Sustainable Materials

(Volume 1)

Bioremediation for Environmental Pollutants

Edited by

Inamuddin
Chemistry Department, Faculty of Science, King Abdulaziz University, Jeddah 21589, Saudi Arabia

Sustainable Materials

(Volume 1)

Bioremediation for Environmental Pollutants

Editor: Dr. Inamuddin

ISBN (Online): 978-981-5123-49-4

ISBN (Print): 978-981-5123-50-0

ISBN (Paperback): 978-981-5123-51-7

First published in 2023.

need for a court order if at any point you breach any terms of this License Agreement. In no event will any delay or failure by Bentham Science Publishers in enforcing your compliance with this License Agreement constitute a waiver of any of its rights.

3. You acknowledge that you have read this License Agreement, and agree to be bound by its terms and conditions. To the extent that any other terms and conditions presented on any website of Bentham Science Publishers conflict with, or are inconsistent with, the terms and conditions set out in this License Agreement, you acknowledge that the terms and conditions set out in this License Agreement shall prevail.

Bentham Science Publishers Pte. Ltd.
80 Robinson Road #02-00
Singapore 068898
Singapore
Email: subscriptions@benthamscience.net

BENTHAM SCIENCE

CONTENTS

Deepesh Tiwari, Athar Hussain, Sunil Kumar Tiwari, Salman Ahmed, Mohd. Wajahat Sultan and Mohd. Imran Ahamed

Grace N. Ijoma, Weiz Nurmahomed, Tonderayi S. Matambo, Charles Rashama and *Joshua Gorimbo*

PREFACE

Modern civilization is experiencing an environmental catastrophe as a result of prevailing pollution issues worldwide. Scientists, researchers, environmentalists, engineers, planners, as well as the developing nations, all need to address this problem to some extent through the introduction of microbes in order to recycle waste into useful forms that other well-beings can utilize efficiently. Varied microbial traits offer a strong substitute to get around the serious issues as they can withstand practically any environmental situation because of their incredible metabolic activity. Hence, the extensive nutritional capacities of microbes can be exploited in the bioremediation of environmental contaminants. Moreover, these microorganisms can be used in bioremediation to eliminate, degrade, detoxify, and immobilize a variety of physical and chemical pollutants. Enzymes are exploited in the destruction and transformation of pollutants like heavy metals, hydrocarbons, pesticides, oil and dyes.

The book *Bioremediation for Environmental Pollutants - Part 1* focuses about the bioremediation of heavy metals, pesticides, textile dyes removal, petroleum hydrocarbon, micro plastics and plastics.

Inamuddin
Chemistry Department, Faculty of Science
King Abdulaziz University
Jeddah 21589, Saudi Arabia

List of Contributors

Abudukeremu Kadier	Laboratory of Environmental Science and Technology, The Xinjiang Technical Institute of Physics and Chemistry, Key Laboratory of Functional Materials and Devices for Special Environments, Chinese Academy of Sciences, Urumqi, 830000, China Center of Materials Science and Optoelectronics Engineering, University of Chinese Academy of Sciences, Beijing, 100049, China
Ankit Kumar	Department of Chemistry, University of Delhi, Delhi 110007, India
Ali Moertopo Simbolon	Research Centre for Environmental and Clean Technology, National Research and Innovation Agency Republic of Indonesia, Kawasan Puspitek gd 820, Serpong, 15314, Tangerang Selatan, Indonesia
Alok Kumar Panda	School of Applied Sciences, Kalinga Institute of Industrial Technology, Deemed to be University, Bhubaneswar-751024, Delhi – 110 042, India
Athar Hussain	Department of Civil Engineering, CBP Government Engineering College Jaffarpur, New Delhi-73, India
Bekti Marlena	Research Centre for Environmental and Clean Technology, National Research and Innovation Agency Republic of Indonesia, Kawasan Puspitek gd 820, Serpong, 15314, Tangerang Selatan, Indonesia
Bhaskarjyoti Gogo	Department of Biotechnology, Assam Royal Global University, Guwahati-781035, Assam, India,
Deepesh Tiwari	Department of Environmental Engineering, Delhi Technological University, Delhi, India
Dongsheng Song	Laboratory of Environmental Science and Technology, The Xinjiang Technical Institute of Physics and Chemistry, Key Laboratory of Functional Materials and Devices for Special Environments, Chinese Academy of Sciences, Urumqi, 830000, China College of Resources and Environment, Xinjiang Agricultural University, Urumqi, 830052, China
Debajit Borah	Department of Biotechnology, Assam Royal Global University, Guwahati-781035, Assam, India,
Evans C. Egwim	Department of Biochemistry, Federal University of Technology Minna, Nigeria
Gargi Dutta	Department of Biotechnology, Assam Royal Global University, Guwahati-781035, Assam, India,
Geetanjali	Department of Chemistry, Kirori Mal College, University of Delhi, Delhi – 110 007, India
Gokul Ram Nishad	Department of Chemistry, Govt. Digvijay PG Autonomous College, Rajnandgaon-491441, Chhattisgarh, India
Grace N. Ijoma	Institute for the Development of Energy for African Sustainability (IDEAS) Research Unit, University of South Africa (UNISA), Florida Campus, Private Bag X6, Johannesburg 1710, South Africa
J. Ranjitha	CO_2 Research and Green Technologies Centre, VIT University, Vellore-14, Tamilnadu, India

Japhet G. Yakubu	Department of Microbiology, Federal University of Technology Minna, Nigeria
Joshua Gorimbo	Institute for the Development of Energy for African Sustainability (IDEAS) Research Unit, University of South Africa (UNISA), Florida Campus, Private Bag X6, Johannesburg 1710, South Africa
Kamlesh Kumar Nigam	Department of Chemistry, Institute of Science, Banaras Hindu University, Varanasi 221005, UP, India
Lakshmi Unnikrishnan	CIPET : School for Advanced Research in Polymers (SARP) - APDDRL - Bengaluru 7P, Hi-Tech Defence and Aerospace Park (IT Sector), Jalahobli, Bengaluru North, Near Shell R&D Centre, Devanahalli, Bengaluru - 562 149, India
Mohd. Wajahat Sultan	Department of Geology, Aligarh Muslim University, Aligarh, India
Mohd. Imran Ahamed	Department of Chemistry, Aligarh Muslim University, Aligarh, India
Manish Kumar Singh	Krishi Vigyan Kendra, Rajnandgaon-491441, Chhattisgarh, India
Nanik Indah Setianingsih	Research Centre for Environmental and Clean Technology, National Research and Innovation Agency Republic of Indonesia, Kawasan Puspitek gd 820, Serpong, 15314, Tangerang Selatan, Indonesia
Oluwafemi A. Oyewole	Department of Microbiology, Federal University of Technology Minna, Nigeria
Priti Gupta	Department of Chemistry, Manav Rachna University, Faridabad, Haryana – 121001, India
Praveen Kumar Yadav	Chemical and Food BND Group, Indian Reference Materials Division, CSIR-National Physical Laboratory, Dr. K.S. Krishnan Marg, New-Delhi 110012, India Academy of Scientific and Innovative Research (AcSIR), Ghaziabad, U.P 201002, India
Peng-Chen Ma	Laboratory of Environmental Science and Technology, The Xinjiang Technical Institute of Physics and Chemistry, Key Laboratory of Functional Materials and Devices for Special Environments, Chinese Academy of Sciences, Urumqi, 830000, China Center of Materials Science and Optoelectronics Engineering, University of Chinese Academy of Sciences, Beijing, 100049, China
Priyanka Singh	Department of Chemistry, Govt. Digvijay PG Autonomous College, Rajnandgaon-491441, Chhattisgarh, India
R. Gayathri	CO_2 Research and Green Technologies Centre, VIT University, Vellore-14, Tamilnadu, India
Rustiana Yuliasni	Research Centre for Environmental and Clean Technology, National Research and Innovation Agency Republic of Indonesia, Kawasan Puspitek gd 820, Serpong, 15314, Tangerang Selatan, Indonesia
Rupesh Kumar	Department of Biotechnology, Assam Royal Global University, Guwahati-781035, Assam, India
Ram Singh	Department of Applied Chemistry, Delhi Technological University, Delhi – 110 042, India

S. Swarupa Tripathy   Chemical and Food BND Group, Indian Reference Materials Division, CSIR-National Physical Laboratory, Dr. K.S. Krishnan Marg, New-Delhi 110012, India

Salman Ahmed   Department of Geology BFIT Group of Institution, Dehradun, India

Setyo Budi Kurniawan   Department of Chemical and Process Engineering, Faculty of Engineering and Built Environment, National University of Malaysia (UKM), 43600 UKM Bangi, Selangor, Malaysia

Shamima Begum   Department of Biotechnology, Assam Royal Global University, Guwahati-781035, Assam, India,

Shishir Kumar Singh   Academy of Scientific and Innovative Research (AcSIR), Ghaziabad, U.P 201002, India
Environmental Science and Biomedical Metrology Division, CSIR- National Physical Laboratory, Dr. K.S. Krishnan Marg, New-Delhi 110012, India

Siti Rozaimah Sheikh Abdullah   Department of Chemical and Process Engineering, Faculty of Engineering and Built Environment, National University of Malaysia (UKM), 43600 UKM Bangi, Selangor, Malaysia

Sonalee Das   CIPET : School for Advanced Research in Polymers (SARP) - APDDRL - Bengaluru 7P, Hi-Tech Defence and Aerospace Park (IT Sector), Jalahobli, Bengaluru North, Near Shell R&D Centre, Devanahalli, Bengaluru - 562 149, India

Soumyaranjan Senapati   School of Applied Sciences, Kalinga Institute of Industrial Technology, Deemed to be University, Bhubaneswar-751024, India,

Sreelipta Das   School of Applied Sciences, Kalinga Institute of Industrial Technology, Deemed to be University, Bhubaneswar-751024, India,

Sunil Kumar Tiwar   Department of Mechanical Engineering, CBP Government Engineering College Jaffarpur, New Delhi-73, India

Telli Alia   Laboratoire de protection des écosystèmes en zone aride and semi aride, Université de KASDI Merbah, BP 511 la route de Ghardaia, Ouargla 30000, Algérie

Tonderayi S. Matambo   Institute for the Development of Energy for African Sustainability (IDEAS) Research Unit, University of South Africa (UNISA), Florida Campus, Private Bag X6, Johannesburg 1710, South Africa

Younus Raza Beg   Krishi Vigyan Kendra, Rajnandgaon-491441, Chhattisgarh, India,

Vikram Singh   Department of Chemistry, NREC College, Khurja, Uttar Pradesh – 203131, India,

Vijayalakshmi Shankar CO_2 Research and Green Technologies Centre, VIT University, Vellore-14, Tamilnadu, India

Weiz Nurmahomed   Institute for the Development of Energy for African Sustainability (IDEAS) Research Unit, University of South Africa (UNISA), Florida Campus, Private Bag X6, Johannesburg 1710, South Africa

CHAPTER 1

Microbial Remediation of Heavy Metals

R. Gayathri[1], **J. Ranjitha**[1] and **V. Shankar**[1,*]

[1] *CO$_2$ Research and Green Technologies Centre, VIT University, Vellore-14, Tamilnadu, India*

Abstract: Chemical elements with an atomic mass unit ranging from 63.5 – 200.6 (relative atomic mass) and a relative density exceeding 5.0 are generally termed as heavy metals. Since they are non-biodegradable inorganic contaminants, physical and chemical methods of degradation are ineffective. Heavy metals cannot be degraded easily due to their physical and chemical properties, such as the rate of oxidation & reduction reactions, rate of solubility, formation of complexes with other metal ions, *etc.* They are flexible, and easily accumulated in the environment. In the case of bioaccumulation, they are highly lethal to the organisms. The process of removal of toxic and hazardous material from the environment using plants and microorganisms is termed bioremediation. The disposal of toxic contaminants using plants is termed phytoremediation. Microbial bioremediation consists of the removal of toxic elements with the application of microorganisms during which the toxic substance is converted into either end products or nontoxic and non-hazardous forms or recovery of metals.

Keywords: Bioremediation, By-products, Hazardous, Heavy metals, Microbe.

INTRODUCTION

Environmental pollution is currently a major problem on a global scale. The ecosystem is severely contaminated due to the rise in industries and increased population. Urbanization with improved standard of life resulted in the reduction of quality of the ecosystem with high pollution within the past 100 years [1]. Air, water, the soil has been contaminated heavily nowadays due to the usage of pesticide, fertilizers, mining, tannery effluents, smelting, electronic appliances, electroplating, paper industries; large scale production of chemicals including solvents, chemical feedstocks, petroleum products, additives, synthetic polymers, pigments and dyes, *etc.*, resulting in the release of heavy metals at a large scale [2]. These contaminants are accumulated in the soil, water, and air resulting in a life-threatening situation for all living organisms [3]. Elevated CO$_2$ levels in the air, fall in natural resources, degradation and release of pollutants such as heavy

*Corresponding Author Vijayalakshmi Shankar: CO$_2$ Research and Green Technologies Centre, VIT University, Vellore-14, Tamilnadu, India, E-mail: vijimicro21@gmail.com

Inamuddin (Ed.)

metals, xenobiotics, toxic gases and chemical substances, *etc.*, into the ecosystem are the primary after-effects of technological & industrial modernization [4]. When compared to various environmental pollutants heavy metals are lethal to biotic factors of the ecosystem. Heavy metals easily tend to accumulate in the soil, & water hence this type of pollution is a major ecological issue. They remain in an unstable form - ionic state and readily react with the surrounding elements [5].

HEAVY METALS

Chemical elements with an atomic mass unit ranging from 63.5 – 200.6 (relative atomic mass) and a relative density exceeding 5.0 are generally termed as heavy metals. They possess high density. Even the lowest concentration of heavy metal shows the highest level of toxicity [6]. They are non-biodegradable inorganic pollutants hence, physical and chemical methods such as are not applicable for their decomposition. Heavy metals cannot be degraded easily due to their physical and chemical properties such as the rate of oxidation & reduction reactions, rate of solubility, formation of complexes with other metal ions, *etc.* they are flexible, easily accumulated in the environment, in case of bioaccumulation they are highly lethal to the organisms [7].

List of Heavy Metals

Aluminum (Al), Antimony (Sb),, Arsenate (As (V)),Arsenic (As), Arsenite (As (III)),Barium (Ba), Bismuth (Bi),Cadmium (Cd), Chromium (Cr), Cr(VI), Cr(III)Cobalt (Co), Copper (Cu), Gold (Au), Iron (Fe), Lead (Pb), Manganese (Mn), Mercury (Hg), Molybdenum (Mo), Nickel (Ni), Selenium (Se), Silver (Ag), Titanium (Ti), Zinc (Zn) are the heavy metals present in both contaminated soil and aquatic environment [8]. The adverse effects of heavy metals on plants & human health are listed in Tables **1** and **2**.

Table 1. Impacts of heavy metals on plants [8].

S. No	Metal Ions	Effects
1.	Zn	Reduced growth, development & senescence, inhibition of metabolism, induced oxidative damage, modifications in enzyme functions, leaf-chlorosis, purple-red leaf, lead to Mn, P deficiencies.
2.	Cd	Injuries similar to chlorosis, growth is inhibited, roots become brown resulting in death, interrupts the transport of other mineral nutrients & inhibits enzyme activity, water imbalance, reduces ATP synthesis, inhibits CO_2 incorporation & chlorophyll functions.

(Table 1) cont.....

S. No	Metal Ions	Effects
3.	Cu	Exhibits cytotoxicity in plants, induction of stress, resulting in tissues injuries, chlorosis of leaves, retarded growth, interruption of metabolism, excess oxidation, influences germination, roots & seed morphology
4.	Hg	Interruption of H_2O flow in plants increased oxidative stress, inhibits functions of mitochondria, influences metabolism & rupture of plasma membrane
5.	Cr	Reduced rate of seed germination & root development, reduced CO_2 incorporation & transportation of electrons, enzymatic functions & ATP synthesis, modification in pigmentation, induced synthesis metabolites & with modified structure
6.	Pb	Inhibition of germination, interference in enzymatic function, retardation & abnormalities of growth in various plant parts, chlorosis, interruption in carboxylation reactions. Inhibition of enzymatic functions. Alterations in H_2O balances, modified membrane permeability. high oxidative stress.
7.	As	Competes with P carrier, leads to P deficiencies, transformed into dimethylarsinic acid (DMA), mono methyl arsenicacid (MMA)
8.	Co	Inhibits the growth of shoots & biomass, impedes the concentration of Fe, chlorophyll, protein, enzyme activity, reduced transpiration, other mineral elements get translocated.
9.	Ni	Necrosis & chlorosis, nutrient imbalance, altered cellular membrane &activities, altered ionic concentration in cytosol, and reduced water absorption.
10.	Mn	Reduced rate of photosynthesis, necrosis, leaf colour change to brown resulting in death. Reduced internodal length, chlorosis,
11.	Fe	Excess free radical formation, irreversible changes in cell morphology, injuries to the plasma membrane, protein & DNA damage, reduced photosynthesis & elevated ascorbate peroxidise function & oxidative stress

Table 2. Impacts of heavy metals on the human health [7].

S, No	Metals	Minimum Threshold Level	Toxic Effects
1.	Ag	0.10	Body parts exposed to Ag (skin & tissues) change into grey or blue-grey colour, breathing difficulties, irritation in throat & lung, and stomach ache
2.	As	0.01	Alters the cellular activities including oxidative phosphorylation and ATP synthesis
3.	Ba	2.0	Results in increased blood pressure, respiratory failure, muscle twitch cardiac arrhythmia and gastrointestinal dysfunction
4.	Cd	5.0	Carcinogenic, mutagenic, endocrine disruptor, lung damage and fragile bones, affects calcium regulation in biological systems
5.	Cr	0.1	Severe Hair loss
6.	Cu	1.3	Renal failure, Brain damage, and high concentration lead to anaemia, chronic gastrointestinal irritation & abnormalities, and liver cirrhosis.

(Table 2) cont.....

S, No	Metals	Minimum Threshold Level	Toxic Effects
7.	Hg	2.0	Brain damage & neural impairments, drowsiness, memory loss, fatigue, temper outbursts, restlessness, insomnia, disturbed vision Autoimmune diseases, depression, hair loss, tremors, respiratory & renal failure
8.	Ni	0.2	Dermal allergies, itching, cancer in the respiratory path and system due to chronic continuous inhalation, hair loss, neurotoxic, immunotoxin & genotoxic, head doses result to infertility,
9.	Pb	15	children exposed to high concentrations may experience growth impairments. Low IQ, severe hair loss short-term memory loss, learning disabilities and coordination problems, risk of cardiovascular disease
10.	Se	50	Intake of 300 µg/day impacts functions of the endocrine system, NK cells. liver damage & affects the gastrointestinal
11.	Zn	0.5	Fatigue & dizziness.

Sources of Heavy Metals

The two major heavy metal sources are natural and anthropogenic sources.

Natural Sources

Natural sources include weathering of minerals from parent rocks, erosion, climatic factors, pH changes, volcanic activities, emission from biogenic sources, particles emitted from vegetation, decomposition & volatilization of the plant bodies and forest fire. Heavy metal leaching, corrosion of metals, resuspension of sediment into the soil & water, evaporation of metals form the water resources. Sea spray and aerosols from the oceans [2, 7].

Anthropogenic Sources

Artificial release of heavy metals due to human activities involves extraction of ores, electroplating, smelting, mining, utilization of pesticides, biosolids, manure and N, P, K fertilizers discharge and atmospheric depositions. Agricultural activities like liming of soil, perpetual irrigation, pigments, paints, plastic stabilizers, fly ash & ash form the combustion of Coal & petrol, nuclear power station. Steel, textiles, microelectronics, plastic, wood preservation and paper processing industries. Release of effluents from tanneries and factories, municipal & industrial wastewater, medical waste & surgical instruments and automobile batteries, lubricants, detergents, landfills & their leachate, industrial spill & leaks, incineration, *etc* [2, 3, 7, 8].

HEAVY METAL ACCUMULATION IN ECOSYSTEM.

Heavy metals and their extracts are highly retained in the soil, due to the cationic affinity towards the soil humus. They are widely accumulated in the dispersed form in the contaminated soil and the water bodies [8]. Due to human activities, the natural biogeochemical cycle of these elements is disturbed. Hence, get piled up in the environment. The deposition of heavy metals is governed by the following physical & chemicals reaction including, oxidation, reduction, absorption, adsorption, precipitation, complex formation, methylation, demethylation, dissolution, ionic exchange resulting in the speciation of the metal contaminants, *etc*. These elements inhibit the biodegradability of other elements present in the soil [3, 9].

BIOREMEDIATION

The process of removal of toxic and hazardous material from the environment using plants and microorganisms is termed as bioremediation. The disposal of toxic contaminants using plants is termed as phytoremediation. Microbial bioremediation consists of the removal of toxic elements with the application of microorganisms during which the toxic substance is converted into either end products or nontoxic and non-hazardous forms or recovery of metals. Bioremediation is grouped into bio-stimulation, and bioaugmentation. Bioremediation involves various processes such as biomineralization, bioaccumulation, bio-simulation, biotransformation, metal-microbe interactions, bioleaching, bioventing, biosorption, bioreactor, land farming, composting and phytoremediation in order to remove the toxic and hazardous contaminants. Bioremediation can be categorized into *in-situ* or *ex-situ* bioremediation based on the type of method adapted for their treatment when the bioremediation is carried out in the original site of contamination then it is termed as *in situ* bioremediation. In *ex-situ* bioremediated the contaminants are transferred to some other place for treatment [1 - 7].

Principles of Bioremediation

Conventional remediation includes 1. Contaminated sediment layers are physically removed (dredging), to isolate the contaminated sediment layer, it is coated with some clean materials (capping), burning of the contaminated materials (incineration). In contrast to conventional system's bioremediation is comparatively cost-effective [10]. In bioremediation, the microorganisms either directly utilize the heavy metals and contaminants as a nutrient source required for their growth or degrade them into secondary substrates *via* breaking their chemical bonds to obtain energy. Electron transfer takes place for obtaining energy. In general, the following elements act as electron acceptors for microorganisms, including O_2, CO_2, sulfate, iron and nitrate, *etc* [3].

Factors Affecting Bioremediation

1. Less expensive.

2. Permanent & effective results.

3. Commercial availability of materials required.

4. The technique selected must be generally acceptable.

5. Should be applicable to a wide range & mixed waste including organic substances.

6. Must be highly potent for detoxification.

7. Must be active and effective under the increased concentration of metal ions.

8. It should be highly capable of immobilization.

9. It should actively reduce the quantity of contaminants [11 - 13].

Bioremediation Strategies

1. Natural attenuation is the natural process of removal or conversion of toxic elements into nontoxic form through an innate microbial community without human intervention.

2. Biostimulation is a part of bioremediation where fertilizers/nutrients are added artificially to enhance the indigenous microbial growth and its activity.

3. Bioremediation occurring naturally in the rhizosphere inhabiting microorganisms is termed as rhizoremediation.

4. Co-metabolism is the process of exploitation of contaminants *via* symbiotic association of microorganism with plants plant -microbe due to degrading potency.

5. Acclimatization is a supplementary technique applied to treat wastewater.

6. Engineered bioremediation involves screening & selection of natural, pollutant-resistant microorganism & genetically engineered microbial strains for effective removal of contaminants [10, 14].

Categories of Bioremediation

Remediation of heavy metals contaminated soil are grouped into three categories in the treatment of hazardous wastes. Category I (*In-situ* remediation), Category II (*In-situ* harsh soil restrictive measures), Category III (In situ or Ex situ harsh soil destructive measures). These two broad categories fixed by USEPA for soil contamination, Category I- source control, Category II- containment remedies. The construction of vertical engineered barrier (VEB), caps & liners is uncomposed within the contamination remedies in order to prevent further spreading /movement of the plume containment. Additionally, the general approach of 5 categories includes extraction, isolation, immobilization, physical separation, and toxicity reduction. Integration of more than two techniques would be more economical [15].

Mode of Operation

Bioremediation can be carried out in both Aerobic and anaerobic phases. *Bacillus, Flavobacterium, Mycobacterium, Nitrosomonas, Pseudomonas, Penicillium, and Xanthobacter* are microorganisms commonly used for bioremediation.

Aerobic

Mycobacterium, Sphingomonas, Rhodococcus and *Pseudomonas* are the most commonly used aerobic bacteria in the bioremediation technology. They actively degrade hydrocarbons, and pesticides, especially groups such as alkanes & aromatic compounds and utilize the contaminants as C and energy sources.

Anaerobic

Anaerobic microbes are used to degrade polychlorinated biphenyls present in the river, chloroform, trichloroethylene and for solvents-dechlorination. Algae, bacteria, fungi and yeast have been reported for successful bioremediation of heavy metals & radio nuclei. *Bacillus, Pseudomonas, P. aeruginosa, Streptomyces, Aspergillus, Rhizopus,* and *Saccharomyces, Streptoverticullum* are highly potent microbial strains used for heavy metal bioremediation. In exclusion, the target site is free from metal ions, in extrusion, the elimination of metal ions from the cell as a result of chromosomal and plasmid movements. Accommodation is the formation of a metal complex with intracellular proteins and cell components, Biotransformation is the reduction of heavy metals into non-toxic forms, methylation and demethylation are the mechanisms used by the above-mentioned fungal species for heavy metal bioremediation. For survival under metallic stress, several mechanisms have been evolved in these microorganisms. A series of reactions take place in the detoxification of heavy metals which includes extracellular efflux of metal ions, formation of intracellular metallic/ionic complexes, and final stage with the reduction of toxic metal into non-toxic form [3].

INTERACTIONS BETWEEN HEAVY METALS AND MICROORGANISMS

Bioremediation using microorganisms includes bioaccumulation, biosorption, biotransformation, immobilization, mobilization and biomineralization processes. Comparatively Biosorption process plays a vital in the environmental remediation of heavy metals. This process includes metabolism which is independent, reversible, and passive uptake of the heavy metals by the adsorption of biological materials to the cell surface. Bioaccumulation is a complicated process that involves the active accumulation of heavy metals in the cellular components. They interact with biomolecules to obtain a stable form resulting in the formation of biotoxins (Fig. **1**). Heavy metals are taken up by the microorganism *via* active/passive methods. The binding capacity of the microbial cells to the heavy metal is due to the presence of biomolecules such as lipids, carbohydrates and proteins which possess functional groups like phosphate, hydroxyl, carboxylate and amino groups. These functional groups are responsible for the adsorption and absorption reactions of the microbial cells to the charged particles [7, 16]. Adsorption of heavy metals to the microbial surface takes place through the following factors net negative charge of the microbial cell and metallic cation charges, ionic exchange between heavy metals and teichoic acid & peptidoglycan of the cell wall. Nucleation reaction resulting in precipitation, and formation of metallic/ionic complexes with O_2 and N ligands. The teichoic acid & peptidoglycan in the cell wall accelerate the adsorption reaction, hence Gram-positive bacteria highly adsorbs to the heavy metals while gram negative bacteria lack in heavy metal adsorption [17 - 20]. Microorganisms sequester metal ions in two routes 1. Intracellular sequestration, 2. Extracellular Sequestration.

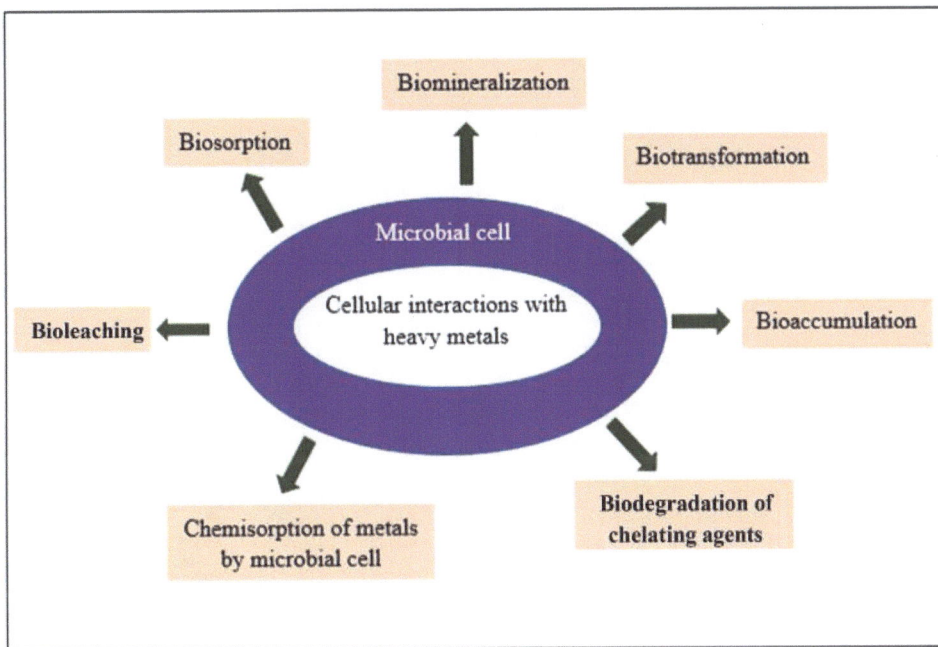

Fig. (1). Interactions of heavy metals with microbial cell in using various methods.

Intracellular Sequestration

Intracellular sequestration is including the absorption of metal complexes inside the cytoplasm. This process is facilized by the formation of ligands /metal binding sites on the surface of microbial cell walls and then slowly transported into the cytoplasmic membrane. The following microbial strains exhibit intracellular sequestration of metal ions during the remediation process Cadmium and Zinc ions by *Pseudomonas putida* (which uses protein carrier molecules), Cadmium ions using glutathione by *Rhizobium leguminosarum* [21 - 23].

Extracellular Sequestration

Extracellular sequestration is the process of bioaccumulation, where metal ions are accumulated into the components of cells present in the periplasm/ formation of insoluble metal complexes. The following microorganism is well known for their extracellular sequestration of heavy metals including Cu ions by *Pseudomonas syringae* (in periplasm), Zn ions by *Synechocystis* PCC 6803 (periplasm). *Geobacter sp* and *Desulfuromonas spp.* reduces the toxicity level of various metals. Mn and Cr ions by *G. metallireducens*, Cr ions by *G. sulfurreducens*. H_2S by *P. aeruginosa*, *Klebsiella planticola*. Pb ions by *Vibrio harveyi* [24 - 29].

TYPES OF BIOREMEDIATION

Based on the procedures adapted, bioremediation is categorised into two types such as *in-situ* bioremediation and *ex-situ* bioremediation.

In-situ Bioremediation

The contaminants are treated at the subsurface level, O_2 and nutrients are supplied to promote the growth of indigenous microorganism consortium for the degradation of heavy metals. This method is applicable for both saturated and unsaturated soil and ground water. It includes intrusion of nutrients and energy sources diffused in water. To improve the rate and quality of bioremediation, the chemotactic capabilities of microbial consortium have to be enhanced. It is divided into intrinsic bioremediation and engineered *in situ* bioremediation

Intrinsic Bioremediation

In this method, the innate abilities of the microbial consortium are stimulated by supplying nutrients and electron acceptors *e.g.* O_2 to degrade the heavy metal contaminants. It is the natural attenuation process.

Engineered In Situ Bioremediation

This method involves the inclusion of selected strains of microorganisms into the site of contamination. Stimulation and enhancement of the physical and chemical parameters to favour microbial growth, thereby accelerating the process of bioremediation. This method is adopted where the environmental conditions are not favourable for the indigenous microbial consortium. Few genetically modified microbial strains are capable of withstanding such conditions.

Advantages

Economically feasible, more efficient, utilization of harmless microbes, reduces the risk of further exposure, an alternate to pump and treat method for remediating aquifers and soil with organic and toxic contaminants *e.g.* Cr (VI), complete conversion of contaminants into simple non- toxic form/end products like CO_2, H_2O, C_2H_6. Treatment for dissolved as well as sorbed contaminants is possible with accelerated *in-situ* bioremediation and is time saving when compared to pump-and-treat method. Cost effective, even inaccessible areas of the plume can be treated with bioremediation.

Limitations

Based on the contamination site, incomplete conversion of pollutions into end products are possible. In the case of the intermediate cease of the bioconversion process the products formed are highly toxic than the original compounds. Recalcitrant are non-biodegradable. Injection well may get blocked/ obstructed if improper operation takes place, ultimately resulting in increased microbial growth due to accumulation of supplied nutrients & electron acceptor/donor. Some innate microbial consortiums may get inhibited due to a high concentration of toxic elements. Microbial strains adapted to contaminated environments are required since they cannot be produced in recent spills.

Ex-situ Bioremediation

The contaminants are treated above ground level. Contaminants are removed from on-site and transported to the treatment area. The excavation method is used for soil and pump and treat method for polluted groundwater. After excavation, the soil is aerated to stimulate the indigenous microbes to degrade the contaminants. *Ex-situ* bioremediation is categorised into the slurry-phase system and solid-phase system

Slurry-Phase System

Slurry-Phase bioremediation involves the use of bioreactors. The contaminated soil is aerated, and blended with water inside the bioreactor. In this stage, the rubbles and stones are removed. In the following step, the soil will be suspended in a fixed amount of water for the slurry formation. Based on the degree or concertation of the contaminants present, the amount of water added will vary. After the completion of this procedure, the soil is taken out and dried with the help of a centrifuge, pressure filters and vacuum filter. Further treatment depends upon the nature of the soil. In the final stage, the resulting fluids will be examined and further treated,

Solid-Phase System

Solid-Phase bioremediation involves the soil placed in form of piles. It contains organic waste from plants, animals, industries and MSW (Municipal Solid Waste). The soil bio-piles are aerated with a network of pipelines to facilitate ventilation and promote the growth of microbial consortium and require wide space & time when compared to slurry-phase system. This process includes composting, soil bio-piles land farming, *etc.*

Soil Bio-Piles

In this method, the contaminated soil is excavated into the treatment site where they are piled up at 3-10 feet heights. The microbial growth is promoted by aeration and supply of nutrients and moisture. These compost piles contain both aerobic as well as anaerobic microorganisms. Air is injected into the piles through a network of pipelines embedded in them. Regular tilling or ploughing help to aerate the piles. Engineered cells in the form of aerated compost piles are used to treat surface contaminants which involve volatilization and leaching.

Land farming involves spreading and continuous treatment such as ploughing and tilling of the contaminated soil at a regular interval till the pollutants are completely degraded. Tilling and ploughing will facilitate aeration and nutrients are supplied to stimulate the growth of the innate microbial consortium.

Composting

In composting method, bulking agents are added to the contaminated soil to facilitate the supply of optimum level of water and air required for microbial growth. The contaminated soil is mixed and aerated in the treatment vessel in case of mechanically agitated composting. For window composting methods, the soil is made into lengthy piles called windows and ploughed or blended with the help of tractors like 25% compost for 75% contaminated is the standard ratio generally followed but can be altered based upon the type of soil, characteristics and concentration of the contaminants. This method has increased rate of clean-up when compared to others.

Advantages

It will be appropriate for nearly all types of organic contaminants and it can be accessed from the site investigation data.

Limitations

This method is not suitable for heavy metals and additional processing is required for soil with low permeability *e.g.*, clay & silt.

MICROBIAL REMEDIATION

Remediation by Bacteria

Bacteria act as a potential bio-sorbent for heavy metals. They actively take up the heavy metals from the contaminated site. Residual microbial biomass from the fermentation factories can be utilized for this process. *E.g. Bacillus, Streptomyces, Citrobacter, Pseudomonas, etc.* These metal ions are extensively utilized and transformed into secondary substrates or reduced into a non-toxic state. Bacteria have evolved several mechanisms at genetic as well as biochemical levels for sequestering heavy metals. A specific strain like *Pseudomonas* and *Bacillus* sp. are highly potent for heavy metal reduction and used on a large scale for remediation of contaminated sites due to their high affinity towards heavy metal and their binding potential. Functional groups present in the biomolecules of bacterial cell wall protein play an important role in metal binding and it includes the following functional groups like amide, carboxyl, phosphonate, hydroxyl and sulfonate groups. The defense mechanism of the microbial cell facilitates the surface level changes of the cell wall and mass attraction of heavy metals, and their further reduction reactions. There is a significant difference between normal bacterial cells and cells exposed to Cr (VI) metal ions. The impact had been reflected on their cell surface when exposed to heavy metals. Smooth cellular surfaces with an elongated cell shape has been reported in non-exposed bacterial cells, whereas in Cr (VI) exposed bacterial extreme modifications were noticed in form of irregular cell surface resulting in clumping of these bacterial cells.

Bacterial cell surface has anionic charges and is drawn towards catatonically charged heavy metals. G +ve bacteria are provided with a thick cell wall consisting of peptidoglycan, teichoic and teichuronic acids and G -ve bacteria the cell wall is not thick and lacks these materials. These materials are responsible for their binding capacity. Hence, G +ve bacteria will potentially sequester heavy metals comparatively than G-ve bacteria. Significant biosorption of Cr (IV) through cell surface in novel haloalkaliphilic bacterium has been reported. reported (Karthik *et al.* 2017a). Active biosorption of Cr (IV) metal ions has been recorded *via* intracellular as well as extracellular mechanisms. Functional groups were found to be responsible for both adsorption & absorption along with active immobilization of metal ions to the bacterial cell. Functional groups observed in this reaction include amines, amide and alkanes. Generally, bacteria used for bioremediation involve chemolithotrophs and soil bacterial (Fig. **2**). *Bacillus licheniformis, Bacillus firmus, Bacillus coagulans, Bacillus megaterium, Enterobacter sp. JI, Bacillus licheniformis, Bacillus licheniformis, Escherichia coli, Pseudomonas fluorescens, Salmonella typhi, Bacillus cereus, Desulfovibrio desulfuricans, Enterobacter cloacae, Kocuria rhizophila, Micrococcus luteus, Lactobacillus sp., Pantoea agglomerans, Alcaligenes sp., Ochrobactrum intermedium, Cupriavidus metallidurans* (Tables **3** and **4**) are potent bacterial strains for adsorption of heavy metal ions [2, 30].

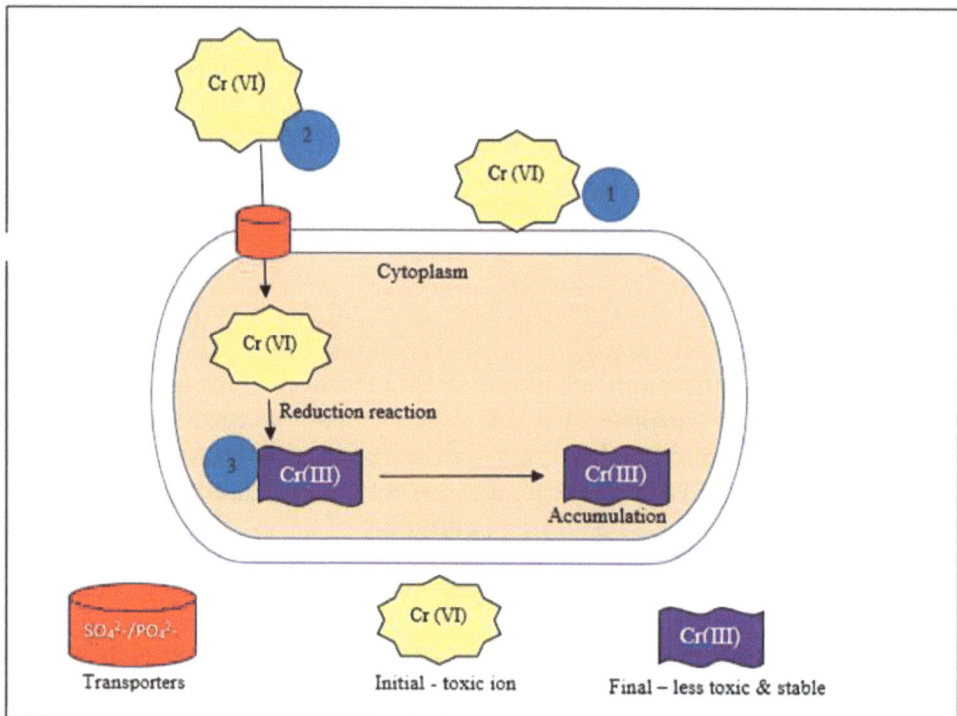

Fig. (2). Bacterial remediation of Cr (VI) ions *via* intracellular reduction including 4 stages. **1.** Interaction & biosorption. **2.** Transportation of Cr (VI) **3.** Reduction of Cr (VI). **4.** Bioaccumulation of Cr (VI).

Table 3. Bacterial Bioremediation of heavy metals under Optimum temperature & pH conditions.

S. No	Factors	Range	Microorganism Species & Strains	Categories	Metals
1.	Temperature in °C	20-40	*Acinetobacter sp., Aspergillus niger, Bacillus cereus, Botrytis cinereal, Enterobacter aerogenes, Rhizopus oryzae., Ochrobactrum sp,, Paenibacillus sp,, Rhizobium sp., Vigribacillus sp, Chlorella vulgaris*	Mesophilic	-
		45	*Chlorella vulgaris*	Moderate thermophilic	Cd (II), Ni (II)
		55	*Bacillus cereus- TA2, Bacillus cereus- TA4*	Moderate thermophilic	Cr (VI)
		60	*Geobacillus thermantarcticus, Anoxybacillus amylolyticus*	Moderate to hyperthermophiles	Cd (II), Cu (II), Co (II) Mn (II)
		70-80	-	Hyperthermophiles	-
2.	pH	10	*Ochrobactrum bacterial strain CSCr-3*	Alkaliphiles	Cr (VI)
		10.5	*Halomonas genus*	Alkaliphiles	Cr (VI)

Table 4. Efficient removal of heavy metals by various microorganisms including bacteria, bacteria & algae [2].

S. No	Microorganism	Range	Incubation Time	Metal Ions
1.	*Micrococcus luteus*	408 mg/g	After 12 h	Cu
2.	*Desulfovibrio desulfuricans*	98.2 mg/g	After 168 h	Cu
3.	*Aspergillus niger*	15.6 mg/g	After 1 h	Cu
4.	*Cellulosimicrobium funkei AR8*	100, 150 and 200 µg/ml	-	Cr (VI),
5.	*Chlorella vulgaris*	69%- 80% of 2.5 ppm	-	Ni (II), Cu (II)

Microbial Remediation by Fungi (Mycoremediation)

Fungi are omnipresent microorganisms and they can survive in any environmental habitat due to their adaptive metabolic mechanism depending on the availability of C and N resources. Fungi have the potency to withstand and convert toxic heavy metals into non-toxic stable forms. The cell of fungal hyphae is made of carbohydrates, amino acids, triglycerides, phosphate groups and inorganic elements in an ionic state. Due to the higher surface, cell ratio has a higher affinity to contact with surrounding enzymatic and physical materials. Fungi sequester heavy metals *via* intracellular precipitation, valence transformation, ion exchange, and formation of metal complexes. The process of bioremediation with the application of fungal species to remediate, decompose, or convert toxic elements in the ecosystem is termed as Mycoremediation (termed by Stamets). Application of the mycelium of fungi to sieve toxic contaminates and microbe present in the soil water *via* stimulating the enzymatic function of the microbial cells is termed as Mycofiltration. Fungi such as saprophytes, endophytes & mycorrhizae possess the ability to remediate and restore soil-water environment & balance in the microbial community.

The fungal mycelia produce extracellular enzymes & acids to disintegrate the basic units of polymeric cell wall materials like lignin & cellulose. The essential part of mycoremediation is the selection of appropriate fungal strains to achieve maximum efficiency for remediating individual target contaminants. It has been reported that the fungi mycorrhizae play a vital role in remediating aluminium present in soil & Melia plant roots by secreting glomalin protein. The following strains of fungi have been reported for efficient recovery of heavy metals from the contaminated site. It includes *Penicillium spp., Trametes versicolor Cladosporium resinae, Aspergillus niger, Funalia trogii, Rhizopus arrhizus, Aureobasidium pullulans, Ganoderma lucidum, etc. Aspergillus versicolor* shows high potency to bioaccumulate heavy metal ions under optimal conditions and it can be utilized in bioremediation of contaminated sites. It is highly recommended for remediation of Cr contaminated fields. Pb ions can be effectively removed by using the fungal strain *Aspergillus fumigates* and it is experimentally proved by Ramasamy *et al.* (2011) [10, 31 - 33]. Fungal biomass for heavy metal biosorption can be cultivated with the help of fermentation methods. Fungal species are cheap and can be obtained from the waste products left out in the enzyme industries. Various fungal machineries such as extracellular and intracellular precipitation, transfer of electrons from the valence shell and active absorptions are involved in the biosorption of heavy metals. The fungal cell wall is composed of carbohydrates that are rich in mannuronic and guluronic acids with carboxyl groups in large quantities facilitating heavy metals sequestration [34]. The difference between lead exposed and non-exposed fungal cells has been reported by Jacob *et al.* (2017). The impact of heavy metal stress has resulted in the structural modification of the fungal cell wall. There is increased pressure in the cytosol of the fungal cell as a stimulated stress response by the external Pb stress in the culture media. *Agaricus bisporus, Bjerkandera adusta, Pleurotus pulmonarius, Trametes versicolor, Pleurotus tuberregium, Lentinula edodes, Pleurotus ostreatus,* and *Irpex lacteus,* the above mentioned fungal species showed identical results under the same stress condition (Table **5**). This reaction illustrates the potency of fungal species to biosorb heavy metals [35]. The following fungal species are highly potent in heavy metal biosorption. It includes *Aspergillus flavus, Lepiota hystrix, Penicillium cirtinum, Mucar rouxii, Aspergillus brasiliensis, Aspergillus terreus, Rhizopus oryzae Aspergillus niger, Trichoderma longibrachiatum, Ganoderma lucidum, Pleurotus sapidus, Saccharomyces cerevisiae, Pleurotus platypus* [2].

Table 5. Rate of Mycoremediation of various heavy metals [10].

S. No	Microorganism	Maximum Activity Showed at pH	Concentration in mg/L	Metal Ions	Recovered Quantity in %
1.	*Aspergillus versicolor*	6	50	Cr (VI)	99.89
2.	*Aspergillus versicolor*	6	50	Ni (II)	30.05
3.	*Aspergillus versicolor*	5	50 mg/L	Cu (II)	29.06
4.	*Aspergillus fumigatus*	-	100 mg/L	Pb (II)	85.41

Microbial Remediation by Algae (Phycoremediation)

Algae are autotrophic microorganism that utilizes radiant energy to perform photosynthesis. The rate of algal growth is comparatively higher than any other photosynthetic organism. It utilizes the radiant energy and converts it into biochemical form required for its growth. Based on their size and morphology they are categorised into microalgae and macroalgae. Microalgae are unicellular microscopic photoautotrophs and macroalgae algae are multicellular photoautotrophs growing at the sea bed commonly termed as seaweeds. Algae are adaptable to any aquatic habitat such as moist soil, marine and freshwater bodies. As a bio sorbent of heavy metals algae remains unexposed comparatively to bacteria & fungi. Algae have evolved many metabolisms to utilize and bioaccumulate heavy metals [36 - 38]. The cell surface plays a vital role in the algal biosorption of heavy metals. Similarly, to bacteria and fungi, the cell surface gets to bind with metal ions. To enhance the binding capacity of the algae, the cell surface has to be modified. Currently, many techniques are available to modify the algal cell surface. The binding potential of marine algae to the heavy metal is due to the presence of plenty of biopolymers. Algal species have the ability to sequester heavy metals by both metabolic dependent/independent processes. The most potent macroalgae for heavy metal sequestration are the red and brown algae. Polysaccharides located in the cell wall surface are provided with binding sites for metal ions. The functional groups such as carboxyl and amino groups will form a coordinate bonds with the heavy metal ions with the help of O_2 and N atoms. The composition of cell walls varies based on the type of algal species. The cell wall of red algae contains protein ranging from 36-50% and green algae with 10-70%. Number binding in the cell wall is determined by the ratio of cell wall components. Electrostatic attraction takes place between the heavy metals and non-protonated functional groups such as O_2, sulphate and carboxyl groups, present in the algal cell wall [39, 40].

Cyanoremediation

Cyanoremediation is one of the bioremediation techniques which involves the application of cyanobacteria for heavy metal removal and is also used as an agent for pollution control. Green algae & blue green algae have been proven to be a potent microorganisms that can be utilized for the removal of heavy metal contamination in water bodies [41 - 44] Indigenous as well as genetically modified strains are used for this purpose. The single celled-blue algae the Synechocysis sp. PCC6803 showed bioaccumulation of Arsenic metal at a maximum rate of 1.0 and 0.9 g/kg DW in 0.5 mM for a duration of 14 days in the arsenate & arsenite solution. Rapid oxidation of arsenite into arsenate resulting in active bioaccumulation by *Synechocysis* sp was noticed in 2.37µM arsenite concentration. Hence, it is experimentally proven that Synechocysis sp has the potency for efficient removal of arsenic elements and recommended for remediating arsenic polluted water bodies. Apart from this, multicellular algae such as *Oscillatoria spp., Synechoccus spp., Calothrix spp., Nostoc spp., and Anabaena spp.,* are highly suitable for cyanoremedaition of heavy metals [45]. *Cladophora fascicularis* is an effective biosorbing marine green alga that removes lead ions in polluted water. Based on the study reports of Lee & Saunders *et al.* (2012), the bioaccumulation of heavy metal vanadium and arsenic has been increased upto 8% of dry biomass when *Hydrodictylon, Oedogonium* and *Rhizoclonium* species were utilized to remediate Coal treated wastewater from the power generating units (Table **6**).

Table 6. Effective bioaccumulation & biosorption of heavy metals by cyanobacteria/ algae at various concentrations of heavy metal ion [10].

S. No	Microorganism	Under the concentration	Recovery rate	Metal ions	References
1.	*Oscillatoria sp., Synechococcus sp., Nodularia sp., Nostoc sp.* and *Cyanothece sp.* Pure culture	5 ppm	69.5 - 99.6%	-	Dubey *et al.* (2011)
	Mixed culture	5 ppm	91.6 - 100%,	-	Dubey *et al.* (2011)

(Table 6) cont.....

S. No	Microorganism	Under the concentration	Recovery rate	Metal ions	References
2.	*Spirogyra*	5 mg/L	-	Pb2+	Lee and Chang (2011)
		5 mg/L	98.23%	Cr	
		5 mg/L	(89.6%	Cu	
		5 mg/L	99.73%	Fe	
		5 mg/L	(99.6%	Mn	
		5 mg/L	98.16%	Se	
		5 mg/L	(81.53%	Zn	
		5 mg/L	-	Cu2+	
3.	*Cladophora*	-	-	Pb2+	Lee and Chang (2011)
		-	-	Cu2+	
4.	*Spirulina sp*	5 mg/L	98.3%	Cr	Lee and Chang (2011)
		5 mg/L	81.2%	Cu	
		5 mg/L	(98.93%	Fe	
		5 mg/L	(99.73%	Mn	
		5 mg/L	98.83%	Se	
		5 mg/L	79%	Zn	
5.	*Hydrodictylon, Oedogonium* and *Rhizoclonium sps*	-	1,543 mg/kg DW	V	Mane and Bhosle, (2012)
		-	137 mg/kg DW	As	

Tran *et al.* 2016 conducted an experiment to evaluate the biosorption ability of cyanobacteria gelatin colonies for the sequestration of Pb^{++}, Cu^{++} and Cd^{++} metal ions in the water. Active biosorption was reported. Algae can be used to treat waste water as well as to produce biofuel. The algal strains used for lipid and biofuel production can be used for active sequestration of heavy metals in contaminated wastewater [38]. Proper maintenance including immobilization, pre-treatment, optimum physiochemical conditions and selection of suitable genetically engineered algal strains will endorse heavy metal biosorption [46]. The following algal strain is highly potent to bioaccumulate the heavy metals. *Chlorealla sorokintana, Spirogyra sp., Dunaliella sp., Spirogyra sp., Palmaria palmate, Sargassum wighti, Spirulina maxima, Sargassum sp, Cystoseira barbata, Spirogyra hyaline, S. neglecta, Micrasterias denticulate, Ulva lactuca,, Cladophora sp, Scenedesmus obliqus, Cladophora, Nitella opaca hutchinsiae, Chlorella vulgaris, Eucheuma denticulatum, Chara aculeolate, etc.*

EFFECTS OF HEAVY METALS ON THE MECHANISM OF MICROBE

Heavy metals are commonly essential micronutrients needed for all living organisms within their required limit. Increased concentration of these elements will result in abnormalities by showing various degrees of toxicity at molecular, biochemical, genetical, cellular, and physiological levels [2]. Increased concentration of heavy metals will affect the microorganism at two stages, first stage is microbial metabolism will be inhibited due to the impact of heavy metals. Various cellular functions are blocked including mitosis, enzymatic functions and proteins are denatured and the second stage is genetic level changes due to alteration in genetic components. The process of transcription is inhibited and genetic mutations will occur (Fig. **3**). Microbial remediation is a complicated process (Table **7**). Microorganism either hinders or inhibits the bioaccumulation of heavy metals by the living organism through various metabolic activities including the secretion of enzymes, inorganic and organic acids *e.g.* H_2SO_4 and citric acid, redox reactions, formation of complexion agents *e.g.*, cyanide [47]. Biotic and abiotic components extremely affect the microbial remediation of heavy metals by disturbing their metabolism and reducing the growth rate. Microorganism are highly adaptable to their surroundings and can withstand unfavourable conditions. Proper selection of microbial strain will result in effective bioremediation of heavy metals. Microorganisms are limited to physical, chemical, biological and climatic factors.

Table 7. Impacts of increased heavy metal concentration on microorganism [2].

S. No	Microorganism	Impacts	Heavy Metals	References
1.	*Aspergillus Micrococcus sp.*	Inhibition of growth & metabolism	Cr (VI), Ni (II)	(Congeevaram *et al.*, 2007)
2.	*Aspergillus, Penicillium & Cephalosporium sp.*	Reduced growth & metal removing efficiency	Cu, Cd & Pb	(Hemambika *et al.*, 2011)
3.	*Cellulosimicrobium funkei AR8*	Reduction in metal removing efficiency	Cr (VI)	(Karthik *et al.*, 2017b)
4.	*Chlorella vulgaris*	Bioremediation efficiency is reduced	Ni (II), Cu (II)	(Mehta and Gaur, 2001)

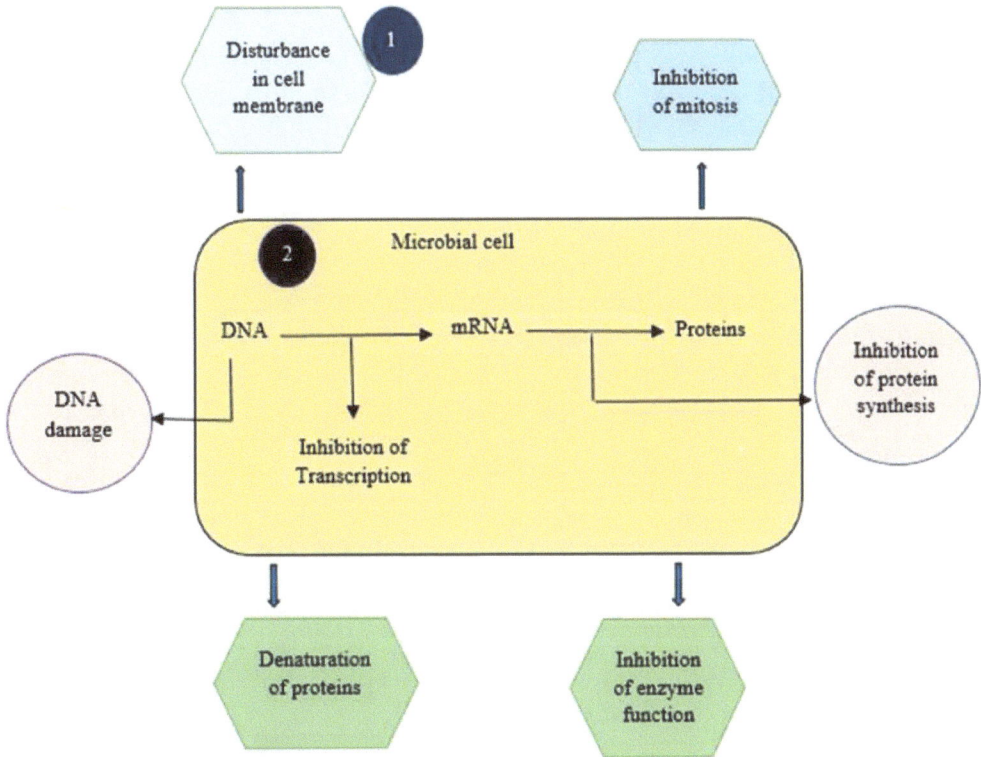

Fig. (3). Influence of heavy metal stress on microbial cells in two routes. 1. Inhibiting the metabolism. 2. Alteration in the genetic material.

Biological Factors

Microbial remediated is highly influenced by biotic factors such as alteration in the cell shape, size, cellular surface, composition of cell wall and production of extracellular products. The concentration of biomass determines the efficiency of heavy metal bioremediation. In the lowest concentration, there is sufficient intercellular space between the microbial cells and the rate of biosorption will be high. The microbial consortium will freely interact with metal ions, to form metal complexes or bio-transform & bioaccumulate heavy metals effectively in the lowest concentration. If the equilibrium stage is attained by the microbial biomass, then there will be an increased concentration gradient in addition to adsorption resulting in the entry of heavy metals into the microbial cell. In case of increased microbial biomass concentration, the metal microbe interaction *via* the binding sites of cellular surface is inhibited [48]. Shell effects on the external surface of the microorganism take place due to elevated concentration of microbial cells. In such cases, the intercellular space gets compressed due to the clumping of microbial cells and the amount of metal absorbed will be reduced [49]. The biosorbent ability of *Scenedesmus abundans* has been reported to be decreased when there is an increased concentration of microbial cells used in Cd (II) & Cu (II) remediation. Similar results had been reported for the removal of heavy metals using various algal strains. Extreme biosorption was recorded at the minimum concentration of biomass. *Bacillus subtilis, Saccharomyces cerevisiae, Microcystis* and *Pseudomonas aeruginosa* also showed the highest activity in the lowest concentration [50 - 53]. Microbial Consortium is the biological factor affecting the efficiency of bioremediation. Different microbial strains like *Mycobacteria, Flavobacteria, Corynebacteria, Aeromonas, Acinetobacter, Pseudomonas, Chlorobacteria, Streptomyces, Aeromonas, Bacilli, Cyanobacteria and Arthrobacter species, etc.,* inhabiting the soil/aquatic ecosystem [10]. Macrobenthos diversiform is diverse form of plants and animals present in domestic wastewater that possess the ability to reduce/ deteriorate chemical oxygen demand (COD) Biological Oxygen Demand (BOD), turbidity, NH3, NO_2^- etc [10, 54].

Physical and Chemical Factors on Microbial Remediation

Temperature

Temperature affects bioremediation in two ways. Direct impact on the physical and chemical state of the pollutants and interrupting the microbial metabolism [2]. Microorganism requires optimum temperature to perform microbial metabolism under normal conditions. Elevated temperature tends to denture the enzymes and proteins resulting in death or reduced activity [55]. Temperature exceeding the optimum level will influence the ribosomes thereby restricting the protein translation resulting in the fall of protein synthesis. Thermophilic bacteria can survive and effectively remediate contaminants in elevated temperatures but under arid conditions the growth of fungal biomass declines [56]. Low temperature will interrupt the growth rate due to disturbance in the transportation of substrate into the cell membrane & its movement. Alterations in membrane fluidity will affect the rate and speed of movement of particles through the cell membrane. Microorganisms will be dead in case of extreme temperature fluctuations (increase/decrease) and disturbance in the diffusion of gas into the cell membrane [2]. The optimum temperature reported for bioremediation is 20 – 40°C for mesophilic bacteria, 45°C for moderate thermophilic bacteria and 55-80°C for hyperthermophiles.

pH

The hydrogen ion concentration plays a vital role in the growth and enzyme regulation, formation of metallic complexes, regulating the functional groups of cellular surfaces. Enzymes are essential for active biotransformation of heavy metals, any alteration in the pH will interrupt the enzymatic function thereby influencing the biosorption of heavy metals. At optimum pH maximum reduction of heavy metal has been reported. The pH fluctuation will directly affect the solubility and bioavailability of heavy metals, a decrease of pH in the media resulted in reduced bioaccumulation. The optimum pH for microbial growth is 5.5 - 8.5 for active bioremediation. Increased pH affects the solubility of metal and its bioavailability. Elevation of pH from 6-7 at 1.3 mM phosphate solution showed a fall in the cadmium ion solubility. The function groups of the cell wall are influenced by alteration in the pH of the media, *e.g.* Acidic/alkaline medium. Thus, the interaction between the metal ion at the binding site to heavy metals is hindered or inhibited. Under low pH, the hydronium ion tends to diminish the negative charge of functional groups present in the cell wall material. Thus, the ability to biosorb and reduce heavy metal at high [H+] ion concentration *via* microbial cell wall is decreased [57]. Increased pH also influences the effectiveness of microbial remediation. The hydroxyl ion in the alkaline media will interact with heavy metals leading to the complexation of metal-hydroxyl ions *via* proton substitution. Under alkaline condition, the microbial log phase tend to elevate resulting in the reduction of heavy metal remediation [58]. The following microorganism has been reported for maximum bioremediation at optimum pH. It includes *Bacillus sp, Acinetobacter junii, Cellulosimicrobium funkei, Escherichia coli, Micrococcus luteus, Pannonibacter phragmitetus, Pleurotus platypus, Lentinula edodes, Pseudochrobactrum saccharolyticum, Vigribacillus sp. Pseudomonas aeruginosa* and Trichoderma sp.

Characteristics of Pollutants

Bioremediation is influenced by the nature of the contaminant. Includes the type, physical, and chemical state of the pollutants. (i) Physical states are liquid, semi-solid, gaseous/volatile, solid, *etc*. (ii) Chemical states are inorganic, organic, hazardous, non-hazardous (iii) Types of materials such as heavy metals, hydrocarbons, pesticides, solvents that are chlorinated (iv) level of toxicity like high, moderate, low and harmless/non-toxic.

Structure of the Soil

It includes the soil type, texture, particle size, and soil profile (*e.g.* clay, sand & silt). Ensuring the availability of H_2O, air and nutrients is mandatory. It will directly reflect on the effectiveness of bioremediation in case of contaminated soil and also determines the type of techniques to be adapted for effective treatment [10].

Nutrients

For normal metabolic activities and cell development nutrients such as N, P, K is required. In general, these lack in the contaminated sites, due to its unavailability microorganism inhabiting that site will be interrupted. In such cases, biostimulation techniques are highly recommended for the efficient removal of contaminants. The ratio for carbon to other nutrients for effective removal of heavy metals & other contaminants includes C: N - 10:1, C: P-30:1, for treatment of soil the ratio varies such as C: N- 2:1 [59].

Redox Potential

Redox potential highly influences the movement and speciation of heavy metal. It affects the interaction of heavy metals and their reaction such as biosorption and desorption, metal ion complexation and speciation [60].

Oxygen Content

Oxygen is the key factor for determining the mode of reaction to be aerobic or anaerobic. The quantity of O_2 at a sufficient level is essential for various biochemical reactions [10, 61].

Ionic Concentration

The concentration of metal ion also influences the efficiency of cellular activities in microbial remediation. Under elevated ionic concentration the microbial remediation escalates followed by a gradual fall in the efficiency to treat the heavy metals [62]. Minimum concentration of metal ions is essential for regulating microbial metabolism but increased concentration will create stress on the cells resulting in retardation of cellular actives [63]. Based on the heavy concentration in the treatment of the medium, the bioremediation potential varies. The bioremediation efficiency has been reported to be reduced in the following microorganism due to elevated heavy metal concentration. It includes *Aspergillus, Micrococcus sp. Penicillium* and *Cephalosporium sp, Cellulosimicrobium funkei* AR8, *Chlorella vulgaris*. This is due to heavy metal toxicity resulting in reduced microbial biomass production, interruption in cellular activities, and enzyme inhibition & denaturation [64]. Increased metal ion concentration will influence the microbial metabolism in the following routes such as permanent impairment or changes in the membrane integrity due to metal-microbe interactions; inhibition/denaturation of enzymes in the cell membrane due to accumulation of metal ions; genetic modifications including DNA damage and mutation [65, 66].

Climatic Factors

Environmental factors such as climate play a vital in the growth of the microbial community. There is an indirect influence of climatic conditions on the microbial community *via* alterations in physical & chemical parameters acting upon microbial metabolism [67, 68]. The synthesis of extracellular enzymes gets modified by these parameters through alterations in the microbial community inhabiting the soil [56]. Increased carbon dioxide in the atmosphere along with temperature greatly influences the microbial community. It has been reported that there is a rise in bacterial cell and a fall in fungal biomass significantly under increased CO_2 concentration [69, 70].

Light

Radiant energy is vital for the process of photosynthesis. Many photoautotrophs depend on radiant energy for typical cellular activities. Rate photosynthesis also reflects on the bioremediation and heavy metal sequestration in some microorganisms, especially algae. The maximum efficacy of the algae to sequester the heavy metals is based on the light/heat intensity due to climatic changes. Since algae are hypertensive even mild alteration in the climatic condition can cause permanent changes in the cellular activities, thereby influencing the rate of bioremediation [71 - 73].

Moisture

Moisture is the water content present in solid/liquid phase. The Relative permittivity (dielectric constant) is determined by the moisture content present in the soil ranging from 25-28% [10].

VARIOUS BIOREMEDIATION METHODS

Bioventing

Bioventing is a process used for stimulating the native aerobic bacteria to remediate the soil contamination by proving nutrients and O_2 *via* wells in located the soil. It is a type of sub-surface *in-situ* bioremediation. O_2 is strictly provided at the required amount and the rate of O_2 flow is low in order to reduce the escape/leakage of volatile gases and contaminants. Naturally contaminates are bioremediation under the aerobic condition, which involves active participation of innate microbial consortium inhabiting the treatment material and site. Under anaerobic conditions, the bioremediation process is promoted by the supply of O_2 at a minimal level. Bioremediation of aquifers by implying redox potential, dynamics adsorption of heavy metals indirectly results in the treatment of soil for heavy metal contaminants in case of sub-surface *in-situ* bioremediation. The treated water is consumable & can be used for irrigation as well [10, 75].

Bioaugmentation

It is the process where either native or selected pre-cultured microbial strains are used for improving the deterioration rate of contaminants. Competence of exogenic microbial cultures with native microbial colonies occurs infrequently. Bioaugmentation with effective results can be achieved by integrating it with biosimulation technique. Integration of biostimulation with potent bacterial strains, nutrients & carbon source, and biosurfactants, with optimum moisture & temperature, will provide maximum results [10, 74 - 76].

Biodegradation

It is a natural process occurring in the ecosystem which involves the microbial decomposition of organic waste into simple organic/inorganic elements or end products such as CO_2, NH_3, H_2O, O_2, N, C, *etc*. This process has been developed into remediation techniques to achieve active elimination of organic/inorganic wastes, remediating contaminated sites and treatment of hazardous materials. To overcome the limitations, appropriate knowledge of microbial growth metabolism, nutrients & other parameters is required. Geological, biochemical, proteomic, genomic studies provide all information required. Under aerobic conditions, O_2 is required in adequate amounts in order to act as an electron acceptor in the subsurface degradation [10, 75, 77 - 79].

Microbial Induced Calcite Precipitation

More environment friendly & low-cost techniques are under demand in current bioremediation technologies to remediate heavy metal contamination. MICP act as a substitute for solving such issues. MICP products are highly adsorbed to the heavy metal surface. The metal ions present within the ionic radius near the elements including Cu^{2+}, Pb^{2+}, Ca^{2+}, Sr^{2+}, and Cd^{2+} are actively amalgamated into the calcite crystals during precipitation of calcite *via* substitution reaction. Kocuria flava CR1 is an innate calcitrant bacterial strain that showed 95% of Cu removal based on the MICP process after its isolation from a mining area. It is a highly potent bacterial strain that can be used for environmental-friendly, heavy metal remediation in Cu polluted spots [80].

Biosparging

It is the process in which indigenous microbial consortia are stimulated by proving aeration under pressurised conditions below the level of the water table in order to achieve the maximum rate of biodegradation [81]. This process elevates the blending of water with soil in the saturated zone. Construction of small sized air injection points is advantageous and favours the designing & system construction with easy installation in a cost-effective way. This method is highly suitable for remediating the contaminated oil fields, petroleum- hydrocarbons associated heavy metals contaminants. In Texas, this process is used for cleansing the arsenic-hydrocarbon contaminates in the aquifer at an oilfield. Air supply depends on the type of soil, contaminants to be removed, mode of treatment and the site structure. Optimum pH and temperature are mandatory for the active biochemical reaction taking place in the microbial cell [82]. Fe^{2+} particles interrupt the permeability of soil in the saturated zone during the operation. It has been experimentally recorded that there is a positive correlation existing between the rate of bioremediation and air supply [83].

Biostimulation

Biostimulation is a process where artificially nutrients and aeration are provided to enhance the ability of indigenous microbial consortium to degrade chemical substances. It is generally used in *in-situ* bioremediation [75, 76]. N, P, O_2, CH_3, C_6H_6O, C_7H_8, are supplied to induce the rate of bioremediation in heavy metals [84, 85]. Isolation of a bacterial consortia from Cr polluted site shows plasmid-mediated Cr resistance and reduced it *via* enzyme activity. Strain with improved activity can be obtained through modification at the molecular level [86]. The capabilities of microorganisms segregated as a sample for culture taken from heavy metal contaminated waste disposal site showed the maximum rate of bioremediation for Cu, Fe, & Cd at an elevated concentration of 100 mg/L with a recovery rate of 99.6, 100 and 98.5%, respectively [87]. The heavy metal bioremediation potential of sulfate reducing bacteria (RBS) was tested in a column reactor, the results showed complete removal of heavy metals including Zn, Cd, Cu, As (V) & Cr (VI) along with 50-60% sulfate reduction [88]. Even though the maximum results are obtained some organisms cannot tolerate elevated heavy metal concentrations or get killed by toxins formed from microbial activity [86].

Biomineralization

Biomineralization is a naturally occurring process in living organism that produces minerals in their tissues/cells. All living organisms are capable of synthesising inorganic elements by metabolic activities [91]. Algae and diatoms store silicates and carbonates are stored in invertebrates and shell-forming marine forms, Ca, P and CO_3 are sequestered by vertebrates, Cu, Fe and Au are mineralized in bacteria. Fungal species are known to be actively involved in biodegradation and biomineralization *via* metal fungi interactions. Fungi biomineralize inorganic elements with the help of organic protein which provide a nucleation site in this process. Cu containing mineral precipitate has been produced by fungal hyphae in form of calcium carbonate by precipitation reaction. The fungal mycelia of *Aspergillus niger* and *Paecilomyces javanicus* has a special feature that it can tolerate and biomineralize P ions amalgamated with uranium [89, 90]. *Sporosarcina ginsengisoli CR5* bacteria have the ability to withstand As (III) heavy metal at the highest concentration. Under 50mM concentration, the bacteria actively synthesized calcite-precipitating enzyme urease to accumulate AS III. This bacterial strain is highly recommended for the recovery of Arsenic polluted sites [92]. For active CaP calcium phosphate biomineralization MSP- mesoporous silica particles are produced with help of hyaluronic acid which provides reaction site for mineral sedimentation Microbial biomineralization is a vital part of biogeochemical cycles for heavy metal sequestration [93]. *Bacillus* sp. KK1 showed tolerance and active bioremediation of Pb ions in mine tailing *via* bacterially induced calcite precipitation [94].

Biosorption

Biosorption is the process of metabolism-independent passive uptake, ad/absorption of heavy metals by the microbial cells. Interactions of metals ion with microbial cells occur at the level of cell wall /periplasm (surface/sub-surface) level. Biomass that are alive or dead possess the ability to biosorb heavy metals. Binding sites/ligands include functional groups present in the microbial cell wall. Biosorption is influenced by diffusion, heterogenous surface and pH. It can be enhanced by improving the metal binding site for specific /target metals to be treated to achieve the maximum rate of metal removal. When the metal-resistant *Ralstonia eutropha* strain was genetically engineered with mouse metallothionein in the outer membrane of microbial cells the resulting strain showed active sequestration of Cd^{2+} ions. Similarly, the *E. coli* fimbriae were modified with metal binding peptides for active uptake of ZnO & metal binding motif for Cd^{2+}, Hg^{2+} ions. *Staphylococci sp.* with external binding peptides for Cd^{2+} and Hg^{2+} ions [95].

Biotransformation

Many microbial strains' potential modifies organic and inorganic substances inside the cell and this process is termed as biotransformation. Various microorganisms reduce heavy metals ion through enzymatic action and convert them from soluble to insoluble states. The majority of the bacteria strains potentially reduce soluble Cr (III), into insoluble precipitate Cr $(OH)_3$. Segregation of Cr (VI) reducing microbial strains such as *Pseudomonas aeruginosa* S128, *Arthrobacter* sp, few anaerobic bacteria & algae from the Cr polluted sites showed effective sequestration of Cr (VI) ions nearly 67% during a period of 30 to 40 days along with nitrate & electron donor supplement. This indicates that the molasses hinders the transportation of Cr and NO^{3-} into the core acquirers. Other than enzymatic reduction of heavy metals into insoluble, indirect reduction also takes place in sedimentary & subsurface conditions. Metal/sulfate-reducing bacteria actively immobilize the pollutants in the indirect method. This can be achieved by combining either oxidization of organic compounds or H_2 to reduce iron Fe (III), Mn (IV) and S (IV) in sulfate from SO_4. The reduced metals finally precipitated in form of hydroxide minerals, reduced oxide or coprecipitate with Fe (III) during the redox reaction of Fe (II) [95].

Microbial Leaching

Microbial leaching is a technique used for the recovery of valuable metals such as Cu, Co, Zn, Pb, Au, Ag, As, Sb, Hg, Ni, Mo, *etc.*, from ores, mining site, heavy metal remediation, with the application of microorganism. Sulfur oxidizing bacteria *Thiobacillus thiooxidans* & *Thiobacillus ferrooxidans* are well known for microbial leaching by oxide formation & sulfur reduction. Fungal strains include *Aspergillus niger*, *Penicillium simplicissimum* for Cu, Sn with a recovery rate of 65% and 96% for Pb, Al, Zn Ni, *via* metabolic activities by dissolving metals [95]. Muti-metal resistant bacteria were experimentally tested for active recovery of the following metals Fe+3, Al+3, Cr6+, Ni2+, Mn2+ and Zn2+. Bacterial strains segregated for experiment were cultured under 30°C at a pH of 7± 0.2 for a period of 24 h, under nutrient agar media with a different ionic concentration of heavy metal concentrations ranging from 25, 50, 100 mg mL^{-1}. $C_6H_{12}O_6$ is used as C source (Table **8**). *Brevibacterium casei sp.* showed 99% of Cr (VI) under pH 7 at 30°C within 12 h and 67% of Mn under 6.5 pH and 36°C with C source with a duration of 9 days [96].

Table 8. Culture conditions for bacterial isolates [97].

S. No	Metals	Source	Media With Metal Concentration in mg mL-1	pH	Temperature	Experimental Concentration in mg mL-1	Incubation Period
1.	Cr (VI)	K2Cr2O7	100 mg mL-1	7± 0.2	30C	25, 50, 100	24 h
2.	Zn	ZnSO4	100 mg mL-1	7± 0.2	30C	25, 50, 100	24 h
3.	Ni	NiCl2,	100 mg mL-1	7± 0.2	30C	25, 50, 100	24 h
4.	Fe	FeSO4	100 mg mL-1	7± 0.2	30C	25, 50, 100	24 h
5.	Al	Al2(SO4)3	100 mg mL-1	7± 0.2	30C	25, 50, 100	24 h

Microbial Remediation with the Use of Chelating Agents

Chelators are chemical substances used to form stable water-soluble metallic complexes. Chelating agents establish the bond between metal ions and other molecules. It controls the metal ion mobility. Ethylenediamine tetra acetic acid (EDTA) and Nitrilotriacetic acid (NTA) are highly capable of forming metallic complexes with heavy metal ions. In general, they used to cleanse the nuclear fuels in industries. The metal-chelate complexes formed at the end of the cleaning process are released into the ecosystem which later on penetrate the soil & reach the ground water table. In the water insoluble mineralization of metal oxides due to precipitation reaction takes place. After decomposition of chelators, these metal ions are permanently immobilized/ mineralized which reduces the risk of further spreading of contaminants. Most of the chelating agents are decomposed by innate microorganisms [95].

Siderophore-Mediated Incorporation of Metal Ions

Siderophores are iron (Fe^{3+})-chelating agents produced by microorganisms including fungi & bacteria. It shows great affinity to iron particles and is involved in the active transport of iron particle through the cell membrane. A wide range of microbes have been experimentally proven to be active siderophore producers. *Microbacterium flavescens* produces siderophores to acquire adequate iron as a nutrient material. This metabolic activity might be a vital route for the entry and transfer of heavy metals in various tropical levels. The following microorganism produces sidephores for active iron biosorption. It includes Yeast (*Rhodotorula pilimanae*), Fungi (*Ustilago sphaerogena, Fusarium roseum*) and Bacteria (*Streptomyces pilosus, Streptomyces coelicolor, Burkholderia cepacian, E. coli,*

Bacillus anthracis, B. subtilis, Vibrio cholerae, Azotobacter vinelandii, Pseudomonas aeruginosa, Yersinia pestis, Shigella sp., Salmonella sp., Pseudomonas sp., Bordetella sp., Staphylococcus sp., and *Mycobacterium tuberculosis,etc*) [95, 97].

Application of Biopolymers in Heavy Metal Remediation

They are polymers macromolecules produced by living cells. It has metal binding properties due to immediate covalent bond formation. It includes exopolysaccharides, biosurfactants and cyclodextrins. They possess functional groups such as hydroxyl, phosphoryl, & carboxyl groups. These materials can be synthesised in the polluted soil using metal resistant microbial strains or directly introduced into washing solvents. These molecules are further improved *via* genetic engineering to achieve maximum effects of bioremediation [95].

Exopolysaccharides

Exopolysaccharides are macromolecules, possessing acid functional groups resulting in a negative charge. Metallic interaction may lead to folding due to its flexibility & extreme hydration. Certain exopolysaccharides may result in flocculation due to metal interaction leading to the formation of insoluble metallic products which provide the nucleus for additional precipitation. The low solubility & mobility helps in effective bioremediation. Many microorganisms synthesize free or capsulated polysaccharides and a majority of them are especially for metallic interactions. Metallic interactions, active binding and adsorption of 3.3 lg cadmium by *Arthrobacter* and 22 lg copper for *Klebsiella* were recorded experimentally. These molecules also perform active desorption of metal ions bound to the mineral surface & transportation of meatal complexes *via* the porous media. The G-ve soil bacterial strain 9702-M4 synthesizes a polysaccharide that produces Cd2+ complexes and a weak bond is established with the surface of sand particles. Nearly 57% (~0.08 ppm) of Cd was biosorbed with the sand particles per 1500 pore volume [95].

Biosurfactants

Biosurfactants are amphiphilic compounds consisting of polar heads & non-polar tails. The head contains P and tails with lipid groups. Biosurfactants are actively synthesized by diverse microorganisms. It helps to form metal complexes. These molecules reduce the force of attraction amongst particles at surface level when present in a liquid. This facilitates bioremediation function *via* concentration of molecules at the surface level. Metal ions present near sorbed molecules reaches the equilibrium stage by the formation of metal complexes due to enhanced activity by surfactants. Critical micelle concentration (CMC) is an adequate maximum concentration of 1–200 mg/L where biosurfactants form micellar bodies *via* aggregation, thereby assimilating chelated metals ion actively enclosed inside the micellar bodies. These micelles are about 50 nm in diameter. *Pseudomonas aeruginosa* produces rhamnolipid micelle with negative charge. It leads to metal complexation including the metal ions Zn2+, Cd2+, & Pb2+ in a solution. The minimum metal-mobilizing potency of rhamnolipid in sand-loam soil was 0.11% organic substances, indicating its strength to lead desorption. After metal reabsorption or recovery, this biosurfactant can be reused [95].

Cyclodextrins

Cyclodextrins are oligosaccharides made up of macrocyclic glucose units produced as a result of enzyme action on starch during microbial decomposition of starchy tissues. The glucose subunits are bonded with the help of glycoside bond. The formation of complexes to non-polar molecules is facilitated through a less polarized cavity present in the middle of the ring leading to the development inclusion in a ratio of 1:1. The outer ring is hydrophilic in nature and makes the complex water soluble. The metal biding potency of cyclodextrin was facilitated by the presence of carboxyl group in the outer and carboxymethyl-b-cyclodextrin (CMCD). This CMCD form complexation with Cd^{2+} ions at a rate of 103.66 for 1 g/l CMCD, cyclodextrins reduces the surface tension of H_2O along with the absence of CMC. Since they are tiny, they cannot be excluded from the soil pores. They are eco-friendly and suitable for the pump-treat method for heavy metals bounded with organic contaminants [95].

Dissimilatory Metal Reduction

Most of the microbial community acquires energy through enzymatic activities redox reaction including electron donor/ acceptors like $C_6H_{12}O_6$, acetate or H_2 coupled with NO_3, O_2, Fe^{3+}, $SO_4^{2+/-}$. Numerous metallic ions are recorded to be terminal electron acceptors. It includes As (V) as arsenate, Se (VI) as selenite, Cr (VI). It helps in the reduction and separation of metal ions from solid surfaces [95].

Role of Enzymes in Bioremediation

In the microbial remediation of heavy metals, enzymes play a key role in metal removal/recovery. Sequential enzymatic reactions favour metal sequestration and metal-microbe interactions. They act as a catalyst for enhancing reactions including precipitation, redox reaction, polymerization & extraction of H+ ion *E.g.*, urease, peroxidase, by bacteria [2].

GENETICALLY MODIFIED MICROORGANISM FOR HEAVY METAL REMEDIATION

Genetically engineered microorganisms (GEMs) are applied in order to enhance the rate of heavy metal bioremediation. Under the certain circumstance, the innate microorganism shows a moderate rate of metal recovery or fails to withstand heavy metal stress or the population gets declined due to increased concentration of heavy metals. In such cases, GEM are adapted for active removal of contaminants. *R.eutropha CH34, Alcaligenes eutrophus AE104 (pEBZ141), Rhodopseudomonas palustris, M.huakuii subsp. Rengei strain B3, Astragalus sinicus* are a few genetically modified microbial strains used for Cr, Hg, Cd^{2+} heavy metal removal. Suicidal genetically engineered microorganisms (S-GEMS) are generally used in the bioremediation process to avoid risk and effective removal of pollutants. Adaptation to the natural ecosystem and sustaining the contaminant stress without any molecular / genetic mutation is a major risk factor for recombinant microbial strains (Table **9**). The release of GEMs into the environment still remains a challenge. Continuous monitoring of remediation site is mandatory for GEMs. Sometimes, these strains may potentially affect the indigenous microbial consortium [1].

Table 9. Effective remediation of heavy metals of GEM bacteria [7].

S. No	Heavy Metal	Initial Conc. (ppm)	Removal Efficiency (%)	Genetically Engineered Bacteria	Expressed Gene
1.	As	**0.05**	100	*E. coli strain*	Metalloregulatory protein ArsR
2.	Cd2+	-	-	*E. coli strain*	SpPCS
3.	Cr6+	1.4–1000	100	*Methylococcus capsulatus*	CrR
4.	Cr	-	-	*P. putida strain*	Chromate reductase (ChrR)
5.	Cd2+, Hg	-	-	*Ralstonia eutropha CH34, Deinococcus radiodurans*	merA
6.	Hg	-	-	*E. coli strain*	Organomurcurial lyase
	Hg	7.4	96	*E. coli JM109*	Hg2+ transporter
	Hg	-	-	*Pseudomonas K-62*	Organomercurial lyase
	Hg	-	-	*Achromobacter sp AO22*	mer

(Table 9) cont.....

S. No	Heavy Metal	Initial Conc. (ppm)	Removal Efficiency (%)	Genetically Engineered Bacteria	Expressed Gene
7.	Ni	145	80	*P. fluorescens 4F39*	Phytochelatin synthase (PCS)

CONCLUSION

Heavy metals were one of the major group of pollutants persisting in the environment and it is highly toxic & life threatening. Bioremediation is an effective eco-friendly technique for pollutant removal. Application of microorganism that was suitable for the removal of particular heavy metals can be a promising method to clean-up the contaminated as well as to recover the valuable metals in the different ionic states further subjected for recycling/reuse. Based on the type of microbial mechanism involved, physicochemical parameters of the contaminated sites, selection of the most suitable microbial strain or consortium will augment the remediation process. It acts as an effective technique for pollutant removal & it requires the combination of detailed analysis of the contaminated site including vital information such as type of metal pollutant, physiochemical properties of the site, selection of most appropriated microbes & type of remediation process.

CONSENT FOR PUBLICATION

Not applicable.

CONFLICT OF INTEREST

The authors declare no conflict of interest, financial or otherwise.

ACKNOWLEDGEMENT

Declared none.

LIST OF ABBREVIATIONS

1. BOD – biological oxygen demand

2. COD – chemical oxygen demand

3. MICP- microbially induced calcite precipitation

4. IQ – intelligence quotient

5. Nk Cells – Natural killer cells.

REFERENCES

[1]   Verma S, Kuila A. Bioremediation of heavy metals by microbial process. Environmental Technology & Innovation 2019; 14: 100369.
        [http://dx.doi.org/10.1016/j.eti.2019.100369]

[2]   Jacob JM, Karthik C, Saratale RG, *et al.* Biological approaches to tackle heavy metal pollution: A survey of literature. J Environ Manage 2018; 217: 56-70.
        [http://dx.doi.org/10.1016/j.jenvman.2018.03.077] [PMID: 29597108]

[3]   Kulshreshtha A, Agrawal R, Barar M, Saxena S. "A Review on Bioremediation of Heavy Metals in Contaminated Water," IOSR. J Environ Sci Toxi Food Technol 2014; 8(7): 44-50.

[4]   Tyagi M, Manuela M, Da Fonseca R, De Carvalho CR. Bioaugmentation and biostimulation strategies to improve the effectiveness of bioremediation processes. Biodegradation 2011; 22(2): 231-41.
        [http://dx.doi.org/10.1007/s10532-010-9394-4]

[5]   Garbisu C, Alkorta I. Basic concepts on heavy metal soil bioremediation. Europ J Miner Proces Env Protect 2003; 3(1): 58-66.

[6]   Carolin CF, Kumar PS, Saravanan A, Joshiba GJ, Naushad M. Efficient techniques for the removal of toxic heavy metals from aquatic environment: A review. J Environ Chem Eng 2017; 5(3): 2782-99.
        [http://dx.doi.org/10.1016/j.jece.2017.05.029]

[7]   Dixit R, Wasiullah , Malaviya D, *et al.* Bioremediation of Heavy Metals from Soil and Aquatic Environment: An Overview of Principles and Criteria of Fundamental Processes. Sustainability (Basel) 2015; 7(2): 2189-212.
        [http://dx.doi.org/10.3390/su7022189]

[8]   Nagajyoti PC, Lee KD, Sreekanth TVM. Heavy metals, occurrence and toxicity for plants: a review. Environ Chem Lett 2010; 8(3): 199-216.
        [http://dx.doi.org/10.1007/s10311-010-0297-8]

[9]   Chen M, Xu P, Zeng G, Yang C, Huang D, Zhang J. Bioremediation of soils contaminated with polycyclic aromatic hydrocarbons, petroleum, pesticides, chlorophenols and heavy metals by composting: Applications, microbes and future research needs. Biotechnol Adv 2015; 33(6): 745-55.
        [http://dx.doi.org/10.1016/j.biotechadv.2015.05.003] [PMID: 26008965]

[10]   Mani D, Kumar C. Biotechnological advances in bioremediation of heavy metals contaminated ecosystems: an overview with special reference to phytoremediation. Int J Environ Sci Technol 2014; 11(3): 843-72.
        [http://dx.doi.org/10.1007/s13762-013-0299-8]

[11]   Sarma H. Metal hyperaccumulation in plants: a review focussing on phytoremediation technology. J Environ Sci Technol 2011; 4(2): 118-38.
        [http://dx.doi.org/10.3923/jest.2011.118.138]

[12]   Dhankher OP, Doty SL, Meagher RB, Pilon-Smits E. Biotechnological approaches for phytoremediation.Plant biotechnology and agriculture. Oxford: Academic Press 2011; pp. 309-28.

[13]   Paliwal V, Puranik S, Purohit HJ. Integrated perspective for effective bioremediation. Appl Biochem Biotechnol 2012; 166(4): 903-24.
        [http://dx.doi.org/10.1007/s12010-011-9479-5] [PMID: 22198863]

[14]   Gupta R, Mohapatra H. Microbial biomass: an economical alternative for removal of heavy metals from waste water. Indian J Exp Biol 2003; 41(9): 945-66.
        [PMID: 15242288]

[15]   Raymond A. Wuana1 and Felix E. Okieimen2., Review Article Heavy Metals in Contaminated Soils: A Review of Sources, Chemistry, Risks and Best Available Strategies for Remediation. International Scholarly Research Notices 2011; 1-20.

[16]   Scott JA, Karanjkar AM. Repeated cadmium biosorption by regenerated *Enterobacter aerogenes*

biofilm attached to activated carbon. Biotechnol Lett 1992; 14(8): 737-40.
[http://dx.doi.org/10.1007/BF01021653]

[17]   Monachese M, Burton JP, Reid G. Bioremediation and tolerance of humans to heavy metals through microbial processes: a potential role for probiotics? Appl Environ Microbiol 2012; 78(18): 6397-404.
[http://dx.doi.org/10.1128/AEM.01665-12] [PMID: 22798364]

[18]   Beveridge TJ, Murray RG. Sites of metal deposition in the cell wall of *Bacillus subtilis*. J Bacteriol 1980; 141(2): 876-87.
[http://dx.doi.org/10.1128/jb.141.2.876-887.1980] [PMID: 6767692]

[19]   Beveridge TJ, Fyfe WS. Metal fixation by bacterial cell walls. Can J Earth Sci 1985; 22(12): 1893-8.
[http://dx.doi.org/10.1139/e85-204]

[20]   Mueller JG, Chapman PJ, Pritchard PH. Creosote-contaminated sites. Their potential for bioremediation. Environ Sci Technol 1989; 23(10): 1197-201.
[http://dx.doi.org/10.1021/es00068a003]

[21]   Bernard E, Igiri , Stanley I R, *et al.* Toxicity and bioremediation of heavy metals contaminated ecosystem from tannery wastewater: a review. Journal of Toxicology 2018; 2018: 16.

[22]   Higham DP, Sadler PJ, Scawen MD. Cadmium-binding proteins in *Pseudomonas putida*: pseudothioneins. Environ Health Perspect 1986; 65: 5-11.
[PMID: 3709466]

[23]   Lima AIG, Corticeiro SC, de Almeida Paula Figueira EM. Glutathione-mediated cadmium sequestration in Rhizobium leguminosarum. Enzyme Microb Technol 2006; 39(4): 763-9.
[http://dx.doi.org/10.1016/j.enzmictec.2005.12.009]

[24]   Gavrilescu M. Removal of heavy metals from the environment by biosorption. Eng Life Sci 2004; 4(3): 219-32.
[http://dx.doi.org/10.1002/elsc.200420026]

[25]   Bruschi M, Florence G. New bioremediation technologies to remove heavy metals and radionuclides using Fe(III)-, sulfate-and sulfur-reducing bacteria. In: Singh SN, Tripathi RD, Eds. Environmental Bioremediation Technologies. 2006; pp. 35-55.

[26]   Luptakova A, Kusnierova M. Bioremediation of acid mine drainage contaminated by SRB Hydrometallurgy 2005; 77: 97-102.
[http://dx.doi.org/10.1016/j.hydromet.2004.10.019]

[27]   White VE, Knowles CJ. Effect of metal complexation on the bioavailability of nitrilotriacetic acid to Chelatobacter heintzii ATCC 29600. Arch Microbiol 2000; 173(5-6): 373-82.
[http://dx.doi.org/10.1007/s002030000157] [PMID: 10896217]

[28]   Sharma PK, Balkwill DL, Frenkel A, Vairavamurthy MA. A new *Klebsiella planticola* strain (Cd-1) grows anaerobically at high cadmium concentrations and precipitates cadmium sulfide. Appl Environ Microbiol 2000; 66(7): 3083-7.
[http://dx.doi.org/10.1128/AEM.66.7.3083-3087.2000] [PMID: 10877810]

[29]   Wang CL, Ozuna SC, Clark DS, Keasling JD. A deep-sea hydrothermal vent isolate, *Pseudomonas aeruginosa* CW961, requires thiosulfate for Cd tolerance and precipitation. Biotechnol Lett 2002; 24(8): 637-41.
[http://dx.doi.org/10.1023/A:1015043324584] [PMID: 20725529]

[30]   Siddiquee S, Rovina K, Azad SA, Naher L, Suryani S, Chaikaew P. Heavy metal contaminants removal from wastewater using the potential filamentous fungi biomass: a review. J Microb Biochem Technol 2015; 7(6): 384-93.
[http://dx.doi.org/10.4172/1948-5948.1000243]

[31]   Dudhane M, Borde M, Jite PK. Effect of aluminium toxicity on growth responses and antioxidant activities in Gmelina arborea Roxb. inoculated with AM fungi. Int J Phytoremediation 2012; 14(7): 643-55.

[http://dx.doi.org/10.1080/15226514.2011.619230] [PMID: 22908633]

[32]   Taştan BE, Ertuğrul S, Dönmez G. Effective bioremoval of reactive dye and heavy metals by *Aspergillus versicolor*. Bioresour Technol 2010; 101(3): 870-6.
[http://dx.doi.org/10.1016/j.biortech.2009.08.099] [PMID: 19773159]

[33]   Ramasamy RK, Congeevaram S, Thamaraiselvi K. Evaluation of isolated fungal strain from e-waste recycling facility for effective sorption of toxic heavy metals Pb(II) ions and fungal protein molecular characterization-a Mycoremediation approach. Asian J Exp Biol 2011; 2(2): 342-7.

[34]   Thatoi H, Das S, Mishra J, Rath BP, Das N. Bacterial chromate reductase, a potential enzyme for bioremediation of hexavalent chromium: a review. J Environ Manag 2014; 146: 383-99.

[35]   Rhodes CJ. Mycoremediation (bioremediation with fungi) e growing mushrooms to clean the earth. Chem Spec Bioavailab 2014; 26: 196-8.

[36]   Saratale GD, Saratale RG, Chang JS, Govindwar SP. Fixed-bed decolorization of Reactive Blue 172 by *Proteus vulgaris* NCIM-2027 immobilized on *Luffa cylindrica* sponge. Int Biodeterior Biodegrad 2011; 65: 494-503.

[37]   Vijayaraghavan K, Balasubramanian R. Is biosorption suitable for decontamination of metal-bearing wastewaters? A critical review on the state-of-theart of biosorption processes and future directions. J Environ Manag 2015; 160: 283-96.

[38]   Zeraatkar AK, Ahmadzadeh H, Talebi AF, Moheimani NR, McHenry MP. Potential use of algae for heavy metal bioremediation, a critical review. J Environ Manag 2016; 181: 817-31.

[39]   Flores-Chaparro CE, Chazaro Ruiz LF, Alfaro de la Torre MC, Huerta-Diaz MA, Rangel-Mendez JR. Biosorption removal of benzene and toluene by three dried macroalgae at different ionic strength and temperatures: algae biochemical composition and kinetics. J Environ Manag 2017; 193: 126-35.

[40]   Ajjabi LC, Chouba L. Biosorption of Cu2þ and Zn2þ from aqueous solutions by dried marine green macroalga Chaetomorpha linum. J Environ Manag 2009; 90: 3485-9.

[41]   Deng L, Su Y, Su H, Wang X, Zhu X. Sorption and desorption of lead (II) from wastewater by green algae Cladophora fascicularis. J Hazard Mater 2007; 143(1-2): 220-5.
[http://dx.doi.org/10.1016/j.jhazmat.2006.09.009] [PMID: 17049733]

[42]   Singhal RK, Joshi S, Tirumalesh K, Gurg RP. Reduction of uranium concentration in well water by Chlorella (*Chlorella pyrendoidosa*) a fresh water algae immobilized in calcium alginate. J Radioanal Nucl Chem 2004; 261(1): 73-8.
[http://dx.doi.org/10.1023/B:JRNC.0000030937.04903.c4]

[43]   Tripathi RD, Dwivedi S, Shukla MK, *et al.* Role of blue green algae biofertilizer in ameliorating the nitrogen demand and fly-ash stress to the growth and yield of rice (Oryza sativa L.) plants. Chemosphere 2008; 70(10): 1919-29.
[http://dx.doi.org/10.1016/j.chemosphere.2007.07.038] [PMID: 17854856]

[44]   Yin XX, Wang LH, Bai R, Huang H, Sun GX. Accumulation and transformation of arsenic in the blue-green alga Synechocysis sp. PCC6803. Water Air Soil Pollut 2012; 223(3): 1183-90.
[http://dx.doi.org/10.1007/s11270-011-0936-0]

[45]   Fiset JF, Blais JF, Riveros PA. Review on the removal of metal ions from effluents using seaweeds, alginate derivatives and other sorbents. Rev Sci Eau 2008; 21(3): 283-308.
[http://dx.doi.org/10.7202/018776ar]

[46]   Favara P, Gamlin J. Utilization of waste materials, non-refined materials, and renewable energy in *in situ* remediation and their sustainability benefits. J Environ Manag 2017; 204: 730-7.

[47]   Gadd GM. Metals, minerals and microbes: geomicrobiology and bioremediation. Microbiol Reading Engl 2010; 156: 609-43.
[http://dx.doi.org/10.1099/mic.0.037143-0]

[48]   Abbas SH, Ismail IM, Mostafa TM, Sulaymon AH. Biosorption of heavy metals: a review. J Chem Sci

Technol 2014; 3: 74-102.

[49]   Fadel M, Hassanein NM, Elshafei MM, Mostafa AH, Ahmed MA, Khater HM. 2017; Biosorption of manganese from groundwater by biomass of Saccharomyces cerevisiae. HBRC J 13: 106-13.

[50]   Romera E, Gonzalez F, Ballester A, *et al.* Comparative ~ study of biosorption of heavy metals using different types of algae. Bioresour Technol 2007; 98: 3344-53.

[51]   Pradhan S, Rai LC. Optimization of flow rate, initial metal ion concentration and biomass density for maximum removal of Cu2þ by immobilized Microcystis. World J Microb Biotechnol 2000; 16: 579-84.

[52]   Tarangini K, Satpathy GR. Optimization of heavy metal biosorption using attenuated cultures of *Bacillus subtilis* and Peseudomonas aeruginosa. J Environ Res Dev 2009; 3.

[53]   Fadel M, Hassanein NM, Elshafei MM, Mostafa AH, Ahmed MA, Khater HM. Biosorption of manganese from groundwater by biomass of Saccharomyces cerevisiae. HBRC J 2017; 13: 106-13.

[54]   Mangunwardoyo W, Sudjarwo T, Patria MP. Bioremediation of effluent wastewater treatment plant Bojongsoang Bandung Indonesia using consortium aquatic plants and animals. Int J Res Rev Appl Sci 2013; 14(1): 150-60.

[55]   Thomson B, Hepburn CD, Lamare M, Baltar F. Temperature and UV light affect the activity of marine cell-free enzymes. Biogeosci Discuss 2017; 2017: 1-13.
[http://dx.doi.org/10.5194/bg-14-3971-2017]

[56]   Sowerby A, Emmett B, Beier C, *et al.* Microbial community changes in heathland soil communities along a geographical gradient: interaction with climate change manipulations. Soil Biol Biochem 2005; 37(10): 1805-13.
[http://dx.doi.org/10.1016/j.soilbio.2005.02.023]

[57]   Arivalagan P, Singaraj D, Haridass V, Kaliannan T. Removal of cadmium from aqueous solution by batch studies using *Bacillus cereus*. Ecol Eng 2014; 71: 728-35.
[http://dx.doi.org/10.1016/j.ecoleng.2014.08.005]

[58]   Govarthanan M, Mythili R, Selvankumar T, Kamala-Kannan S, Rajasekar A, Chang Y-C. Bioremediation of heavy metals using an endophytic bacterium *Paenibacillus* sp. RM isolated from the roots of *Tridax procumbens*. 3 Biotech 2016; 6: 242.

[59]   Atagana HI, Haynes RJ, Wallis FM. Optimization of soil physical and chemical conditions for the bioremediation of creosote-contaminated soil. Biodegradation 2003; 14(4): 297-307.
[http://dx.doi.org/10.1023/A:1024730722751] [PMID: 12948059]

[60]   Popenda A. Effect of redox potential on heavy metals and as behavior in dredged sediments. Desalin Water Treat 52: 3918-27.
[http://dx.doi.org/10.1080/19443994.2014.887449]

[61]   Ramos JL, Marqués S, van Dillewijn P, *et al.* Laboratory research aimed at closing the gaps in microbial bioremediation. Trends Biotechnol 2011; 29(12): 641-7.
[http://dx.doi.org/10.1016/j.tibtech.2011.06.007] [PMID: 21763021]

[62]   Karthik C, Oves M, Thangabalu R, Sharma R, Santhosh SB, Indra Arulselvi P. Cellulosimicrobium funkei-like enhances the growth of Phaseolus vulgaris by modulating oxidative damage under Chromium(VI) toxicity. J Adv Res 2016; 7: 839-50.

[63]   Singh R, Chadetrik R, Kumar R, *et al.* Biosorption optimization of lead(II), cadmium(II) and copper(II) using response surface methodology and applicability in isotherms and thermodynamics modeling. J Hazard Mater 2010; 174: 623-34.

[64]   Oves M, Saghir Khan M, Huda Qari A, Nadeen Felemban M, Almeelbi T. Heavy metals: biological importance and detoxification strategies. J Bioremediat Biodegrad 2016; 7: 334.

[65]   Fashola M, Ngole-Jeme V, Babalola O. Heavy metal pollution from gold mines: environmental effects and bacterial strategies for resistance. Int J Environ Res Public Health 2016; 13(11): 1047.

[http://dx.doi.org/10.3390/ijerph13111047] [PMID: 27792205]

[66]   Ayangbenro A, Babalola O. A new strategy for heavy metal polluted environments: a review of microbial biosorbents. Int J Environ Res Public Health 2017; 14(1): 94.
[http://dx.doi.org/10.3390/ijerph14010094] [PMID: 28106848]

[67]   Nie M, Pendall E, Bell C, *et al.* Positive climate feedbacks of soil microbial communities in a semi-arid grassland. Ecol Lett 2013; 16: 234-41.
[http://dx.doi.org/10.1111/ele.12034]

[68]   Srivastava J, Naraian R, Kalra SJS, Chandra H. Advances in microbial bioremediation and the factors influencing the process. Int J Environ Sci Technol 2014; 11(6): 1787-800.
[http://dx.doi.org/10.1007/s13762-013-0412-z]

[69]   Castro HF, Classen AT, Austin EE, Norby RJ, Schadt CW. Soil microbial community responses to multiple experimental climate change drivers. Appl Environ Microbiol 2010; 76: 999-1007.
[http://dx.doi.org/10.1128/AEM.02874-09]

[70]   Frey SD, Drijber R, Smith H, Melillo J. Microbial biomass, functional capacity, and community structure after 12 years of soil warming. Soil Biol Biochem 2008; 40: 2904-7.
[http://dx.doi.org/10.1016/j.soilbio.2008.07.020]

[71]   Bwapwa JK, Jaiyeola AT, Chetty R. Bioremediation of acid mine drainage using algae strains: a review. South Afr J Chem Eng 2017; 24: 62-70.
[http://dx.doi.org/10.1016/j.sajce.2017.06.005]

[72]   Elbaz-Poulichet F, Dupuy C, Cruzado A, Velasquez Z, Achterberg EP, Braungardt CB. Influence of sorption processes by iron oxides and algae fixation on arsenic and phosphate cycle in an acidic estuary (Tinto river, Spain). Water Res 2000; 34: 3222-30.

[73]   Brake SS, Hasiotis ST, Dannelly HK. Diatoms in acid mine drainage and their role in the formation of iron-rich stromatolites. Geomicrobiol J 2004; 21: 331-40.
[http://dx.doi.org/10.1080/01490450490454074]

[74]   Robinson C, Brömssen M, Bhattacharya P, *et al.* Dynamics of arsenic adsorption in the targeted arsenic-safe aquifers in Matlab, south-eastern Bangladesh: Insight from experimental studies. Appl Geochem 2011; 26(4): 624-35.
[http://dx.doi.org/10.1016/j.apgeochem.2011.01.019]

[75]   Tyagi M, da Fonseca MMR, de Carvalho CCCR. Bioaugmentation and biostimulation strategies to improve the effectiveness of bioremediation processes. Biodegradation 2011; 22(2): 231-41.
[http://dx.doi.org/10.1007/s10532-010-9394-4] [PMID: 20680666]

[76]   Cheng SS, Hsieh TL, Pan PT, *et al.* Study on biomonitoring of aged TPHcontaminated soil with bioaugmentation and biostimulation (Conference paper). 10th International *in situ* and on-site bioremediation symposium 2009.2009.

[77]   Chauhan A, Jain RK. Biodegradation: gaining insight through proteomics. Biodegradation 2010; 21(6): 861-79.
[http://dx.doi.org/10.1007/s10532-010-9361-0] [PMID: 20422258]

[78]   Jeffries TC, Seymour JR, Newton K, Smith RJ, Seuront L, Mitchell JG. Increases in the abundance of microbial genes encoding halotolerance and photosynthesis along a sediment salinity gradient. Biogeosciences 2012; 9(2): 815-25.
[http://dx.doi.org/10.5194/bg-9-815-2012]

[79]   Rayu S, Karpouzas DG, Singh BK. Emerging technologies in bioremediation: constraints and opportunities. Biodegradation 2012; 23(6): 917-26.
[http://dx.doi.org/10.1007/s10532-012-9576-3] [PMID: 22836784]

[80]   Achal V, Pan X, Zhang D. Remediation of copper-contaminated soil by Kocuria flava CR1, based on microbially induced calcite precipitation. Ecol Eng 2011; 37(10): 1601-5.
[http://dx.doi.org/10.1016/j.ecoleng.2011.06.008]

[81]   Adams JA, Ready KR. Extent of benzene biodegradation in saturated soil column during air sparging. Ground Water Monit Remediat 2003; 23(3): 85-94.
        [http://dx.doi.org/10.1111/j.1745-6592.2003.tb00686.x]

[82]   Kumar C, Mani D. Advances in bioremediation of heavy metals: a tool for environmental restoration. Saarbru¨cken: LAP Lambert Academic Publishing AG & Co. KG 2012.

[83]   Machackova J, Wittlingerova Z, Vlk K, Zima J. Major factors affecting *in situ* biodegradation rates of jet-fuel during large-scale biosparging project in sedimentary bedrock. J Environ Sci Health Part A Tox Hazard Subst Environ Eng 2012; 47(8): 1152-65.
        [http://dx.doi.org/10.1080/10934529.2012.668379] [PMID: 22506708]

[84]   Ma X, Novak PJ, Ferguson J, *et al.* The impact of H2 addition or dechlorinating microbial communities. Bioremediat J 2007; 11(2): 45-55.
        [http://dx.doi.org/10.1080/10889860701369490]

[85]   Baldwin BR, Peacock AD, Park M, *et al.* Multilevel samplers as microcosms to assess microbial response to biostimulation. Ground Water 2008; 46(2): 295-304.
        [http://dx.doi.org/10.1111/j.1745-6584.2007.00411.x] [PMID: 18194316]

[86]   Kanmani P, Aravind J, Preston D. Remediation of chromium contaminants using bacteria. Int J Environ Sci Technol 2012; 9(1): 183-93.
        [http://dx.doi.org/10.1007/s13762-011-0013-7]

[87]   Fulekar MH, Sharma J, Tendulkar A. Bioremediation of heavy metals using biostimulation in laboratory bioreactor. Environ Monit Assess 2012; 184(12): 7299-307.
        [http://dx.doi.org/10.1007/s10661-011-2499-3] [PMID: 22270588]

[88]   Cruz Viggi C, Pagnanelli F, Cibati A, Uccelletti D, Palleschi C, Toro L. Biotreatment and bioassessment of heavy metal removal by sulphate reducing bacteria in fixed bed reactors. Water Res 2010; 44(1): 151-8.
        [http://dx.doi.org/10.1016/j.watres.2009.09.013] [PMID: 19804893]

[89]   Geoffrey M. GADD, Geomycology: biogeochemical transformations of rocks, minerals, metals and radionuclides by fungi, bioweathering and bioremediation. Mycological research 2007; 111(2007): 3-49.

[90]   Li Q, Gadd GM. Biosynthesis of copper carbonate nanoparticles by ureolytic fungi. Appl Microbiol Biotechnol 2017; 101(19): 7397-407.
        [http://dx.doi.org/10.1007/s00253-017-8451-x] [PMID: 28799032]

[91]   Kim S, Park CB. Bio-inspired synthesis of minerals for energy, environment and medicinal applications. Adv Funct Mater 2013; 23(1): 10-25.
        [http://dx.doi.org/10.1002/adfm.201201994]

[92]   Achal V, Pan X, Zhang D. Bioremediation of strontium (Sr) contaminated aquifer quartz sand based on carbonate precipitation induced by Sr resistant Halomonas sp. Chemosphere 2012; 89(6): 764-8. b
        [http://dx.doi.org/10.1016/j.chemosphere.2012.06.064] [PMID: 22850277]

[93]   Chen Z, Li Z, Lin Y, Yin M, Ren J, Qu X. Biomineralization inspired surface engineering of nanocarriers for pH-responsive, targeted drug delivery. Biomaterials 2013; 34(4): 1364-71.
        [http://dx.doi.org/10.1016/j.biomaterials.2012.10.060] [PMID: 23140999]

[94]   Govarthanan M, Lee KJ, Cho M, Kim JS, Kamala-Kannan S, Oh BT. Significance of autochthonous *Bacillus* sp. KK1 on biomineralization of lead in mine tailings. Chemosphere 2013; 90(8): 2267-72.
        [http://dx.doi.org/10.1016/j.chemosphere.2012.10.038] [PMID: 23149181]

[95]   Tabak HH, Lens P, van Hullebusch ED, *et al.* Developments in bioremediation of soils and sediments polluted with metals and radionuclides – 1. Microbial processes and mechanisms affecting bioremediation of metal contamination and influencing metal toxicity and transport-. Rev Environ Sci Biotechnol 2005; 4(3): 115-56.
        [http://dx.doi.org/10.1007/s11157-005-2169-4]

[96]   Dasa AP, Swaina S, Pradhanb N, Sukla L B. Bioremediation and bioleaching potential of multi-metal resistant microorganism. International seminar on mineral processing technology (MPT-2013)CSI--IMMT. Bhubaneswar. 2013.

[97]   Weinberg ED. Iron availability and infection. Biochim Biophys Acta 2009; 1790(7): 600-5.

CHAPTER 2

Removal of Heavy Metals using Microbial Bioremediation

Deepesh Tiwari[1], Athar Hussain[2,*], Sunil Kumar Tiwari[3], Salman Ahmed[4], Mohd. Wajahat Sultan[5] and Mohd. Imran Ahamed[6]

[1] *Department of Environmental Engineering, Delhi Technological University, Delhi, India*

[2] *Department of Civil Engineering, CBP Government Engineering College Jaffarpur, New Delhi-73, India*

[3] *Department of Mechanical Engineering, CBP Government Engineering College Jaffarpur, New Delhi-73, India*

[4] *Department of Geology BFIT Group of Institution, Dehradun, India*

[5] *Department of Geology, Aligarh Muslim University, Aligarh, India*

[6] *Department of Chemistry, Aligarh Muslim University, Aligarh, India*

Abstract: The unorganized dumping of effluents along with different wastes directly into the water and soil has resulted in the rise of the concentration of many harmful metals, chemicals, and other gases in the environment. Widely known heavy metals triggering pollution issues are Lead (Pb), Chromium (Cr), Mercury (Hg), Cadmium (Cd), Copper (Cu), Arsenic (As) and Selenium (Se), as these heavy metals are generally found in the effluents of fertilizers, metallurgy, electroplating, and electronics industries. A number of physical-chemical reactions such as acid-base, oxidation-reducing, precipitation- dissolution, solubilization and ion-exchange processes occur and affect metal speciation. The physical methods used for heavy metals removal include magnetic separation, electrostatic separation, mechanical screening method, hydrodynamic classification, gravity concentration, flotation, and attrition scrubbing. The chemical methods used for eliminating heavy metals are chemical precipitation, coagulation and flocculation processes and the heavy metals are therefore removed as sludge. Electro-deposition, membrane filtration, electro-flotation and electrical oxidation are the various electrochemical treatment methods that are used to remove heavy metals from wastewater. Bioremediation is a biological method of eliminating toxins from the environment by using biological microbial bacteria such as *Pseudomonas* and *Sphingomonas*. Examples of bioremediation technologies include field farming, bioleaching, phytoremediation, bioventing, bioreactor, bio-stimulation and composting. Bioremediation is a natural process and is quite applicable as a waste treatment process for contaminated soils. The microbes present in the solution or soil can degrade the pollutants. It can also prove to be less expensive than

---

* **Corresponding author Athar Hussain:** Associate Professor, Department of Civil Engineering, CBP Government Engineering College Jaffarpur, New Delhi-73, India; E-mails: salmanahmed.alig@gmail.com, athariitr@gmail.com and deep1212001@gmail.com

Inamuddin (Ed.)

other technologies that are used for clean-up of hazardous waste and are also useful for the destruction of a wide variety of contaminants as many hazardous compounds can be transformed into harmless products.

Keywords: Bioremediation, Floatation, Hazardous, Heavy metals, Membrane filtration.

INTRODUCTION

The industrialization has grown at a faster rate since the 1950s and is considered as the growth of the economy with pollution on the earth increasing at a higher rate than the threshold predictions of WHO [1]. The knowledge of its adverse impact on health and the ecosystem leading to the unorganized dumping of effluents and waste directly into the water and soil was negligible. It resulted in the rise of the concentration of many harmful metals, chemicals, and gases that are exposed to the environment, and later on, also causes some global pandemics. After the first Global Environmental Summit in 1995, awareness of the environment and its safety had increased. Still many small and medium-size manufacturing units are not following any norms for waste disposal that are issued by the government. The problem of the rising concentration of pollutants appears to be growing for many years. Heavy metal is one of such pollutants, majorly for water and soil. Heavy metal is found naturally in the earth's crust and so it still exists in the water and soil around us but within limits [3]. Sometimes their exposure to the environment increases naturally. But after the rapid growth in industrialization and mining activities, exposure of these heavy metals to our surroundings has been increased and some of these also crossed the safety limits at various places. For example, in the case of the Bengal epidemic, there was a rise of Arsenic (As) due to the direct discharge of industrial waste in the river, causing an increase in its concentration in drinking water and soil resulting in the chronic disease Arsenicosis in the residents of that town [2].

Mining and manufacturing activities are major causes of an increase in heavy metal concentration on the surface in most of the regions. Rock weathering plays the second most abundant role. These two factors take around 97% contribution to heavy metal pollution. Other sources can be fertilizers, pesticides, and waste discharge [1]. To overcome this problem, the disciplines for waste disposal from manufacturing industries, mining activities, construction works, and others, must be regulated by the government [4]. There is a need for recycling and bioremediation of heavy metals that are one of the highly toxic pollutants. This article will cover heavy metals and their remediation processes and will mainly focus on the bioremediation process.

HEAVY METALS: SOURCES AND EFFECTS

The metals that possess a high density of approximately greater than 4000 kg.m^{-3}are termed as heavy metals and have high molecular weight with respect to other metals with their capability of forming colorful complexes (Cotton S., 2013; Pourret, 2018). Widely known heavy metals triggering pollution issues are Lead (Pb), Chromium (Cr), Mercury (Hg), Cadmium (Cd), Copper (Cu), Arsenic (As), Selenium (Se), *etc.* These heavy metals are generally found in the effluents of fertilizers, metallurgy, electroplating and electronics industries [5]. Mercury and lead are also used in laboratories, batteries and other scientific equipment. Some of the heavy metals like iron, zinc in a fixed concentration, are essential to our body for growth and immunity; but after a certain limit, or indifferent oxidation state, they can start interfering with regular functions in the body of a living entity and may cause serious health diseases. The heavy metals originating from the solid waste landfill leachate disposed of sewage, min tailing leachate, liquid wastes disposed of in deep wells, seepage of industrial effluents through lagoons or leakages ultimately join the groundwater reservoirs and contaminate the groundwater.

Distinct physical-chemical reactions such as acid-base, oxidation-reducing, precipitation- dissolution, solubilization and ion-exchange processes, occur and affect metal speciation and also typically influence the mobility of these metal pollutants. Abnormalities with other dissolved elements, pH, Eh, sorption and ion exchange capability, and the quantity of organic matter, usually accelerate the rate and intensity of these reactions. The availability of Eh, pH, temperature and moisture also influence the environmental toxicity, stability, and reactivity of heavy metals that eventually affect the speciation of these toxic metals. The diffusion of some heavy metals in soils can be decided by the use of special solvents, which solubilize the various phases [6].

The detrimental characteristics of such toxic metals are to eliminate the critical structure of cells of almost any living creature, decrease the rate of functions in that organism and modify the transport structure or work of enzymes, proteins or membranes [7]. Continuous exposure to these heavy metals can affect the various parts of the human body such as exposure to arsenic affects the human skin. Exposure to mercury affects the bones and similarly, the presence of lead affects the central nervous system and liver. Similarly, continuous exposure to cadmium affects blood and enzymes. Therefore, the presence of these heavy metals causes toxicity in some states or at higher concentrations. There is a need to get an understanding of the eco-friendly process that can eliminate these heavy metals from the food chain or the environment. Various sources of heavy metals as mentioned in the literature are summarized in Table **1**.

Table 1. Origins of heavy metals and public health risks.

Heavy Metals	Natural Sources	Anthropogenic Sources	Adverse Effects	References
Zn	Minerals (sulfides, oxides, silicates)	Dyes, metal alloys, galvanise, factory pollutants, containers, fertilizers and pesticides, plastics.	Abdominal discomfort, nausea, diarrhoea and vomiting, gastrointestinal problems, fatigue, irritability and restlessness., anemia	[8]
Pb	Galena mineral	Pipeline, gasoline, diesel, dyes, herbicide., Battery plants,	Neurotoxic	[8]
As	Rocks, Metal ores, Minerals(oxyhydroxide or sulfide)	Smelting and refining metal ores, arsenic-containing chemicals, pesticides, manufacturing glass, semiconductors, pharmaceutical substances	skin manifestation, neurological effects, vascular diseases, chronic lung disease, cancers of skin, lungs, liver, kidney and bladder	[9]
Cd	Zn and Pb minerals, phosphate rocks	Mining residue, galvanizing, battery plants, paints, automotive emission	Pulmonary, cardiovascular and renal disturbances	[8]
Hg	Volcanic eruption, geothermal, oceanic, terrestrial emissions	Combustion of fossil fuels, Au-ag refining, concrete, paper mills processing, ammonia, incineration of garbage, amalgamation of dental implants.	Neurotoxicity, fetal brain toxicity, renal toxicity, cardiomyopathy, immune malfunction and abnormal blood sugar	[10]

HEAVY METALS OCCURRENCES

This phrase of heavy metal encompasses the metal or metal category with a density of more than 4 ± 1 g / cm^3. The concept is also known and often applicable to toxins broadly dispersed in land and freshwater ecosystems [11]. The other most prevalent heavy metals comprise arsenic, copper, boron, mercury, lead, zinc, cadmium, chromium, copper and platinum metals [12]. The Earth's crust includes these heavy metals as natural components and propagates in the ecosystem through anthropogenic practices.

They are regarded as persistent contaminants in the environment as they cannot be degraded or eliminated easily. However, over time these heavy metals enter the human body through the food chain with food, air, and water and bio-accumulates over a while [13]. The possible sources for the occurrence of these heavy metals can be categorized as natural sources, anthropogenic sources, and industrial activities. Natural Sources mainly include the occurrence of heavy metals in the earth crust or mineral rocks and stones. Due to the seasoning and rock weathering, these heavy metals come on the surface of the earth and spread in some particular regions like the Mendip Region in Great Britain, where the soil found contains heavy metals such as lead, cadmium, and zinc. These are also enriched in the earth's crust in that area [13].

The main anthropogenic origins of heavy metals occurred in the environment are the result of human activity such as mining including various industrial and other hazardous waste management operation. The same can be explained with example like the fact that even after 5-15 years of hard rock mines operation and closure with depletion of the minerals the heavy metals that come into the environment cannot be eliminated even after hundreds of years after the termination of mining operations [14]. Besides the mining, mercury is emitted through cosmetics and chloroalkali industry using mercury to process sodium hydroxide into the environment [12]. The heavy metals are emitted in the environment both in the elemental form as well as in the form of organic and inorganic compounds. The sources of heavy metal emissions through human activities are various industries as point sources along with foundries, smelters, combustion by-products and transportation [13]. Sewage and surface water discharge areas and ceiling areas; incinerators and crematoria and motor cars are the numerous practices in which heavy metals are discharged into the environment. Such anthropogenic operations releasing heavy metals into the atmosphere are the electroplating, smelting, dentists, hospitals, storing timber, drum fixing, waste management and storage, metal handling and care of livestock and poultry, sheep and cattle, scrap metal yard, tanning plants, chemical manufacturers, accumulator manufacturing and use, mercury lamps, thermometer, battery and other miscellaneous equipment [12].

HEAVY METAL REMOVAL STRATEGIES

Physical Methods

The heavy metals can be removed by using the physical separation techniques as these techniques primarily remove the particulate forms of metals, discrete particles, or metal-bearing particles. The physical method includes magnetic separation, electrostatic separation, mechanical screening method, hydrodynamic classification, gravity concentration, flotation, and attrition scrubbing [15]. These physical methods that can be used for the removal of heavy metals from soil or water are based on the filtration process or the pressure-resistant and also the charge resistant, which can be briefly discussed below as:

Ultrafiltration is a process where the permeable membrane is used to separate the heavy metals, macromolecules and suspended solids of pore size ranging from 5–20 nm from inorganic solution. The separating heavy metals possess a molecular weight ranging from 1000–100,000 Da [16].

Electrodialysis (ED) is another membrane separation process that is generally used for the removal of heavy metals. The principle behind the removal of heavy metals is that utilizing the electrical potential, the ionized species in the solution are passed through an ion-exchange membrane. The membranes used in the process possess anionic or cationic characteristics and are manufactured in thin plastic sheets. These ionic species present in solution when passed through the various cell compartments get bifurcated as the cations present in the solution migrate towards the cathode and the anions migrate towards the anode. This leads to crossing the anion exchange and cation-exchange membranes [17].

Coagulation-flocculation is another chemical process and mechanism that is utilized for the removal of heavy metals. It is based on charge neutralization on particles and works on the principle of zeta potential (ζ). This zeta potential measurement is used as the criteria to define the electrostatic interaction between pollutants and coagulant-flocculant agents [18]. In the coagulation process, the chemicals such as alum and ferric salts are generally added to the solution containing pollutants that reduces the net surface charge of the colloidal particles and stabilizes them by electrostatic repulsion process [19]. The flocculation is an agitation process of liquid being carried out after adding for a period of around 25-30 minutes. The charge on the particles neutralises over this time period and the particles constantly increase the particle size by agglomeration and reaction with inorganic polymers produced by the inserted organic polymers [20].

The ion exchange process is another very commonly used technique in which the soluble ions are separated from the liquid phase to a solid phase by the addition of materials containing ions of opposite charge. The method is most widely used in the water treatment industry. This process is very cost-effective as it utilizes low-cost materials and is very convenient in operation and maintenance. The water has a low concentration of heavy metals that can be treated efficiently and heavy metals from aqueous solutions can be removed effectively and efficiently [21, 22].

Reverse Osmosis (RO) is an essential method based on the pollutant separating principle by pressuring a solution by virtue of a semi-permeable membrane, which regulates the solute on the other, thus enabling the treated solution to be moved. The semi-permeable membrane permits the solvent to flow through but not to metals. Thus, the most semi-permeable membranes used only for reverse osmosis have a denser barrier coating in the matrix [23].

Chemical Methods

Chemical treatment is one of the most commonly applied approaches for eliminating heavy metals from untreated wastewater because of its comfort and efficiency of use. In traditional chemical precipitation techniques, the insoluble precipitation of these heavy metals in form of hydroxides, sulphides, carbonates and phosphates are removed. The process is based on the principle where insoluble metal precipitates are produced by the reaction of dissolved metals in the precipitant and solution. During this precipitation process very fine particles are generally generated and therefore to increase the size of these particles in the solution itself the chemical precipitants, coagulation and flocculation processes are generally used and the heavy metals are therefore removed as sludge [24].

Electrolysis is another important technology utilized for the removal of heavy metals from wastewater effluents. The process is carried out by passing the electric current through a metal-bearing aqueous solution containing a cathode plate and an insoluble anode. The electricity in the process is generated by the movements of electrons from one element to another.

Electrochemical is also another technique used by virtue of the acidic medium to eliminate heavy metals from the wastewater. Heavy metal accumulation is performed through hydroxides in a slightly acidic or neutralized solution. Electrodeposition, membrane filtration, electroflotation and electrical oxidation are the various electrochemical treatment methods that are used to treat wastewater [25].

Biological Methods

Bioremediation is a method of eliminating toxins from the environment by treating them with biological microbial. Relevant examples of bioremediation technologies include field farming, bioleaching, phytoremediation, bioventing, bioreactor, biostimulation and composting. The aerobic method includes the use of aerobic bacteria such as *Pseudomonas*, *Sphingomonas* that are being recognized for their degradative abilities. However, the microbes such as *Rhodococcus* and *Mycobacterium* are generally used to degrade pesticides and hydrocarbons, both alkanes and polyaromatic compounds. Many of these bacteria use the contaminants as the sole source of carbon and energy. Consequently, the anaerobic approach involves the use of bacterial species to eliminate different contaminants. In recent years, aerobic bacteria also gained attention for use in the bioremediation of trichloroethylene solvent dechlorination, polychlorinated biphenyls in river sediments and chloroform [26].

Phytoremediation

Phytoremediation is a term that refers to making use of plants to remediate the polluted soils by the process of extracting or detoxifying the pollutants. The technique is very effective, nonintrusive, economical, aesthetically pleasing and is a socially accepted technology [27]. Plants can be useful to remove contamination of heavy metals, especially from soils. Several processes are used for metal removal using plants. Another important technique Phytoextraction is used that involves the mechanism such as the uptaking and migration of metal contaminants present in soil take place through the plant's roots into plant systems above ground [28]. Cleaning of the soil's environmental toxins can be prevented through the use of the radical plant or seedlings from water waste and the process is termed as Phyto-filtration [29, 30].

Phytostimulation is another important technique generally used for the augmentation of microbe activity to decompose biological compounds through exudation from plant roots [31]. Another very effective method known as phytostabilization is generally used to absorb pollutants from the soil by making use of plant roots and thereafter it retains these pollutants in the rhizosphere. Thereafter, the rhizosphere is separated and stabilized, thus making them harmless and ultimately preventing these pollutants from dispersing in the environment [32]. The process of removing the pollutants by changing them into a vapor state and releasing them into the atmosphere through making use of plants is known as phytovolatilization [30, 33]. The plant enzymes are also used in the breakdown of organic pollutants and changing them from hazardous to non-hazardous forms is carried out by a process known as phytodegradation [33]. Rhizofiltration is yet another procedure used to eliminate hazardous chemicals or contaminants from groundwater by virtue of filtration and is based on the process of the aggregation of rhizospheric by floras [34].

Bioremediation

Contaminants present in the soil, sediments, and water including metals, hydrocarbons, toxicants, agrochemicals, *etc.* can be cleaned or stabilized by doing biologically encoded changes in their oxidation states by passing it through mediator film containing microbes. This process is defined as Bioremediation [35]. Some inorganic compounds like heavy metals cannot be simplified by general species of microbes. They need specific microbes for the type of contaminating compounds depending on their active metabolizing capabilities. For example, Fe is a necessitating requirement of all bacteria, but Fe^{2+} is an essential micronutrient for only anaerobic bacteria [36].

Among the above-discussed methods for remediation of contaminated soil, water, and atmosphere, Physico-chemical methods are advantageous as they provide a swift process with simple procedures with a range of input loads and controls. Sometimes chemical methods are the only or most suitable treatment methods for some inorganic compounds. Although, such type of methods possesses certain major drawbacks that include the cost of operation being very high while making use of costly chemicals with their after-effects on the environment. It also requires high-energy consumption in the handling of toxic sludge disposal which again is very uneconomical. On the other hand, biological treatment methods are sustainable and low-cost methods with the availability of a large variety of adsorbents on this planet, but they provide a slow treatment rate, need a larger area, and are sometimes unable to treat certain inorganic compounds [23]. In biological methods, bioremediation is a method that does not require much large area and is capable of treating inorganic and toxic compounds like heavy metals by using some specified microbes [35]. So, here bioremediation methods will be discussed in detail.

Mechanism of Bioremediation

The bioremediation processes are carried out by making use of microorganisms that mineralizes these organic pollutants and the end-products obtained are carbon dioxide and water. However, the intermediate metabolic products in the process can be utilized by the microorganisms as a sole substrate for their cell growth. Therefore, these microorganisms utilize these chemicals and eradicate the heavy metals present in the soil for their growth and development. These microorganisms possess the capability of dissolving these metals and reducing or oxidizing them to transition metals. The diverse processes such as oxidizing, binding, immobilizing, volatilizing methods exist by which microbes reinstate the environment and the transformation of heavy metals takes place by any such method [35]. The microorganism's structure is such that the cell wall consists of various macromolecules, such as polysaccharides and proteins. It also contains several charged functional groups that include the thioether, phenol, carboxyl, imidazole, sulfhydryl, ester, hydroxyl sulfate, amino, and carbonyl, amide groups [37 - 39]. These positively charged metals in the solutions gravitate to these functional groups for adsorption. The adsorption capacity of these microorganisms can be increased and can be best exploited by cultivating these microorganisms and influencing the cell wall composition [1]. The organic pollutants may disrupt the membranes but sometimes the microorganisms are acclimatized that develop a defense mechanism in the cells and form an outer cell-membrane-protective material [40].

Bioremediation by Biosorption

The process of holding anything, either by absorption or adsorption is termed as sorption. However, in case the ability of microorganisms to bind metals from an aqueous solution is defined as 'biosorption' and the microorganisms carrying out the process are called bio-sorbents. The biosorption process has a great scope and wide applications in industries as the removal of heavy metals can be effectively carried out by using the microorganisms. As a biological substratum, marine algae can be used as a metal accumulator of dissolved metal ions because it is capable to absorb metal ions from aqueous solutions through many micro-organisms [41]. The Zn (II) and Cd (II) elimination can be achieved with *Saccharomyces cerevisiae*, which functions as a biosorbent by means of the ion substitution process.

These microbial may ingest heavy metals without using any energy from linking sites in their cell structure because there are some bacterial cell walls contain different reactive compounds. The extracellular polymeric compounds have major effects on the properties of both the acid and base and are quite effective in carrying out metal adsorption therefore they are of particular importance [42]. The biomass concentration, temperature, pH, ion strength, other solvent ions and particle size are the valuable variables that can impact the bioassorption of metals [43]. It can be well utilized by both living and dead biomass and biosorption can occur in any case as it is independent of cell metabolism [8].

Bioremediation by Bioaccumulation

The energy for the cell metabolism cycle degrades heavy metals. Bioaccumulation involves bioremediation of both the mixed active and inactive forms of hazardous metals. The textile waste-water effluents contain heavy metals that can be effectively treated by using *Cunninghamella elegans* as a promising sorbent. The use of fungi as biocatalyst helps in eradicating the heavy metals toxicity by converting them h into less toxic compounds. The genus of fungi including *Stachybotrys* sp., *Klebsiella, Oxytoca Klebsiella, Oxytoca Phlebia* sp., *Pleurotus pulmonarius, Allescheriella* sp., and *Botryosphaeria Rhodina* have the capability of metal-binding. For the removal of Pb (II) from the contaminated sites by a biosorption process, fungal species such as *Aspergillus parasasitica* and *Cephalosporium aphidicola* may be implemented. The fungi species, such as *Hymenoscyphus ericae, Neocosmospora vasinfecta* and *Verticillium terrestrial*, can eliminate Hg efficiently by biotransforming Hg to Hg (II) and finally by reducing its toxic effects [35]. The bioaccumulation process consists of both intra- and extracellular processes in which the passive uptake of heavy metals is limited and does not play a very well-defined role [44].

Comparison of Biosorption and Bioaccumulation Process

Comparison between Biosorption and Bioaccumulation can be done based on various characteristics. Biosorption process is cheaper because the biomass used in it is generally found in industrial waste, so only the transportation and production cost of biosorbent is included, whereas bioaccumulation cost is high because of the presence of living cells is required, that need to get feeded. The sorption process is affected by the pH of the solution, but it can be carried in a wider pH range, but the bioaccumulation process is sensitive as a slight change in pH can affect the microbial. The selectivity of metals in the case of bioaccumulation is better than in biosorption. However, the rate of removal is faster with biosorption mechanisms. Reuse and regeneration of biosorbents can be done to use in multiple cycles and also recovery of heavy metals is possible, whereas it is limited in bioaccumulation due to intercellular accumulation and hence the biomass recovered from it cannot be used for other purposes. The bioaccumulation process demands high energy for cell growth and is lower for biosorption [5].

Biotechnological Intervention in Bioremediation Processes by the Microbial Approach

Kang *et al.*, 2016 [45] in a study reported the synergistic effect of bacterial consortia for the removal of Pb, Cd and Cu from contaminated soil through the bioremediation process using strains such as *Viridibacillus arenosi* B-21, *Sporosarcina soli* B-22, *Enterobacter cloacae* KJ-46 and *E. cloacae* KJ-47. The bacterial mix consortia possess greater efficiency and resistance to heavy metals remediation in contrast to single strain cultivation in 48 hours. The Pb, Cd and Cu reported an efficiency of 98.3%, 85.4% and 5.6%, respectively. The different mechanisms used for the application of microbial bioremediation are (i) Intracellular metal-binding proteins and peptides such as metallothioneins (MTs) and phytochelatins, along with compounds such as catecholic bacterial siderophores, as opposed to fungi which produce hydroxamate siderophores or toxic cell-wall sequestration of metals (ii) Modification of biochemical pathways to inhibit metal absorption (iii) Transformation of metals into innocuous enzyme forms (iv) Reduction by complex efflux mechanisms of intracellular metals [46].

By using recombinant DNA technology, the genetic content of genetically engineered microorganisms may be customised. A character-specific productive strain is added to the biorestructuring of activated soil and water in the sludge. It gives the advantage of cultivating diverse microbial strains that can be used as bio-remediators under different environmentally demanding conditions and endure adverse stressful circumstances [47]. However, the genetic engineering of rhizospheric bacteria and endophytes is perceived to be one of the most advanced and interesting innovation to rebuild polluted sites with metals for plant-associated soil depletion of contaminants [35]. It has also been reported that *Escherichia coli* and *Moreaxella* sp. are such bacteria that while expressing phytochelatin 20 on the cell surface can even accumulate 25 times more Cd or Hg as compared to wild-type strains [48].

However, sustaining the recombinant bacteria population in the soil is one of the major hindrances to utilizing these GEMs in hostile field conditions under various environmental conditions. This may also be because of the competition that exists between these microorganisms from native bacterial populations [49]. In comparison, only a few bacterial strains such as *Escherichia coli. Pseudomonas putida, Bacillus subtilis, etc.* have been used with molecular approaches. This means that other microorganisms must be studied by molecular interference for use in heavy metal bioremediation [35].

The Ability of Microorganisms to Bioremediate Heavy Metals

Bacteria Remediation Capacity of Heavy Metal

The biosorption abilities in microbial biomass differ and it also varies considerably among the microbes. However, the pretreatment and experimental conditions of these microbial cells alter the biosorption capability of removing any pollutant. The microbes can be used as effective biosorbents because of their ubiquity, size and potential to develop under temperate climates and their resilience to environmental conditions [50].

The functional groups such as phosphate groups, carbonyl groups, hydroxyl groups are a component of the negatively charged biomolecules of microbial cell wall surfaces that binds readily to heavy metal ions. Bacterial functional groups, including sulphate uronic acid ($SO4^{2-}$) and carboxylic groups (RCOOH), can also easily carry out ion exchanges [8]. The peptidoglycan layers of gram-positive bacteria cell walls contain the glutamic acid, amino acids and alanine along with meso-di-aminopimelic acid and teichoic acid. Enzymes such as lipopolysaccharides, lipoproteines, glycoproteins and phospholipids are located within the cell wall of gram-negative bacteria [51].

Jaysankar *et al.* [52] in a study used *Brevibacterium iodonium, Bacillus pumilus, Pseudomonas aeruginosa,* and *Alcaligenes faecalis* as a mercury resistant bacteria and has been successfully used for the removal of cadmium (Cd) and lead (Pb) from the soil. It has been reported that the cadmium (Cd) concentration can be reduced from 1000 mg/L to 17.4 mg/L using the *P. aeruginosa* and *A. faecalis* species with removal efficiency to be 70% and 75%, respectively. However, using

P. aeruginosa and *A. faecalis* the Cd can be reduced to 19.2 mg/L in a 72 hours duration. Similarly, by using the species of *Brevibacterium iodonium* and *Bacillus pumilus*, the Pd concentration can be reduced from 1000 mg/L to 1.8 mg/L indicating the removal efficiency greater than 87% and 88% in 96 hours duration. They also revealed that the *Acinetobacter* sp and *Arthrobacter* sp. consortium of bacteria could decrease by 78% with 16 mg / L of metal ion concentration.

Abioye *et al.*, 2018 [53] used the species *Bacillus subtilis*, *Aspergillus niger*, *B. megaterium*, and *Penicillium* sp. *B. Megaterium* and explored the biosorption of Pb, Cr and Cd in industrial waste distilleries. The highest Pb reduction from 2.13 mg/l to 0.03 mg/L has been reported by using *Bacillus subtilis*. However, the highest Cr concentration has been found to be reduced from 1.38 to 0.08 mg/L by using *A. niger* followed by *Penicillium* sp. reducing the Cr concentration from 1.38 to 0.13 mg/L. However, the Cd concentration has been reported to be reduced from 0.4 mg/L to 0.03 mg/L using *B. subtilis* followed by *B. megaterium*, which can reduce the Cd concentration from 0.04 mg/L to 0.06 mg/L after 20 days period [53]. The selective genetically engineered bacteria used for heavy metals removal from the soil is summarized in Table **2**.

Table 2. Studies on heavy metal removal using bioremediation technique by various investigators.

Heavy Metal	Initial Concentration (ppm)	Removal Efficiency	Mediated Bacteria	References
Cr(IV)	100	100%	*Bacillus cereus* strain XMCr-6	[48]
Cu			*Kocuria flava*	[8]
Au			*Stenotrophomonas* spp.	[8]
As	0.05	100%	*Escherichia coli* strain	[49]
Cr^{6+}	1.4-1000	100%	*Methylococcus capsulatus*	[54]
Cr			*Pseudomonas putida*	[55]
Cd^{2+}, Hg			*Ralstonia eutropha* CH34, *Deinococcus diodurans*	[56]
Hg	7.4	96%	*Escherichia coli* JM109	[57]
Ni	145	80%	*P. fluorescens* 4F39	[58, 59]
Cd, Zn, Cu	0.5 mM* each	50%	*Pseudomonas veronii*	[8, 60]

*all three Cadmium, Zinc and Copper are taken as mixtures, each for an experiment carried at 32°C.

Fungi Remediation Capacity of Heavy Metal

Another important biological species fungus can be effectively used as a biosorbents for the removal of a wide variety of toxic metals. It poses an excellent capability for the uptake of heavy metals and their recovery [61]. These active and lifeless fungal cells have been considered to play an important part in the elasticity of inorganic compounds across many studies [62 - 64]. In a study that assessed the removal efficiency of Cr present in tannery wastewater by using *Aspergillus* sp. It has been reported that in a bioreactor system with a synthetic substrate medium around 85% of chromium can be removed effectively at a pH value of 6 as compared to a 65% removal of Cr from the tannery effluent. The lesser efficiency in tannery effluent indicates that the other organic pollutants present in the effluent can hinder the growth of the organisms [65]. Park *et al.*, 2005 [66] observed that the dead fungal biomass of Rhizopus oryzae, *Saccharomyces cerevisiae, Penicillium chrysogenum* and *Aspergillus niger* can be used to convert toxic Cr (VI) to less toxic or nontoxic Cr (III). Luna *et al.*, 2016 [67] in a study reported that using *Candida sphaerica* species that produce biosurfactants with Fe, Zn and Pb removal efficiency of 95%, 90% and 79% respectively. Until isolation from the soil, these surfactants might form complexes with metal ions and interact directly with heavy metals. The various biological species used for heavy metals removal are shown in Table 3.

Table 3. Heavy metals removal by using various biological species.

Family	Biological Species	Heavy Metals Removal	References
Fungus	*Aspergillus versicolor*	Cr, Cu, Ni, Hg	[68]
	Aspergillus niger	Ni, Cu, Cr	[69]
	Aspergillus sp	Cr, Ni	[70]
	Sphaerotilus natans	Cr, Cu	[71]
	A. lentulus	Cu	[72]
	Saccharomyces cerevisiae	Cr	[73]
	Saccharomyces cerevisiae, Rhizopus arrhizus	Cu, Cd, Zn	[74]
Yeast	*Saccharomyces cerevisiae*	Pb, Cd	[75]
	Candida parapsilosis	Hg	[76]
	Coriolus versicolor, Phanerochaete chrysosporium	LiP	[77]

Remediation Capacity of Heavy Metal by Algae

Algae is an alternative autotrophic biota that can function in low nutrients and provide adequate biomass relative to other microbial biosorbents. The evidence of algae as for heavy metal removal with a high sorption capacity as biosorbents exists in the literature [37]. The study was carried out using an aqueous solution under various conditions by utilizing *Chlorella vulgaris* dead cells for productive ion elimination, differing in pH, concentration, and contact time of Cd^{2+}, Cu^{2+}, and Pb^{2+} ions. The obtained results were found to be very promising in the removal of metal ions. The biomass of C. vulgaris can be efficiently used as biosorbent and about 95%, 97,7% and 99,4% of the Cd^{2+}, Cu^{2+} and Pb^{2+} ions can be separated efficiently from each metal ion 's 50 $mgdm^{-3}$ mixed solution, respectively [38].

In another study being carried out by Mustapha and Halimoon, 2015 [78] observed that using algae as biosorbent the metal ions can be effectively treated with 15.3–84.6% higher efficiency as compared to other microbial biosorbents. However, this can be achieved through ion exchange mechanisms. It has also been reported that heavy metals such as Cd, Ni and Pb can be effectively removed by brown marine algae that can remediate the chemical groups on their surfaces such as carboxyl, sulfonate, amino and sulfhydryl groups. The algal family species (Table **4**) were used for the removal of heavy metals.

Table 4. The removal of Heavy metals by various microbial-species.

Family	Species	Heavy Metals	References
Algae	*Spirogyra* spp. and *Cladophora* spp.	Pb(II), Cu(II)	[79]
	Spirogyra spp. and *Spirullina* spp.	Cr, Cu, Fe, Mn	[80]
	Lessonia nigrescens	As(V)	[81]
	Gelidium (Agar extraction)	Cd(II)	[82]
	Sargassum natans, Ascophyllum nodosum	Au, Co	[74]

Heavy Metal Removal Using Biofilms

Several studies being carried out by different investigators for the removal of heavy metals from the soil through the application of biofilms have been reported. The usage of biofilm also functions as a specialist bioremediation instrument and as a bio-stabilizer. These biofilms possess a very high tolerance capacity and can work effectively and accurately under the high level of poisonous inorganic elements, even at an acute toxicity. Grujić *et al.*, 2017 [39] conducted a study and observed that the use of *Rhodotorula mucilaginosa* can remove heavy metals with efficiency varying from 4.79 to 10.25% for planktonic cells and the efficiency of removal may vary from 91.71 to 95.39% for biofilm cells. These biofilms are comprised of microorganisms consortia that are encased in an extracellular matrix of polysaccharides, exudates and detritus. However, these biofilms can also be constituted of a single bacterial species [83].

Plant Approach

The application of Phytoremediation technology by accumulating the pollutants or their metabolites in plant tissues possesses some constraints. This may be due to the fact that it shortens the plant life and the pollutants are released in the troposphere through volatilization. The problem can be curbed by minimizing metal tolerance through manipulation, degradation and accumulation potential of plants against various inorganic pollutants [35]. This can be best carried out by making use of some designer plant approach. This can be done by introducing the specific bacterial genetic factor for metal degradation in plant tissues that can enhance the degradation of metals inside these plant tissues [84]. The plant resistance can be imposed by the metal detoxifying chelators, such as metallothineins and phytochetain, by enhanced absorption, transportation and aggregation of different heavy metals [85]. However, through the volatilization of Hg and Se, the transgenic plants carry bacterial reductase that can enhance the process while accumulating the arsenic in plant shoots [86].

Therefore, vegetation like poplar, willow and Jatropha can be effectively utilized for phytoremediation as they are rising rapidly as well as high-biomass-yielding vegetation that can produce enormous energy. Also, the transfer of these metals from soil or water by utilizing the above processes never solves the problem. This is because the burning of metal-contaminated plant material or biomass for its use in energy production will again release these heavy metals into the atmosphere. This will ultimately result in the release of these heavy metals from soil or water into the air. Therefore the biomass or plants accumulating these heavy metals should be disposed of with proper disposal methods at the proper site so that they do not create any harm to the environment. However, the manipulation of desired plant species as discussed earlier can solve the problem up to a certain extent. The multigene plant species can eradicate pollutants as it ensures that the biomass collected is used entirely to further support the population [87].

Advantages of Bioremediation

1. The soil contaminated with heavy metals can be effectively treated by using the bioremediation process. This is a natural process and is quite applicable as a waste treatment process for contaminated soils. The microbes present in the solution or soil can degrade the pollutants also when the contaminant is degraded, the biodegradative population declines.

2. Less energy is required as compared to other technologies.

3. Bioremediation can prove less expensive than other technologies that are used for clean-up of hazardous waste.

4. Bioremediation is useful for the destruction of a wide variety of contaminants. Many compounds that are legally considered to be hazardous can be transformed into harmless products.

5. Instead of transferring contaminants from one environment medium to another, *e.g.* from land to water or air, the destruction of target pollutants is possible and is able to flourish at one uninhabitable site.

Disadvantages of Bioremediation

1. Biological processes are often highly specific. Important site factors required for success include the presence of metabolically capable microbial populations, suitable environmental growth conditions, and appropriate levels of nutrients and contaminants.

2. Bioremediation often takes longer than other treatment options, such as excavation and removal of soil or incineration.

3. Contaminants may be present as solids, liquids and gases.

4. A dynamic process, difficult to predict future effectiveness.

5. Bioremediation is limited to those compounds that are biodegradable. Not all compounds are susceptible to rapid and complete degradation.

CONCLUSION

Manufacturing industries, mining activities, construction works, and other such activities are major causes of an increase in heavy metal concentration on the surface in most of the regions. The heavy metals originating from the solid waste landfill leachate disposed of sewage, min tailing leachate, liquid wastes disposed of in deep wells, seepage of industrial effluents through lagoons or leakages ultimately join the groundwater reservoirs and contaminate the groundwater. The continuous exposure to these heavy metals can affect the various parts of the human body. There is a need to get an understanding of the eco-friendly process that can eliminate these heavy metals from the food chain or from the environment. The physical methods include the magnetic separation, electrostatic separation, mechanical screening method, hydrodynamic classification, gravity concentration, flotation, and attrition scrubbing can be used for the removal of heavy metals from soil or water. Chemical treatment is one of the most commonly applied approaches for eliminating heavy metals from untreated wastewater because of its comfort and efficiency of use. Bioremediation is a biological method of eliminating toxins from the environment by treating them with biological microbial. Biosorption process is cheaper because the biomass used in it is generally found in industrial waste, so only the transportation and production cost of biosorbent is included, whereas bioaccumulation cost is high because of the presence of living cells is required, that need to be fed. By using recombinant DNA technology, the genetic content of genetically engineered microorganisms may be customised. *Acinetobacter* sp and *Arthrobacter* sp. consortium of bacteria could decrease by 78% with 16 mg / L of metal ion concentration. Algae is an alternative autotrophic biota that can function in low nutrients and provide adequate biomass relative to other microbial biosorbents. Biological processes are often highly specific. Important site factors required for success include the presence of metabolically capable microbial populations, suitable environmental growth conditions, and appropriate levels of nutrients and contaminants.

CONSENT FOR PUBLICATION

Not applicable.

CONFLICT OF INTEREST

The authors declare no conflict of interest, financial or otherwise.

ACKNOWLEDGEMENT

Declared none.

REFERENCES

[1]   Zhou Q, Yang N, Li Y, *et al.* Total concentrations and sources of heavy metal pollution in global river and lake water bodies from 1972 to 2017. Glob Ecol Conserv 2020; 22: e00925.
[http://dx.doi.org/10.1016/j.gecco.2020.e00925]

[2]   Mazumder DN, Ghosh A, Majumdar K, Ghosh N, Saha C, Mazumder RN. Arsenic contamination of ground water and its health impact on population of district of Nadia, West Bengal, India. Indian J Community Med 2010; 35(2): 331-8.
[http://dx.doi.org/10.4103/0970-0218.66897] [PMID: 20922118]

[3]   Cotton S. Lanthanide and Actinide Chemistry. John Wiley & Sons 2013.

[4]   Pourret O. On the Necessity of Banning the Term "Heavy Metal" from the Scientific Literature. Sustainability (Basel) 2018; 10(8): 2879.
[http://dx.doi.org/10.3390/su10082879]

[5]   Zabochnicka-Świątek M, Krzywonos M. Potentials of Biosorption and Bioaccumulation Processes for Heavy Metal Removal. Pol J Environ Stud 2014; 23(2).

[6]   Kulshreshtha A, Agrawal R, Barar M, Saxena S. A Review on Bioremediation of Heavy Metals in Contaminated Water. IOSR J Environ Sci Toxicol Food Technol 2014; 8(7): 44-50.

[http://dx.doi.org/10.9790/2402-08714450]

[7] Tak HI, Ahmad F, Babalola OO. Advances in the application of plant growth-promoting rhizobacteria in phytoremediation of heavy metals. Rev Environ Contam Toxicol 2013; 223: 33-52.
[PMID: 23149811]

[8] Coelho L M, Rezende H C, Coelho L M, de Sousa P A, Melo D F, Coelho N M. Bioremediation of Polluted Waters Using Microorganisms. Adv bioremediation wastewater polluted soil 2015; 10: 60770.

[9] Singh N, Kumar D, Sahu AP. Arsenic in the environment: effects on human health and possible prevention. J Environ Biol 2007; 28(2) (Suppl.): 359-65.
[PMID: 17929751]

[10] Liu G, Cai Y, O'Driscoll N, Feng X, Jiang G. Overview of Mercury in the Environment.Environ Chem Toxicol Mercur. 2012; pp. 1-12.

[11] Duffus JH. "Heavy metals" a meaningless term? (IUPAC Technical Report). Pure Appl Chem 2002; 74(5): 793-807.
[http://dx.doi.org/10.1351/pac200274050793]

[12] Mohammed AS, Kapri A, Goel R. Heavy Metal Pollution: Source, Impact, and Remedies.Biomanagement of metal-contaminated soils. Springer 2011; pp. 1-28.
[http://dx.doi.org/10.1007/978-94-007-1914-9_1]

[13] Fuge R, Glover SP, Pearce NJG, Perkins WT. Some observations on heavy metal concentrations in soils of the Mendip region of north Somerset. Environ Geochem Health 1991; 13(4): 193-6.
[http://dx.doi.org/10.1007/BF01758636] [PMID: 24203102]

[14] Peplow D. Environmental Impacts of Mining in Eastern Washington. Washington: University of Washington Water Center 1999.

[15] Dermont G, Bergeron M, Mercier G, Richer-Laflèche M. Soil washing for metal removal: A review of physical/chemical technologies and field applications. J Hazard Mater 2008; 152(1): 1-31.
[http://dx.doi.org/10.1016/j.jhazmat.2007.10.043] [PMID: 18036735]

[16] Vigneswaran S, Ngo HH, Chaudhary DS, Hung Y-T. Physicochemical Treatment Processes for Water Reuse.Physicochemical treatment processes. Springer 2005; pp. 635-76.
[http://dx.doi.org/10.1385/1-59259-820-x:635]

[17] L.; CHEN, Q. Industrial Application of UF Membrane in the Pretreatment for RO System. Membr Sci Technol 2003; 4: 9. [J].

[18] López-Maldonado EA, Oropeza-Guzman MT, Jurado-Baizaval JL, Ochoa-Terán A. Coagulation–flocculation mechanisms in wastewater treatment plants through zeta potential measurements. J Hazard Mater 2014; 279: 1-10.
[http://dx.doi.org/10.1016/j.jhazmat.2014.06.025] [PMID: 25036994]

[19] Benefield L D, Judkins J F, Weand B L. Process Chemistry for Water and Wastewater treatment. Prentice-Hall, Englewood Cliffs, 1982.

[20] Tripathy T, Rajan De B. Flocculation: A New Way to Treat the Waste Water. J Phys Sci 2006; 10: 93-127.

[21] Dizge N, Keskinler B, Barlas H. Sorption of Ni(II) ions from aqueous solution by Lewatit cation-exchange resin. J Hazard Mater 2009; 167(1-3): 915-26.
[http://dx.doi.org/10.1016/j.jhazmat.2009.01.073] [PMID: 19231079]

[22] Hamdaoui O. Removal of copper(II) from aqueous phase by Purolite C100-MB cation exchange resin in fixed bed columns: Modeling. J Hazard Mater 2009; 161(2-3): 737-46.
[http://dx.doi.org/10.1016/j.jhazmat.2008.04.016] [PMID: 18486328]

[23] Gunatilake SK. Methods of Removing Heavy Metals from Industrial Wastewater. Methods 2015; 1(1): 14.

[24]   Ku Y, Jung IL. Photocatalytic reduction of Cr(VI) in aqueous solutions by UV irradiation with the presence of titanium dioxide. Water Res 2001; 35(1): 135-42.
[http://dx.doi.org/10.1016/S0043-1354(00)00098-1] [PMID: 11257867]

[25]   Shim HY, Lee KS, Lee DS, *et al.* Application of Electrocoagulation and Electrolysis on the Precipitation of Heavy Metals and Particulate Solids in Washwater from the Soil Washing. J Agric Chem Environ 2014; 3(4): 130-8.
[http://dx.doi.org/10.4236/jacen.2014.34015]

[26]   Manahan SE. Environmental Chemistry: Boca Raton. Florida: Lewis Publ 2000; p. 898.

[27]   Alkorta I, Hernández-Allica J, Becerril JM, Amezaga I, Albizu I, Garbisu C. Recent Findings on the Phytoremediation of Soils Contaminated with Environmentally Toxic Heavy Metals and Metalloids Such as Zinc, Cadmium, Lead, and Arsenic. Rev Environ Sci Biotechnol 2004; 3(1): 71-90.
[http://dx.doi.org/10.1023/B:RESB.0000040059.70899.3d]

[28]   Małachowska-Jutsz A, Gnida A. Mechanisms of Stress Avoidance and Tolerance by Plants Used in Phytoremediation of Heavy Metals. Arch Environ Prot 2015; 41(4): 104-14.

[29]   Mesjasz-Przybyłowicz J, Nakonieczny M, Migula P, *et al.* Uptake of Cadmium, Lead Nickel and Zinc from Soil and Water Solutions by the Nickel Hyperaccumulator Berkheya Coddii. Acta Biol Cracov Ser; Bot 2004; 46: 75-85.

[30]   Rahman MA, Reichman SM, De Filippis L, Sany SBT, Hasegawa H. Phytoremediation of Toxic Metals in Soils and Wetlands: Concepts and Applications.Environmental remediation technologies for metal-contaminated soils. Springer 2016; pp. 161-95.
[http://dx.doi.org/10.1007/978-4-431-55759-3_8]

[31]   Gaiero JR, McCall CA, Thompson KA, Day NJ, Best AS, Dunfield KE. Inside the root microbiome: Bacterial root endophytes and plant growth promotion. Am J Bot 2013; 100(9): 1738-50.
[http://dx.doi.org/10.3732/ajb.1200572] [PMID: 23935113]

[32]   Lone MI, He Z, Stoffella PJ, Yang X. Phytoremediation of heavy metal polluted soils and water: Progresses and perspectives. J Zhejiang Univ Sci B 2008; 9(3): 210-20.
[http://dx.doi.org/10.1631/jzus.B0710633] [PMID: 18357623]

[33]   Ali H, Khan E, Sajad MA. Phytoremediation of heavy metals—Concepts and applications. Chemosphere 2013; 91(7): 869-81.
[http://dx.doi.org/10.1016/j.chemosphere.2013.01.075] [PMID: 23466085]

[34]   López-Chuken UJ. Hydroponics and Environmental Clean-Up. *Hydroponics–a Stand. Methodol. plant.* Biol Res 2012; 181.

[35]   Dixit R, Wasiullah , Malaviya D, *et al.* Bioremediation of Heavy Metals from Soil and Aquatic Environment: An Overview of Principles and Criteria of Fundamental Processes. Sustainability (Basel) 2015; 7(2): 2189-212.
[http://dx.doi.org/10.3390/su7022189]

[36]   Ahemad M. Remediation of metalliferous soils through the heavy metal resistant plant growth promoting bacteria: Paradigms and prospects. Arab J Chem 2019; 12(7): 1365-77.
[http://dx.doi.org/10.1016/j.arabjc.2014.11.020]

[37]   Abbas SH, Ismail IM, Mostafa TM, Sulaymon AH. Biosorption of Heavy Metals: A Review. J Chem Sci Technol 2014; 3(4): 74-102.

[38]   Goher ME. AM, A. E.-M.; Abdel-Satar, A. M.; Ali, M. H.; Hussian, A.-E.; Napiórkowska-Krzebietke, A. Biosorption of Some Toxic Metals from Aqueous Solution Using Non-Living Algal Cells of *Chlorella vulgaris*. J Elem 2016; 21(3).

[39]   Grujić S, Vasić S, Radojević I, Čomić L, Ostojić A. Comparison of the Rhodotorula mucilaginosa Biofilm and Planktonic Culture on Heavy Metal Susceptibility and Removal Potential. Water Air Soil Pollut 2017; 228(2): 73.

[http://dx.doi.org/10.1007/s11270-017-3259-y]

[40]   Sikkema J, de Bont JA, Poolman B. Mechanisms of membrane toxicity of hydrocarbons. Microbiol Rev 1995; 59(2): 201-22.
[http://dx.doi.org/10.1128/mr.59.2.201-222.1995] [PMID: 7603409]

[41]   Zouboulis AI, Matis KA. Removal of metal ions from dilute solutions by sorptive flotation. Crit Rev Environ Sci Technol 1997; 27(3): 195-235.
[http://dx.doi.org/10.1080/10643389709388499]

[42]   Guiné V, Spadini L, Sarret G, *et al.* Zinc sorption to three gram-negative bacteria: combined titration, modeling, and EXAFS study. Environ Sci Technol 2006; 40(6): 1806-13.
[http://dx.doi.org/10.1021/es0509811] [PMID: 16570601]

[43]   Volesky B. Sorption and Biosorption.St. Lambert, Quebec: BV Sorbex. Inc 2004.

[44]   Gadd G, White C. Microbial treatment of metal pollution? a working biotechnology? Trends Biotechnol 1993; 11(8): 353-9.
[http://dx.doi.org/10.1016/0167-7799(93)90158-6] [PMID: 7764182]

[45]   Kang CH, Kwon YJ, So JS. Bioremediation of heavy metals by using bacterial mixtures. Ecol Eng 2016; 89: 64-9.
[http://dx.doi.org/10.1016/j.ecoleng.2016.01.023]

[46]   Jan AT, Azam M, Ali A, Haq QMR. Prospects for Exploiting Bacteria for Bioremediation of Metal Pollution. Crit Rev Environ Sci Technol 2014; 44(5): 519-60.
[http://dx.doi.org/10.1080/10643389.2012.728811]

[47]   Sayler GS, Ripp S. Field applications of genetically engineered microorganisms for bioremediation processes. Curr Opin Biotechnol 2000; 11(3): 286-9.
[http://dx.doi.org/10.1016/S0958-1669(00)00097-5] [PMID: 10851144]

[48]   Dong G, Wang Y, Gong L, *et al.* Formation of soluble Cr(III) end-products and nanoparticles during Cr(VI) reduction by Bacillus cereus strain XMCr-6. Biochem Eng J 2013; 70: 166-72.
[http://dx.doi.org/10.1016/j.bej.2012.11.002]

[49]   Kostal J, Yang R, Wu CH, Mulchandani A, Chen W. Enhanced arsenic accumulation in engineered bacterial cells expressing ArsR. Appl Environ Microbiol 2004; 70(8): 4582-7.
[http://dx.doi.org/10.1128/AEM.70.8.4582-4587.2004] [PMID: 15294789]

[50]   Igiri B E, Okoduwa S I R, Idoko G O, Akabuogu E P, Adeyi A O, Ejiogu I K. Toxicity and Bioremediation of Heavy Metals Contaminated Ecosystem from Tannery Wastewater: A Review. J Toxicol 2018. 2018: 2568038.
[http://dx.doi.org/10.1155/2018/2568038]

[51]   Ayangbenro A, Babalola O. A New Strategy for Heavy Metal Polluted Environments: A Review of Microbial Biosorbents. Int J Environ Res Public Health 2017; 14(1): 94.
[http://dx.doi.org/10.3390/ijerph14010094] [PMID: 28106848]

[52]   De J, Ramaiah N, Vardanyan L. Detoxification of toxic heavy metals by marine bacteria highly resistant to mercury. Mar Biotechnol (NY) 2008; 10(4): 471-7.
[http://dx.doi.org/10.1007/s10126-008-9083-z] [PMID: 18288535]

[53]   Abioye OP, Oyewole OA, Oyeleke SB, Adeyemi MO, Orukotan AA. Biosorption of lead, chromium and cadmium in tannery effluent using indigenous microorganisms. Braz J Biol Sci 2018; 5(9): 25-32.
[http://dx.doi.org/10.21472/bjbs.050903]

[54]   Al Hasin A, Gurman SJ, Murphy LM, Perry A, Smith TJ, Gardiner PHE. Remediation of chromium(VI) by a methane-oxidizing bacterium. Environ Sci Technol 2010; 44(1): 400-5.
[http://dx.doi.org/10.1021/es901723c] [PMID: 20039753]

[55]   Ackerley DF, Gonzalez CF, Keyhan M, Blake R II, Matin A. Mechanism of chromate reduction by the *Escherichia coli* protein, NfsA, and the role of different chromate reductases in minimizing oxidative

stress during chromate reduction. Environ Microbiol 2004; 6(8): 851-60.
[http://dx.doi.org/10.1111/j.1462-2920.2004.00639.x] [PMID: 15250887]

[56]   Brim H, McFarlan SC, Fredrickson JK, *et al.* Engineering Deinococcus radiodurans for metal
         remediation in radioactive mixed waste environments. Nat Biotechnol 2000; 18(1): 85-90.
         [http://dx.doi.org/10.1038/71986] [PMID: 10625398]

[57]   Zhao XW, Zhou MH, Li QB, *et al.* Simultaneous mercury bioaccumulation and cell propagation by
         genetically engineered *Escherichia coli.* Process Biochem 2005; 40(5): 1611-6.
         [http://dx.doi.org/10.1016/j.procbio.2004.06.014]

[58]   López A, Lázaro N, Morales S, Marqués AM. Nickel Biosorption by Free and Immobilized Cells of
         Pseudomonas Fluorescens 4F39: A Comparative Study. Water Air Soil Pollut 2002; 135(1/4): 157-72.
         [http://dx.doi.org/10.1023/A:1014706827124]

[59]   Sriprang R, Hayashi M, Ono H, Takagi M, Hirata K, Murooka Y. Enhanced accumulation of Cd2+ by
         a *Mesorhizobium* sp. transformed with a gene from Arabidopsis thaliana coding for phytochelatin
         synthase. Appl Environ Microbiol 2003; 69(3): 1791-6.
         [http://dx.doi.org/10.1128/AEM.69.3.1791-1796.2003] [PMID: 12620871]

[60]   Vullo DL, Ceretti HM, Daniel MA, Ramírez SAM, Zalts A. Cadmium, zinc and copper biosorption
         mediated by Pseudomonas veronii 2E. Bioresour Technol 2008; 99(13): 5574-81.
         [http://dx.doi.org/10.1016/j.biortech.2007.10.060] [PMID: 18158237]

[61]   Dursun AY, Uslu G, Cuci Y, Aksu Z. Bioaccumulation of copper(II), lead(II) and chromium(VI) by
         growing *Aspergillus niger.* Process Biochem 2003; 38(12): 1647-51.
         [http://dx.doi.org/10.1016/S0032-9592(02)00075-4]

[62]   Tiwari S, Singh SN, Garg SK. Microbially Enhanced Phytoextraction of Heavy-Metal Fly-Ash
         Amended Soil. Commun Soil Sci Plant Anal 2013; 44(21): 3161-76.
         [http://dx.doi.org/10.1080/00103624.2013.832287]

[63]   Karakagh RM, Chorom M, Motamedi H, Kalkhajeh YK, Oustan S. Biosorption of Cd and Ni by
         inactivated bacteria isolated from agricultural soil treated with sewage sludge. Ecohydrol Hydrobiol
         2012; 12(3): 191-8.
         [http://dx.doi.org/10.1016/S1642-3593(12)70203-3]

[64]   Vankar PS, Bajpai D. Phyto-remediation of chrome-VI of tannery effluent by Trichoderma species.
         Desalination 2008; 222(1-3): 255-62.
         [http://dx.doi.org/10.1016/j.desal.2007.01.168]

[65]   Srivastava S, Thakur IS. Isolation and process parameter optimization of *Aspergillus* sp. for removal
         of chromium from tannery effluent. Bioresour Technol 2006; 97(10): 1167-73.
         [http://dx.doi.org/10.1016/j.biortech.2005.05.012] [PMID: 16023341]

[66]   Park D, Yun YS, Hye Jo J, Park JM. Mechanism of hexavalent chromium removal by dead fungal
         biomass of *Aspergillus niger.* Water Res 2005; 39(4): 533-40.
         [http://dx.doi.org/10.1016/j.watres.2004.11.002] [PMID: 15707625]

[67]   Luna JM, Rufino RD, Sarubbo LA. Biosurfactant from Candida sphaerica UCP0995 exhibiting heavy
         metal remediation properties. Process Saf Environ Prot 2016; 102: 558-66.
         [http://dx.doi.org/10.1016/j.psep.2016.05.010]

[68]   Taştan BE, Ertuğrul S, Dönmez G. Effective bioremoval of reactive dye and heavy metals by
         Aspergillus versicolor. Bioresour Technol 2010; 101(3): 870-6.
         [http://dx.doi.org/10.1016/j.biortech.2009.08.099] [PMID: 19773159]

[69]   Javaid A, Bajwa R, Manzoor T. Biosorption of Heavy Metals by Pretreated Biomass of *Aspergillus
         niger.* Pak J Bot 2011; 43(1): 419-25.

[70]   Congeevaram S, Dhanarani S, Park J, Dexilin M, Thamaraiselvi K. Biosorption of chromium and
         nickel by heavy metal resistant fungal and bacterial isolates. J Hazard Mater 2007; 146(1-2): 270-7.
         [http://dx.doi.org/10.1016/j.jhazmat.2006.12.017] [PMID: 17218056]

[71]   Ashokkumar P, Loashini VM, Bhavya V. Effect of PH, Temperature and Biomass on Biosorption of Heavy Metals by Sphaerotilus Natans. Int J Microbiol Mycol 2017; 6(1): 32-8.

[72]   Shipra J, Dikshit SN, Pandey G. Comparative Study of Agitation Rate and Stationary Phase for the Removal of Cu2+ by A. Lentulus. Int J Pharma Bio Sci 2011; 2(3)

[73]   Parvathi K, Nagendran R, Nareshkumar R. Effect of pH on Chromium Biosorption by Chemically Treated *Saccharomyces cerevisiae*. J Sci Industr Res 2007; 66: 675-79.

[74]   Kuyucak N, Volesky B. Biosorbents for recovery of metals from industrial solutions. Biotechnol Lett 1988; 10(2): 137-42.
[http://dx.doi.org/10.1007/BF01024641]

[75]   M. H.; Benedito, C. Potential Application of Modified *Saccharomyces cerevisiae* for Removing Lead and Cadmium. J Bioremediat Biodegrad 2015; 6(2).

[76]   Muneer B. JIqbal, M.; Shakoori, F. R.; Shakoori, A. R. Tolerance and Biosorption of Mercury by Microbial Consortia: Potential Use in Bioremediation of Wastewater. Pak J Zool 2013; 45(1).

[77]   Jebapriya GR, Gnanadoss JJ. Bioremediation of Textile Dye Using White Rot Fungi: A Review. Int J Curr Res Rev 2013; 5(3): 1.

[78]   Mustapha MU, Halimoon N. Microorganisms and Biosorption of Heavy Metals in the Environment: A Review Paper. J Microb Biochem Technol 2015; 7(5): 253-6.
[http://dx.doi.org/10.4172/1948-5948.1000219]

[79]   Lee YC, Chang SP. The biosorption of heavy metals from aqueous solution by Spirogyra and Cladophora filamentous macroalgae. Bioresour Technol 2011; 102(9): 5297-304.
[http://dx.doi.org/10.1016/j.biortech.2010.12.103] [PMID: 21292478]

[80]   Mane P C, Bhosle A B. Bioremoval of Some Metals by Living Algae Spirogyra Sp. and Spirullina Sp. from Aqueous Solution. Int J Environ Res 2012; 6(2): 571-6.

[81]   Hansen HK, Ribeiro A, Mateus E. Biosorption of arsenic(V) with Lessonia nigrescens. Miner Eng 2006; 19(5): 486-90.
[http://dx.doi.org/10.1016/j.mineng.2005.08.018]

[82]   Vilar VJP, Botelho CMS, Boaventura RAR. Equilibrium and kinetic modelling of Cd(II) biosorption by algae Gelidium and agar extraction algal waste. Water Res 2006; 40(2): 291-302.
[http://dx.doi.org/10.1016/j.watres.2005.11.008] [PMID: 16380148]

[83]   Balcázar JL, Subirats J, Borrego CM. The role of biofilms as environmental reservoirs of antibiotic resistance. Front Microbiol 2015; 6: 1216.
[http://dx.doi.org/10.3389/fmicb.2015.01216] [PMID: 26583011]

[84]   Van Aken B, Tehrani R, Schnoor JL. Endophyte-Assisted Phytoremediation of Explosives in Poplar Trees by Methylobacterium Populi BJ001 T.Endophytes of Forest Trees. Springer 2011; pp. 217-34.
[http://dx.doi.org/10.1007/978-94-007-1599-8_14]

[85]   Ruiz ON, Daniell H. Genetic engineering to enhance mercury phytoremediation. Curr Opin Biotechnol 2009; 20(2): 213-9.
[http://dx.doi.org/10.1016/j.copbio.2009.02.010] [PMID: 19328673]

[86]   Maestri E, Marmiroli M. Genetic and Molecular Aspects of Metal Tolerance and Hyperaccumulation.Metal toxicity in plants: perception, signaling and remediation. Springer 2012; pp. 41-63.
[http://dx.doi.org/10.1007/978-3-642-22081-4_3]

[87]   Abhilash PC, Powell JR, Singh HB, Singh BK. Plant–microbe interactions: novel applications for exploitation in multipurpose remediation technologies. Trends Biotechnol 2012; 30(8): 416-20.
[http://dx.doi.org/10.1016/j.tibtech.2012.04.004] [PMID: 22613174]

CHAPTER 3

Bioremediation of Heavy Metal in Paper Mill Effluent

Priti Gupta[1,*]

[1] *Department of Chemistry, Manav Rachna University, Faridabad, Haryana – 121001, India*

Abstract: The pulp and papermaking industry, being a large consumer of natural resources, *i.e.*, wood and water, has become one of the largest sources of pollution to the environment. Wastewater generated during various stages of the pulp and paper-making process continues to be toxic in nature even after secondary treatment. The effluent water contains not only various toxic chemicals such as volatile organic compounds but also heavy metals like copper, mercury, iron, zinc aluminium, *etc*. Even at very low concentrations, most of the heavy metals are toxic and deadly in nature. Prolonged exposure to heavy metals causes various diseases in humans and animals either through skin contact, inhalation, or *via* consuming food materials.

Treatment of pulp and paper industry wastewater by conventional methods is not efficient due to its complex nature. These conventional methods, either physical, biological, chemical or a combination of these methods are also not environmentally safe and economically viable. Complete degradation of heavy metals is not possible by the application of a single method. The generation of a huge volume of toxic sludge is an ongoing and major problem. Therefore bioremediation methods are preferred as they are highly efficient, cost effective, eco-friendly in nature, there is no secondary waste created in the environment and metabolize the highly toxic heavy metals into degradable, less toxic components with the help of microbes.

This chapter focuses on Micro-Bioremediation methods using algae, fungi, yeasts and bacteria as the most preferred medium to treat wastewater generated by the pulp and paper industry. These are further also used to reduce toxic organic compounds.

Keywords: Accumulative, Adsorption, Anaerobic, Bioaccumulation, Biosor-bents, Bioremediation, Carcinogenic, Chelation, Degradation, Efficiency, Effluent, Genetic engineering, Hazardous, Heavy metals, Organic halides, Persistent, Physicochemical, Phytoremediation, Substrate, Transformation.

---

* **Corresponding author Priti Gupta:** Assistant Professor, Department of Chemistry, Manav Rachna University, Faridabad, Haryana – 121001 India; E-mail: me.pritigupta@gmail.com

Inamuddin (Ed.)

INTRODUCTION

Urban industrialization has long been identified as a major source of aquatic environment contamination through atmospheric deposition as well as wastewater discharge. All types of manufacturing industries contribute to this. The paper and pulp industry being one of the most volatile and important industries in the world, is by-far the sixth largest polluting industry (after cement, oil, textile, leather, and steel industries). The process of producing paper from pulp produces various discharges of a variety of pollutants including organic and inorganic pollutants like gases, liquid, and solid wastes into the environment [1]. The global volume of toxic effluent released into the water bodies by paper and pulp industries is approximated to be around $178 - 303$ m^3.

The process of wood pulping and paper products production generate a considerable amount of pollutants. When these effluents are discharged in either an untreated condition or in the poorly treated conditions in the water bodies, this causes a severe problem of water pollution. Usage of a higher quantity of water between 20,000 and 60,000 gallons per ton of product results in the generation of a large amount of water [2].

The process of pulping and bleaching are energy intensives and consume a considerable volume of chemicals and fresh water. Usage of raw material, pulping method, bleaching process as well as the size of Paper and Pulp mills decide the extent of wastewater production and, therefore, the level of pollution and toxicity [3].

The distribution of wastewater generated by the Paper and Pulp mills encompasses the whole of the surrounding ecosystem. In aquatic life, it may lead to reproductive irregularities and damage the genetic and immune system in invertebrates and zooplankton, which are a source of nutrients and food for fish [4]. Such polluted water when released into the water bodies also causes slime growth, thermal impacts, scum formation, colour problems, and loss of aesthetic beauty in the environment [5].

Paper mill generated waste water contains significantly higher concentrations of BOD (biological oxygen demand), COD (chemical oxygen demand) as well as phosphorous. Electrical conductivity, temperature and colour value also increases in such water [6]. The effluent water not only contains volatile organic compounds like alcohols, phenols, methanol, acetone, surfactants, dyes, acids and alkalies but also carries heavy metals such as zinc, mercury, iron, copper and aluminium [7]. These heavy metals are highly toxic, carcinogenic as well as mutagenic in nature even though present in very low concentrations [8, 9]. They are commonly present in paper mill effluents and being hazardous in nature causes serious problems to the sewage network pipelines. Hazardous impact of heavy metals on biological processes is complex in nature and depends upon various factors such as species, solubility and concentration of the metal as well as on various characteristics such as pH of the influent [10]

Given the intensity of the hazards posed, various methods have been studied and reported for the remediation of wastewater. Physical methods such as adsorption, microfiltration, *etc.* and chemical treatment methods such as coagulation, sedimentation, oxidation, ozonation, *etc.*, or combinations of these methods are prevalent, however, these methods are not cost-effective. Additionally, these also increase the BOD, and COD level of the treated effluent along with the generation of a large amount of sludge [11].

It is therefore essential to develop not only a cost-effective as well as eco-friendly and efficient method for the remediation of the paper pulp mill effluent. Biological methods provide popular and attractive alternative methods that utilize the metabolic potential of microbes to clean up the environment [12, 13]. These include various aerobic and anaerobic treatments using bioreactors.

PAPER & PULP INDUSTRY: GLOBAL OUTLOOK ON UTILITY AND GROWTH

There has been a widespread application of paper products for packaging. With the increasing awareness of long term use of plastic, and increased use of hygiene products that include toilet papers for sanitation, paper towels during travel, and disposable wipes for babies the use of paper products has further elevated the consumption of paper products many folds. Increased demand along with increase in spending capacity of middle class across the globe adds to it. The paper & pulp market was valued at USD 518.83 Billion in 2019 and is expected to reach USD 679.72 Billion by 2027 [14].

Increased e-commerce leading to packaging and shipment will also drive the global paper & pulp market with increased demand for packaging. The paper & pulp industry continues to engage professionals in various areas of research activities such as - maximizing yield, increase in efforts to reduce the use of energy, ways to attain sustainable environmental goals, *etc.* The industry is also shifting focus towards replacing the larger share of products that are currently made out of plastic materials.

PAPER & PULP INDUSTRY: GLOBAL OUTLOOK ON HAZARDS

Unlike many other production units or manufacturing industry, the paper and pulp industry is a heavy consumer of energy and emits an equally large amount of pollutants which includes greenhouse gases, water pollutants, *etc.* The major constituent of waste generated is in the form of wastewater that causes severe damage to the aquatic life, causes disruption in the food chain, and also leads to health hazards. The primary pollutants released from the industry are a mix of water, solid, and gaseous discharges that mix into the human ecosystem and cause severe dangers to mankind.

The solid wastes are majorly an outcome of the bleaching and pulp process, whereas, unprocessed water during the process gets mixed with surrounding water bodies and pollutes them infinitely. Gases like chlorine dioxides, nitrogen oxides, hydrogen and sodium sulphides, *etc.* are also produced during the process. These gases also form a major component of pollutants from the paper and pulping process.

PAPER MAKING PROCESSES AND WASTEWATER GENERATION

Globally, the processes applied in paper and pulp industry vary from region to region, and purpose to purpose. Depending upon the kind of wood stock available and the quality of paper to be produced will largely define the concentration of the chemicals to be applied. The quality of paper refers to its utility, for *e.g.* for books, newspaper, shopping bags, magazines *etc.* Whatever process may be applied, there will be streaming of wastewater due to this. The volume and properties of pulp and paper mills wastewater and related pollutants depend upon various factors, such as production scale, the raw materials used, and applied production technologies [15]. Also, the nature of wastewater depends upon the quantity of fresh water utilized during the process. Different compositions of wastewater streams emerge from specific kinds of operations of pulping, bleaching or paper making [16].

A typical paper manufacturing is a five step process (Fig. 1). Each step can be carried out using a variety of methods. Therefore, the final effluent is a combination of wastewaters generated from each of the five different unit processes and the methods employed therein. These five basic unit processes are:

- Debarking
- Pulping (also called delignification)
- Bleaching
- Washing
- Papermaking

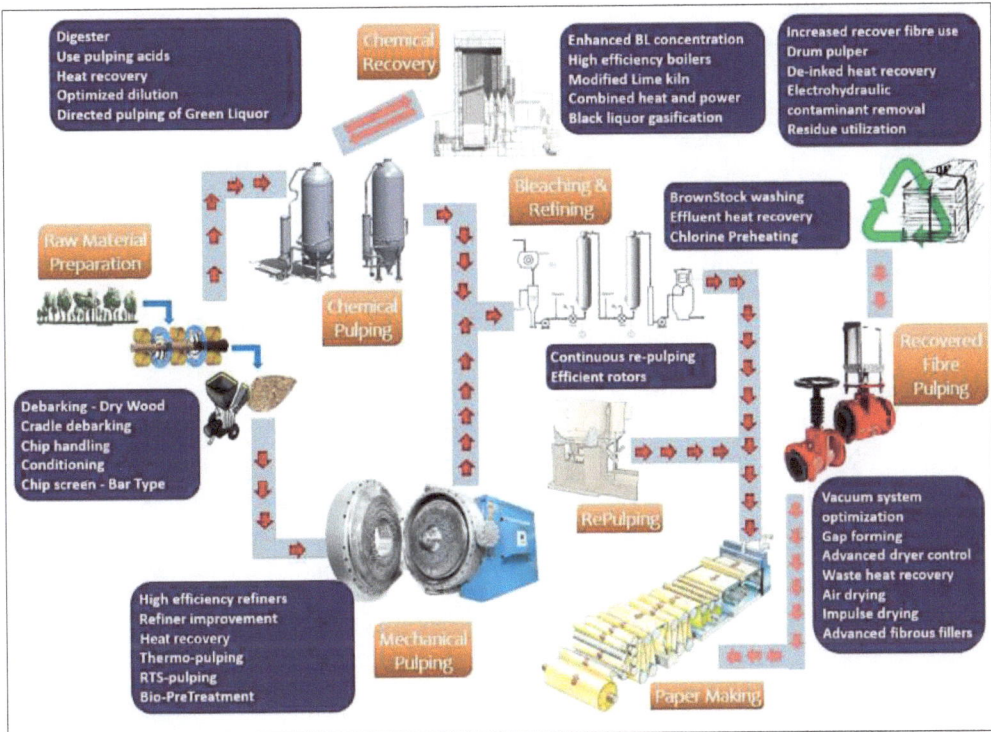

Fig. (1). Stages of the paper making process.

Debarking

Following the sequence of manufacturing steps, one of the first streams of contaminated water generation is associated with the logs soaking and bark removal process [17, 18]. Depending upon the raw material used such as softwood, hardwood, agro-residues, *etc.* the processed water may contain differential concentrations of resin acids, tannins, *etc.* These are usually transferred basis on their presence in the bark [19]. The debarked logs are then moved into a chipper, where they are reduced into small pieces or chips ready for pulping.

Pulping

Pulping is the stage in which chips are turned into pulp. This process aims to produce a cellulose-rich 'pulp' by removing the majority of lignin and hemicellulose content from the raw material. Pulping can be carried out by various methods, such as Mechanical, Chemical, Kraft, Sulfite pulping, *etc.* Each of these processes utilizes a separate raw material and in separate quantities.

These then form deciding factors for the nature and quantity of transfer of long chain fatty acids and resin acids into processed waters.

Mechanical Pulping

This process is also referred to as Thermo-Mechanical Pulping (TMP) and is carried out at high temperature conditions. By this process, the yield of the pulp is as high as 90–95% [20] but the quality of the pulp produced is of low grade, highly colored, and contains short fibers.

Chemical Pulping

In this process, the wood chips are subject to heating with appropriate chemicals in an aqueous solution. A high temperature and pressure conditions are maintained to break chips into a fibrous mass. The original wood material produces 40-50% of the pulp through this process. Chemical pulping is carried out in two media: alkaline and acidic.

Kraft Process

In this process, wood chips are introduced into a pressure vessel large enough in size, also referred to as digester at a temperature of 155–175 ^{0}C in an aqueous medium containing Na_2S_2 (white liquor) and NaOH. This serves as a good medium to dissolve a considerable amount of lignin from the lignocelluloic biomass.

Sulfite Process

In this process, wood chips are mixed with an acidic medium of Sulfurous Acid (H_2SO_3) and bisulfide ions (HSO^{3-}) to dissolve lignin [21].

Bleaching

Bleaching of pulp is an important part of the process and has many objectives. The primary one is to increase the brightness of the pulp so that it can be used in paper products such as printing grades, *etc.* The brown pulp obtained after the pulping process is introduced with bleaching agents such as chlorine, chlorine dioxide, hydrogen peroxide, oxygen, ozone, *etc.* either individually or in combination. The bleaching process determines the quality, gradation and colour of the paper to be produced.

The bleaching process is the main contributor to introducing contamination in wastewater. Upon usage of chlorine-based chemicals used for bleaching, a large volume of chlorinated toxic compounds are produced. The wastewater produced during this stage of bleaching contains various compounds which are toxic and colored. This includes EOXs (extractable organic halides), PCBs (polychlorinated biphenyls), chlorophenols, AOXs (Adsorbable Organic Halides), and extremely toxic DDT (Dichloro-diphenyl-trichloroethane) in small proportions, and PCDDs (polychlorinated dibenzodioxins) [22 - 25]. By this step chlorination of lignin, phenols and resin acids also takes place and they get transformed into highly toxic Xenobiotics.

Over the recent times, modern mills have been developed that attempt to minimize these contaminations. For making it chemical free "total chlorine-free" (TCF) methods are employed for bleaching in these mills [26].

Washing

Washing is done for the removal of the bleaching agents from the pulp. To extract colour and bleaching agents from pulp, generally, an alkali (caustic soda) is used. Due to this reason process is also known as the Alkali Extraction stage (E1) [27]. The washing process results in making effluents more alkaline in nature [28].

Stock Preparation and Paper Making Process

Stock preparation is a much-required step to convert raw stock into finished stock (furnish) preparation is conducted for the manufacturing of paper. Furnish is then prepared by blending different pulps, dilution and the addition of chemicals [29]. The stock preparation process is a series of steps that adapts from one step to another, such as fiber disintegration, cleaning, fiber modification, and storage and mixing. Paper is finally produced by mixing the bleached and washed pulp with a variety of appropriate fillers such as clay, alum, titanium dioxide, calcium carbonates, dyes, *etc.* in a beater tank. The process releases water with a mix of various components. This wastewater produced is popularly called white water which contains talc, very fine fibers, alum, *etc* [30].

Paper is finally produced by consuming pulp. However, during the process, the short fibers are transferred to the wastewater and not retained within the paper [31]. These remaining fibres and other materials (viz. filler, ink) are required to be separated from the wastewater. This separation is carried out through the process of decantation. The sediment material obtained from this process is inducted into the press where it gets converted into sludge. When this sludge is obtained from the production of virgin wood fiber, it is called primary sludge. However, if sludge originated from removing inks from postconsumer fiber, it is referred to as secondary sludge. When wastewater is treated by the activated sludge process, secondary sludge is formed (Fig. **2**). Overall, 45% of wastewater sludge is generated during paper production. This is rich in various compounds such as carbohydrate polymers *e.g.* Cellulose and hemicellulose, lignin, and other extractives such as lipids *etc.* These also contain potentially toxic compounds like resin acids, chlorinated organics, and heavy metals [32]. Effluents or wastewater sludge that is formed from paper making process contains heavy metals (HMs) and are of umpteen concern from risk to ecosystem [33]. These industrial effluents when constantly polluting our environment pose a serious threat to human and aquatic life. Effluents contain organic compounds and heavy metals in excess quantity and constantly exert negative influences on mother earth, mankind and the environment. It causes a hazardous impact on all forms of biotics and abiotics that are continuously exposed to potentially toxic heavy metals [34].

HEAVY METALS AT GLANCE

Heavy metals is no more an alien term today, rather a commonly known term due to its obvious hazardous impact on the environment. It refers to relatively high density metals or a metalloid element that ranges from 3 -7.5 gram per cubic centimetre [35]. Heavy metals such as, arsenic, antimony, cadmium, chromium, lead, mercury, and selenium have higher atomic weight and are five times more in density as compared to water.

Even at very low concentration most of the heavy metals are serious toxicants such as carcinogens. Heavy metals are one of the many and most prominent reasons for environmental pollution such as water, air and soil pollution. They are ubiquitous and non-biodegradable in nature [36]. The impact of polluted wastewater effluents containing heavy toxic metals may effects humans causing neurotoxic, acute chronic or sub-chronic, teratogenic, mutagenic, or carcinogenic [37].

In conclusion, heavy metals are commonly available and commercially used in industries. These, in general, are very toxic to animals as well as to aerobic and anaerobic processes. They cause an adverse effect on human health and the environment.

One of the most common ways to decrease the impact of Heavy metals present in industrial effluents is to ensure wastewater treatment at the source. Although, treatment of wastewater does not expedite the complete removal of contaminant and generally heavy metal removal is less than 50%. Due to this reason even after waste water treatment, effluent is loaded with a significant quantity of heavy metal. The sludge produced in the treatment plant also contains a higher concentration of metal content [38].

Adverse Effect of Heavy Metal Contamination

One of the characteristics of heavy metals is that they are highly soluble in water. This characteristic makes them get easily absorbed by biological systems and hence their separation from water becomes difficult. Heavy metals have long-lasting existences in the environment as they cannot be further disintegrated into smaller forms, therefore arsenic, copper, cadmium, lead, chromium, nickel, and zinc [39] make it gradually to the environment through metal evaporation from water resources to soil and ground. These deposited elements then start accumulating into the biosphere through absorption [40].

Contaminated wastewater effluents having heavy metals and whether untreated or partially treated are released into receiving water bodies, there is a higher probability of a variety of health and environmental impacts. Within aquatic ecosystems, heavy metals negatively impact the number of living organisms and their growth. The presence of heavy metal pollutants has an adverse effect on soil; and plants growing on such soils. When these plants are ingested by humans or animals, they enter into the food chain. Processes like Bio magnification and Bioaccumulation lead to severe detrimental effects [41].

Soil

Heavy metals cannot be separated easily by physical methods from wastewater and when they enter the soil, creates interference with the plant roots and also affects soil respiration. Plants uptake lead and cadmium along with other nutrients and their accumulation may get affected by the concentration, time of exposure and climatic factor. This, as a result, affects the quality and yield of crops, which in turn enters into food cycle and the entire food industry becomes prone to waste water pollution [42]. Vegetables and fruits are more susceptible to insect attacks when grown in such heavy metal-contaminated soils [43].

Microbial Population

Microbes play a critical role in the recycling of nutrients in the environment. Plant respiration and metabolism processes are adversely affected by heavy metal contamination. It, therefore, may dangerously disturb the ecosystem functioning resulting in threatened long-term soil activity and productivity. Heavy metals tend to pose adverse impact on microbial cells and they may obstruct cellular growth and maintenance as well. For example, arsenic can lead to enzyme deactivation [44]. Mercury, cadmium and lead cause denaturing of protein, hindering cell division, altering transcription process, *etc* [45 - 52].

Plants

Heavy metals and plants have complex relationships. Trace concentrations of heavy metals are essential for healthy growth and essential physiological functions [53]. However, when the concentration exceeds the requirement, heavy metals result in disfunctioning of biochemical and physiological processes. This also inhibits the process of photosynthesis and respiration, degeneration of main cell organelles, malnutrition in plants and sometimes even leads to the death of plants [54, 55]. For example yellowing of wheat and paddy leaves is caused by hexavalent chromium. Generally, adverse harmful impacts of heavy metal contamination on plants are not observed immediately due to the accumulation of these toxic metals in various parts of plants such as walls and vacuoles of cells, intercellular spaces of roots, barks of plants, *etc.* Arsenic is harmful to legumes, rice and onions [56]. Lead, cadmium, chromium and nickel accumulation causes growth inhibition, *etc.*

Animals

Heavy metals play vital roles in various oxidation-reduction reactions when consumed within permissible limits [57]. The essential heavy metals exert biochemical and physiological functions in animals as they are important constituents of several key enzymes. When the consumption or accumulation of these heavy metal concentrations exceeds the permissible limit it may disturb various physiological and biochemical functions in animals. On the basis of various studies reproductive and teratogenic effects on animals have also been reported. The harmful effects reported by heavy metals are mutagenicity, immunosuppression, teratogenicity, carcinogenicity *etc.* For example, chromium being highly toxic; if enters through inhalation or dermal route causes lung cancer, ulcers in the stomach and small intestines and also impacts the reproduction system. Mercury has an adverse impact on the central nervous system, moreover, cellular degeneration, osteoporosis (skeletal damage), kidney stones formation, *etc.* in animals may be caused by cadmium poisoning [58].

Humans

Heavy metals may enter the human body through food, water, air or absorption through the skin and their constant accumulation causes an adverse impact on human health. It has been reported that individual metals exhibit specific signs of their toxicity when they enter as a contaminant mixed with other sources. However, when volatile vapours of metals such as cadmium, lead, arsenic, mercury, zinc, and copper aluminium are inhaled, the signs associated with poisoning are gastrointestinal disorders, diarrhoea, tremor, hemoglobinuria causing a rust-red colour to stool, ataxia, paralysis, vomiting and convulsion, depression and pneumonia [59].

Heavy metals by virtue of their characteristic properties will have health hazards if not available in appropriate proportions (Table **1**). Various metals will have an affinity to impact specific body parts of human beings, although the impact largely depends upon the concentration of heavy metals available (Fig **2**).

Table 1. Sources of Heavy metal and its impact on Human health [60 - 69].

Heavy Metal	Sources of Contamination	Impact of Contamination
Arsenic [60]	Industrial sources like smelting from micro-electronic industry. Contamination from pesticides, preservatives, fungicides, herbicides and paints	Blood vessels destruction, pulmonary disease, acute or chronic toxicity
Lead [61]	Industrial sources such as lead-acid batteries, gasoline, plumbing materials, alloys, house paint, lead bullets, Mining, smelting, and informal processing and recycling of electric and electronic waste, toys, *etc.*	Abdominal pain, allergies, autism, brain damage, dyslexia, fatigue, hallucinations. Headache, hyperactivity, hypertension, mental retardation, kidney damage, birth defects, loss of appetite, muscular weakness, paralysis, psychosis, renal dysfunction, sleeplessness, vertigo, weight loss,
Mercury [62]	Industrial sources such as the battery, wastewater discharge	brain functions, irritability, kidney, memorization problems and behavioural changes in visibility, hearing, shyness,
Cadmium [63]	Industrial sources such as coatings, pigment alloys, paints, batteries and plastics. Potential deposition of cadmium smelters that enter sewage and groundwater	lungs irritation, respiratory irritation, irritation in the digestive system resulting in vomiting and diarrhoea,
Chromium [64]	Industrial sources like tanning, electroplating, metallurgy, wood preservation, pulp and paper production, pigments and paint production, chemical production	anaemia, damage to sperm and male reproductive system, Irritation in nose lining, stomach and small intestines, ulcers in small intestine, nose and stomach
Copper [65]	Industrial sources such as production of copper pipes, cables, wires, copper cookware, *etc.* Activities that lead to assimilation in soil followed by plant uptake.	Abdominal pain, cardiac arrhythmia, diarrhoea, flatulence, generalized rash, vomiting [66]
Manganese [67]	Gasoline Fumes	cognitive disorder, gait disorder, neurotoxicity, postural instability, tremor

(Table 1) cont.....

Heavy Metal	Sources of Contamination	Impact of Contamination
Nickel [68]	Industrial sources like battery production, jewellery plated with Nickel, parts of machine, metallic objects plated with nickel, steel manufacturing, smoking tobacco and cigarette, electrical wires and other parts	chronic bronchitis, cancer of the lung, reduced lung function, nasal sinus [69]

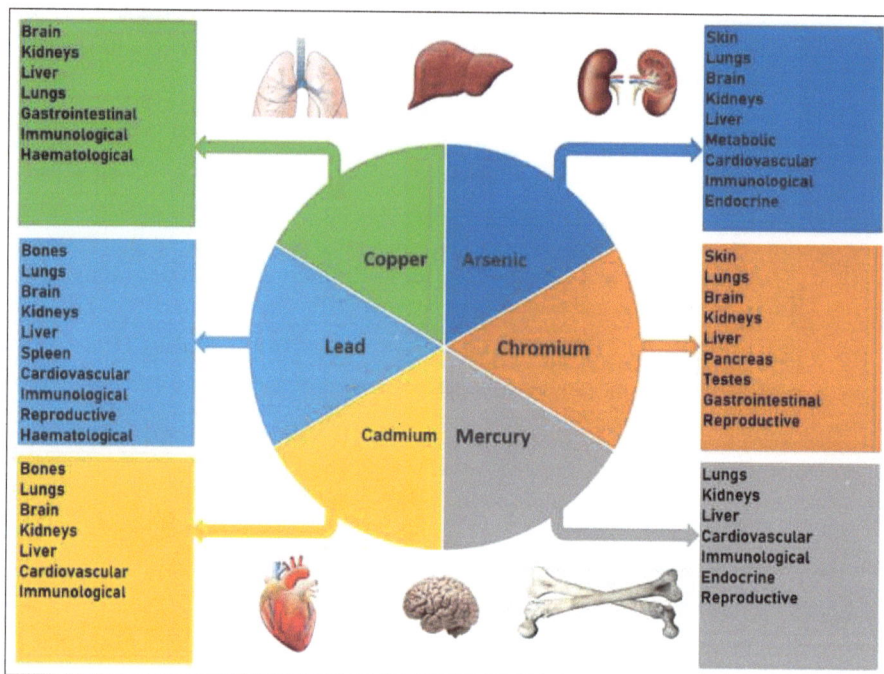

Fig. (2). Heavy metal impact on human body parts.

Remediation Technologies for the Treatment of Heavy Metal Contaminated Wastewater Effluent

There are several technologies that exist for the remediation of heavy metals - contaminated wastewater. These techniques have some definite outcomes such as substantial degradation of the pollutants, extraction of pollutants which may be either treated further or disposed of, separation and recycling of non-contaminated materials for further treatment and confinement of the polluted material to avoid its exposure to larger environment area [70, 71].

Various approaches have been suggested and used to degrade heavy metals present in wastewater effluent of paper mills (Fig. **3**). These technologies are categorised into various classes:

- Chemical Treatment Technologies
- Biological/ Biochemical/Biosorptive Treatment Technologies
- Physico-Chemical Treatment Technologies

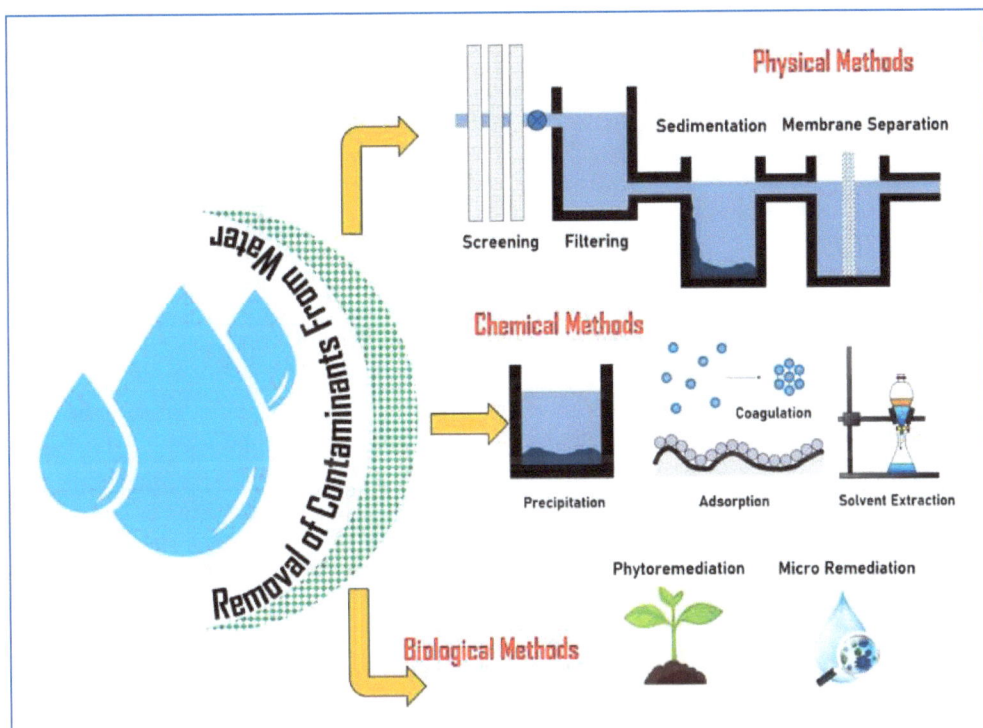

Fig. (3). Various technologies of waste water remediation.

These physicochemical methods such as reverse osmosis, adsorption, ion exchange, chemical precipitation *etc.* are not even environment-friendly in nature as well as economically viable [72]. Moreover, there isn't a single concrete method that affirms the complete removal or disintegration of heavy metals and their complexes leading to the formation of secondary wastes at the treatment site [73]. Conventional physicochemical methods do not exhibit much efficiency in case of low metal toxicity and they are expensive too [74].

Table **2** below is the comparative study between biological and physiochemical methods of heavy metal removal or degradation.

Table 2. Comparative study of biological and physiochemical methods [75].

Biological Methods	Physicochemical Methods
Methods based on bioleaching, biotransformation, biosorption, phytoremediation, interaction between plants and microbes, bio mineralization, rhizo-remediation are classified under these techniques.	>Methods based on oxidation/reduction process, electrochemical treatment, reverse osmosis, membrane technology, filtration/evaporation, ion exchange, *etc.* are classified under physicochemical techniques.
To eliminate hazardous pollutants, they utilize the biological mechanisms which are inherent in microbes, plants, and derived components like enzymes. They restore the ecosystem to its original condition.	>Comparatively, these methods are expensive, as a full instrumental setup, skilled workers, chemicals and continuous power supply, *etc.* are essentials to proceed.
These methods possess the ability to remove heavy metals from the polluted site effectively even at low concentration as well as eco-friendly in nature.	>These methods have a hazardous impact on human health and environment as they produce various toxic gases, therefore, there is high requirement of post-treatment monitoring of effluents.
During the use of these methods chances of ill effects on on-site workers decreases as they get minimal exposure to the contaminants.	>Physiochemical methods affect treatment site with secondary pollution.
On the basis of the nature of metal that needs to be removed, there are different biological systems requirements. There is no universal application to do so. Also, as compared to physiochemical processes biological processes are slow in nature.	>They are ineffective or expensive in areas where there is very low heavy metal concentration/toxicity, but a fast process as compared to Biological methods.

BIOREMEDIATION: AN INNOVATIVE AND USEFUL APPROACH

The bioremediation process refers to the application of living organisms, like microbes and bacteria, in the removal of contaminants, pollutants, and toxins from soil, water, and other environments either by degradation or conversion of larger toxicants into lesser or non-toxicants. Multiple physicochemical techniques are known to be developed for detoxifying the heavy metal concentration which is increasing day by day in the environment but in this regard, bioremediation technique is considered an innovative and promising approach for the removal of heavy metals in polluted water because of many reasons:

- They are eco-friendly in nature
- There is no secondary waste created in the environment
- Metabolize the highly toxic heavy metals into degradable, less toxic components and bio-products such as microbes, *etc.*

The biological remediation method utilizes the biological mechanisms inherent in microbes and plants and their derivatives for heavy metals detoxification present in the environment [76]. Not only plants and microorganisms are helpful in the detoxification process but many other waste materials such as industrial by-products, agricultural wastes also work.

Given the widespread application of these techniques, it is important to learn about them in a little more detail. Given below are some descriptive details for some of these.

Industrial by-Products

These are typically low-cost waste materials from different industries such as paper, steel, tea, pharmaceutical, fertilizer, sugar, aluminium, *etc.* and can be effectively used as biosorbents. Therefore, with the application of these zero-cost industrial wastes, treatment of wastewater effluent from paper mill can solve the problem of not only disposal rather detoxification as well [77].

Agricultural Wastes

Agricultural and plant wastes can be very helpful in the removal of heavy metal ions and with almost no cost. There are many agricultural wastes such as rice and wheat husks, rice bran, wheat bran, and activated carbon obtained from different plant wastes that may be used as adsorbents. It is reported that wastes containing a high percentage of cellulose and lignin have a high potential for metal binding [78].

Phytoremediation Methods and its Types

Phytoremediation is a combination of two terms. The first one is a Greek word *phyto* (plant), and second is a Latin word *remedium* (restoring balance). Such techniques where heavy usage of plants is applied for the treatment of wastewater to restore the balance of contaminants, is called phytoremediation [79]. Several plants have been investigated for the removal of heavy metals that get added as contaminants in wastewater.

- *Spartina maritima* is used to remove metals like lead, arsenic, zinc and copper
- *Arundo donax* for cadmium and zinc
- *Plectranthus amboinicus* for lead
- *Eichhornia crassipes* (water hyacinth) are used for removing chromium, copper, cadmium, iron, and zinc
- *Carex pendula* for lead *etc.*

The basis on various processes involved, phytoremediation can be further categorized into:

- **Phytovolatization**: In this process metal absorbed gets converted into volatile form and subsequently gets released into the atmosphere from the leaf surface.
- **Rhizofiltration**: Process in which absorption, concentration, and precipitation of heavy metals uptake by plant roots.
- **Phytoextraction**: Process in which the extraction and accumulation of contaminants in harvestable plant tissues including roots and surface shoots.
- **Phytodegradation**: In this process, the complex organic molecules degrade into simple molecules and these molecules get incorporated into plant tissues.
- **Phytostimulation**: Plant-assisted bioremediation: In this process, the breakdown of organic contaminants in the soil takes place *via* enhanced microbial activity in the plant root zone or rhizosphere.
- **Phytostabilization**: This method involves absorption and precipitation of contaminants, majorly metals, by plants' roots and their exudates and thus reducing their mobility and spreading.

Microbial Biosorbents

Microorganisms have the capability of tolerating unfavourable conditions which leads to their usage as biosorbents for removing heavy metals from waste water. Microbes used for this purpose include bacteria, yeast, algae, and fungi.

MICROBIAL BIOREMEDIATION METHODS

Microbial bioremediation is defined as a mechanism in which degradation of the hazardous larger molecules into smaller ones takes place by the action of microbes.

Survival of microbes in the heavy metal contaminated environment takes place *via* two ways:

- Their capability of binding the metal onto the cellular wall or EPS (extrapolymeric substances)
- To resist higher metal concentration in their surroundings.

Microbes undergo some specific metabolic modifications which lead to the detoxification of the contaminant. Therefore, it can be safely stated that different microbes such as bacteria, algae, and fungi adopt and develop various detoxifying mechanisms for the removal of heavy metals from their surrounding environment [80].

Microbes can be deployed in various ways for treating wastewater and for the removal of heavy metals from various contaminated sites. A detailed overview of each of these can be seen below (Fig. **4**).

These can be listed as [81 - 83]:

- Biosorption
- Biotransformation
- Bioaccumulation,
- Bio-leaching

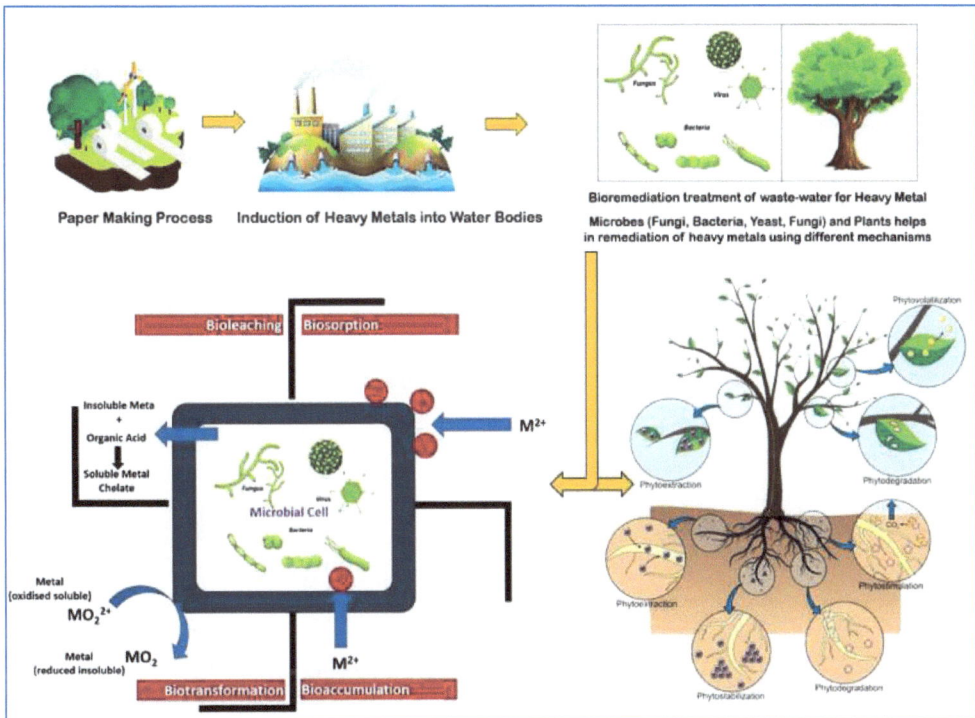

Fig. (4). Different remediation techniques used in removing heavy metals from different contaminated sites.

Biosorption

Biosorption refers to a simple metabolically passive physicochemical process involved in the nonspecific binding of heavy metals (biosorbate) to the surface of the microorganisms [84].

How Does Biosorption Works?

Biosorption is a complex process in which the binding of ions with the functional groups present on the surface of the biosorbent takes place. This binding of metals with biosorbent may be physical or chemical. Physical binding includes electrostatic interaction or Van der Waals forces whereas chemical one may be either through the displacement of either bound metal cations (ion exchange) or protons. Other than these interactions process like reduction, chelation, precipitation and complex formation also occurs (Fig. **5**) [85].

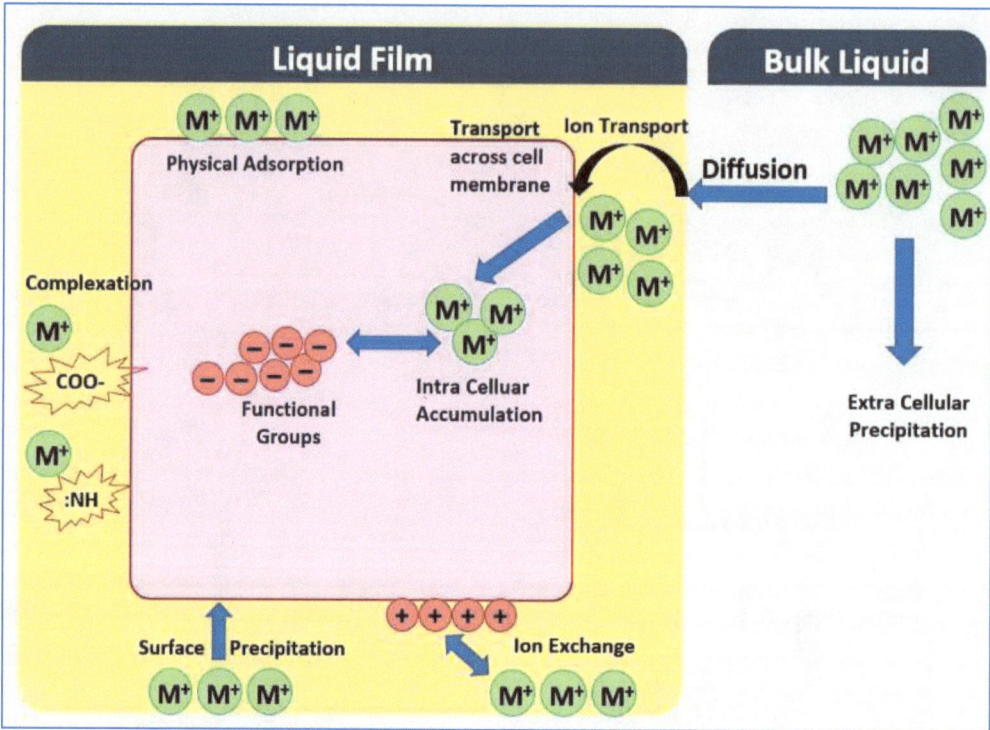

Fig. (5). Processes that occur during Biosorption.

The presence of chemical/functional groups such as amine, imidazole, amide, thioether, carbonyl, sulfonate, *etc.* attract appropriate metal ions and combine with them.

Important Factors Governing Biosorption Mechanism

There are various factors that control and characterize the biosorption mechanism like weight, ionic radius, and oxidation state of the metal species; type and availability of the binding site, structure and nature of biosorbents, *etc.* The combined effects of these factors influence the metal speciation *i.e.* the nature and characteristics of new metal formed after the absorption process.

Types of Biosorption

Two different types of biosorption are known and studied

- *Passive Biosorption* - in the passive biosorption process there is no chemical bond formation that takes place and also lower interaction energy is observed, which is why the process is reversible in nature. There is no effect of outer environmental conditions such as temperature, pH and ionic concentration on the process.
- *Active Biosorption* – This process depends on cell metabolism and is comparatively slow [86].

Examples of Efficient Biosorbents

A large number and types of microbes have been reported as biosorbents for the removal of metal ions efficiently [87].

Cell wall in general is responsible for surface binding sites and binding strength for different metal ions, out of different microbial species some are reported as efficient biosorbents *e.g.* yeast, microalgae, bacteria, and fungi

- **Bacteria** due to their small size and ability to sustain and grow under controlled conditions are also considered efficient biosorbents [88]. For example, the *cyanobacterium Synechococcus* are efficient for the removal of Zn and Cd metals while *Bacillus, Pseudomonas, Streptomyces, Escherichia, and Micrococcus* species of bacteria work on heavy metal in lakes.
- **Algae** are also known to be efficient biosorbents. Algaes such as *Phaeophyta* and *Rhodophyta* are well-known metal binders. *Chlorella vulgaris, Cladophora crispate, Anabaena sp., and Synechococcus sp.* are known to work best on Cr metal and absorb through the walls [89].
- **Filamentous Fungi** and yeasts are also deployed to work towards treating wastewater for removing heavy metals and are very efficient microbiosorbents.
- **Extracellular Polymeric Substances (EPS)** are the polymer microbes comprised of polysaccharides, proteins, and other small moieties, they are produced and released outside the cell wall. EPS exhibit characteristic property to adsorb heavy metals so they are generally used for the wastewater treatment process. During this adsorption phenomenon cations of metals interact with oppositely charged regions of EPS [90].

Advantages

The process has the capability of adsorbing heavy metals even from dilute solutions and is highly efficient to remove heavy metal owing to toxicity at *ppb* levels. This process has many advantages such as: simplicity in operation, high efficiency yield, and additional nutrients are not required, sludge generation is very less, low operational cost, biosorbent regeneration and chemical oxygen demand (COD) of water remain the same. Most conventional techniques fail on these important parameters [91].

Biotransformation

Biotransformation is one of the most important metal detoxification mechanisms among microbes in which conversion of highly toxic metal into less toxic or easily recoverable forms takes place. Microbial transformations are majorly classified into two forms *i.e.*:

- Redox conversions of inorganic forms;
- Conversions from inorganic to organic form and vice-versa are typically methylation and demethylation.

Biotransformation involves reactions such as oxidation, reduction, methylation, and demethylation. The oxidation of metals like Fe, Mn, S and As, work as an energy source for microbes [92]. On the other hand, during the reduction process of metals like AS, Cr, Ur and Se, microbes use them as an electron acceptor for anaerobic respiration. Certain microorganisms impart metal resistance through reduction mechanisms that are not coupled to respiration such as aerobic and anaerobic reduction of Cr (IV) to Cr (III); reduction of Se (VI) to elemental selenium [93]; U (VI) to U (IV) [94]; and reduction of Hg (II) to Hg [95, 96]. Such biotransformation processes are the main constituents of biological geochemical cycles of heavy metals, and they are used in microbial bioremediation of wastewater.

Fungi can facilitate biotransformation of metals (Fig. 6) efficiently by undergoing certain reactions (*i.e.* oxidation, reduction, methylation and dealkylation) and can lead to metal detoxification. Fungal biomass is easily available and cost effective, thereby making it economically *via* ble. It finds vast application in several industrial fermentation processes.

Bioaccumulation

Bioaccumulation is a process similar to a heavy metal transportation system, in which heavy metal from the outer surface of the cell moves within the cell cytoplasm [97]. The process is considered to be energy dependent and needs very high energy. The reason is the inflow of heavy metals through the membrane of bacteria by different mechanisms which requires energy to proceed. These mechanisms include carrier mediated transport channels, ion pump channels, endocytosis channels, complex and lipid permeation, *etc.* Multiple active channels of transport have been reported for a variety of fungal and bacterial species such as cadmium and lead uptake by *Citrobacter sp.*, silver influx by *Thiobacillus ferrooxidans* and Chromium influx by *Bacillus subtilis* [98]. The Table **3** shows some specific metal uptake volumes.

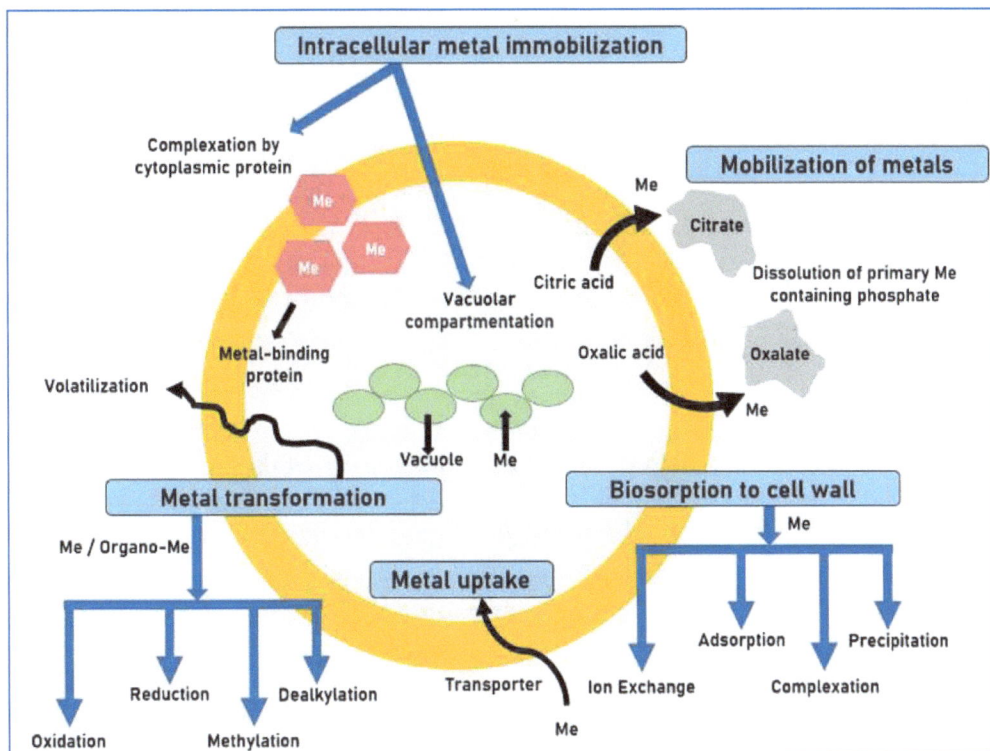

Fig. (6). Schematic of fungi-metal interactions.

Table 3. Heavy metal uptake by different microbial species [98].

Species	Co^{2+} (mg/g)	Cu^{2+} (mg/g)	Zn^{2+} (mg/g)	Cd^{2+} (mg/g)
Debaryomyces senii	4.31	3.72	3.27	9.25
Candida blankie	9.54	6.61	5.35	16.54
Aspergillus ussamii	5.03	4.5	2.56	7.95
Aspergillus niger	4.34	3.96	2.41	7.52

(Table 3) cont.....

Species	Co^{2+} (mg/g)	Cu^{2+} (mg/g)	Zn^{2+} (mg/g)	Cd^{2+} (mg/g)
Aspergillus awamori	5.17	4.32	2.51	8.15
Rhizopus delemar	18.82	17.16	10.31	33.12
Saccharomyces cerevisiae	10.73	7.44	6.12	18.6

Bioleaching

Bioleaching can be defined as "the solubilization of metals from solid substrates either directly by the metabolism of leaching bacteria or indirectly by the products of metabolism" [99]. The Bioleaching process has recently gained preference from industry segments as well as researchers. The process is environment friendly and cost effective also.

Chemolithoautotrophic bacteria like *Acidothiobacillus sp* are used to carry out the process. This bacteria is strong enough to oxidize Sulphur and Fe ions in their elemental form within the acidic environment. The capacity to solubilize heavy metals is favourable as well. Mixed culture of bacteria that can oxidize Fe or S ions was deployed to study the extraction of cadmium, mercury, lead, copper and other heavy metals by the process of bioleaching. Bioleaching and bio-oxidation combined together are also popularly known as Biomining [100, 101].

Organic acids with comparatively lower molecular weight are released as a result of the metabolic activity of microorganisms. Heavy metals and soil containing heavy metals are easily dissolvable in these organic acids. Microbes are further efficient and use nutrients, *i.e.* substrates, mixed with energy. They produce organic acids that in turn support cadmium leaching. The rate of the leaching process appreciates by approximately 36% when takes place in the presence of beneficial nutrients for metabolic activity, however, the rate is only 9% when no nutrients are available. The prokaryotic microbes contributes to modifying the valence of heavy metals by undergoing redox reactions, resulting in heavy metals with altered properties [102, 103].

FACTORS AFFECTING MICROBIAL REMEDIATION OF HEAVY METALS

Various biotic and abiotic factors affect the behavior and the growth of microbial cells which ultimately affects various biological processes occurring in a microbial community. Efficacy of process gets largely reduced due to a lack of information regarding the factors influencing the microbial processes and a non effective implementation [104]. The changing environment does not impact microbes a lot and they get easily adapted to the changing environment due to their inherent ability, although with certain limitations. Therefore, for an effective implementation; improvement in the microbiological action becomes important and hence, there is a requirement for in-depth understanding of microbial ecology which may result in a successful bioremediation process [105]. The various factors which affect the microbial processes are classified as biological factors or biotic factors, physicochemical characteristics of the environment (abiotic factors), and, climatic conditions, *etc.* Among all the factors which influence microbial metabolic rate, physicochemical and climatic conditions are major.

Physicochemical factors include a set of parameters such as temperature, pH, ionic strength, redox potential (Eh), solubility, presence or absence of electron acceptors and donors, *etc.* In the biosorption process, the initial step in microbial remediation of heavy metal detoxification majorly depends upon pH. An optimum pH is often different for different microorganisms and microbial growth may get affected due to undesirable pH variance [106].

Temperature also plays important role in the process of heavy metal adsorption. The rate of adsorbate diffusion increases across the external boundary layer with an increase in temperature. Therefore, the solubility of heavy metals increases which result in an improvement in the bioavailability of heavy metals [107].

An increase in temperature also favours microbial metabolism and enzyme activity which accelerate the bioremediation process. Heavy metal removal efficiency is also affected by species and concentration of heavy metals. Different heavy metals have different solubility such as Zn, Ni, and Cu are easily dissolved, whereas Pb^{2+} and Cr are less soluble. Furthermore, the presence of metal ions in wastewater also affects the enrichment of microorganisms [108].

CHALLENGES

Removal of heavy metals from paper mill wastewater is still a major challenge as an enormous volume of wastewater is generated. Even after using the most suitable and appropriate techniques, the rate at which wastewater gets generated is far higher than the rate at which this gets treated. For sustainable development of industry and the environment, the pulp and paper wastewater must undergo detoxification treatment before it gets discharged into the river streams and eventually contaminates the overall environment. During the process of paper making, pulping is reported as the most significantly polluting step. This step is also responsible for a considerable release of the effluents from paper mill and is a big challenge for the global community.

The traditional methods of treatment adopted by some have only shown minimal results in the treatment of multiple complex pollutants such as Adsorbable Organic Halides (AOX). These organic halides are released in wastewater majorly from paper industries. They are persistent in the environment due to their accumulative characteristic and causes several environmental and health problems. Although certain biological treatments have shown high efficiency, less environmental impact and economical as well, however, such versatile biological agents when applied on a larger scale for the treatment of wastewater still had some limitations. Some of these challenges are like requirement of favourable environmental conditions for microorganism's growth, lack of optimum substrate specificities, limited sources of biocatalyst, *etc.*

Other than these issues there are some more drawbacks associated with these methods:

- Relatively larger area is required for biological treatments
- Time-consuming processes
- Local concerns such as the unpleasant odour released from the bacterial activities
- Prevention of horizontal gene transfer or swapping of genetic material and uncontrolled growth

CONCLUSION AND FUTURE ASPECTS

Even though the world is moving towards electronic data exchange and the demand for paper generation is assumed to deplete, it actually continues to grow year on year. This means the paper and pulp industries will continue to be the backbone for all our printing needs. It is therefore important that there are enough optimal solutions and control methods should be made available.

There is a grave need for ongoing research to create controls that will help minimize the release of pollutants and make paper making process more effective. The pulp and paper industry has been considered a major contributor of pollutants to the environment and moreover a larger consumer of wood, water and energy. So, detoxification of these wastewater produced from paper mills is a thrust need for the world community and a major concern, before its discharge in the environment. However, several efforts have been done in this area for generating less pollutants in the form of wastewaters. Efficient bio remedial techniques have been adapted by many industries which help in reducing the amount and load of the emissions to the environment. Biological treatments are efficient for removing heavy metals present in wastewater effluents and offer a great opportunity to reduce cost (both capital and operating), have less energy consumption, and are environment friendly. However, these methods are comparatively slow. The availability of microorganisms used for remediation methods is also not sufficient enough to meet the consumer's demand because of its low yields. Furthermore, they do not meet the desired standard for industrial processes. Keeping this in mind and for achieving the desired standard norms to discharge wastewater there should be successful implementation of microbes (for bioremediation techniques) in optimal conditions such as temperature, pH, reaction media and substrate specificities.

For improving metal binding abilities, modification in the outer membrane proteins of bacteria with potential bioremediation properties are very helpful. This is probably a way to enhance their capacity for the biotransformation of toxic heavy metals. The focus of future studies should be on the factors involved in improving *in-situ* bioremediation strategies using Genetically Engineered Microorganisms (GEM). Genetic engineering gives better opportunities for obtaining microorganisms having a high potential for environmental clean-up as compared to indigenous microbes. These also exhibit high adaptation to adverse heavy -metal- polluted conditions.

The focus on GEM as well as their acknowledgement and acceptability for bioremediation of heavy metal removal among the public also needs to gain appropriate attention for future studies. Other than these techniques, the application of bioreactors must be put into action for improving removal efficacy and treatment time. The techniques majorly used in bioremediations are aerobic in nature and has their limitations; therefore, more anaerobic processes are required for the degradation of heavy metals and other contaminants in the anaerobic region.

CONSENT FOR PUBLICATION

Not applicable.

CONFLICT OF INTEREST

The author declares no conflict of interest, financial or otherwise.

ACKNOWLEDGEMENTS

Declared none.

REFERENCES

[1]   Ali M, Sreekrishnan TR. Aquatic toxicity from pulp and paper mill effluents: a review. Adv Environ Res 2001; 5(2): 175-96.
[http://dx.doi.org/10.1016/S1093-0191(00)00055-1]

[2]   Chandra R, Singh R. Decolourisation and detoxification of rayon grade pulp paper mill effluent by mixed bacterial culture isolated from pulp paper mill effluent polluted site. Biochem Eng J 2012; 61: 49-58.
[http://dx.doi.org/10.1016/j.bej.2011.12.004]

[3]   Nagarathnamma R, Bajpai P. Decolorization and detoxification of extraction-stage effluent from chlorine bleaching of kraft pulp by Rhizopus oryzae. Appl Environ Microbiol 1999; 65(3): 1078-82.
[http://dx.doi.org/10.1128/AEM.65.3.1078-1082.1999] [PMID: 10049866]

[4]   Aprianti T, Miskah S, Selpiana R, Komala R, Hatina S. Komala, and S. Hatina, *"Heavy metal ions adsorption from pulp and paper industry wastewater using zeolite/activated carbon-ceramic composite adsorbent,"*. AIP Conf Proc 2018; 2014: 020127.
[http://dx.doi.org/10.1063/1.5054531]

[5]   Pokhrel D, Viraraghavan T. Treatment of pulp and paper mill wastewater—a review. Sci Total Environ 2004; 333(1-3): 37-58.
[http://dx.doi.org/10.1016/j.scitotenv.2004.05.017] [PMID: 15364518]

[6]   Gaete H, Larrain A, Bay-Schmith E, Baeza J, Rodriguez J. Ecotoxicological assessment of two pulp mill effluent, Biobio River Basin, Chile. Bull Environ Contam Toxicol 2000; 65(2): 183-9.
[http://dx.doi.org/10.1007/s001280000113] [PMID: 10885995]

[7]   Mishra S, Mohanty M, Pradhan C, Patra HK, Das R, Sahoo S. Physico-chemical assessment of paper mill effluent and its heavy metal remediation using aquatic macrophytes—a case study at JK Paper mill, Rayagada, India. Environ Monit Assess 2013; 185(5): 4347-59.
[http://dx.doi.org/10.1007/s10661-012-2873-9] [PMID: 22993029]

[8]   Tchounwou PB, Yedjou CG, Patlolla AK, Sutton DJ. Heavy metal toxicity and the environment. EXS 2012; 101: 133-64.
[http://dx.doi.org/10.1007/978-3-7643-8340-4_6] [PMID: 22945569]

[9]   Jaishankar M, Tseten T, Anbalagan N, Mathew BB, Beeregowda KN. Toxicity, mechanism and health effects of some heavy metals. Interdiscip Toxicol 2014; 7(2): 60-72.
[http://dx.doi.org/10.2478/intox-2014-0009] [PMID: 26109881]

[10]   Gikas P. Single and combined effects of nickel (Ni(II)) and cobalt (Co(II)) ions on activated sludge and on other aerobic microorganisms: A review. J Hazard Mater 2008; 159(2-3): 187-203.
[http://dx.doi.org/10.1016/j.jhazmat.2008.02.048] [PMID: 18394791]

[11]   Mehmood K, Rehman , Wang , *et al.* Treatment of Pulp and Paper Industrial Effluent Using Physicochemical Process for Recycling. Water 2019; 11(11): 2393.
[http://dx.doi.org/10.3390/w11112393]

[12]   Chuphal Y, Kumar V, Thakur IS. Biodegradation and Decolorization of Pulp and Paper Mill Effluent by Anaerobic and Aerobic Microorganisms in a Sequential Bioreactor. World J Microbiol Biotechnol 2005; 21(8-9): 1439-45.
[http://dx.doi.org/10.1007/s11274-005-6562-5]

[13]   Winter B, Fiechter A, Zimmermann W. Degradation of organochlorine compounds in spent sulfite bleach plant effluents by actinomycetes. Appl Environ Microbiol 1991; 57(10): 2858-63.
[http://dx.doi.org/10.1128/aem.57.10.2858-2863.1991] [PMID: 1746946]

[14]   Synthetic Paper Market by Raw Material (BOPP, HDPE), Application (Printing, Paper Bags, Labels), End-use Industry (Paper, Packaging), and Region (APAC, Europe, North America, Middle East & Africa, South America) - Global Forecast to 2025. Appl Environ Microbiol 2020; 57: 2858-63.
[http://dx.doi.org/10.1128/aem.57.10.2858-2863.1991]

[15]   Tiku DK, Kumar A, Sawhney S, Singh VP, Kumar R. Effectiveness of treatment technologies for wastewater pollution generated by Indian pulp mills. Environ Monit Assess 2007; 132(1-3): 453-66.
[http://dx.doi.org/10.1007/s10661-006-9548-3] [PMID: 17295114]

[16]   Kamali M, Khodaparast Z. Review on recent developments on pulp and paper mill wastewater treatment. Ecotoxicol Environ Saf 2015; 114: 326-42.
[http://dx.doi.org/10.1016/j.ecoenv.2014.05.005] [PMID: 24953005]

[17]   Menezes GB, De Galvão TCB, Moo-Young HK. Pulp and Paper Mill Effluents Management. Water Environ Res 2010; 82(10): 1560-7.

[http://dx.doi.org/10.2175/106143010X12756668801617]

[18]   Bajpai P. Generation of Waste in Pulp and Paper Mills 2015; 9-17.
        [http://dx.doi.org/10.1007/978-3-319-11788-1_2]

[19]   Hubbe MA, Metts JR, Hermosilla D, *et al.* Wastewater Treatment and Reclamation: A Review of Pulp
        and Paper Industry Practices and Opportunities. BioResources 2016; 11(3): 7953-8091.
        [http://dx.doi.org/10.15376/biores.11.3.Hubbe]

[20]   Smook G. Handbook for Pulp and Paper Technologists. Angus Wilde Publications, Inc.: Canada 1989.

[21]   Abdelaziz OY, Brink DP, Prothmann J, *et al.* Biological valorization of low molecular weight lignin.
        Biotechnol Adv 2016; 34(8): 1318-46.
        [http://dx.doi.org/10.1016/j.biotechadv.2016.10.001] [PMID: 27720980]

[22]   Moo-Young HK. Pulp and Paper Effluent Management. Water Environ Res 2007; 79(10): 1733-41.
        [http://dx.doi.org/10.2175/106143007X218566]

[23]   Singh , Chowdhary P, Singh J, Chandra R. Pulp and paper mill wastewater and coliform as health
        hazards: A review. Microbiol Res Int 2016; 4: 28-39.

[24]   Lacorte S, Latorre A, Barcelo D, Rigol A, Malmqvist A, Welander T. *"Organic compounds in paper-
        mill process waters and effluents. Trends in Analytical Chemistry, 22(10), 725-737,"* TrAC. Trends
        Analyt Chem 2003; 22: 725-37.
        [http://dx.doi.org/10.1016/S0165-9936(03)01009-4]

[25]   Rocha-Santos T, Duarte AC. A critical overview of the analytical approaches to the occurrence, the
        fate and the behavior of microplastics in the environment. Trends Analyt Chem 2015; 65: 47-53.
        [http://dx.doi.org/10.1016/j.trac.2014.10.011]

[26]   Bajpai PT. *"Environmentally Benign Approaches for Pulp Bleaching,"* Nether LandsEl. Sci 2012;
        (Jan): 97-134.

[27]   Santos R, Hart P. Brown Stock Washing – A Review of the Literature. Tappi J 2014; 13: 9-19.
        [http://dx.doi.org/10.32964/TJ13.1.9]

[28]   Chaudhry S, Paliwal R. Techniques for Remediation of Paper and Pulp Mill Effluents: Processes and
        Constraints.BT - Handbook of Environmental Materials Management. Cham: Springer International
        Publishing 2018; pp. 1-19.

[29]   Biermann C. Stock Preparation and Additives for Papermaking 1996; 190-208.
        [http://dx.doi.org/10.1016/B978-012097362-0/50012-1]

[30]   Rao G, Dutta RK, Ujwala D. Strength characteristics of sand reinforced with coir fibres and coir
        geotextiles. Electron J Geotech Eng 2005; 10.

[31]   Jenkins J, Chojnacky DC, Heath L, Birdsey RA. National Scale Biomass Estimators for United States
        Tree Species. For Sci 2003; 49: 12-35.

[32]   Raj A, Reddy MMK, Chandra R, Purohit HJ, Kapley A. Biodegradation of kraft-lignin by *Bacillus* sp.
        isolated from sludge of pulp and paper mill. Biodegradation 2007; 18(6): 783-92.
        [http://dx.doi.org/10.1007/s10532-007-9107-9] [PMID: 17308883]

[33]   Kumar V, Thakur I, Shah M. Bioremediation Approaches For Pulp And Paper Wastewater Treatment:
        Recent Advances and Challenges 2020; 1-63.
        [http://dx.doi.org/10.1007/978-981-15-1812-6_1]

[34]   Paranthaman SR, Karthikeyan B. Bioremediation of paper mill effluent on growth and development of
        seed germination (Vigna mungo). CIBTech J Biotechnol 2015; 4(1): 22-6.

[35]   Alissa EM, Ferns GA. Heavy metal poisoning and cardiovascular disease. J Toxicol 2011; 2011: 1-21.
        [http://dx.doi.org/10.1155/2011/870125] [PMID: 21912545]

[36]   Wu X, Cobbina SJ, Mao G, Xu H, Zhang Z, Yang L. A review of toxicity and mechanisms of
        individual and mixtures of heavy metals in the environment. Environ Sci Pollut Res Int 2016; 23(9):

8244-59.
[http://dx.doi.org/10.1007/s11356-016-6333-x] [PMID: 26965280]

[37]   Duruibe J. O. C, and J. Egwurugwu, *"Heavy Metal Pollution and Human Biotoxic Effects,".* Int J Phys Sci 2007; 2: 112-8.

[38]   Sharma S, Bhattacharya A. Drinking water contamination and treatment techniques. Appl Water Sci 2017; 7(3): 1043-67.
[http://dx.doi.org/10.1007/s13201-016-0455-7]

[39]   Appenroth KJ. What are "heavy metals" in Plant Sciences? Acta Physiol Plant 2010; 32(4): 615-9.
[http://dx.doi.org/10.1007/s11738-009-0455-4]

[40]   Sahni SK. Hazardous metals and minerals pollution in India. A Position Paper, August 2011.New Delhi, India: Indian Natl. Sci. Acad. 2011; pp. 1-22.

[41]   Saidi M. Experimental studies on effect of heavy metals presence in industrial wastewater on biological treatment. Int J Environ Sci 2010; 1: 666-76.

[42]   Kapoor J, Jabin S, Bhatia H. Optimization of coagulation-flocculation process for food industry waste water treatment using polyelectrolytes with inorganic coagulants J Indian Chem Soc 2015; 92: 1697-703.

[43]   Kamran S, Shafaqat A, Samra H, *et al.* Heavy Metals Contamination and what are the Impacts on Living Organisms. Greener Journal of Environmental Management and Public Safety 2013; 2(4): 172-9.
[http://dx.doi.org/10.15580/GJEMPS.2013.4.060413652]

[44]   Singh J, Kalamdhad AS. Effects of heavy metals on soil, plants, human health and aquatic life. Int J Res Chem Environ 2011; 1(2): 15-21.

[45]   Igiri BE, Okoduwa SIR, Idoko GO, Akabuogu EP, Adeyi AO, Ejiogu IK. Toxicity and Bioremediation of Heavy Metals Contaminated Ecosystem from Tannery Wastewater: A Review. J Toxicol 2018; 2018: 1-16.
[http://dx.doi.org/10.1155/2018/2568038] [PMID: 30363677]

[46]   Hodson M. Effects of Heavy Metals and Metalloids on Soil Organisms, in Heavy Metals in Soils: Trace Metals and Metalloids in Soils and Their Bioavailability. Environ Pollut 2013; 22: 141-60.

[47]   Ding Z, Wu J, You A, Huang B, Cao C. Effects of heavy metals on soil microbial community structure and diversity in the rice (*Oryza sativa* L. subsp. Japonica, Food Crops Institute of Jiangsu Academy of Agricultural Sciences) rhizosphere. Soil Sci Plant Nutr 2017; 63(1): 75-83.
[http://dx.doi.org/10.1080/00380768.2016.1247385]

[48]   Xie Y, Fan J, Zhu W, *et al.* Effect of Heavy Metals Pollution on Soil Microbial Diversity and Bermudagrass Genetic Variation. Front Plant Sci 2016; 7: 755. [Online] Available from: https://www.frontiersin.org/article/10.3389/fpls.2016.00755

[49]   Sobolev D, Begonia M. Effects of heavy metal contamination upon soil microbes: lead-induced changes in general and denitrifying microbial communities as evidenced by molecular markers. Int J Environ Res Public Health 2008; 5(5): 450-6.
[http://dx.doi.org/10.3390/ijerph5050450] [PMID: 19151442]

[50]   Giller KE, Witter E, McGrath SP. Heavy metals and soil microbes. Soil Biol Biochem 2009; 41(10): 2031-7.
[http://dx.doi.org/10.1016/j.soilbio.2009.04.026]

[51]   Abdu N, Abdullahi AA, Abdulkadir A. Heavy metals and soil microbes. Environ Chem Lett 2017; 15(1): 65-84.
[http://dx.doi.org/10.1007/s10311-016-0587-x]

[52]   Oliveira A, Pampulha ME. Effects of long-term heavy metal contamination on soil microbial characteristics. J Biosci Bioeng 2006; 102(3): 157-61.

[http://dx.doi.org/10.1263/jbb.102.157] [PMID: 17046527]

[53]   Tangahu BV, Sheikh Abdullah SR, Basri H, Idris M, Anuar N, Mukhlisin M. A Review on Heavy Metals (As, Pb, and Hg) Uptake by Plants through Phytoremediation. Int J Chem Eng 2011; 2011: 1-31.
[http://dx.doi.org/10.1155/2011/939161]

[54]   Schmidt U. Enhancing Phytoextraction. J Environ Qual 2003; 32(6): 1939-54.
[http://dx.doi.org/10.2134/jeq2003.1939] [PMID: 14674516]

[55]   Hattori H. Influence of heavy metals on soil microbial activities. Soil Sci Plant Nutr 1992; 38(1): 93-100.
[http://dx.doi.org/10.1080/00380768.1992.10416956]

[56]   Schwartz C, Echevarria G, Morel JL. Phytoextraction of cadmium with Thlaspi caerulescens. Plant Soil 2003; 249(1): 27-35.
[http://dx.doi.org/10.1023/A:1022584220411]

[57]   Organization WH. World health organization = organisation mondiale de la santé 1952. Available from: https://apps.who.int/iris/handle/10665/240673

[58]   Pandey G, Sharma M. Heavy metals causing toxicity in animals and fishes 2014; 2: 17-23.

[59]   McCluggage D. Heavy Metal Poisoning.NCS Magazine. Columbus: The Bird Hospital 1991.

[60]   Sauvé S. Time to revisit arsenic regulations: comparing drinking water and rice. BMC Public Health 2014; 14(1): 465.
[http://dx.doi.org/10.1186/1471-2458-14-465] [PMID: 24884827]

[61]   Wani A L, Ara A, Usmani J A. Lead toxicity: A reviewInterdisciplinary Toxicology. Slovak Toxicology Society 2015; 8: pp. 55-64.
[http://dx.doi.org/10.1515/intox-2015-0009]

[62]   Rahimzadeh MR, Rahimzadeh MR, Kazemi S, Moghadamnia AA. Cadmium toxicity and treatment: An updateCaspian Journal of Internal Medicine. Babol University of Medical Sciences 2017; 8: pp. 135-45.
[http://dx.doi.org/10.22088/cjim.8.3.135]

[63]   Unaegbu M, Engwa GA, Abaa Q, *et al.* Heavy metal, nutrient and antioxidant status of selected fruit samples sold in Enugu, Nigeria. Int J Food Contam 2016; 3: 7.
[http://dx.doi.org/10.1186/s40550-016-0031-9]

[64]   Abdul Ghani AG. Effect of chromium toxicity on growth, chlorophyll and some mineral nutrients of brassica juncea L. Egypt Acad J Biol Sci H Bot 2011; 2(1): 9-15.
[http://dx.doi.org/10.21608/eajbsh.2011.17007]

[65]   H UD, Services H. Agency for Toxic Substances and Disease Registry-ATSDR 1999.(Accessed: Oct. 27, 2020). Available from: https://pesquisa.bvsalud.org/portal/resource/pt/lis-LISBR1.1-128

[66]   Health Effects of Excess Copper 2000.(Accessed: Oct. 27, 2020). Available from: https://www.ncbi.nlm.nih.gov/books/NBK225400/

[67]   Miah MR, Ijomone OM, Okoh COA, *et al.* The effects of manganese overexposure on brain health. Neurochem Int 2020; 135: 104688.
[http://dx.doi.org/10.1016/j.neuint.2020.104688] [PMID: 31972215]

[68]   Jaishankar M, Tseten T, Anbalagan N, Mathew B B, Beeregowda K N. Toxicity, mechanism and health effects of some heavy metalsInterdisciplinary Toxicology. Slovak Toxicology Society 2014; 7: pp. 60-72.
[http://dx.doi.org/10.2478/intox-2014-0009]

[69]   Nickel: health effects, nickel allergy-Metalpedia. (accessed Oct. 27, 2020). Available from: http://metalpedia.asianmetal.com/metal/nickel/health.shtml

[70]   Remediation Approaches. Reclamation of Contaminated Land 2004; 125-49.
        [http://dx.doi.org/10.1002/0470020954.ch8]

[71]   Scullion H, Collings DG. International Talent Management. In: Scullion H, Collings DG, Eds. Global
        Staffing. London: Routledge 2006: p 30.
        [http://dx.doi.org/10.4324/9780203643433]

[72]   Emenike CU, Jayanthi B, Agamuthu P, Fauziah SH. Biotransformation and removal of heavy metals: a
        review of phytoremediation and microbial remediation assessment on contaminated soil. Environ Rev
        2018; 26(2): 156-68.
        [http://dx.doi.org/10.1139/er-2017-0045]

[73]   Sharma S, Tiwari S, Hasan A, Saxena V, Pandey L M. Recent advances in conventional and
        contemporary methods for remediation of heavy metal-contaminated soils 3 Biotech 2018; 8: 216.
        [http://dx.doi.org/10.1007/s13205-018-1237-8]

[74]   Gunatilake S. Methods of Removing Heavy Metals from Industrial Wastewater J Multidiciplinary Eng
        Sci Stud 2015; 1.

[75]   Kumar L, Bidlan R, Sharma J, Bharadvaja N. Biotechnological management of water quality: A mini
        review. Biosc Biotech Res Comm 2019; 12(1).

[76]   Rajendran P, Muthukrishnan J, Gunasekaran P. Microbes in heavy metal remediation. Indian J Exp
        Biol 2003; 41(9): 935-44.

[77]   De Gisi S, Lofrano G, Grassi M, Notarnicola M. Characteristics and adsorption capacities of low-cost
        sorbents for wastewater treatment: A review. Sustainable Materials and Technologies 2016; 9: 10-40.
        [http://dx.doi.org/10.1016/j.susmat.2016.06.002]

[78]   Hossain MA, Ngo HH, Guo WS, Setiadi T. Adsorption and desorption of copper(II) ions onto garden
        grass. Bioresour Technol 2012; 121: 386-95.
        [http://dx.doi.org/10.1016/j.biortech.2012.06.119] [PMID: 22864175]

[79]   Salt DE, Blaylock M, Kumar NPBA, *et al.* Phytoremediation: a novel strategy for the removal of toxic
        metals from the environment using plants. Nat Biotechnol 1995; 13(5): 468-74.
        [http://dx.doi.org/10.1038/nbt0595-468] [PMID: 9634787]

[80]   Muthu M, Wu HF, Gopal J, Sivanesan I, Chun S. Exploiting Microbial Polysaccharides for
        Biosorption of Trace Elements in Aqueous Environments—Scope for Expansion via Nanomaterial
        Intervention. Polymers (Basel) 2017; 9(12): 721.
        [http://dx.doi.org/10.3390/polym9120721] [PMID: 30966021]

[81]   Gadd GM. Bioremedial potential of microbial mechanisms of metal mobilization and immobilization.
        Curr Opin Biotechnol 2000; 11(3): 271-9.
        [http://dx.doi.org/10.1016/S0958-1669(00)00095-1] [PMID: 10851150]

[82]   Lim PE, Mak KY, Mohamed N, Noor AM. Removal and speciation of heavy metals along the
        treatment path of wastewater in subsurface-flow constructed wetlands. Water Sci Technol 2003; 48(5):
        307-13.
        [http://dx.doi.org/10.2166/wst.2003.0337] [PMID: 14621178]

[83]   Lin CC, Lin HL. Remediation of soil contaminated with the heavy metal (Cd2+). J Hazard Mater
        2005; 122(1-2): 7-15.
        [http://dx.doi.org/10.1016/j.jhazmat.2005.02.017] [PMID: 15943924]

[84]   Kanamarlapudi SLRK, Chintalpudi VK, Muddada S. Application of Biosorption for Removal of
        Heavy Metals from Wastewater. In: Derco J, Vrana B. Biosorption. InTech: Rijeka, Croatia, 2018: pp.
        69-116.

[85]   Mrvčić J, Stanzer D, Šolić E, Stehlik-Tomas V. Interaction of lactic acid bacteria with metal ions:
        opportunities for improving food safety and quality. World J Microbiol Biotechnol 2012; 28(9): 2771-
        82.

[http://dx.doi.org/10.1007/s11274-012-1094-2] [PMID: 22806724]

[86]   Chander K, Klein T, Eberhardt U, Joergensen R. Decomposition of carbon-14-labelled wheat straw in repeatedly fumigated and non-fumigated soils with different levels of heavy metal contamination. Biol Fertil Soils 2002; 35(2): 86-91.
        [http://dx.doi.org/10.1007/s00374-002-0443-y]

[87]   Tapia JM, Muñoz JA, González F, Blázquez ML, Ballester A. Mechanism of adsorption of ferric iron by extracellular polymeric substances (EPS) from a bacterium Acidiphilium sp. Water Sci Technol 2011; 64(8): 1716-22.
        [http://dx.doi.org/10.2166/wst.2011.649] [PMID: 22335116]

[88]   Shamim S. Biosorption of Heavy metals. In: Derco J, Vrana B. Biosorption. InTech: Rijeka, Croatia, 2018: pp. 21-49.

[89]   Nourbakhsh M, Saḡ Y, Özer D, Aksu Z, Kutsal T, Çaḡlar A. A comparative study of various biosorbents for removal of chromium(VI) ions from industrial waste waters. Process Biochem 1994; 29(1): 1-5.
        [http://dx.doi.org/10.1016/0032-9592(94)80052-9]

[90]   Huckle JW, Morby AP, Turner JS, Robinson NJ. Isolation of a prokaryotic metallothionein locus and analysis of transcriptional control by trace metal ions. Mol Microbiol 1993; 7(2): 177-87.
        [http://dx.doi.org/10.1111/j.1365-2958.1993.tb01109.x] [PMID: 8446025]

[91]   Chojnacka K. Biosorption and bioaccumulation – the prospects for practical applications. Environ Int 2010; 36(3): 299-307.
        [http://dx.doi.org/10.1016/j.envint.2009.12.001] [PMID: 20051290]

[92]   Dt O, Aa A, Oe O. Heavy Metal Concentrations in Plants and Soil along Heavy Traffic Roads in North Central Nigeria. J Environ Anal Toxicol 2015; 5(6).
        [http://dx.doi.org/10.4172/2161-0525.1000334]

[93]   Shah , Ahmad N, Masood K, Peralta-videa J, Ahmad F. Heavy Metal Toxicity in Plants. Plant Adaptation and Phytoremediation 2010; 71-97.

[94]   Siddiquee S, Rovina K, Azad SA. Heavy Metal Contaminants Removal from Wastewater Using the Potential Filamentous Fungi Biomass: A Review. J Microb Biochem Technol 2015; 7(6).
        [http://dx.doi.org/10.4172/1948-5948.1000243]

[95]   von Canstein H, Kelly S, Li Y, Wagner-Döbler I. Species diversity improves the efficiency of mercury-reducing biofilms under changing environmental conditions. Appl Environ Microbiol 2002; 68(6): 2829-37.
        [http://dx.doi.org/10.1128/AEM.68.6.2829-2837.2002] [PMID: 12039739]

[96]   Brim H, McFarlan SC, Fredrickson JK, *et al.* Engineering Deinococcus radiodurans for metal remediation in radioactive mixed waste environments. Nat Biotechnol 2000; 18(1): 85-90.
        [http://dx.doi.org/10.1038/71986] [PMID: 10625398]

[97]   Ramaswamy V. Cresence VM, Rejitha JS. Listeria--review of epidemiology and pathogenesis. J Microbiol Immunol Infect 2007; 40(1): 4-13.

[98]   Vasudevan N, Rajaram P. Bioremediation of oil sludge-contaminated soil. Environ Int 2001; 26(5-6): 409-11.
        [http://dx.doi.org/10.1016/S0160-4120(01)00020-4] [PMID: 11392759]

[99]   Rulkens WH, Grotenhuis JTC, Tichý R. Methods for Cleaning Contaminated Soils and Sediments BT - Heavy Metals: Problems and Solutions. Berlin, Heidelberg: Springer Berlin Heidelberg 1995; pp. 165-91.

[100]   Kantachote D, Panwichian S, Wittayaweerasak B, Mallavarapu M. Removal of heavy metals by exopolymeric substances produced by resistant purple nonsulfur bacteria isolated from contaminated shrimp ponds. Electron J Biotechnol 2011; 14(4): 2.
        [http://dx.doi.org/10.2225/vol14-issue4-fulltext-2]

[101] Pereira, dr, Silva-Correia j, Caridade SG, *et al.* Development of gellan gum-based microparticles/hydrogel matrices for application in the intervertebral disc regeneration. Tissue Eng Part C 2011; 17(10): 961-72.

[102] Beolchini F, Ubaldini S, Passariello B, *et al.* Bioremediation of Dredged Sediments Polluted by Heavy Metals. Adv Mat Res 2007; 20: 307-10.
[http://dx.doi.org/10.4028/0-87849-452-9.307]

[103] Li T, Zheng X, Dai Y *et al.* Mapping near-surface air temperature, pressure, relative humidity and wind speed over Mainland China with high spatiotemporal resolution. Adv Atm Sci 2014; 13: 1127-35.
[http://dx.doi.org/10.13140/2.1.4690.7849]

[104] Lovley DR. Cleaning up with genomics: applying molecular biology to bioremediation. Nat Rev Microbiol 2003; 1(1): 35-44.
[http://dx.doi.org/10.1038/nrmicro731] [PMID: 15040178]

[105] Watanabe K. Microorganisms relevant to bioremediation. Curr Opin Biotechnol 2001; 12(3): 237-41.
[http://dx.doi.org/10.1016/S0958-1669(00)00205-6] [PMID: 11404100]

[106] Laurenti M, García Blanco F, López-Cabarcos E, Rubio-Retama J. Detection of heavy metal ions using a water-soluble conjugated polymer based on thiophene and *meso* -2,3-dimercaptosuccinic acid. Polym Int 2013; 62(5): 811-6.
[http://dx.doi.org/10.1002/pi.4369]

[107] Wilcke W, Kiesewetter M, Musa Bandowe BA. Microbial formation and degradation of oxygen-containing polycyclic aromatic hydrocarbons (OPAHs) in soil during short-term incubation. Environ Pollut 2014; 184: 385-90.
[http://dx.doi.org/10.1016/j.envpol.2013.09.020] [PMID: 24100048]

[108] Kapoor A, Viraraghavan T. Heavy metal biosorption sites in *Aspergillus niger.* Bioresour Technol 1997; 61(3): 221-7.
[http://dx.doi.org/10.1016/S0960-8524(97)00055-2]

CHAPTER 4

Bioremediation of Pesticides

Praveen Kumar Yadav[1,2,*], Kamlesh Kumar Nigam[3], Shishir Kumar Singh[2,4], Ankit Kumar[5] and **S. Swarupa Tripathy[1]**

[1] *Chemical and Food BND Group, Indian Reference Materials Division, CSIR- National Physical Laboratory, Dr. K.S. Krishnan Marg, New-Delhi 110012, India*

[2] *Academy of Scientific and Innovative Research (AcSIR), Ghaziabad, U.P 201002, India*

[3] *Department of Chemistry, Institute of Science, Banaras Hindu University, Varanasi 221005, UP, India*

[4] *Environmental Science and Biomedical Metrology Division, CSIR- National Physical Laboratory, Dr. K.S. Krishnan Marg, New-Delhi 110012, India*

[5] *Department of Chemistry, University of Delhi, Delhi 110007, India*

Abstract: Increasing population has raised the demand for food grains, which compels the producers for the heavy use of pesticides to meet the demand for sufficient production of food grains. Heavy utilization of pesticides polluted soil, water, plant, animal, food grains, *etc*. Additionally, that much utilization of pesticides has also created several legal and illegal contaminated sites across the world, which are continuously polluting the environment. There are several methods available for pesticide treatment, but the bioremediation method has been more promising than the others. Bioremediation of pesticides is carried out through either *ex situ* or *in situ* methods using different organisms like bacteria, fungi and higher plants. The pesticides degradation using bacteria, fungi and higher plants is called bacterial degradation, mycodegradation and phytodegradation, respectively. Present review discusses different methods, mechanisms and recent tools used for the bioremediation of pesticides.

Keywords: Bacteria, Bioremediation, Food grains, Fungi, Mycodegradation and phytodegradation, Pesticides.

INTRODUCTION

In this contemporary world, increasing population, rapid industrialization, and advancement in technologies have increased the productivity of agricultural activity. But it is well said that every development needs a cost, so, this proverb

* **Corresponding author Praveen Kumar Yadav:** Chemical and Food BND Group, Indian Reference Materials Division, CSIR- National Physical Laboratory, Dr. K.S. Krishnan Marg, New-Delhi 110012, India & Academy of Scientific and Innovative Research (AcSIR), Ghaziabad, U.P 201002, India; E-mail: kpraveen.yadav414@gmail.com

Inamuddin (Ed.)

exactly fits modern agriculture. For example, these days, millions of tons ofpesticides are being produced to improve agricultural productivity, to control the harmful effects of various organisms (bacteria fungi, insects) as well as the effect of some herbs on the economical crop's productivity [1].

Though the use of pesticides is not new for the world, it has been utilised for pest control for centuries. For example, Romans, Greeks, and Sumerians also used different types of chemicals like mercury, copper, sulphur, arsenic or plant extracts to control the pests. Though, at that time, these chemicals were not very effective due to primitive chemistry and poor application methods. However, after World War II, several chemical compositions such as aldrin, endrin, DDT (dichlorodiphenyltrichloroethane), 2,4-D (2,4-dichlorophenoxyacetic acid), BHC (benzene hexachloride), and dieldrin were introduced which triggered their rapid use in agriculture due to their certain properties like low cost, effective application and easy to use which made these pesticides more popular [2].

So, the extensive production and application of pesticides have made them the most broadly distributed pollutant in the environment. It is estimated that more than 95% of these pesticides are waste and accumulate in the environment and only 5% reach the target organisms. Therefore, the high leaching capacity of pesticides into the groundwater and accumulation in the soil is the main environmental concern. Additionally, the disposal of outdated pesticide stocks is a major source of pesticide contamination in the environment that has left many long-term contaminated sites in the world. There are several officially recognized sites of pesticides that have been reported from the different parts of the world *e.g.* Poland [3], Argentina, Chile [4], USA [5], Spain [6], Brazil, The Netherlands [7], India [8], China [9], Canada, *etc.* In spite of these sites, several illegal disposal sites are reported. For example, the Santiago del Estero, Argentina [10, 11] has been reported as a known illegal pesticide disposal site where more than 30 tons of pesticides (chlordane (CLD), DDT, lindane (g-HCH), aldrin, methoxychlor (MTX)) along with several heavy metals [Cd(II), Cr(VI), Cu(II)] disposal is carried out. However, there are several studies which report the residues of pesticides in water [12], air [13], food commodities [14], fishes [15], milk [16], soil [11], and even in human blood and adipose tissue [17] as well. Therefore, the availability of pesticides from the environment to human beings could directly affect the plants, animals, microorganisms and human health as well [18, 19] (Fig. **1**).

Hence, in order to reduce the various pollutants and pesticides from the environment, several new techniques such as recycling, landfills, pyrolysis of pesticides, *etc.* have been developed. But these techniques were not completely successful because they also produced some toxic intermediates [20].

Additionally, these techniques were very expensive and sophisticated, especially in the case of pesticide treatment [21].

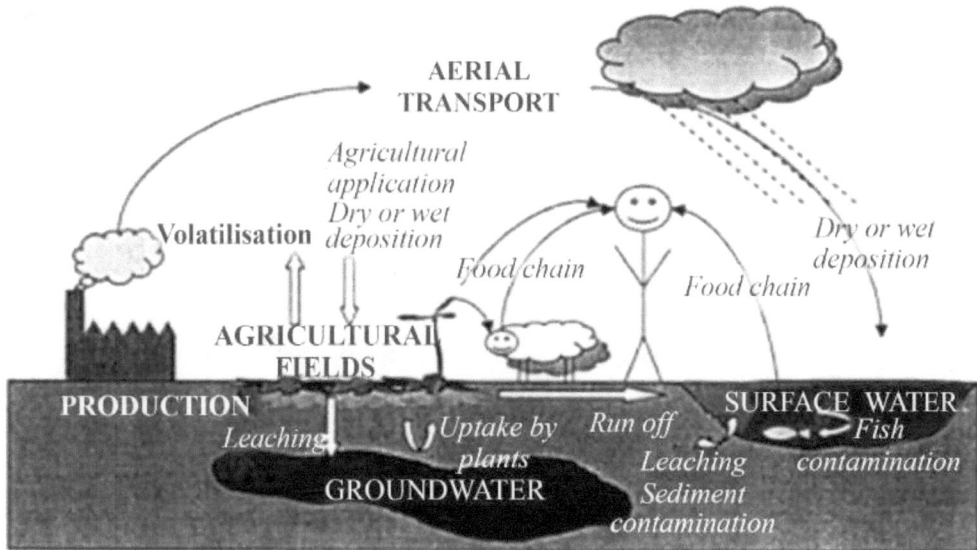

Fig. (1). Environmental fate of pesticides [19].

In the early decades, research has been focused on the development of eco-friendly and cost-effective methods for the cleaning of environmental pollutants using various microbial species and the process of pollutant treatment using microbes called bioremediation. This approach has shown the upper hand over the traditional physicochemical methods due to its low invasive and soil curative property [22]. Bioremediation has been proven as a sustainable technology that is utilised for the tracking of the environmental release of anthropogenic pollutants.

Pesticides

A pesticide is any substance either chemical, biological or herbal constituents that can be utilised to control, destroy or mitigate pests *e.g.* weeds, insects, nematodes, mites, *etc* [23]. Though in different periods, different definitions have been explained for the pesticide but its crux remains unchanged. Pesticides include herbicide, insecticide, fungicide and several other constituents utilised for pest management [24].

Basically, on the basis of the composition, pesticides are of three types *i.e.* chemical and biological and botanical pesticides (Table **1**). The structure-based classification of pesticides is also made which included carbamates, organochlorine, nitrogen-based pesticides, organophosphorus, *etc* [25, 26]. The global consumption of pesticides has reached 2 million tonnes in which Europe contributes 45%, the USA 24% and the rest of the world contributes 25% of the total consumption of pesticides [27]. Additionally, in the case of Asian countries, China showed the highest percentage of pesticide consumption followed by Korea, Japan and India. Moreover, India is the biggest producer of pesticides as well.

Table 1. Composition based classification of pesticides.

Pesticides	Examples	Sources
Organochlorines	DDT, aldrin, lindane, chlordane, mirex	Chemical synthesis
Organophosphates	Malathion, methyl parathion, diazinon	Chemical synthesis
Carbamates	Sevin, carbaryl	Chemical synthesis
Cyclodienes	Aldrin, chlordane, heptachlor, endrin	Chemical synthesis
Botanical	Neem, rotinone, linalool, limonene, ryania	Plant source
Biological	Dispel, foray, thuricide	Microorganism, virus or their metabolic products

However, the global sales of pesticides have been constant, since 1990s. Europe has been the largest global consumer of pesticides followed by Asia. Some other countries like the United States, China, Brazil, France, and Japan have been the biggest global producers and consumers of pesticides. Globally, pesticides are mostly utilised in vegetable and food crops. The developed countries mainly use herbicides mostly for the maize crop [24].

The pesticides get degraded into another but less toxic degradation product than the original pesticides. But there are several microorganisms available in the environment which can further be transferred or completely degrade the pesticides or their degradation products by the microbial consortium. Though, the metabolic dead-end products and remaining pesticides (called persistent xenobiotics) accumulate in the soil and make it humus soil and hence, enter the food chain. The excess use of pesticides could affect both environment and human health as well. So, these two factors (*i.e.* environmental and human health) are used in the impact evaluation of pesticides. Fig. (**2**) illustrates the impact of pesticides.

The environmental fate of pesticides strongly depends on the soil sorption processes, which control their bioavailability and transfer as well [28]. Soil as well as groundwater contamination, basically, occurs due to the bulk use of pesticides on the farm, their accidental release and bulk handling at the farmyard [29]. The degradation and the persistence of pesticides on the soil control their potential toxicity, efficiency and persistence as well [30]. The biodegradation processes are influenced by several factors like temperature, moisture, physicochemical properties of the soil, and availability of different nitrogen or carbon sources, which can alter the population of microbes and their activity as well [24]. On the other hand, the illegal and inappropriate storage or disposal of obsolete products, and liquid and solid wastes have become an environmental liability. Therefore, most inappropriate waste management cases are not conveyed to the suitable authority. Mostly, developing countries are worst affected by the environmental and biological impact of outdated pesticides due to mismanagement of pesticide waste [31]. So, the lack of effective or clear obsolete pesticides management strategy for long tenure has stockpiled the obsolete pesticides in the developing countries [32].

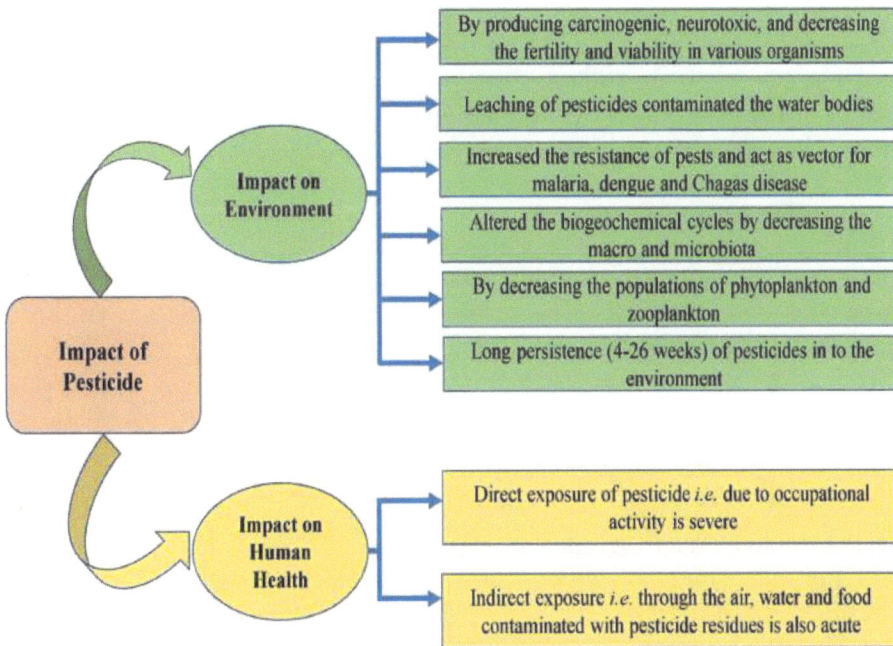

Fig. (2). Impact of pesticide on human health and environment.

Bioremediation of Pesticides

In order to avoid the problems related to pesticide waste management, and reduce the effect on the environment and human health, several safe, efficient and economical methods or technologies have been developed. Basically there are two basic treatment techniques *i.e.* chemical treatment where powerful transient species (-OH radical) is utilised for oxidation of waste and physical treatments *i.e.* adsorption and percolator filters are utilised for pesticide waste treatment. Some other techniques such as photodegradation of pesticides using TiO_2 [33] and high temperature incineration of pesticide waste in special furnaces are also utilised, currently, in pesticide waste treatment. For high temperature incineration, mostly the pesticides are packaged and transported to the country having special disposable facilities for hazardous wastes. And the cost of 3 to 4 thousand USD/ton is estimated for these whole operations [26]. But these methods have some disadvantages viz. the use of expensive techniques and chemical catalysts (TiO_2) *etc.* Additionally, the treatment of pesticide waste can lead to the formation of secondary pollutants *e.g.* alkaline hydrolysis of pesticides (hydrolysis of organophosphates), which needs controlled experimental conditions otherwise it can produce complexes with metal already present in the water.

So, an alternative pesticide treatment technique (*i.e.* bioremediation) has been developed which has shown the upper hand over the traditional methods and overcome their limitations as well. In this method, microorganisms are used to convert the organic pollutants into simple and nontoxic components. Consequently, this technique is globally utilised for pesticide treatment at large scale due to its cost effective and eco-friendly nature [34 - 37].

Type of Bioremediation

Bioremediation is actually the degradation of pollutant material using bio-organisms and hence, the technique involves the biodegradation of pesticide materials which is stimulated either *in-situ* or *ex-situ* in bioreactors or compost heaps.

In-situ Bioremediation

In this method of bioremediation, the biodegradation of pollutants is carried out within the affected areas *e.g.* sediment, soil and groundwater or surface water environments. Hence, it is called the *in-situ* bioremediation process. *In-situ* treatment techniques offer direct contact between the dissolved and/or adsorbed contaminants, and microorganisms for biotransformation [38]. Additionally, this technique also offers cost effective, eco-friendly technique as well as produces less dust and volatile pollutants. Moreover, some other potential advantages of *in-situ* techniques include concurrent treatment of polluted groundwater and soil, minimal site disruption, public exposure and site personnel exposure. Additionally, the genetically engineered microorganisms are also utilised for the simulation of remediation process whenever the microorganisms lack their biodegradation capacity during the *in-situ* bioremediation process. Though, these methods also suffer from certain disadvantages such as a seasonal change in microbial activity due to the variation in environmental conditions, time taking method, and use of treatment additives such as surfactants, nutrients and oxygen.

Ex-situ Bioremediation

In *ex-situ* bioremediation process, the pesticide waste is collected from the contaminated site and their treatment is done in different places in order to improve their biodegradation. Here, the degradation process is generally aerobic which needs solid or slurry phase systems for the treatment of polluted soil or sediments. In *ex-situ* process, the treatment is based on the physical nature of the contaminant and divided into two types: (i) Slurry phase bioremediation where the treatment of solid-liquid suspensions takes place in bioreactors and (ii) Solid phase bioremediation included land farming (soil treatment units) and soil piles, compost heaps and engineered bio piles. There are several contaminated sites having explosives such as Tetryl, TNT and RDX *etc.* called the military sites. So, in military sites, the composting techniques have proven important for the remediation of these sites. These techniques are also having certain disadvantages and limitations but the main disadvantage is the involvement of the high-cost and time-consuming techniques.

Additionally, the bioremediation process is further divided into the aerobic and anaerobic bioremediation processes.

Aerobic Bioremediation

The aerobic bacteria utilize the oxygenase enzyme for the degradation of organic pesticides but the aromatic molecules having electron-withdrawing groups (chloro, nitro and azo groups) do not show the degradation due to the presence of electron-withdrawing group on aromatic ring which hinders the attack of electrophile. Therefore, the bioremediation of these pesticides occurs *via* the co-metabolism process only. During the co-metabolism process, microbes catalyse the hydrolysis of pesticides and transform them into less toxic and degradable components. The enzymes utilised in the co-metabolism process are, basically, transferase, reductase, oxidase and hydrolytic enzymes. In the case of contaminated soil bioremediation, the availability of pollutants to the microorganism is a very critical part of the soil bioremediation. Additionally, the pollutant availability depends on their adsorption affinity to the soil as well as their solubility in water also. Another significant factor is moisture content which offers a very critical role in microbial degradation. For example, moisture content of 50% could significantly increase the degradation rates. Additionally, the pesticide concentration in the soil also affects the rate of degradation during the low microbial population. The contaminated soil can be remediated with help of using specific seed cultures, bio-augmentation and inoculation of the soil or bioreactor. It is found that anaerobic bioremediation shows better degradation for some pesticides like heptachlor, DDT and lindane than aerobic degradation [39].

Anaerobic Bioremediation

In the anaerobic bioremediation process, the degradation of aromatic compounds takes place *via* nucleophilic reaction on aromatic ring followed by the substitution of functional groups like nitro, chloro or azo present on the ring along with the reduction of ring by anaerobic bacteria. Whereas, if the aromatic ring is having the electron donating groups *i.e.* amino groups, it reduces the capability of anaerobic bacteria toward the degradation of an aromatic compound, while aerobic degradation works well in this case. Additionally, if there is no functional group present on the ring, it also decreases the capability of the anaerobic bacteria for their degradation. So, the hydrocarbons are less susceptible toward anaerobic degradation. Hence, there are several degradation reactions that are used for the anaerobic degradation of less susceptible pesticides *e.g.* (i) oxidation of sulphur to SO_2 and oxidation of hydrocarbon to produce epoxide which is resistant to microbial degradation and toxic for the microbes as well (ii) addition of hydroxyl group replacing the hydrogen atom (iii) methylation or demethylation of pesticides such as arsenic pesticides, (iv) replacement of sulphur with oxygen (v) chlorine migration (vi) reduction of a nitro (NO_2-) group to amino group (NH_3), (vii) dehalogenation of pesticides, *etc* [40].

There are three basic approaches for the bioremediation of pesticides, where the biodegradation of pesticides takes place. These approaches are as follows:

  I. Phytodegradation of pesticides
 II. Fungal degradation of pesticides
III. Bacterial degradation of pesticides

Mycodegradation of Pesticides

Phytoremediation is associated with the application of green plants, associated microorganisms, agronomic techniques, and soil amendments for the removal of environmental pollutants. It is a suitable, easy handling, low-cost, eco-friendly method used for bioremediation of polluted soil and sub-soil water with the help of plants. Plants have a special ability (*i.e.* the selective and unique uptake capabilities of root systems, translocation, bioaccumulation and degradation/storage of contaminants) to degrade the organic pollutants, through which they can accumulate and metabolise the pollutants and this process is called the phytodegradation. Phytoremediation method is basically triggered by three systems *i.e.* biological, solar-driven, pump -and-treat systems [41]. The partial recovery of pesticides from contaminated water and soil with the help of buffer zones of nontarget plants and vegetation filter strips [42] offers very valuable and cost-effective technology as well as it can reduce the herbicide runoff also. The potential of detoxification of higher plants, which is also called the green livers of earth, can provide the base for the development of treatment technologies for pollutant removal from soil and water [43, 44].

There are two methods of phytoremediation: (I) direct remediation of the environment *via* root uptake, detoxification by phytotransformation and storage of nontoxic elements in plant tissues, (II) indirect remediation *i.e.* by the release of enzymes or exudates which can enhance degradation [45, 46]. To make the remediation process more effective, the plants need to be grown in the presence of target elements. Additionally, the plant must have pollutant resistant capacity, pollutant removal, or transformation capability as well.

The vegetative filter strips have been utilising for the treatment of agricultural runoff water since very long time [47]. For example, the use of hybrid-poplar buffer strips showed the effective removal efficiency for atrazine from runoff water or agricultural percolation [48]. The poplar (*Populus deltoides x nigra)*, transforms the herbicides into polar ammine and dealkylates the herbicides as well [49, 50].

Grasses and semiaquatic plants can also be very helpful in pesticide remediation. These plants decrease the transport of pollutants and increase the deposition of pollutants. In this way, they provide a long time to the plants for uptake, transformation and metabolism of pollutants and hence, prevent them from entering to the water system. The grass common cattails (*Typha latifolia*) have been utilised for the removal of simazine from contaminated water [45]. Rhizosphere soil of several plants is also utilised for pesticide treatment due to the presence of bacteria in the rhizosphere [51].

Basically, two types of bacteria (*i.e.* endophytic and rhizospheric bacteria) involve in the degradation of pesticides and these bacteria are present either inside or outside the plant roots which makes a favourable environment for the enhancement of the pesticide degradation process. Endophytic bacteria are basically non-pathogenic bacteria that occur naturally in the plant tissue. It enhances plant growth, provides food to plant host and helps in the biodegradation of contaminants as well [52, 53]. Whereas, the rhizospheric bacteria provide a unique treatment zone (*i.e.* rhizosphere) with great potential for pesticide remediation [54]. High biodegradation in the rhizosphere is possibly due to a large microbial population, improved oxygen, soil moisture, nutrient conditions and/or co-metabolism as well. The roots of plant adsorb pesticides onto the surface because when the roots become dead, it produces organic matter into the soil which increases the pesticide adsorption over the soil surface where microbial degradation of pesticides take place [55]. However, these days, DNA recombinant technology is utilised in place of plant-associated bacteria to develop transgenic plant which releases bacterial enzyme and hence offers improved catabolic activity for toxic elements and tolerance level in plants [56]. Using this technology in several plants like *Lactuca sativa, Amaranthus caudate, Phaseolus vulgaris and Nasturtium officinale* can give very fruitful results in detoxification and degradation of contaminants (like dimethoate and malathion) during cultivation periods [57].

Mycodegradation of Pesticides

Bioremediation of pesticides using the fungi is known as mycoremedeation. Fungi are omnipresent and utilised as a potential tool for the biodegradation of organic contaminants. *Phanerochaete chrysosporium* also called the white rot fungi are the most widely used fungi that release a very complex and extracellular enzyme that has been utilised for the long range of organic compound treatment. The enzyme released by the white rot fungi is a peroxidase enzyme *i.e.* manganese peroxidase (MnP), lignin peroxidase (LiP), laccase, *etc.* which catalyses the degradation of various complex organic compounds. These enzymes oxidize the organic pollutants into carbon dioxide in the absence of nitrogen. There are several species of white rot like *Hypholoma fasciculare, Stereum hirsutum and Coriolus versicolor* which have shown significant degradation potential against various organic compounds such as metalaxyl, atrazine, terbuthylazine and diuron. Additionally, the white rot fungi have also shown the potential degradation capacity for anthracene, lindane, chlordane, pyrene, di- and tribenzoic acids, DDT (Dichlorodiphenyltrichloroethane), and several polychlorinated biphenyls (PCBs). Though, the rate of mycodegradation of several organic contaminants is slower in soil. However, increasing the amount of fungus has shown the enhanced degradation of several organic pollutants. There are some fungal isolates *i.e. Polyporus* sp., *Trametes* sp., *Nigroporus* sp. and some unidentified species like U11 and F33 which were also utilised for the biodegradation purpose of different organic compounds. The fungal isolates also have different enzymes; dehydrogenase, lignin peroxidase, laccase, peroxidase, esterase, hydrolase, manganese peroxidase, *etc.* and able to degrade several pesticides including DDT, endosulfan, atrazine, methamidophos, chlorpyrifos, 2, 8-DCDD (2,8- dichlorodibenzo-p-dioxin), dieldrin, lindane, methyl parathion, cypermethrin, heptachlor, *etc* [58]. Though the rate of fungal degradation of pesticides depends on soil pH, oxygen level, nutrient availability, temperature, moisture content, *etc.* during the degradation process, different pesticides possess different processes such as oxidation, demethylation, hydroxylation, dioxygenation, dechlorination, esterification, dehydrochlorination, *etc.*

Bacterial Degradation of Pesticides

The complete bio-degradation of the pesticide includes the oxidation of the parent compound for the production of oxygen and carbon dioxide. This approach produces carbon and energy for the growth and reproduction of microbes. A particular enzyme released by a decaying cell or extracellular enzyme catalyses each degradation step. If an effective enzyme is not present, the pesticide degradation by either an internal or external enzyme will cease in either stage. One common cause for any pesticide to persist is the absence of an appropriate enzyme. If a suitable microorganism is missing from soil or if biodegradable microbial populations are decreased by pesticide toxicity, a particular microorganism may be introduced or added into the soil to allow the current community to increase their activity [59]. Bacterial degradation and detoxification of various toxic components present in the different environmental mediums are the effective ways to treat the polluted areas. Isolating indigenous bacteria which can metabolize pesticides is an effective way to *in situ* detoxify them [60]. Microbial degradation is based on a wide range of environmental factors, not just on decaying enzymes. Temperature, water content, nutrients, pH and pesticide or soil metabolite concentrations in soil can also be used to restrict pesticide degrading microorganisms that require further analysis with respect to the total microbial populations and their biochemistry [59]. The microorganisms easily degrade certain pesticides; some have proved to be recalcitrant [61, 62]. Many plasmids have been recorded for the pesticides degrading genes in soil-bacteria. Such genes that encode degradable enzymes have been well established and these plasmids are classified as catabolic plasmids, which have the potential to degrade other compounds. In the species of *Flavobacterium*, *Pseudomonas*, *Actinobacteria*, *Moraxella Klebsiella*, *Arthrobacter* and *Alcaligenes* several catabolic plasmids are identified [63]. A number of bacterial groups including genera of *Alcaligene*, *Flavobacterium*, *Pseudomonas* and *Rhodococcus*, metabolize pesticides. *Actinomycetes* have a significant biotransformation and biodegradation ability for pesticides. The gram-positive group has found that diminish pesticides with wide-range chemical structures, including striazines, organochlorines, carbamates, triazinones, organophosphonates, organophosphates, sulfonylureas and acetanilides. There are small numbers of xenobiotic pesticides which may be metabolised by single isolates, but bacterial association is also required to degrade completely. Throughout this group of bacteria, pesticide co-metabolism is sometimes observed. Similar to the degradation of pesticide by Gram-negative bacteria, there are fewer details about molecular pathways of actinomycetes in the biotransformation of pesticides. A lack of suitable molecular genetic tools has seriously impeded progress in this area. The methomyl inhibits the acetyl cholinesterase enzyme which hydrolyses the neurotransmitter acetyl choline and is a common class of oxide carbamates used in insect and nematode pest control.

S-methyl N- (methylcarbamoyloxy) thioacetimidate is the IUPAC name of methomyl. Methomyl has been listed as a highly toxic and dangerous pesticide by the European Chemical Classification (ECC), Environment Protection Agency (EPA) and the World Health Organization (WHO) [64], and as the soil absorption affinity for this pollutant is relatively small, it can easily contribute to soil and water pollution. Microorganisms played a crucial role in their environmental degradation. Many bacteria capable of degrading carbamate pesticides were isolated worldwide from the soil. Stenotrophomonas maltophilia has been reported to degrade various xenobiotic compounds [65] and purify high molecular weight polycyclic aromatic hydrocarbons (PAHs). Hence, it has shown great bioremediation potential as well. Individually, S. maltophilia M1 strain has shown great degradation capability towards the methomyl pesticide. Additionally, this strain is having two plasmids (PMa and PMb), between them, PMb is probably utilised for the breakdown of the gene which carries methomyl pesticide. Moreover, the plasmid can be transferred to the other strains so that its degrading capability for methomyl could be increased. These two plasmids were used for transforming DH5α strain of *Escherichia coli*. Only PMb plasmid have been transported to that strain which was permitted to nourish on methomyl whereas no gene responsible for the degradation of the PMa plasmid had been identified. Transformed strains (*E. coli* DH5α strain M2) were capable of hydrolysing methomyl (100 ppm) in media of M9 and continued to operate following repeated subculture. Moreover, in contrast to M1 strain, the performance of transformed strain M2 was less. The result may be the reason that strain M2 has been able to degrade methomyl after processing with PMb plasmid that converts M1's degrading gene(s), but the degradation rate was less than M1, which could be because of the shorter replication rate for the plasmid transmitting to the host DH5α strain M2 than the original host [60].

Many biochemical methods have shown the ability of some of the soil bacterium to use aldicarb, a Carbamate insecticide that was known to cause adverse health effects by inhibition of the activity of acetyl cholinesterase in neuromuscular junction [66] by hydrolysis and bacterial enzyme-esterase degradation. The most degrading result in aldicarbs was observed in S. maltophilia [67]. Esterase and amidase may lead to aldicarb enzymatic hydrolysis, but the enzyme associated was observed to be esterase when it degraded by S. maltophilia through the determination of cytoplasmic enzyme activity of the S. maltophil.

Mechanisms Involved in Bioremediation

Though the diversity and complexity in pollutants are high; the microbial diversity is always higher than the pollutants with significant degradation capability [68, 69]. The pollutant biodegradation process in the environment is the result of metabolic cooperation which involves the transfer of products as well as the substrates within a well-coordinated microbial community. These microorganisms interact with contaminants *via* different means such as chemical and physical both ways which lead to the complete or partial degradation of pollutants. Actinomycetes, fungi and bacteria are the main degraders among the microbial communities [70]. Generally, fungi biotransform the pollutants into less toxic components, which are further biotransformed or completely degraded by the bacteria into the environment [71].

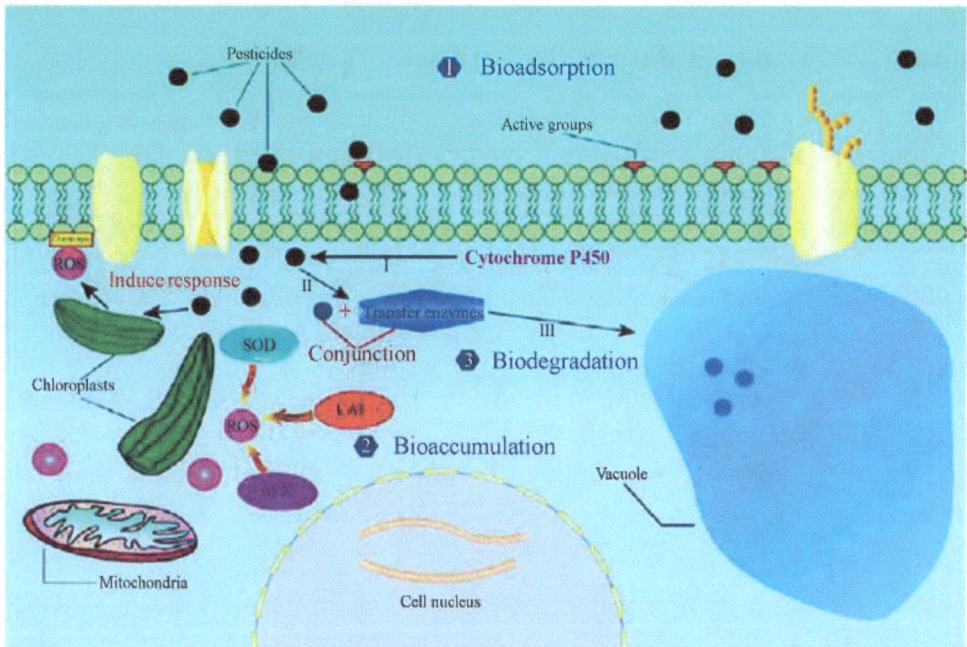

Fig. (3). Mechanism of biodegradation of pesticides [78].

Bacteria and fungi are extracellular enzyme-producing microorganisms that have shown a very significant role biodegradation of diverse pollutants and enzyme metabolism affects their degradation as well. Among all, esterase, transferase and cytochrome P_{450} are the main enzymes involved in the bioremediation process [72]. Actually, the pesticide degradation is a multi-step process that includes the following steps: (i) the activation of pesticides *via* hydroxylation, oxidation and reduction reactions to get the less toxic element; (ii) transfer of enzyme, present in cytosol, on the activated (having functional groups) pesticides; (iii) transportation of these pesticides into vacuoles with the help of transporters [73 - 75]. However, the mechanism of biodegradation (Fig. **3**) of different mostly depends on the nature of the mechanism of microbes and pesticides. In short, enzymes play a critical role in the process of bioremediation and bioaccumulation of contaminants [76].

The application of enzymes in pesticide degradation offers unique and innovative treatment technology which is more effective and significant than the chemical methods. The mechanism involved in pesticide degradation offers, basically, two modes of action *i.e.* either pesticides are activated by enzymes or they act according to the particular enzymes. They can be utilised in target organisms as well as in a wider environment also for bioremediation purpose [77].

Genetic Modification in Bioremediation Tools

The gene editing technology offers critical applications for microorganisms in numerous fields such as agriculture, medical food and feed, *etc* [78, 79]. It has also shown very fruitful application in bioremediation processes like conversion of more toxic components into less toxic ones, elimination of xenobiotics and complete degradation of pesticides. This technology is used to add, remove or alter the genetic material at specific locations in the genome. The manipulation of DNA in a genome is done by using the engineered nuclease enzyme called a molecular scissor. These molecular scissors have shown a broad range of application *i.e.* in animals, plants and microorganisms, *etc* [80]. The editing process includes the development of a guide sequence that can attach to the target sequence of the gene of interest along with the enzyme (which cuts the genetic sequence at particular site) followed by the deletion or insertion of the required target.

However, clustered regularly interspaced short palindromic repeats (CRISPR-Cas), zinc-finger nucleases (ZFNs) and transcription activator-like effector nucleases (TALENs) are the three main gene editing tools that are most widely used in the gene editing process [81, 82]. Basically, these tools collectively create a double stranded break (DSB) in the target gene sequence, repaired *via* homology–directed repair (HRD) and error-prone non- homologous end joining (NHEJ) pathway. The TALENs and ZFNs use artificial restriction enzymes, which cut the specific target DNA sequence by TAL effector DNA binding domain and Zinc finger DNA binding domain, respectively. The main aim of these tools is to develop the best microbe with more complex genes and maximum quality. But there are several limitations of using these two gene editing tools such as off target and lethal mutations, risk of environmental release (accidentally or intentionally) of genetically modified organism, *etc* [83].

However, the CRISPR-Cas is simple, cost effective and easy to use in comparison to the other gene editing tools like TALENs and ZFNs. In this tool, the CRISPR locus is an assembly of spacer sequences and alternating repeats which offers the chronological history of the plasmids and viruses invaded in a given bacterial strain. Basically, three steps are involved in CRISPR-based immunity to protect the bacteria from invading viruses and these steps are adaptation, expression, and interference. The first step is the adaptation step in which the foreign DNA fragments are introduced into the CRISPR locus. In the second step *i.e.* expression, the transcription process of the CRISPR locus offers an RNA template that copies the complementary protospacer sequences present in the invading DNA. Finally, the invading DNA is cleaved and inactivated by the Cas effector protein in the interference step. The protospacer adjacent motif (PAM) present in the target DNA helps the immune system identify and distinguish the invader DNA from genomic DNA introduced in the CRISPR array [81]. The CRISPR-Cas technology has shown several advantages such as it helps to evaluate the gene interaction and the relationship between genetics and phenotype [84]. This genetic tool also suffers from certain limitations such as genomic disintegration, off target mutation and hindrance in applicability [85]. However, unlike the TALENs and ZFNs, CRISPR-Cas offers more specific target binding.

CONCLUSION

Heavy utilisation of pesticides and mismanagement in pesticide handling lead to environmental and biological toxicity. Therefore, the remediation of contaminated soil, surface water and different contaminated sites are important where bioremediation plays a very crucial role in order to provide a clean and sustainable environment. Bioremediation offers the best pesticide treatment technique than the other chemical methods, *etc.* due to its cost effectiveness and eco-friendly nature. Bioremediation of pesticides can be done either by *ex-situ* or by *in-situ* methods. Anaerobic and aerobic process of bioremediation is also discussed. Additionally, the bioremediation process involves the degradation of pesticides in which three main biocomponents are used such as fungi, bacteria and higher plants. Remediation of pesticides using the higher plants, fungi and bacteria are called phytodegradation, myco-degradation and bacterial degradation, respectively. Additionally, the mechanism involved in the bioremediation process as well as the application of genetic engineering is also discussed, in this review.

CONSENT FOR PUBLICATION

Not applicable.

CONFLICT OF INTEREST

The authors declare no conflict of interest, financial or otherwise.

ACKNOWLEDGEMENTS

The authors are thankful to the Director NPL for his encouragement. One of the authors Mr. Praveen Kumar Yadav is thankful to AcSIR and UGC also for providing fellowship to carry out his Ph.D. work.

REFERENCES

[1]   Liu YH, Chung YC, Xiong Y. Purification and characterization of a dimethoate-degrading enzyme of *Aspergillus niger* ZHY256, isolated from sewage. Appl Environ Microbiol 2001; 67(8): 3746-9.
        [http://dx.doi.org/10.1128/AEM.67.8.3746-3749.2001] [PMID: 11472959]

[2]   Damalas CA. Understanding benefits and risks of pesticide use. Sci Res Essays 2009; 4(10): 945-9.

[3]   Gałuszka A, Migaszewski ZM, Manecki P. Pesticide burial grounds in Poland: A review. Environ Int 2011; 37(7): 1265-72.
        [http://dx.doi.org/10.1016/j.envint.2011.04.009] [PMID: 21531026]

[4]   Barra R, Colombo JC, Eguren G, Gamboa N, Jardim WF, Mendoza G. Persistent organic pollutants (POPs) in eastern and western South American countries. In Reviews of environmental contamination and toxicology . New York, NY: Springer 2006; pp. 1-33.
        [http://dx.doi.org/10.1007/0-387-30638-2_1]

[5]   Phillips TM, Lee H, Trevors JT, Seech AG. Full-scale *in situ* bioremediation of hexachlorocyclohexane-contaminated soil. *Journal of Chemical Technology & Biotechnology: International Research in Process*. Environmental & Clean Technology 2006; 81(3): 289-98.

[6]   Concha-Graña E, Turnes-Carou MI, Muniategui-Lorenzo S, López-Mahía P, Prada-Rodríguez D, Fernández-Fernández E. Evaluation of HCH isomers and metabolites in soils, leachates, river water and sediments of a highly contaminated area. Chemosphere 2006; 64(4): 588-95.
        [http://dx.doi.org/10.1016/j.chemosphere.2005.11.011] [PMID: 16403559]

[7]   Van Liere H, Staps S, Pijls C, Zwiep G, Lassche R, Langenhoff A. Full scale case: successful in situ bioremediation of a HCH contaminated industrial site in central Europe (The Netherlands). In Forum book . 2003; pp. 128-32.

[8]   Singh KP, Malik A, Sinha S. Persistent organochlorine pesticide residues in soil and surface water of northern Indo-Gangetic alluvial plains. Environ Monit Assess 2007; 125(1-3): 147-55.
        [http://dx.doi.org/10.1007/s10661-006-9247-0] [PMID: 16957856]

[9]   Zhu Y, Liu H, Xi Z, Cheng H, Xu X. Organochlorine pesticides (DDTs and HCHs) in soils from the outskirts of Beijing, China. Chemosphere 2005; 60(6): 770-8.
        [http://dx.doi.org/10.1016/j.chemosphere.2005.04.018] [PMID: 15972227]

[10]   Chaile AP, Romero N, Amoroso MJ, Hidalgo MDV, Apella MC. Organochlorine pesticides in Sali River. *Tucumán-Argentina.* Rev Boliv Ecol 1999; 6: 203-9.

[11]   Fuentes MS, Benimeli CS, Cuozzo SA, Amoroso MJ. Isolation of pesticide-degrading actinomycetes from a contaminated site: Bacterial growth, removal and dechlorination of organochlorine pesticides. Int Biodeterior Biodegradation 2010; 64(6): 434-41.
        [http://dx.doi.org/10.1016/j.ibiod.2010.05.001]

[12]   Kumari B, Madan VK, Kathpal TS. Status of insecticide contamination of soil and water in Haryana, India. Environ Monit Assess 2007; 136(1-3): 239-44.
        [http://dx.doi.org/10.1007/s10661-007-9679-1] [PMID: 17406996]

[13] Lammel G, Ghim YS, Grados A, Gao H, Hühnerfuss H, Lohmann R. Levels of persistent organic pollutants in air in China and over the Yellow Sea. Atmos Environ 2007; 41(3): 452-64.
[http://dx.doi.org/10.1016/j.atmosenv.2006.08.045]

[14] Bajpai A, Shukla P, Dixit BS, Banerji R. Concentrations of organochlorine insecticides in edible oils from different regions of india. Chemosphere 2007; 67(7): 1403-7.
[http://dx.doi.org/10.1016/j.chemosphere.2006.10.026] [PMID: 17140628]

[15] Malik A, Singh KP, Ojha P. Residues of organochlorine pesticides in fish from the Gomti river, India. Bull Environ Contam Toxicol 2007; 78(5): 335-40.
[http://dx.doi.org/10.1007/s00128-007-9188-5] [PMID: 17618385]

[16] Zhao G, Xu Y, Li W, Han G, Ling B. PCBs and OCPs in human milk and selected foods from Luqiao and Pingqiao in Zhejiang, China. Sci Total Environ 2007; 378(3): 281-92.
[http://dx.doi.org/10.1016/j.scitotenv.2007.03.008] [PMID: 17408724]

[17] Ridolfi AS, Álvarez GB, Girault MER. Organochlorinated contaminants in general population of Argentina and other Latin American Countries. In Bioremediation in Latin America. Cham: Springer 2014; pp. 17-40.
[http://dx.doi.org/10.1007/978-3-319-05738-5_2]

[18] Amakiri MA. Microbial degradation of soil applied herbicides. Nig J Microl 1982; 2: 17-21.

[19] Schwitzguébel JP, Meyer J, Kidd P. Pesticides removal using plants: phytodegradation versus phytostimulation. In Phytoremediation rhizoremediation. Dordrecht: Springer 2006; pp. 179-98.
[http://dx.doi.org/10.1007/978-1-4020-4999-4_13]

[20] Norris A. Past Alaska Master Gardener Manual. cited 2012 February, 10th 2011.

[21] Paul D, Pandey G, Pandey J, Jain RK. Accessing microbial diversity for bioremediation and environmental restoration. Trends Biotechnol 2005; 23(3): 135-42.
[http://dx.doi.org/10.1016/j.tibtech.2005.01.001] [PMID: 15734556]

[22] Kidd P, Barceló J, Bernal MP, *et al.* Trace element behaviour at the root–soil interface: Implications in phytoremediation. Environ Exp Bot 2009; 67(1): 243-59.
[http://dx.doi.org/10.1016/j.envexpbot.2009.06.013]

[23] İdiz N, Karakus A, Dalgıç M. The forensic deaths caused by pesticide poisoning between the years 2006 and 2009 in Izmir, Turkey. J Forensic Sci 2012; 57(4): 1014-6.
[http://dx.doi.org/10.1111/j.1556-4029.2012.02085.x] [PMID: 22372492]

[24] Zhang W, Jiang F, Ou J. Global pesticide consumption and pollution: with China as a focus. Proceedings of the International Academy of Ecology and Environmental Sciences 2011; 1(2): 125.

[25] Vaccari DA, Strom PF, Alleman JE. Environmental biology for engineers and scientists. New York: Wiley-Interscience 2006; Vol. 7: p. 242.

[26] Ortiz-Hernández ML. Biodegradación de plaguicidas organofosforados por nuevas bacterias aisladas del suelo. Universidad Autónoma del Estado de Morelos, 2002. (Doctoral dissertation, Dissertation).

[27] Abhilash PC, Singh N. Pesticide use and application: An Indian scenario. J Hazard Mater 2009; 165(1-3): 1-12.
[http://dx.doi.org/10.1016/j.jhazmat.2008.10.061] [PMID: 19081675]

[28] Besse-Hoggan P, Alekseeva T, Sancelme M, Delort AM, Forano C. Atrazine biodegradation modulated by clays and clay/humic acid complexes. Environ Pollut 2009; 157(10): 2837-44.
[http://dx.doi.org/10.1016/j.envpol.2009.04.005] [PMID: 19419808]

[29] Singh BK, Walker A. Microbial degradation of organophosphorus compounds. FEMS Microbiol Rev 2006; 30(3): 428-71.
[http://dx.doi.org/10.1111/j.1574-6976.2006.00018.x] [PMID: 16594965]

[30] Worrall F, Fernandez-Perez M, Johnson AC, Flores-Cesperedes F, Gonzalez-Pradas E. Limitations on

the role of incorporated organic matter in reducing pesticide leaching. J Contam Hydrol 2001; 49(3-4): 241-62.
[http://dx.doi.org/10.1016/S0169-7722(00)00197-2] [PMID: 11411399]

[31]   Ortiz-Hernández ML, Sánchez-Salinas E, Dantán-González E, Castrejón-Godínez ML. Pesticide biodegradation: mechanisms, genetics and strategies to enhance the process. Biodegradation-life of Science. 2013; pp. 251-87.

[32]   Dasgupta S, Meisner C, Wheeler D. Stockpiles of obsolete pesticides and cleanup priorities: A methodology and application for Tunisia. The World Bank 2009.
[http://dx.doi.org/10.1596/1813-9450-4893]

[33]   Ferrusquía-García CJ. Evaluación de la degradación de metil paratión en solución usando fotocatálisis heterogénea. Revista Latinoamericana de Recursos Naturales 2008; 4(2): 285-90.

[34]   Vidali M. Bioremediation. An overview. Pure Appl Chem 2001; 73(7): 1163-72.
[http://dx.doi.org/10.1351/pac200173071163]

[35]   Singleton I. Microbial metabolism of xenobiotics: fundamental and applied research. *Journal of Chemical Technology & Biotechnology: International Research in Process.* Environmental AND Clean Technology 1994; 59(1): 9-23.

[36]   Blackburn J, Hafker WR. The impact of biochemistry, bioavailability and bioactivity on the selection of bioremediation techniques. Trends Biotechnol 1993; 11(8): 328-33.
[http://dx.doi.org/10.1016/0167-7799(93)90155-3] [PMID: 7764179]

[37]   M D, A S, N S, A J. Biotechnology and bioremediation: successes and limitations. Appl Microbiol Biotechnol 2002; 59(2-3): 143-52.
[http://dx.doi.org/10.1007/s00253-002-1024-6] [PMID: 12111139]

[38]   Alcalde M, Ferrer M, Plou FJ, Ballesteros A. Environmental biocatalysis: from remediation with enzymes to novel green processes. Trends Biotechnol 2006; 24(6): 281-7.
[http://dx.doi.org/10.1016/j.tibtech.2006.04.002] [PMID: 16647150]

[39]   Fathepure BZ, Youngers GA, Richter DL, Downs CE. *In situ* bioremediation of chlorinated hydrocarbons under field aerobic-anaerobic environments (No. CONF-950483-). Columbus, OH (United States).: Battelle Press 1995.

[40]   Varshney K. Bioremediation of Pesticide of waste at contaminated sites. J Emerg Technol Innov Res 2019; 6(5): 128-34.

[41]   Cobbett CS. Phytochelatins and their roles in heavy metal detoxification. Plant Physiol 2000; 123(3): 825-32.
[http://dx.doi.org/10.1104/pp.123.3.825] [PMID: 10889232]

[42]   Beitz H, Schmidt H, Herzel F. Occurrence, toxicological and ecotoxicological significance of pesticides in groundwater and surface water. Pesticides in ground and surface water. Berlin, Heidelberg: Springer 1994; pp. 1-56.
[http://dx.doi.org/10.1007/978-3-642-79104-8_1]

[43]   Hall JC, Wickenden JS, Yau KY. Biochemical conjugation of pesticides in plants and microorganisms: an overview of similarities and divergences. In:J. C. Hall, R. E. Hoagland, and R. M. Zablotowicz (Eds.) Pesticide Biotransformation in Plants and Microorganisms: Similarities and Divergences. Washington, DC: American Chemical Society 2001; pp. 89–118.

[44]   Nzengung VA, McCutcheon SC. Occurrence, toxicological and ecotoxicological significance of pesticides in groundwater and surface water. Phytoremediation: Transformation and control of contaminants. 2003; pp. 863-85.

[45]   Wilson PC, Whitwell T, Klaine SJ. Metalaxyl and simazine toxicity to and uptake by Typha latifolia. Arch Environ Contam Toxicol 2000; 39(3): 282-8.
[http://dx.doi.org/10.1007/s002440010106] [PMID: 10948277]

[46]   Marcacci S, Schwitzguébel JP. Using plant phylogeny to predict detoxification of triazine herbicides. In Phytoremediation. Humana Press 2007; pp. 233-49.
[http://dx.doi.org/10.1007/978-1-59745-098-0_19]

[47]   Krutz LJ, Senseman SA, Zablotowicz RM, Matocha MA. Reducing herbicide runoff from agricultural fields with vegetative filter strips: a review. Weed Sci 2005; 53(3): 353-67.
[http://dx.doi.org/10.1614/WS-03-079R2]

[48]   Nair DR, Burken JG, Licht LA, Schnoor JL. Mineralization and uptake of triazine pesticide in soil-plant systems. J Environ Eng 1993; 119(5): 842-54.
[http://dx.doi.org/10.1061/(ASCE)0733-9372(1993)119:5(842)]

[49]   Burken JG, Schnoor JL. Phytoremediation: plant uptake of atrazine and role of root exudates. J Environ Eng 1996; 122(11): 958-63.
[http://dx.doi.org/10.1061/(ASCE)0733-9372(1996)122:11(958)]

[50]   Burken JG, Schnoor JL. Uptake and metabolism of atrazine by poplar trees. Environ Sci Technol 1997; 31(5): 1399-406.
[http://dx.doi.org/10.1021/es960629v]

[51]   Stone JK, Bacon CW, White JF. An overview of endophytic microbes: endophytism defined. Microbial endophytes 2000; (5): 29-30.

[52]   Sessitsch A, Reiter B, Pfeifer U, Wilhelm E. Cultivation-independent population analysis of bacterial endophytes in three potato varieties based on eubacterial and Actinomycetes-specific PCR of 16S rRNA genes. FEMS Microbiol Ecol 2002; 39(1): 23-32.
[http://dx.doi.org/10.1111/j.1574-6941.2002.tb00903.x] [PMID: 19709181]

[53]   Davis LC, Erickson LE, Narayanan M, Zhang Q. Modeling and design of phytoremediation. Phytoremediation: Transformation and control of contaminants. 2003; pp. 663-94.

[54]   Karthikeyan R, Davis LC, Erickson LE, *et al.* Potential for plant-based remediation of pesticide-contaminated soil and water using nontarget plants such as trees, shrubs, and grasses. Crit Rev Plant Sci 2004; 23(1): 91-101.
[http://dx.doi.org/10.1080/07352680490273518]

[55]   Kawahigashi H. Transgenic plants for phytoremediation of herbicides. Curr Opin Biotechnol 2009; 20(2): 225-30.
[http://dx.doi.org/10.1016/j.copbio.2009.01.010] [PMID: 19269160]

[56]   Al-Qurainy F, Abdel-Megeed A. Phytoremediation and detoxification of two organophosphorous pesticides residues in Riyadh area. World Appl Sci J 2009; 6(7): 987-98.

[57]   Alexander M. Biodegradation and bioremediation. Gulf Professional Publishing. Academic Press 1999: p. 453.

[58]   Siripong P, Oraphin B, Sanro T, Duanporn P. Screening of fungi from natural sources in Thailand for degradation of polychlorinated hydrocarbons. Am-Eurasian J Agric Environ Sci 2009; 5(4): 466-72.

[59]   Singh DK. Biodegradation and bioremediation of pesticide in soil: concept, method and recent developments. Indian J Microbiol 2008; 48(1): 35-40.
[http://dx.doi.org/10.1007/s12088-008-0004-7] [PMID: 23100698]

[60]   Mohamed MS. Degradation of methomyl by the novel bacterial strain Stenotrophomonas maltophilia M1. Electron J Biotechnol 2009; 12(4): 6-7.
[http://dx.doi.org/10.2225/vol12-issue4-fulltext-11]

[61]   Richins RD, Kaneva I, Mulchandani A, Chen W. Biodegradation of organophosphorus pesticides by surface-expressed organophosphorus hydrolase. Nat Biotechnol 1997; 15(10): 984-7.
[http://dx.doi.org/10.1038/nbt1097-984] [PMID: 9335050]

[62]   Mulchandani A, Kaneva I, Chen W. Detoxification of organophosphate nerve agents by immobilized *Escherichia coli* with surface-expressed organophosphorus hydrolase. Biotechnol Bioeng

1999; 63(2): 216-23.
[http://dx.doi.org/10.1002/(SICI)1097-0290(19990420)63:2<216::AID-BIT10>3.0.CO;2-0] [PMID: 10099598]

[63]   Sayler GS, Hooper SW, Layton AC, King JMH. Catabolic plasmids of environmental and ecological significance. Microb Ecol 1990; 19(1): 1-20.
[http://dx.doi.org/10.1007/BF02015050] [PMID: 24196251]

[64]   Nawaz K, Hussain K, Choudary N, *et al.* Eco-friendly role of biodegradation against agricultural pesticides hazards. Afr J Microbiol Res 2011; 5(3): 177-83.

[65]   Lee EY, Jun YS, Cho KS, Ryu HW. Degradation characteristics of toluene, benzene, ethylbenzene, and xylene by Stenotrophomonas maltophilia T3-c. J Air Waste Manag Assoc 2002; 52(4): 400-6.
[http://dx.doi.org/10.1080/10473289.2002.10470796] [PMID: 12002185]

[66]   Lifshitz M, Shahak E, Bolotin A, Sofer S. Carbamate poisoning in early childhood and in adults. J Toxicol Clin Toxicol 1997; 35(1): 25-7.
[http://dx.doi.org/10.3109/15563659709001161] [PMID: 9022648]

[67]   Saptanmasi B, Karayilanoglu T, Kenar L, Serdar M, Kose S, Aydin A. Bacterial biodegradation of aldicarb and determination of bacterium which has the most biodegradative effect. Turk J Biochem 2008; 33: 209-14.

[68]   Ramakrishnan B, Megharaj M, Venkateswarlu K, Naidu R, Sethunathan N. The impacts of environmental pollutants on microalgae and cyanobacteria. Crit Rev Environ Sci Technol 2010; 40(8): 699-821.
[http://dx.doi.org/10.1080/10643380802471068]

[69]   Abraham WR, Nogales B, Golyshin PN, Pieper DH, Timmis KN. Polychlorinated biphenyl-degrading microbial communities in soils and sediments. Curr Opin Microbiol 2002; 5(3): 246-53.
[http://dx.doi.org/10.1016/S1369-5274(02)00323-5] [PMID: 12057677]

[70]   Briceño G, Palma G, Durán N. Influence of organic amendment on the biodegradation and movement of pesticides. Crit Rev Environ Sci Technol 2007; 37(3): 233-71.
[http://dx.doi.org/10.1080/10643380600987406]

[71]   Diez MC. Biological aspects involved in the degradation of organic pollutants. J Soil Sci Plant Nutr 2010; 10(3): 244-67.
[http://dx.doi.org/10.4067/S0718-95162010000100004]

[72]   Bass C, Field LM. Gene amplification and insecticide resistance. Pest Manag Sci 2011; 67(8): 886-90.
[http://dx.doi.org/10.1002/ps.2189] [PMID: 21538802]

[73]   Ghasemi Y, Rasoul-Amini S, Fotooh-Abadi E. The biotransformation, biodegradation, and bioremediation of organic compounds by microalgae 1. J Phycol 2011; 47(5): 969-80.
[http://dx.doi.org/10.1111/j.1529-8817.2011.01051.x] [PMID: 27020178]

[74]   Kumar A, Singh JS. Cyanoremediation: a green-clean tool for decontamination of synthetic pesticides from agro-and aquatic ecosystems. In Agro-environmental sustainability . Cham: Springer 2017; pp. 59-83.
[http://dx.doi.org/10.1007/978-3-319-49727-3_4]

[75]   Asad MAU, Lavoie M, Song H, Jin Y, Fu Z, Qian H. Interaction of chiral herbicides with soil microorganisms, algae and vascular plants. Sci Total Environ 2017; 580: 1287-99.
[http://dx.doi.org/10.1016/j.scitotenv.2016.12.092] [PMID: 28003051]

[76]   Nie J, Sun Y, Zhou Y, *et al.* Bioremediation of water containing pesticides by microalgae: Mechanisms, methods, and prospects for future research. Sci Total Environ 2020; 707: 136080.
[http://dx.doi.org/10.1016/j.scitotenv.2019.136080] [PMID: 31869621]

[77]   Scott C, Pandey G, Hartley CJ, *et al.* The enzymatic basis for pesticide bioremediation. Indian J Microbiol 2008; 48(1): 65-79.
[http://dx.doi.org/10.1007/s12088-008-0007-4] [PMID: 23100701]

[78] Yadav R, Kumar V, Baweja M, Shukla P. Gene editing and genetic engineering approaches for advanced probiotics: A review. Crit Rev Food Sci Nutr 2018; 58(10): 1735-46.
[http://dx.doi.org/10.1080/10408398.2016.1274877] [PMID: 28071925]

[79] Basu S, Rabara RC, Negi S, Shukla P. Engineering PGPMOs through gene editing and systems biology: a solution for phytoremediation? Trends Biotechnol 2018; 36(5): 499-510.
[http://dx.doi.org/10.1016/j.tibtech.2018.01.011] [PMID: 29455935]

[80] Bier E, Harrison MM, O'Connor-Giles KM, Wildonger J. Advances in engineering the fly genome with the CRISPR-Cas system. Genetics 2018; 208(1): 1-18.
[http://dx.doi.org/10.1534/genetics.117.1113] [PMID: 29301946]

[81] Waryah CB, Moses C, Arooj M, Blancafort P. Zinc fingers, TALEs, and CRISPR systems: a comparison of tools for epigenome editing. In Epigenome Editing. New York, NY: Humana Press 2018; pp. 19-63.
[http://dx.doi.org/10.1007/978-1-4939-7774-1_2]

[82] Arazoe T, Kondo A, Nishida K. Targeted nucleotide editing technologies for microbial metabolic engineering. Biotechnol J 2018; 13(9): 1700596.
[http://dx.doi.org/10.1002/biot.201700596] [PMID: 29862665]

[83] Canver MC, Joung JK, Pinello L. Impact of genetic variation on CRISPR-Cas targeting. CRISPR J 2018; 1(2): 159-70.
[http://dx.doi.org/10.1089/crispr.2017.0016] [PMID: 31021199]

[84] VanderSluis B, Costanzo M, Billmann M, *et al.* Integrating genetic and protein–protein interaction networks maps a functional wiring diagram of a cell. Curr Opin Microbiol 2018; 45: 170-9.
[http://dx.doi.org/10.1016/j.mib.2018.06.004] [PMID: 30059827]

[85] Sun J, Wang Q, Jiang Y, *et al.* Genome editing and transcriptional repression in *Pseudomonas putida* KT2440 *via* the type II CRISPR system. Microb Cell Fact 2018; 17(1): 41.
[http://dx.doi.org/10.1186/s12934-018-0887-x] [PMID: 29534717]

CHAPTER 5

Biosurfactants for Biodégradation

Telli Alia[1,*]

[1] *Laboratoire de protection des écosystèmes en zone aride and semi aride, Université de KASDI Merbah, BP 511 la route de Ghardaia, Ouargla 30000, Algérie*

Abstract: The low toxicity, biodegradability, powerful surface activity, and the functionality under extreme conditions (pH, salinity and temperature) make the surfactants produced by micro-organisms (bacteria, fungi, and yeasts) best surface-active molecules that can replace hazardous and non degradable chemical surfactants in different industries and fields. In recent decades, there has been growing interest in the use of biosurfactants for bioremediation of environmental pollution and biodegradation of various categories of hydrophobic pollutants and waste due to their eco-friendly and low-cost properties. This chapter presents the classification, the characteristics, and the potential uses of biosurfactants in the solubilization and enhancing the biodegradation of low solubility compounds.

Keywords: Biosurfactants, Surface-active, Biodegradation, Application, Environmental pollution, Micro-organisms.

INTRODUCTION

Rapid development, industrialization, urbanization, as well as increased demand for energy and various products, cause the depletion of natural resources and environmental pollution. Different kinds of pollutants and contaminants are released into the environment, such as petroleum products, pharmaceutical compounds, pesticides, organic dyes, and heavy metals [1]. These hazardous compounds lead to serious problems and are a major threat to all living organisms and ecosystems [1, 2].

In order to reduce the harmful effect of these contaminants, environmental pollution requires efficient strategies that would lead to the removal of hazardous compounds from contaminated areas. Several physical and chemical techniques were adopted for reaching a suitable solution [3]. However, these conventional

* **Corresponding author Telli Alia:** Laboratoire de protection des écosystèmes en zone aride et semi aride, Université de KASDI Merbah, BP 511 la route de Ghardaia, Ouargla 30000, Algérie;
E-mails: telli.alia@univ-ouargla.dz and alia.telli@gmail.com

Inamuddin (Ed.)

methods have many drawbacks, especially their high cost, ineffectiveness, and being unsafe for the environment by generating toxic by-products [3 - 5].

There are many problems associated with the use of surfactants. These substances are the main components in different products (pharmaceutic, cosmetic, household detergents, food, lubricating agents, *etc.*) and are used by several industries (paper, textiles and fibers, pharmaceutical industry, agriculture, and petroleum industry) [6 - 8]. These tension-active molecules are also used in soil and water remediation techniques [9]. Industries discharge a wide range of chemical surface-active agents to wastewater treatment facilities [10]. Most of them end up dispersed in different environmental compartments such as soil, water, or sediment [7, 11].

The biological approaches of remediation have recently attracted more attention for the removal and detoxification of diverse pollutants and contaminants from soils and water. These techniques consist of the use of living organisms (plants and micro-organisms) or biological compounds (enzymes, polysaccharides, secondary metabolites of plants and micro-organisms) in the elimination and degradation of contaminants [12 - 15]. These techniques are economical, non-invasive and provide an enduring solution [16]. One of the most deeply driven approaches in recent years is the employment of biological synthesis surfactants. These molecules are secondary metabolites fabricated by micro-organisms that interact with an interface and alter the surface properties such as wettability and other properties in order to use the hydrophobic substrates as a source of carbon [17].

In recent years, biosurfactants have paid more attention because of their many advantages in comparison with their chemically synthesized equivalents. They are environmentally friendly, biodegradable, less toxic, active under extreme conditions and have exceptional surface properties [18], which make them good candidates for enhanced oil recovery [19], controlling oil spills [20, 21], biodegradation and detoxification of oil- [22] and metal-contaminated soils [23 - 27], enhanced degradation of pesticides [28] and organic dyes [29, 30]. In addition to their biotechnological applications, biosurfactants also have therapeutic actions. Indeed, they are used as antimicrobial agents against bacteria, fungi and viruses [31, 32]. They have exhibited anti-cancer and immunomodulatory activities and drug delivery [33 - 35].

This chapter discusses the use of biosurfactants as an efficient and safe alternative to remove and enhance the biodegradation of some kinds of pollutants and hazardous wastes.

BIOSURFACTANTS

Definition and Importance

Surfactants are amphipathic molecules that can decrease surface and interfacial tensions by assembling at the interface of immiscible liquids and elevate the solubility, mobility, bioavailability and ulterior biodegradation of lipophilic or insoluble organic molecules [36].

Emulsion is a heterogeneous medium consisting of two immiscible liquid phases, one of which is dispersed in the form of very fine particles in the other (dispersing phase) [37].

Biological surfactants or biosurfactants are tensio-active metabolites manufactured by a broad range of micro-organisms (bacteria, fungi and yeast) [38, 39].

The chemical and bio-surfactants are surface-active molecules possessing hydrophobic (tail group), and hydrophilic (head group) portions which divide themselves in the middle of two immiscible liquids, with the action of decreasing the surface/interfacial tensions causing the solubility of non-polar molecules in polar solvents [40, 41]. These compounds aggregate at higher concentrations into micelles, minimizing the system's free energy [11]. The hydrophilic moiety of biosurfactants can be carbohydrates, amino acids, a protein, a phosphate group or other substances, while the hydrophobic portion is a long carbon saturated, unsaturated or hydroxylated chain fatty acids or fatty alcohols [42, 43].

The chemical surfactants are still preferred and used because of their availability in commercial quantities [40]. These petroleum-derived compounds have generated environmental pollution, putting public health at risk [44]. The biosurfactants are bio-emulsifiers produced from low-cost substrates (e.g. agro-industrial wastes) by diverse micro-organisms that may be considered a safe and green alternative to chemical surfactants due to their less hazardous effect and biodegradability [45].

Surface Activity

Surface tension is a measure of cohesive forces between liquid molecules presented at the surface [46]. Interfacial tension means tension at the interface of two immiscible liquids like oil and water. At the interface of two dissimilar liquids, the forces acting on the molecules of each of these fluids are not the same as within each phase and form interfacial tension [47]. The surface and interfacial tensions play a significant role in characterizing surface-active molecules. The addition of biosurfactants lowers distilled water surface tension from 72 mN/m to less than 30 mN/m [48].

Critical Micelle Concentration (CMC)

The critical micelle concentration is defined as the minimum concentration necessary to initiate micelle formation [24] spontaneously. The CMC is used to measure the biosurfactant efficiency, which means that a low amount of biosurfactant is needed to decrease the surface tension. Consequently, the efficient biosurfactant has a lower CMC. Low CMC characterizes a large number of biosurfactants in comparison with chemical surfactants. The CMC of biosurfactants varied from 0.1 mg/L for glycolipids to 160 mg/L for lipopeptids [48]. The CMC is influenced by pH, salinity and temperature [24].

Hydrophile-lipophile Balance

The hydrophile–lipophile balance (HLB) has been utilized to typify surfactants. This number exhibits relatively the tendency to solubilize in oil or water and consequently form water-in-oil or oil-in-water emulsions [49].

Emulsion Stability

The emulsification potency is evaluated by the capacity of the surface-active agent to produce turbidity by cause of suspended hydrocarbons in a water assay system. The de-emulsification activity is derived by determining the impact of tensio-active molecules on a standard emulsion by utilizing a surfactant [48].

Classification, Properties and Applications of Biosurfactants

Several criteria can be used to classify biosurfactants: molecular weight, ionic charge and type of secretion (intracellular, extracellular or adhered to microbial cells). In this review, the criterion adopted is that of the chemical structure of the surfactant molecules produced by micro-organisms [43, 50]. The main class of biosurfactants, their structures, properties and producing micro-organisms, as well as their applications, are shown in Table **1**.

Table 1. Different classes of biosurfactants and their applications.

Class	Example	Chemical structure	Properties	Producing microorganism	Applications	References
Glycolipids	Rhamnolipids	One or two molecules of rhamnose linked to one or two molecules of β-hydroxy fatty acid	- Amphipathic molecules - Low toxicity - Biodegradability	*Pseudomonas aeroginosa*, *Serratia rubidaea* SNAU02	- Adjuvant in vaccines - Drug delivery - Cosmetic products - Herbicides - Insecticides - Antimicrobial agents - Food industry - Immunomodulation - Biodegradation and bioremediation of various pollutants	[24,35,39,41,42,43,50-57]
	Trehalolipids	Trehalose linked at C-6 and C-6' to mycolic acid		*Rhodococcus erythropolis Arthrobacter sp.*		
	Sophorolipids	Shophorose (dimeric carbohydrate) bonded to a long chain fatty acid		*Candida* species, *Aspergillus flavus*, *Rhizopus oryzae*		
Lipopeptides and lipoprotiens	Surfactin, Fengycin and Iturin	Cyclic heptapeptides interlinked with β-hydroxy fatty acid (surfactin) Cyclic decapeptides linked to β-hydroxy fatty acid (fengycin) Cyclic heptapeptides bonded with β-hydroxy fatty acid (iturin)	Biocompatibility and digestibility - Great structure diversity (2000 described) - Great stability over extreme conditions (pH, salinity and temperature) - Efficiency in comparison with chemical surfactants (surface and interface activity) - Physiological activity	*Bacillus* species, including *Bacillus subtillis*	- Antimicrobial agents - Cosmetic products - Larvicidal agents - Antitumor agents - Thrombolotic activity - Antiviral activity - Anti-adhesive application - Removal of heavy metals from a contaminated soil, sediment and water	[24,39,41-43,50,52,57-65]
	Viscosin	Cyclic oligopeptide linked to β-hydroxy fatty acid		*Pseudomonas fluorescens*		
	Serrawettin	Cyclic pentapeptide linked to β-hydroxy fatty acid		*Serratia marcescens*		
Phospholipids	Phosphatidic acid	Glycerol esterified with two fatty acids at positions C-1 and C-2 and with phosphate in position C-3		Structures commons to many micro-organisms, for example: *Corynebacterium lepus Acinetobacter* sp. *Rhodococcus erythropolis*	- Drug delivery systems - Food industry - Cosmetic industry - Bioleaching of heavy metals	[39-43,50,52,60,66-68]
	Phosphatidyl-ethanolamine	Glycerol esterified with two fatty acids at positions C-1 and C-2 and with a phosphorylethanolamine in position C-3				

(Table 1) cont.....

Fatty acid and neutral lipids	Spiculisporic acid	γ-butenolide derivatives		*Penicillium spiculisporum Aspergillus* sp. HDf2	- Increasing the tolerance of bacteria to heavy metals - Antitumor agents - Antibacterial activity	[24,52,53,58-72]
	Glycolipids	Monocarbohydrate linked to β-hydroxy fatty acid		*Lactobacillus fermentum Lactococcus lactis*		
Polymeric biosurfactants	Emulsan	polyanionic amphipathic heteropolysaccharide. N- acetyl--galactosamine, N-acetylgalactosamine uronic acid and an unidentified N-acetylamino sugar		*Acinetobacter calcoaceticus* RAG-1	- Biodegradation of oil - Bioremediation of soil - Dispersion of limestone in water	[24,41-43,52,73-76]
	Biodispersan	anionic heteropolysaccharide contains four reducing sugars: glucosamine, 6-methylaminohexose, galactosamine uronic acids and an unidentified amino sugar		*Acinetobacter calcoaceticus* RAG-1 *Acinetobacter calcoaceticus* A2		
	Liposan	83% carbohydrate and 17% protein. Glucose, galactose, galactosamine and galacturonic acid are the components of the carbohydrate portion		*Acinetobacter radioresistens* KA-53		
Particulate biosurfactants	Vesicles	The vesicles are composed of protein, phospholipid and lipopolysaccharide		*Acinetobacter* sp.	- Bioremediation processes - Drug delivery systems	[24,36,41-43,52,77]
	Whole-cell	Various micro-organisms have been dispersed in microemulsions		Microalgae, cyanobacteria and *Candida* species		

APPLICATION OF BIOSUFACTANT IN BIODEGRADATION

Biodegradation is one of the best solutions to reduce the negative impacts of contaminants and wastes on the environment, but the major limit and challenge of this technique is the insolubility or low solubility of the majority of these contaminants in water. To overcome this problem, biosurfactants are a powerful tool for improving the solubilization, the bioavailability of these pollutants and the enhancement of their biodegradation [78].

Biodegradation of Crude Oil and Petroleum Wastes

Crude oil is a complex hydrocarbon formed mostly of alkanes (saturated hydrocarbons), alkenes, alkynes (unsaturated hydrocarbons) and aromatic hydrocarbons [79]. Pollution of soil and water caused by petroleum hydrocarbon causes significant ecological and social problems. The contamination sources are diverse: accidents during fuel transportation; leakage; oil extraction; and inadequate release of waste generated by industries that use oil byproducts in plastics, solvents, pharmaceuticals and cosmetics production [80]. Physical and chemical cleaning processes to decontaminate the oil-polluted areas have been limited in their applications [2, 81]. Diverse species of micro-organisms are used in the techniques of bioremediation and removing petroleum hydrocarbon from contaminated soil and water. The end product of bioremediation is CO_2, H_2O and dead biomass, which is environment friendly [82]. However, the hydrocarbon pollutants' hydrophobicity may limit their availability to microbes. Several works proved the improvement of hazardous oily wastes biodegradation and management by using biosurfactants [83 - 85].

The mechanisms of hydrocarbons biodegradation by micro-organisms consist of increasing substrate bioavailability for micro-organisms or changes in surface hydrophobicity and membrane permeability of micro-organisms' cells [86]. However, Franetti *et al.* [87] determined four phenomena that were accompanying the enhancement of biodegradation by microbial surfactants in little diverse ways: (a) emulsification, (b) micellization, (c) adhesion–deadhesion of microbes to and from hydrocarbons, and (d) desorption of pollutants.

Numerous studies demonstrate the microbial surfactants' potential applications in environmental decontamination. Kang *et al.* [84] evaluated the sophorolipid ability in scrubbing and biodegradation of model hydrocarbon and crude oil in soil. They found that the addition of sopholipid in soil increased the biodegradation of model hydrocarbon (between 85% and 97% in 6 days) and crude oil (between 72% and 80% in 8 weeks) [84]. The addition of mannosylerythritol lipid produced by *Pseudozyma* sp. NII 08165 to *Pseudozyma putida* culture medium containing crude oil improved the degradation of crude oil by *Pseudozyma putida* [88]. Bezza *et al.* [86] investigated the application of a biosurfactant produced by *Bacillus subtilis* CN2 in the degradation of used motor oil polycyclic aromatic hydrocarbon and found a removal rate of 82% in 18 days of incubation. Using surfactin and rhamnolipids could enhance the petroleum hydrocarbon removal from oil contaminated soil with efficiency over 86% [89]. Besides biosurfactants, Fanaei *et al.* [89] also noted that the use of H_2O_2 stimulates and accelerates the biodegradation of petroleum hydrocarbon by over 99%. Saeki *et al.* [90] tested the biosurfactants produced by *Gordonia* sp. strain JE-1058 for the removal of oil spills at sea and the crude from the surface of contaminated sea sand. The obtained results showed the effectiveness of this biosurfactant in the clean-up of oil spills at sea [90]. Sun *et al.* [91] reported the potential action of biosurfactants produced by strain S5 of *Pseudomonas aeroginosa* isolated from coking wastewater in promoting the biodegradation of polycyclic aromatic hydrocarbons.

Removal and Detoxification of Heavy Metals

Environmental pollution arising from heavy metals (Pb, Cd, Hg, Cr, Ni, As, Zn, Cu) is a result of industrialization and has led to serious health issues. Conventional methods of heavy metal removal often result in the generation of secondary waste, which is toxic to the environment [23, 25, 92, 93]. Among the eco-friendly techniques employed to remove heavy metals is the use of micro-organisms or their bioactive molecules. The biosurfactants, bioflocculants and biofilms are among the substances produced by micro-organisms and used to remove toxic heavy metals [25].

The micro-organisms cannot degrade heavy metals, but they participate in metal mobility and bioavailability in different manners, such as lowering soil pH, producing plant growth-promoting and metal-chelating compounds such as siderophores, organic acids, biosurfactants and altering soil redox conditions [3, 16, 58]. The heavy metal could be removed by immobilization (by reduction of metal ions), biofilms, bioprecipitation, accumulation, biosorption and increasing metal availability and up-taking by plants [94]. It has been reported that microbial surfactants are an effective metal complexing agent. Both the cationic and the anionic biosurfactants are efficient in the elimination of the heavy metals from the polluted sites. The mechanisms in both methods are dissimilar. The mechanisms behind metal binding are (1) anionic biosurfactants create complexes with metals in a nonionic form by ionic bonds. For the cationic biosurfactants, metal ions can be removed by the biosurfactants micelles [95].

Miller [92] mentioned that biosurfactants were an alternative with potential for remediation of metal-contaminated soils due to their advantages (small size, wide variety of chemical structures). Mulligan *et al.* [23] showed in their study the possibility of employing biosurfactants for the removal of heavy metals from the sediments. They proved that 0.5% rhamnolipid eliminated 65% of the Cu and 18% of the Zn, while 4% sophorolipid suppressed 25% of the Cu and 60% of the Zn. Surfactin was less effective, removing 15% of the Cu and 6% of the Zn [23]. The rhamnolipid potentiality produced by *Bacilli* sp. strains isolated from heavy metal contaminated soil and water in removing Cr and Zn were studied. The concentrations of these two metals were reduced (from 80 ppm to 41 ppm for Cr and from 40 ppm to up to 2 ppm for Zn) [96]. The anionic biosurfactant produced by *Candida guilliermondii* UCP 0992 was able to remove 98.9% of Zn, 89.3% of Fe and 89.1% of Pb [97]. The biosurfactant and extracellular reductase released by marine *Bacillus* sp. MTCC 5514 participates in the total reduction of 2000 mg/L of hexavalent chromium to trivalent chromium in 96 h [98].

The problem encountered is the competition between beneficial cations existing in soil with metal contaminants for the biosurfactant complexation sites. Therefore, the selectivity of biosurfactants for metals both in solution and in soil systems must be examined prior [58]. Additionally, various factors present in the soil environment may affect the biosurfactant activity like pore size, charge present on soil particles, soil pH, soil composition, particle size, time and type of contamination interfering with the effectiveness of biosurfactants action [93].

Biodegradation of Pesticides

A large amount of pesticides (herbicides, insecticides, fungicides, *etc.*) is used in agriculture. The demand for pesticides is increasing year over year due to the increase in agricultural productivity to meet the global need for food. Pesticides, in particular organochlorine, are one of the enduring organic contaminants which are disturbing by reason of their emergence in different environments. In nature, the pesticide residues are exposed to physical, chemical and biochemical deterioration operations, but due to their elevated steadiness and aqueous solubility, the pesticide residues last in the environment [99]. The extensive use of pesticides in agriculture causes harmful effects on humans through food and drinking water. Also, pollinisators insects and beneficial micro-organisms in soil are affected by the toxicity of pesticides or their residues. Pesticides movement in soil compartments depends on their solubility in water, their adsorption by soil particles, and lasting [100]. The remediation procedures imply physical, chemical, and biological remediation as well as combined methods for the elimination of pollutants. The low solubility of these compounds limits their availability to micro-organisms and consequently their biodegradation. The use of biosurfactants ameliorates the solubility and the availability of pesticides, thus inducing their degradation by micro-organisms [101, 102].

Dubey *et al.* [103] tested the emulsification capacity of two pesticides (monocrotophos and imidacloprid) by biosurfactants fabricated in curd whey by *Pseudomonas aeroginosa* strain–PP2 and Kocuria turfanesis strain-J at extreme environment conditions (pH, salinity and temperature). These authors proved that the microbial surfactants have demonstrated variances in their surface-active characteristics and have a noticeable specificity to emulsify pesticides in utmost environmental conditions [103]. The glycolipid (mannosylerythritol lipid) fabricated by yeast strain Pseudozyma VITJzN01 exhibited excellent surface properties and stability under extreme conditions. This microbial surfactant enhanced the solubility and the biodegradation of lindane (organochlorine pesticide) in a liquid medium at the rate of 36% in 2 days, compared with degradation in 12 days without biosurfactant [104]. The impact of an extract from *Pseudomonas sp.* B0406 strain with tensio-active characteristics on the solubility of two insecticides: endosulfan and methyl parathion, is studied. The surfactant in this extract is an anionic glycolipid that raises the solubility of both insecticides to 0.41 at 0.92 mg/L for endosulfan and 34.58 at 48.10 mg/L for methyl parathion [105]. Quinalphos, an organphospoorus insecticide, was degraded at a rate of 94% by *Pseudomonas aerogina* Q10. This strain of *Pseudomonas aerogina* owned biosurfactant fabrication capacity that makes Quinalphos available to cells [106]. Wang *et al.* [107] examined the capacity of microbial surfactant manufactured by *Pseudomonas* sp. SB to assist the phytoremediation of dichlorodiphenyltrichloroethane (DDT) pesticide polluted soil by two grass species. This study showed that the DDT contents in soil clearly decreased compared with the original soil. The *Pseudomonas* sp. surfactant enhanced the availability and bioremediation of DDT pesticides with removal efficiency greater than 60% [107]. Gaur *et al.* [108] characterized and evaluated the ability of a rhamnolipid produced by *Lysinibacillus sphaericus* IITR51 for the dissolution of hydrophobic pesticides. At the concentration of 90 mg/L rhamnolipid showed enhanced dissolution of α-, β-endosulfan, and γ-hexachlorocyclohexane up to 7.2, 2.9, and 1.8 folds, respectively. This biosurfactant exhibited potential antimicrobial activity against different pathogen bacteria [108].

Biodegradation of Organic Dyes

Dyes can be defined as products that once used to a substrate, give color by a procedure that alters, at least temporarily, any crystal structure of the colored products. More than 10,000 different dyes and pigments are utilized industrially, and more than 7×10^5 tons of synthetic dyes are fabricated annually [109]. Different dyes are used in textile industries in processing units that are liberated in natural water bodies through wastewater. Major classes of synthetic dyes are azo, anthraquinone and triarylmethane dyes [110]. The methods for the detection of dyes are cost intensive and useless because the dyes undergo chemical changes under environmental conditions, and the transformation products may be more poisonous and carcinogenic than the parent compound [111].

Some micro-organisms have the potential to degrade and decolorized the dyes in aerobic and anaerobic conditions. These micro-organisms produce extracellular enzymes such as laccases and peroxidases that participate in the degradation of dyes by oxidation [110]. The potency of some local isolates, biosurfactant-producing bacteria on decolorization of naphtol was examined. One of the six isolated strains showed the best ability to produce biosurfactants and degrade naphthol [112]. The catalytic activity of a combination of biosurfactant-$Cu_5(PO_4)_2 3H_2O$ nanoflowers was investigated for the degradation of cationic dyes. The obtained results showed the high stability and catalytic activity of this complex which may have potential applications in industrial biocatalysis, biosensors and environmental chemistry in the future [30].

CONCLUSION

The biosurfactants are produced by a variety of micro-organisms. These surface-active molecules act in the interface by decreasing the interfacial tension or as biodispesants. These actions of biosurfactants lead to the increase of bioavailability of non-polar substances for their removal and/or degradation. Several advantages like biodegradability, less hazardous and exceptional surface characteristics make biosurfactants the best biological choice and promising future solutions for the remediation of polluted environments. Some problems related to large-scale production must be resolved before generalizing the use of these tension-active compounds. The use of waste substrates in the biosurfactants fabrication procedures may help to surmount at the same time the pollution and the production costs problems.

CONSENT FOR PUBLICATION

Not applicable.

CONFLICT OF INTEREST

The authors declare no conflict of interest, financial or otherwise.

ACKNOWLEDGEMENT

Declared none.

REFERENCES

[1]   Dzionek A, Wojcieszyńska D, Guzik U. Natural carriers in bioremediation: A review. Electron J Biotechnol 2016; 23: 28-36.
[http://dx.doi.org/10.1016/j.ejbt.2016.07.003]

[2]   Samanta SK, Singh OV, Jain RK. Polycyclic aromatic hydrocarbons: environmental pollution and bioremediation. Trends Biotechnol 2002; 20(6): 243-8. http://tibtech.trends.com
[http://dx.doi.org/10.1016/S0167-7799(02)01943-1] [PMID: 12007492]

[3]   Yao Z, Li J, Xie H, Yu C. Review on remediation technologies of soil contaminated by heavy metal. Procedia Environ Sci 2012; 16: 722-9.
[http://dx.doi.org/10.1016/j.proenv.2012.10.099]

[4]   Shrimali M, Singh KP. New methods of nitrate removal from water. Environmental Pollution 2001; 112: 351-59.
[http://dx.doi.org/10.1016/S0269-7491(00)00147-0]

[5]   Sharma S, Tiwari S, Hasan A, Saxena V, Pandey LM. Recent advances in conventional and contemporary methods for remediation of heavy metal-contaminated soils. 3 Biotech 2018; 8: 216.
[http://dx.doi.org/10.1007/s13205-018-1237-8]

[6] Piorr R, Henkel K, Piorr R. Structure and Application of Surfactants. In: Falbe J, Ed. Surfactants in Consumer Products. Springer-Verlag Heidelberg 1987; pp. 5-22.
[http://dx.doi.org/10.1007/978-3-642-71545-7_2]

[7] Yuan CL, Xu ZZ, Fan MX, Liu HY, Xie YH, Zhu T. Study on characteristics and harm of surfactants. J Chem Pharm Res 2014; 6(7): 2233-7. www.jocpr.com

[8] Azarmi R, Ashjaran A. Type and application of some common surfactants. J Chem Pharm Res 2015; 7(2): 632-40. www.jocpr.com

[9] Schramm LL, Stasiuk EN, Marangoni DG. 2 Surfactants and their applications. Annu Rep Sect C Phys Chem 2003; 99: 3-48.
[http://dx.doi.org/10.1039/B208499F]

[10] Rebello S, Asok AK, Mundayoor S, Jisha MS. Surfactants: toxicity, remediation and green surfactants. Environ Chem Lett 2014; 12(2): 275-87.
[http://dx.doi.org/10.1007/s10311-014-0466-2]

[11] Ivanković T, Hrenović J. Surfactants in the Environment. Arh Hig Rada Toksikol 2010; 61(1): 95-110.
[http://dx.doi.org/10.2478/10004-1254-61-2010-1943] [PMID: 20338873]

[12] Rao MA, Scelza R, Scotti R, Gianfreda L. Role of enzyme in the remdiation of polluted environment. J Soil Sci Plant Nutr 2010; 10(3): 333-53. https://scielo.conicyt.cl/pdf/jsspn/v10n3/art08.pdf
[http://dx.doi.org/10.4067/S0718-95162010000100008]

[13] Srinivasan R. Natural polysaccharides as treatment agents for wastewater. In: Mishra A, Clark JH (Eds.) Green Materials for Sustainable Water Remediation and Treatment. RSC Publishing: Combridge 2013; 51-81.
[http://dx.doi.org/10.1039/9781849735001-00051]

[14] Singer AC, Crowley DE, Thompson IP. Secondary plant metabolites in phytoremediation and biotransformation. Trends Biotechnol 2003; 21(3): 123-30.
[http://dx.doi.org/10.1016/S0167-7799(02)00041-0] [PMID: 12628369]

[15] Sharma N, Lavania M, Lal B. Microbes and their secondary metabolites: Agents in bioremediation of hydrocarbon contaminated site. Archives of Petroleum & Environmental Biotechnology 2019; 4(1): 151.
[http://dx.doi.org/10.29011/2574-7614.100051]

[16] Khalid S, Muhammad S, Khan NN, Behzad M, Irshad B, Durnat C. A comparison of technologies for remediation of heavy metal contaminated soils. J Geochem Explor 2016; 182 (Part B): 247-68.
[http://dx.doi.org/10.1016/j.gexplo.2016.11.021]

[17] Abdel-Mawgoud AM, Lépine F, Déziel E. Rhamnolipids: diversity of structures, microbial origins and roles. Appl Microbiol Biotechnol 2010; 86(5): 1323-36.
[http://dx.doi.org/10.1007/s00253-010-2498-2] [PMID: 20336292]

[18] Md F. Biosurfactant: Production and Application. J Pet Environ Biotechnol 2012; 3(4): 4.
[http://dx.doi.org/10.4172/2157-7463.1000124]

[19] Li G, McInerney MJ. Use of biosurfactants in oil recovery. In: Lee SY, Ed. Consequences of Microbial Interactions with Hydrocarbons, Oils, and Lipids: Production of Fuels and Chemicals, Handbook of Hydrocarbon and Lipid Microbiology. Springer: Cham 2016; pp. 1-16.
[http://dx.doi.org/10.1007/978-3-319-31421-1_364-1]

[20] Banat IM. Potentials for use of biosurfactants in oil spills cleanup and oil bioremediation. In: Brebbia CA, Rodriguez GR (Eds) Oil and Hydrocarbon Spills II. WIT Press 2000; pp. 177-85.

[21] Patel S, Homaei A, Patil S, Daverey A. Microbial biosurfactants for oil spill remediation: pitfalls and potentials. Appl Microbiol Biotechnol 2019; 103(1): 27-37.
[http://dx.doi.org/10.1007/s00253-018-9434-2] [PMID: 30343430]

[22] Mnif I, Sahnoun R, Ellouze-Chaabouni S, Ghribi D. Evaluation of *B. subtilis* SPB1 biosurfactants'

potency for diesel-contaminated soil washing: optimization of oil desorption using Taguchi design. Environ Sci Pollut Res Int 2014; 21(2): 851-61.
[http://dx.doi.org/10.1007/s11356-013-1894-4] [PMID: 23818070]

[23]   Mulligan CN, Yong RN, Gibbs BF. Heavy metal removal from sediments by biosurfactants. J Hazard Mater 2001; 85(1-2): 111-25.
[http://dx.doi.org/10.1016/S0304-3894(01)00224-2] [PMID: 11463506]

[24]   Mulligan CN. Environmental applications for biosurfactants. Environ Pollut 2005; 133(2): 183-98.
[http://dx.doi.org/10.1016/j.envpol.2004.06.009] [PMID: 15519450]

[25]   Ayangbenro A, Babalola O. Metl(loid) bioremediation: Stategies employed by microbial polymers. Sustainability (Basel) 2018; 10(9): 3028.
[http://dx.doi.org/10.3390/su10093028]

[26]   Chellaiah ER. Cadmium (heavy metals) bioremediation by *Pseudomonas aeruginosa*: a minireview. Appl Water Sci 2018; 8(6): 154.
[http://dx.doi.org/10.1007/s13201-018-0796-5]

[27]   Yang Z, Shi W, Yang W, *et al.* Combination of bioleaching by gross bacterial biosurfactants and flocculation: A potential remediation for the heavy metal contaminated soils. Chemosphere 2018; 206: 83-91.
[http://dx.doi.org/10.1016/j.chemosphere.2018.04.166] [PMID: 29730568]

[28]   Singh PB, Sharma S, Saini HS, Chadha BS. Biosurfactant production by *Pseudomonas* sp. and its role in aqueous phase partitioning and biodegradation of chlorpyrifos. Lett Appl Microbiol 2009; 49(3): 378-83.
[http://dx.doi.org/10.1111/j.1472-765X.2009.02672.x] [PMID: 19627480]

[29]   Svarajasekar N, Ramasubbu S, Prakash Maran J, Priya B. Cationic dyes sequestration from aqueous using biosurfactant based reverse micelles. In: Regupathi I, Ed. Recent advances in chemical engineering Springer Science+Business Media Singapore. 2016; pp. 67-74.
[http://dx.doi.org/10.1007/978-981-10-1633-2_8]

[30]   Jiao J, Xin X, Wang X, Xie Z, Xia C, Pan W. Self-assembly of biosurfactant–inorganic hybrid nanoflowers as efficient catalysts for degradation of cationic dyes. RSC Advances 2017; 7(69): 43474-82. https://pubs.rsc.org/en/content/articlepdf/2017/ra/c7ra06592b
[http://dx.doi.org/10.1039/C7RA06592B]

[31]   Gudiña EJ, Rocha V, Teixeira JA, Rodrigues LR. Antimicrobial and antiadhesive properties of a biosurfactant isolated from *Lactobacillus paracasei* ssp. *paracasei* A20. Lett Appl Microbiol 2010; 50(4): 419-24.
[http://dx.doi.org/10.1111/j.1472-765X.2010.02818.x] [PMID: 20184670]

[32]   Vollenbroich D, Özel M, Vater J, Kamp RM, Pauli G. Mechanism of inactivation of enveloped viruses by the biosurfactant surfactin from *Bacillus subtilis*. Biologicals 1997; 25(3): 289-97.
[http://dx.doi.org/10.1006/biol.1997.0099] [PMID: 9324997]

[33]   Chiewpattanakul P, Phonnok S, Durand A, Marie E, Thanomsub BW. Bioproduction and anticancer activity of biosurfactant produced by the dematiaceous fungus *Exophiala dermatitidis* SK80. J Microbiol Biotechnol 2010; 20(12): 1664-71.
[http://dx.doi.org/10.4014/jmb.1007.07052] [PMID: 21193821]

[34]   Kuyukina MS, Kochina OA, Gein SV, Ivshina IB, Chereshnev VA. Mechanisms of immunomodulatory and membranotropic activity of trehalolipid biosurfactants (a review). Appl Biochem Microbiol 2020; 56(3): 245-55.https://link.springer.com/content/pdf/10.1134/S0003683820030072.pdf
[http://dx.doi.org/10.1134/S0003683820030072]

[35]   Faivre V, Rosilio V. Interest of glycolipids in drug delivery: from physicochemical properties to drug targeting. Expert Opin Drug Deliv 2010; 7(9): 1031-48.
[http://dx.doi.org/10.1517/17425247.2010.511172] [PMID: 20716018]

[36] Singh A, Van Hamme JD, Ward OP. Surfactants in microbiology and biotechnology: Part 2. Application aspects. Biotechnol Adv 2007; 25(1): 99-121.
[http://dx.doi.org/10.1016/j.biotechadv.2006.10.004] [PMID: 17156965]

[37] Marouf A, Tremblin G. Abrégé de biochimie appliqué. France 2009: p. 484.

[38] Płaza G, Achel V. Biosurfactants: Eco-friendly and innovative biocides against biocorrosion. Internantional Journal of Molecular Sciences 2020; 21(2152): 11.
[http://dx.doi.org/10.3390/ijms21062152]

[39] Roy A. A review on the biosurfactants: properties, types and its applications. J Fundam Renewable Energy Appl 2017; 8(1): 6.
[http://dx.doi.org/10.4172/2090-4541.1000248]

[40] Fenibo EO, Ijoma GN, Selvarajan R, Chikere CB. Microbial surfactants: the next generation multifunctional biomolecules for applications in the petroleum industry and its associated environmental remediation. Micro-organisms 2019; 7, 581: 29.
[http://dx.doi.org/10.3390/microorganisms7110581]

[41] Fenibo EO, Douglas SI, Stanley HO. A review on microbial surfactants: production, classification, properties and characterization. J Adv Microbiol 2019; 18(3): 1-22.
[http://dx.doi.org/10.9734/jamb/2019/v18i330170]

[42] Alejandro DJC-S. Surfactants of microbial origin and its application in foods. Sci Res Essays 2020; 15(1): 11-7.
[http://dx.doi.org/10.5897/SRE2019.6656]

[43] Mukherjee AK, Das K. 2010. Microbial surfactants and their potential applications: An overview. In Ramkrishna Sen. Editor. Biosurfactants. Landes Bioscience and Springer+Business Media, (2010). https://link.springer.com/book/10.1007/978-1-4419-5979-9

[44] Pacwa-Płociniczak M, Płaza GA, Piotrowska-Seget Z, Cameotra SS. Environmental applications of biosurfactants: recent advances. Int J Mol Sci 2011; 12(1): 633-54.
[http://dx.doi.org/10.3390/ijms12010633] [PMID: 21340005]

[45] Karlapudi AP, Venkateswarulu TC, Tammineedi J, *et al.* Role of biosurfactants in bioremediation of oil pollution-a review. Petroleum 2018; 4(3): 241-9.
[http://dx.doi.org/10.1016/j.petlm.2018.03.007]

[46] Zhang Y, Ji X, Lu X. Choline-based deep eutectic solvents for mitigating carbon dioxide emissions. In: Shi F, Ed. Novel Materials for Carbon Dioxide Mitigation Technology Elsevier, 2015. 2015; pp. 87-116.
[http://dx.doi.org/10.1016/B978-0-444-63259-3.00003-3]

[47] Dimri VP, Srivastava RP, Vedanti N. Reservoir Geophysics. Fractal Models in Exploration Geophysics - Applications to Hydrocarbon Reservoirs. In: Handbook of Geophysical Exploration: Seismic Exploration. Elsevier 2012; 41: 89-118.
[http://dx.doi.org/10.1016/B978-0-08-045158-9.00005-1]

[48] Desai JD, Banat IM. Microbial production of surfactants and their commercial potential. Microbiology and Moleculaire Biology Reviews 1997; 61(1): 47-64. https://mmbr.asm.org/content/mmbr/61/1/47.full.pdf

[49] Sheng JJ. Surfactant Flooding. In: Sheng JJ, Ed. Modern Chemical Enhanced Oil Recovery: theory and practice, 239–335 Elsevier, 2011. 2011.
[http://dx.doi.org/10.1016/B978-1-85617-745-0.00007-3]

[50] Hausmann R, Syldatk C. Types and classification of microbial surfactants. 2015. http://www.taylorandfrancis.com

[51] Ryll R, Kumazawa Y, Yano I. Immunological properties of trehalose dimycolate (cord factor) and other mycolic acid-containing glycolipids--a review. Microbiol Immunol 2001; 45(12): 801-11.

[http://dx.doi.org/10.1111/j.1348-0421.2001.tb01319.x] [PMID: 11838897]

[52]   Gautam KK, Tyagi VK. Microbial surfactants: A review. J Oleo Sci 2006; 55(4): 155-66.
       https://www.jstage.jst.go.jp/article/jos/55/4/55_4_155/_pdf
       [http://dx.doi.org/10.5650/jos.55.155]

[53]   Mnif I, Ellouz-Chaabouni S, Ghribi D. Glycolipid biosurfactants, main classes, functional properties
       and related potential applications in environmental biotechnology. J Polym Environ 2018; 26(5):
       2192-206.
       [http://dx.doi.org/10.1007/s10924-017-1076-4]

[54]   Costa SGVAO, Nitschke M, Lépine F, Déziel E, Contiero J. Structure, properties and applications of
       rhamnolipids produced by *Pseudomonas aeruginosa* L2-1 from cassava wastewater. Process Biochem
       2010; 45(9): 1511-6.
       [http://dx.doi.org/10.1016/j.procbio.2010.05.033]

[55]   Lourith N, Kanlayavattanakul M. Natural surfactants used in cosmetics: glycolipids. Int J Cosmet Sci
       2009; 31(4): 255-61.
       [http://dx.doi.org/10.1111/j.1468-2494.2009.00493.x] [PMID: 19496839]

[56]   Mnif I, Ghribi D. Glycolipid biosurfactants: main properties and potential applications in agriculture
       and food industry. J Sci Food Agric 2016; 96(13): 4310-20.
       [http://dx.doi.org/10.1002/jsfa.7759] [PMID: 27098847]

[57]   Seydlová G, Svobodová J. Review of surfactin chemical properties and the potential biomedical
       applications. Open Med (Wars) 2008; 3(2): 123-33.
       [http://dx.doi.org/10.2478/s11536-008-0002-5]

[58]   Maikudi Usman M, Dadrasnia A, Tzin Lim K, Fahim Mahmud A, Ismail S. Application of
       biosurfactants in environmental biotechnology; remediation of oil and heavy metal. AIMS Bioeng
       2016; 3(3): 289-304.
       [http://dx.doi.org/10.3934/bioeng.2016.3.289]

[59]   Kanlayavattanakul M, Lourith N. Lipopeptides in cosmetics. Int J Cosmet Sci 2010; 32(1): 1-8.
       [http://dx.doi.org/10.1111/j.1468-2494.2009.00543.x] [PMID: 19889045]

[60]   Sobrinho HB, Luna JM, Rufino RD, Porto ALF, Sarubbo LA. Biosurfactants: Classification,
       properties and environmental applications. In: Govil JN, Ed. Recent Developments in Biotechnology.
       USA: Studium Press LLC 2013; pp. 303-30.

[61]   Meena KR, Sharma A, Kanwar SS. Microbial lipopeptides and their medical applications. Ann
       Pharmacol Pharm 2017; 2(11): 1111.

[62]   Shaligram NS, Singhal RS. Surfactin- A review on biosynthesis, fermentation, purification and
       applications.   Food   Technol   Biotechnol   2010;   48(2):
       119-34.
       https://pdfs.semanticscholar.org/e1aa/d5c827bfa258264a8eaf8f49072e0d7825e3.pdf?_ga=2.13395751
       8.1289626574.1599258577-1215473906.1559439321

[63]   Maget-Dana R, Peypoux F. Iturins, a special class of pore-forming lipopeptides: biological and
       physicochemical properties. Toxicology 1994; 87(1-3): 151-74.
       [http://dx.doi.org/10.1016/0300-483X(94)90159-7] [PMID: 8160184]

[64]   Khattari Z, Al-Abdullah T, Maghrabi M, Khasim S, Roy A, Fasfous I. Interaction study of lipopeptide
       biosurfactant viscosin with DPPC and cholesterol by Langmuir monolayer technique. Soft Mater
       2015; 13(4): 254-62.
       [http://dx.doi.org/10.1080/1539445X.2015.1085873]

[65]   Matsuyama T, Tanikawa T, Nakagawa Y. Serrawettins and other surfactants produced by *Serratia*. In:
       Soberón-Chávez G, Ed. Biosurfactants, Microbiology monographs 20. Springer-Verlag Berlin
       Heidelberg 2011; pp. 93-120.
       [http://dx.doi.org/10.1007/978-3-642-14490-5_4]

[66]   Shekhar S, Sundaramanickam A, Balasubramanian T. Biosurfactant producing microbes and its potential applications: A Review. Crit Rev Environ Sci Technol 2014.
[http://dx.doi.org/10.1080/10643389.2014.955631]

[67]   Li J, Wang X, Zhang T, *et al.* A review on phospholipids and their main applications in drug delivery systems. Asian Journal of Pharmaceutical Sciences 2015; 10(2): 81-98.
[http://dx.doi.org/10.1016/j.ajps.2014.09.004]

[68]   Alagumuthu M, Dahiya D, Singh Nigam P. Phospholipid—the dynamic structure between living and non-living world; a much obligatory supramolecule for present and future. AIMS Mol Sci 2019; 6(1): 1-19.
[http://dx.doi.org/10.3934/molsci.2019.1.1]

[69]   Ishigami Y, Zhang Y, Ji F. Spiculisporic acid: Functional development of biosurfactant. Chim Oggi 2000; 18(7-8): 32-4.

[70]   Kumla D, Dethoup T, Buttachon S, Singburaudom N, Silva AMS, Kijjoa A. Spiculisporic acid E, a new spiculisporic acid derivative and ergosterol derivatives from the marine-sponge associated fungus Talaromyces trachyspermus (KUFA 0021). Nat Prod Commun 2014; 9(8): 1934578X1400900.https://www.researchgate.net/publications.265862074
[http://dx.doi.org/10.1177/1934578X1400900822] [PMID: 25233594]

[71]   Saravanakumari P, Mani K. Structural characterization of a novel xylolipid biosurfactant from *Lactococcus lactis* and analysis of antibacterial activity against multi-drug resistant pathogens. Bioresour Technol 2010; 101(22): 8851-4.
[http://dx.doi.org/10.1016/j.biortech.2010.06.104] [PMID: 20637606]

[72]   Wang R, Liu TM, Shen MH, *et al.* Spiculisporic acids B–D, three new γ-butenolide derivatives from a sea urchin-derived fungus *Aspergillus* sp. HDf2. Molecules 2012; 17(11): 13175-82.
[http://dx.doi.org/10.3390/molecules171113175] [PMID: 23128094]

[73]   Rosenberg E, Rubinovitz C, Legmann R, Ron EZ. Purification and chemical properties of *Acinetobacter calcoaceticus* A2 biodispersan. Appl Environ Microbiol 1988; 54(2): 323-6. https://aem.asm.org/content/54/2/323.short
[http://dx.doi.org/10.1128/aem.54.2.323-326.1988] [PMID: 16347545]

[74]   Foght JM, Gutnick DL, Westlake DWS. Effect of emulsan on biodegradation of crude oil by pure and mixed bacterial cultures. Appl Environ Microbiol 1989; 55(1): 36-42.https://aem.asm.org/content/aem/55/1/36.full.pdf
[http://dx.doi.org/10.1128/aem.55.1.36-42.1989] [PMID: 16347832]

[75]   Calvo C, Manzanera M, Silva-Castro GA, Uad I, González-López J. Application of bioemulsifiers in soil oil bioremediation processes. Future prospects. Sci Total Environ 2009; 407(12): 3634-40.
[http://dx.doi.org/10.1016/j.scitotenv.2008.07.008] [PMID: 18722001]

[76]   Rosenberg E, Ron EZ. Bioemulsans: microbial polymeric emulsifiers. Curr Opin Biotechnol 1997; 8(3): 313-6.http://biomednet.com/elecref/0958166900800313
[http://dx.doi.org/10.1016/S0958-1669(97)80009-2] [PMID: 9206012]

[77]   Pfammatter N, Hochköppler A, Luisi PL. Solubilization and growth of *Candida pseudotropicalis* in water-in-oil microemulsions. Biotechnol Bioeng 1992; 40(1): 167-72.
[http://dx.doi.org/10.1002/bit.260400123] [PMID: 18601058]

[78]   Kaczorek E, Pacholak A, Zdarta A, Smułek W. The impact of biosurfactants on microbial cell properties leading to hydrocarbon bioavailability increase. Colloids and Interfaces 2018; 2(3): 35.
[http://dx.doi.org/10.3390/colloids2030035]

[79]   Kiran GS, Ninawe AS, Lipton AN, Pandian V, Selvin J. Rhamnolipid biosurfactants: evolutionary implications, applications and future prospects from untapped marine resource. Crit Rev Biotechnol 2015; 1-17.
[http://dx.doi.org/10.3109/07388551.2014.979758] [PMID: 25641324]

[80]   Silva R, Almeida D, Rufino R, *et al.* Applications of biosurfactants in the petroleum industry and the remediation of oil spills. Int J Mol Sci 2014; 15(7): 12523-42.
[http://dx.doi.org/10.3390/ijms150712523] [PMID: 25029542]

[81]   Thavasi R, Jayalakshmi S, Banat IM. Application of biosurfactant produced from peanut oil cake by *Lactobacillus delbrueckii* in biodegradation of crude oil. Bioresour Technol 2011; 102(3): 3366-72.
[http://dx.doi.org/10.1016/j.biortech.2010.11.071] [PMID: 21144745]

[82]   Mandal AK, Sarma PM, Singh B, *et al.* Journal of Sustainable Development and Environmental Protection 2011; 1(3): 5-23.

[83]   Zhang Y, Miller RM. Enhanced octadecane dispersion and biodegradation by a *Pseudomonas rhamnolipid* surfactant (biosurfactant). Applied and Environmantal Microbiology 1992; 58(10): 3276-82. https://aem.asm.org/content/aem/58/10/3276.full.pdf

[84]   Kang SW, Kim YB, Shin JD, Kim EK. Enhanced biodegradation of hydrocarbons in soil by microbial biosurfactant, sophorolipid. Appl Biochem Biotechnol 2010; 160(3): 780-90.
[http://dx.doi.org/10.1007/s12010-009-8580-5] [PMID: 19253005]

[85]   Akbari S, Abdurahman NH. Biosurfactants-a new frontier for social and environmental safety: a mini review. Biotechnology Research nad. Innovation 2018; 2: 81-90.
[http://dx.doi.org/10/1016/j.biori.2018.09.001]

[86]   Bezza FA, Chirwa EMN. Bioremediation of polycyclic aromatic hydrocarbon contamined soil by a microbial consortium through supplementation of biosurfactant produced by *Pseudomonas aeroginnosa* strain. Polycycl Aromat Compd 2016; 36(5): 848-72.
[http://dx.doi.org/10.1080/10406638.2015.1066403]

[87]   Franzetti A, Tamburini E, Banat IM. Applications of Biological Surface Active Compounds in Remediation Technologies 2010.https://link.springer.com/chapter/10.1007/978-1-4419-5979-9_9
[http://dx.doi.org/10.1007/978-1-4419-5979-9_9]

[88]   Sajna KV, Sukumaran RK, Gottumukkala LD, Pandey A. Crude oil biodegradation aided by biosurfactants from Pseudozyma sp. NII 08165 or its culture broth. Bioresour Technol 2015; 191: 133-9.
[http://dx.doi.org/10.1016/j.biortech.2015.04.126] [PMID: 25985416]

[89]   Fanaei F, Moussavi G, Shekoohiyan S. Enhanced treatment of the oil-contaminated soil using biosurfactant-assisted washing operation combined with H_2O_2-stimulated biotreatment of the effluent. J Environ Manage 2020; 271: 110941.
[http://dx.doi.org/10.1016/j.jenvman.2020.110941] [PMID: 32778265]

[90]   Saeki H, Sasaki M, Komatsu K, Miura A, Matsuda H. Oil spill remediation by using the remediation agent JE1058BS that contains a biosurfactant produced by *Gordonia* sp. strain JE-1058. Bioresour Technol 2009; 100(2): 572-7.
[http://dx.doi.org/10.1016/j.biortech.2008.06.046] [PMID: 18692393]

[91]   Sun S, Wang Y, Zang T, *et al.* A biosurfactant-producing *Pseudomonas aeruginosa* S5 isolated from coking wastewater and its application for bioremediation of polycyclic aromatic hydrocarbons. Bioresour Technol 2019; 281: 421-8.
[http://dx.doi.org/10.1016/j.biortech.2019.02.087] [PMID: 30849698]

[92]   Miller RM. Biosurfactant-facilitated remediation of metal-contaminated soils. Environ Health Perspect 1995; 103 (Suppl. 1): 59-62. https://www.ncbi.nlm.nih.gov/pmc/articles/PMC1519337/pdf/envhper00361-0063.pdf
[PMID: 7621801]

[93]   Sarubbo LA, Rocha RB Jr, Luna JM, Rufino RD, Santos VA, Banat IM. Some aspects of heavy metals contamination remediation and role of biosurfactants. Chem Ecol 2015; 31(8): 707-23.
[http://dx.doi.org/10.1080/02757540.2015.1095293]

[94]   Azubuike CC, Chikere CB, Okpokwasili GC. Bioremediation; An eco-friendly sustainable technology

for environmental management. In: Saxena G, Bharagava RN, Eds. Bioremediation of industrial waste for environmental safety. Nature Singapore: Springer 2020; Volume I: Industrial waste and its management.: pp. 19-39.
[http://dx.doi.org/10.1007/978-981-13-1891-7]

[95]   Mulligan CN, Gibbs BF. Types, production and applications of biosurfactants 2004.https://insa.nic.in/writereaddata/UpLoadedFiles/PINSA/Vol70B_2004_1_Art03.pdf

[96]   Karnwal A, Bhardwaj V. Bioremediation of heavy metals (Zn and Cr) using microbial biosurfactant. Journal of Environmental Research and Protection 2014; 11(1): 29-33. www.ecoterra-online.ro

[97]   Sarubbo LA, Brasileiro PF, Silveira GNM, Luna JM, Rufino RD, dos Santos VA. Application of a low cost biosurfactant in the removal of heavy metals in soil. Chem Eng Trans 2018; 64. https://www.aidic.it/ibic2018/final/90sarubbo.pdf

[98]   Gnanamani A, Kavitha V, Radhakrishnan N, Suseela Rajakumar G, Sekaran G, Mandal AB. Microbial products (biosurfactant and extracellular chromate reductase) of marine microorganism are the potential agents reduce the oxidative stress induced by toxic heavy metals. Colloids Surf B Biointerfaces 2010; 79(2): 334-9.
[http://dx.doi.org/10.1016/j.colsurfb.2010.04.007] [PMID: 20483569]

[99]   Odukkathil G, Vasudevan N. Toxicity and bioremediation of pesticides in agricultural soil. Rev Environ Sci Biotechnol 2013; 12(4): 421-44.
[http://dx.doi.org/10.1007/s11157-013-9320-4]

[100] Alvarez A, Saez JM, Davila Costa JS, *et al.* Actinobacteria: Current research and perspectives for bioremediation of pesticides and heavy metals. Chemosphere 2017; 166: 41-62.
[http://dx.doi.org/10.1016/j.chemosphere.2016.09.070] [PMID: 27684437]

[101] Cameotra SS, Makkar RS. Biosurfactant-enhanced bioremediation of hydrophobic pollutants. Pure Appl Chem 2010; 82(1): 97-116.
[http://dx.doi.org/10.1351/PAC-CON-09-02-10]

[102] Mulligan CN. Recent advances in the environmental applications of biosurfactants. Curr Opin Colloid Interface Sci 2009; 14(5): 372-8.
[http://dx.doi.org/10.1016/j.cocis.2009.06.005]

[103] Dubey KV, Charde PN, Meshram SU, Shendre LP, Dubey VS, Juwarkar AA. Surface-active potential of biosurfactants produced in curd whey by *Pseudomonas aeruginosa* strain-PP2 and Kocuria turfanesis strain-J at extreme environmental conditions. Bioresour Technol 2012; 126: 368-74.
[http://dx.doi.org/10.1016/j.biortech.2012.05.024] [PMID: 22683199]

[104] Abdul Salam J, Das N. Enhanced biodegradation of lindane using oil-in-water bio-microemulsion stabilized by biosurfactant produced by a new yeast strain, Pseudozyma VITJzN01. J Microbiol Biotechnol 2013; 23(11): 1598-609.
[http://dx.doi.org/10.4014/jmb.1307.07016] [PMID: 23928846]

[105] García-Reyes S, Yáñez-Ocampo G, Wong-Villarreal A, *et al.* Partial characterization of a biosurfactant extracted from *Pseudomonas* sp. B0406 that enhances the solubility of pesticides. Environ Technol 2017.
[http://dx.doi.org/10.1080/21622515.2017.1363295] [PMID: 28783001]

[106] Nair AM, Rebello S, Rishad KS, Asok AK, Jisha MS. Biosurfactant Facilitated Biodegradation of Quinalphos at High Concentrations by *Pseudomonas aeruginosa* Q10. Soil Sediment Contam 2015; 24(5): 542-53.
[http://dx.doi.org/10.1080/15320383.2015.988205]

[107] Wang B, Wang Q, Liu W, *et al.* Biosurfactant-producing microorganism *Pseudomonas* sp. SB assists the phytoremediation of DDT-contaminated soil by two grass species. Chemosphere 2017; 182: 137-42.
[http://dx.doi.org/10.1016/j.chemosphere.2017.04.123] [PMID: 28494357]

[108]   Gaur VK, Bajaj A, Regar RK. Kamthan, M., Jha R.R., Srivastava J.K., Manickam N., Rhamnolipid from a *Lysinibacillus sphaericus* strain IITR51 and its potential application for dissolution of hydrophobic pesticides. Bioresour Technol 2019; 272: 19-25.
[http://dx.doi.org/10.1016/j.biortech.2018.09.144] [PMID: 30296609]

[109]   Chequer FMD, de Oliveira GAR, Ferraz ERA, Cardoso JC, Zanoni MVB, de Oliveira DP. Textile dyes: Dyeing process and environmental impact. In: Günay M, Ed. Eco-friendly textile dyeing and finishing. Intechopen 2013.
[http://dx.doi.org/10.5772/53659]

[110]   Gangola S, Bhatt P, Chaudhary P, Khati P, Kumar N, Sharma A. Bioremediation of industrial waste using microbial metabolic diversity. In: Bhatt P, Sharma A, Eds. Microbial biotechnology in environmental monitoring and cleanup. IGI Global Publishing 2018: p. 27.
[http://dx.doi.org/10.4018/978-1-5225-3126-5.ch001]

[111]   Ratna & Padhi B.S.. 2012.Pollution due to synthetic dyes toxicity & carcinogenicity studies ans bioremediation Int J Environ Sci 2012; 3(3): 940-54 http://www.ipublishing.co.in/ijesarticles/ twelve/articles/volthree/EIJES31094.pdf

[112]   Priyani N, Tampubolon JM, Munir E. Decoloration of naphtol batik dye by biosurfactant-producing bacteria. SEMIRATA, IOP Conf. Series. J Phys Conf Ser 2018; 1116: 052051.
[http://dx.doi.org/10.1088/1742-6596/1116/5/052051]

CHAPTER 6

Potential Application of Biological Treatment Methods in Textile Dyes Removal

Rustiana Yuliasni[1], Bekti Marlena[1], Nanik Indah Setianingsih[1], Abudukeremu Kadier[2,3,*], Setyo Budi Kurniawan[4], Dongsheng Song[2,5] and Peng-Cheng Ma[2,3]

[1] Research Centre for Environmental and Clean Technology, National Research and Innovation Agency Republic of Indonesia, Kawasan Puspitek gd 820, Serpong, 15314, Tangerang Selatan, Indonesia

[2] Laboratory of Environmental Science and Technology, The Xinjiang Technical Institute of Physics and Chemistry, Key Laboratory of Functional Materials and Devices for Special Environments, Chinese Academy of Sciences, Urumqi, 830000, China

[3] Center of Materials Science and Optoelectronics Engineering, University of Chinese Academy of Sciences, Beijing, 100049, China

[4] Department of Chemical and Process Engineering, Faculty of Engineering and Built Environment, National University of Malaysia (UKM), 43600 UKM Bangi, Selangor, Malaysia

[5] College of Resources and Environment, Xinjiang Agricultural University, Urumqi, 830052, China

Abstract: The most problematic issue related to textile wastewater is dyes. The occurrence of toxic and carcinogenic compounds in textile dyes creates aesthetic problems and affects the aquatic ecosystem. Dyestuff removal methods include physical, chemical, and biological-based technology. For a more environmentally friendly process that is low cost, produces less sludge, and needs a lesser amount of chemicals, biological treatment is preferable technology. To get maximum effectiveness and efficiency, integrations/ hybrids consisting of several technologies are commonly used. This chapter is dedicated to exploring the potential of biological technology to remove dyes from wastewater, especially dyes used in textile industries. This chapter briefly discusses dyes' characteristics, their utilization, and toxicity. Deeper reviews about the biodegradation potential of dyes are elaborated, along with a discussion about biodegradation mechanisms and reviews of either lab-scale or full-scale applications of biological-based technology for dyes treatment. Lastly, this chapter also gives future insight into the biological treatment of dyes.

* **Corresponding author Abudukeremu Kadier:** Laboratory of Environmental Science and Technology, The Xinjiang Technical Institute of Physics and Chemistry, Key Laboratory of Functional Materials and Devices for Special Environments, Chinese Academy of Sciences, Urumqi, 830011, China; Tel: +860991-3677875; E-mail: abudukeremu@ms.xjb.ac.cn

Inamuddin (Ed.)

Keywords: Biological treatment of dyes, Dyes wastewater treatment technology, Gye biodegradation, Textile wastewater, Wastewater containing dyes.

INTRODUCTION

Fresh water pollution from industrial, agricultural, and domestic activities has raised major concern due to its severe impacts on the environment niche [1]. Industrial pollution, especially from textile industry wastewater, has caused a negative impact on both environment and human health [2]. A wet textile industrial process consists of scouring, bleaching, dying, printing, and finishing (Fig. **1**). All these phases need a high quantity of water, chemicals, and energy. Hence, textile wet process commonly generates a large volume of wastewater containing toxic chemicals [3], characterized by concentrated pollutant parameters such as: color, salinity, biological oxygen demand (BOD), chemical oxygen demand (COD), total dissolved solids (TDS), total nitrogen (TN), total phosphorus (TP), and heavy metals, such as chromium (Cr) [4].

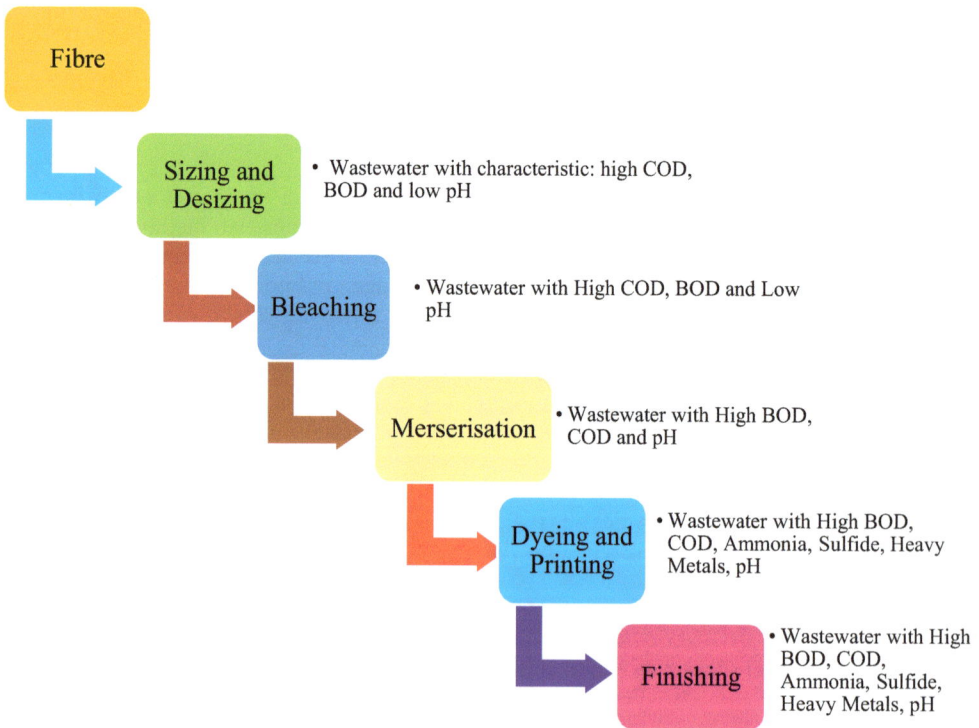

Fibre

Sizing and Desizing
- Wastewater with characteristic: high COD, BOD and low pH

Bleaching
- Wastewater with High COD, BOD and Low pH

Merserisation
- Wastewater with High BOD, COD and pH

Dyeing and Printing
- Wastewater with High BOD, COD, Ammonia, Sulfide, Heavy Metals, pH

Finishing
- Wastewater with High BOD, COD, Ammonia, Sulfide, Heavy Metals, pH

Fig. (1). Textile wet industrial process.

The most problematic issue related to textile wastewater is dyes. The occurrence of toxic and carcinogenic compounds in textile dyes creates aesthetic problems and affects the aquatic ecosystem [5, 6]. For more than 150 years, natural dyes were typically used in textile dyeing [7]. However, since the discovery of synthetic dyes by Henry Perkin in the 17th century, the 21st century has been occupied by many producers in synthetic dyes industry [8]. Dyes are described as complex organic substances (aromatic compounds) that absorb light and give color to the visible range (350 – 700 nm) [7 - 9]. The largest class of dyes is acid dyes. Other types of dyes include reactive, metal complex, direct, basic, mordant, disperse, pigment, vat dyes, anionic, sulphur, solvent, *etc.* Based on dyes consumption data, dyes discharged by the textile processing industry belongs to the class reactive dyes (±35%), acid (±25%), and direct (±15%), all of which are classified as azo dyes [10].

Dyes are highly persistent in the environment and possess the eco-toxic hazards of bioaccumulation. Thus, the release of dyes into the environment can affect man *via* the food chain. The degree of bioaccumulation is normally determined by Bio-concentration values (BCF's). Water soluble dyes, which have log BCF values of <0.5 and low K_{ow} (*i.e.*: acid, reactive and basic dyes), do not bio-accumulate. Bioaccumulation of water soluble dyes increases as they become more insoluble in water (higher K_{ow}) [11]. The acute toxicity of dyes is generally low. However, the chronic effects of dyes, especially azo dyes, are relatively high [12]. Under anaerobic condition, Azo dyes may release aromatic amines that are considered to be carcinogenic [12].

Dyestuff removal methods include physical, chemical and biological based technology. Adsorption and filtration are examples of physical treatment. These methods can achieve 86.8-99% dyes removal [13]. Chemical based technology is considered the most common technology for dyestuff removal. The oldest chemical technique that is still being used is coagulation-flocculation, while advanced oxidation process (AOP) and electrochemical technology are considered to be more advanced methods [14, 15]. AOP and electrochemical technology have short retention time, less sludge production and high removal efficiency [13]. However, the disadvantages of these technologies are high costs in terms of energy and chemicals requirement. For more environmentally friendly processes that are low cost, produce less sludge and need a lesser amount of chemicals, biological treatment is preferable technology [16, 17]. To get a maximum effectivity and efficiency, in terms of pollutants removal performance and costs, integration/ hybrid of physical, chemical and biological treatments is commonly used, *i.e.*: hybrid technology between sequencing batch reactor (biology) and ultrafiltration (physic) or integration technology between electrochemical oxidation/bio-treatment (chemistry/biology) [14 - 18].

Research related to biological dyes degradation, exclusively about azo dye, has gained popularity between 2018 -2020 [1, 19 - 23]. Compared to physical and chemical methods, the biological method has more benefits such as being eco-friendly, low cost, producing less sludge, producing non-hazardous final product and consuming less water [24]. The biological azo dyes degradation is not merely assisted by the microorganism, but also can be facilitated by algae and yeast (via secretion of enzymes) [25, 26]. Microbial assisted azo dyes removal is still relied on a combination of conventional anaerobic - aerobic process [23], where under anaerobic conditions, carcinogenic colorless aromatic amines are generated as inter-mediate products. To achieve a complete biodegradation/ mineralization of aromatic amines, aerobic oxidation should be applied consecutively (Fig. **2**). As for the type of microorganisms, either pure or mixed cultures are generally used [20, 21, 27].

This chapter is dedicated to explore the potential of biological technology to remove dyes from wastewater, especially dyes used in textile industries. In this chapter, the history and classification of dyes will be discussed briefly to give a general idea about their characteristics and utilization. Furthermore, the environmental consequence of dyes will be described, along with brief reviews of current technologies to treat dyes. Lastly, the biodegradation potential of dyes will be elaborated, with more depth discussion about biodegradation mechanisms, types of biodegradation agents and reviews of either lab scale or full scale application of biological based technology for dyes treatment.

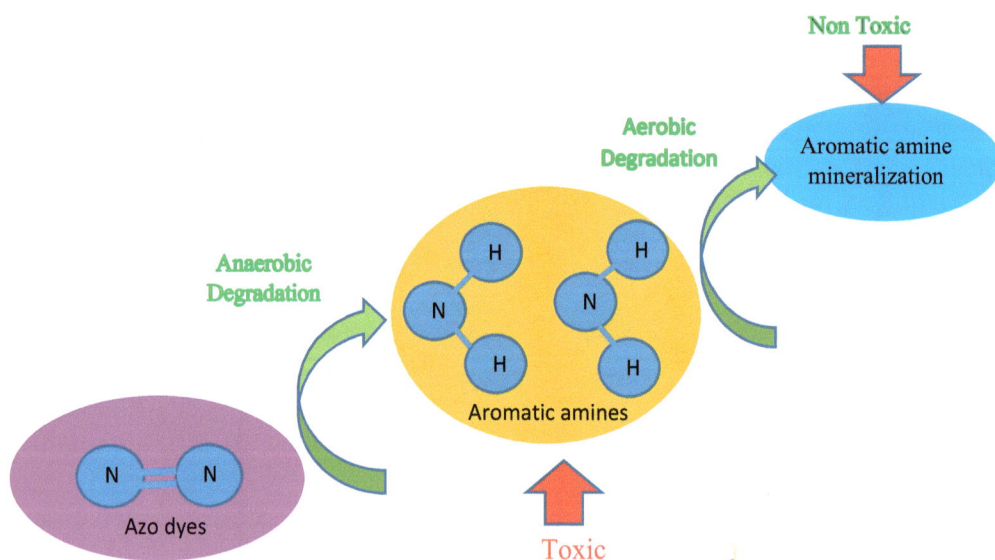

Fig. (2). Illustration of biological azo dyes removal *via* integration anaerobic-aerobic treatment.

HISTORY AND CLASSIFICATION OF DYES

History of Textile Dyes

Plants and a variety of organisms (*i.e.* lichens, insects and shellfish) were the most important source of dyes but the variety of colors was still very limited [7]. To extend color diversity, the mix between two main colors was prepared, for instance, a yellow and blue dyestuffs could generate a green color. The ideal characteristic of dyes is "fast", meaning that dyes could not be easily altered by washing or by air and light exposure. To apply dyes into the fabrics, they are generally dissolved in an aqueous solution and need mordant (consists of metal salts: Aluminum, iron, tin, chromium or copper ions) to improve fastness and need four main steps to ensure dyeing process cost- efficient, namely:(1) Extraction (2) Isolation of dye 3) Dyeing (4) Dye fixing. Dyes classification based on the application method to the fabrics includes direct, vat and mordant. Vat dyes were applied in water-insoluble solution by reducing condition, while mordant need "mordant "that dissolve in the water to produce colored complex ions that are absorbed by the fabrics. Finally, direct dyes are applied directly to the fiber but usually are less light-fast than vat or mordant dyes.

Synthetic dyes began to embrace in the 19th century, discovered by a British chemist, William Perkin. Synthetic dyes were originally invented for medical use. The first synthetic dyes were quinine made of coal tar, a type of liquid which is a byproduct of coal. In 1856, Perkin made another synthetic purple dyes, and followed with artificial red dye, in 1869. The development of synthetic dyes was at the same phase as the growth of industrial fabric production. The peak of utilization of synthetic dyes was in World War I, when the Germans mass produced most of the synthetic dyes used in the textile industry. From then on, most textile industries were colored with synthetic chemical dyes.

Classification of Dyes Based on Industrial Application

Direct Dyes

Direct dyes are used to color paper products because they have stable color during the washing process and have a strong affinity to cellulose fibre [28, 29]. Direct dyes have chromophoric groups include: azo, oxazine and phthalocyanine, with some thiazole and copper complex azo dyes [8]. The example of direct dyes is presented in Table **1**.

Disperse Dyes

Disperse dyes are majority used on polyester, but also can be used on nylon, cellulose acetate and acrylic fibers, despite their poor wet-fastness properties [30]. Azo is the majority of dispersed dyes structures [31]. Disperse azo dyes are the largest group among disperse dyes, accounting for more than 50% of the total disperse dyes group [32]. These dyes belong to the persistent class of dyes, and have recalcitrant nature and non-biodegradable behavior [33]. Some examples of disperse dyes are shown in Table **2**.

Table 1. Chemical structure of some direct dyes.

NO	Chemical structure of direct dyes	Chemical Name
1		Direct Black 38
2		Direct red 28
3		Direct Blue 6

Table 2. Chemical structure of some disperse dyes.

NO	Chemical structure of disperse dyes	Chemical name
1		Disperse violet 26

(Table 2) cont.....

NO	Chemical structure of disperse dyes	Chemical name
2		Disperse red 9
3		C.I disperse blue 56
4		Disperse Brown 1

Vat Dyes

Vat dyes have excellent light. They are also wet fast and mainly soluble in hot water or in Na_2CO_3 solution and are good to be applied on cellulose fiber [34, 35]. Table **3** shows the examples of the vat dyes.

Table 3. Chemical structure of some vat dyes.

NO	Chemical structure of vat dyes	Chemical Name
1		Vat Blue 1 (synthetic indigo)
2		Vat Black 25

(Table 3) cont.....

NO	Chemical structure of vat dyes	Chemical Name
3		C.I Vat Green 1
4		Vat Yellow 1

Basic Dyes

Basic dyes are easily dissolved in water and producing colored cations in solution, due to their positively charged ammonium group. These cations are electrostatically attracted to negatively charged substrates [36, 37]. Basic dyes are generally used in wool, silk, cotton and modified acrylic fibers. Usually, acetic acid is added to the dye bath to help the take up of the *dye* onto the fibre. Table **4** depicts some chemical structures of basic dyes.

Table 4. Chemical structure of some basic dyes.

NO	Chemical structure of basic dyes	Chemical Name
1		Basic Blue 26
2		Basic red 1

(Table 4) cont.....

NO	Chemical structure of basic dyes	Chemical Name
3		Basic green 1
4		Basic Blue 1
5		Basic Brown 1

Acid Dyes

Acid dyes have one or more acid functions, such as SO_3H and COOH and are normally applied at acidic pH. Due to their acid function properties, Acid dyes are suitable for coloring basic fibers, *e.g.* polyamides [8]. Acid dyes are widely used in textiles, leather, papers and pharmaceutical, because of their brightness and high solubility [38]. In textile industries, acid dyes are applied extensively on wool, nylon, and cotton. They contribute around 30-40% of the total consumption of dyes [39]. Table **5** presents some examples of acid dyes.

Sulphur Dyes

Sulphur dyes are one of the most popular types of dyes due to their low cost and good blendability in cellulosic fibers. Sulphur dyes give muted color to fiber [40]. Sulphur dyes constitute 9.1% of total US dye production and 15.8% of the dyes made for use on cellulosic fibers [7]. Aromatic compounds (*i.e*: naphthalene, benzene, nitro, amino) are the precursor for the synthesis of sulphur dye. Aromatic rings in sulphur dyes cause high toxicity [8]. Example of sulphur dyes is presented in Table **6**.

Table 5. Chemical structure of some acid dyes.

No.	Chemical structure of Acid dyes	Chemical Name
1		Acid Red 249 dye
2		Acid Blue 349
3		Acid orange 7

Table 6. Chemical structure of some sulphur dyes.

No.	Chemical structure of Suphur dyes	Chemical Name
1		Sulphur black CI 53185

(Table 6) cont.....

No.	Chemical structure of Suphur dyes	Chemical Name
2	O_2N ... H, N ... OH, O_2N	Sulphur blue dyes, CI 53235
3	(structure with O, O, O)	Phthalic anhydrate

Azo Dyes

Azo dyes are considered as the most popular dyes. It contributes 60-70% of all textile dyes in practice. Azo dyes are mainly applied in dyeing [41]. Azo dyes become a key part of the textile dyeing industry due to their color strength, light fastness, and brighter color intensity [42]. Azo dyes have functional group (-N=--) integrating two symmetrical and/or asymmetrical identical or non-azo alkyl or aryl radicals [42]. Biodegradation of azo dyes can produce amine by products as metabolites, that can be hazardous and carcinogenic [43]. Some azo dyes examples are shown in Table 7.

Table 7. Chemical Structure of some azo dyes.

No.	Chemical Structure of Azo Dyes	Chemical Name
1	H_3C-N, H_3C ... N=N ... COOH	Acid Red 2
2	N=N ... N=N ... CH_3, OH	Disperse yellow 7

(Table 7) cont.....

No.	Chemical Structure of Azo Dyes	Chemical Name
3		Acid orange 20
4		Methyl red

Reactive Dyes

Reactive dyes are very popular due to their high wet fastness, brilliance and range of hues [44]. They are applied mainly on Cellulose fibers (linen, cotton, and rayon), protein fibers (wool and silk) and some nylon [45]. Most of the reactive dyes have azo chromophores and are widely used due to their ability to easily bind to the fibers and their robustness [45]. These dyes are the second largest dyes group [8]. Table **8** shows some examples of reactive dyes.

Table 8. Chemical structure of some reactive dyes.

No.	Chemical Structure of Reactive Dyes	Chemical Name
1		C.I reactive red 3
2		C.I Reactive Blue 19

(Table 8) cont.....

No.	Chemical Structure of Reactive Dyes	Chemical Name
3		C.I Reactive red 17

Dyes Classification Based on Chromophores

Chromophore is described as a group of atoms and electrons forming part of an organic molecule that causes it to be colored. Textile dyes have a unique chemical structure for each color [53, 54]. They can be named according to the chemical structure of their chromophoric group. Table **9** shows some examples of dyes chromophore.

Table 9. Examples of Dye chromophores.

No.	Chromophores	Class
1		Anthraquinone
2		Azo
3		Cyanine
4		Indigoid

(Table 9) cont.....

No.	Chromophores	Class
5		Nitro
6		Quinione-imine
7		Xanthene

ENVIRONMENTAL CONCERN RELATED TO DYES

The most problematic environmental related issue related to dyes is color. Most dyes are visible in the water at a concentration as low as 1 mg/L, while textile industry typically generates wastewater with dye content as high as 200 mg/L, due to the frequently used alkaline dyes [46]. The high color intensity in the water body may affect photosynthetic activity in aquatic life due to reduced light penetration [47]. In addition, many dyes types are considered carcinogenic and hazardous to the aquatic environments [48]. Typically azo dyes are recalcitrant, hence they are un-biodegradable to microorganisms, making it difficult to treat [23]. Recalcitrant characteristic of azo dyes is because of aromatic amines's generation as inter-mediates of anaerobic azo-dye degradation [23].

According to Clarke and Anliker [49], dyes show low toxicity to mammals and aquatic organisms. They surveyed 3000 colorants and found only 2% of the dyes had an LC50 <1.0 mgdmy^3 and over 96% of dyes had an LC50 above 10 mgdmÿ^3. However, environmental anaerobic conditions lead to the conversion of non-toxic dyes into toxic metabolites [50]. 24 dye metabolites, known as aromatic amines, have been enlisted as toxic compounds by European Union (EU). Furthermore, since 1999, EU banned their use in industries [12, 50]. Even though azo dyes have been categorized as toxic and non-toxic, but under the real environmental condition, the potential of toxic dyes is failed to detect. As an example, a non-toxic Acid Orange 7 and Reactive Black 5 can transform into harmful aromatic amines under certain environmental conditions [50].

Apprehending the toxic potential of dye effluents, efforts are being made to develop treatment and management techniques to protect the biodiversity and ecosystem [12].

DYES REMOVAL TECHNIQUES

Dyes treatment technologies, based on physical, chemical and biological methods, have been established [51, 52]. These technologies also can be categorized based on the level of innovation, *i.e.* conventional and advanced methods. All these technologies have their own strength and weakness, depending on many factors, such as wastewater characteristics, removal efficiency, climate, space, cost, operator skill, *etc.* Hence, to give maximum results for full scale application of these technologies, integration of several technologies should be considered. Fig. (**3**) depicts wastewater technologies scheme that is commonly used for full-scale applications.

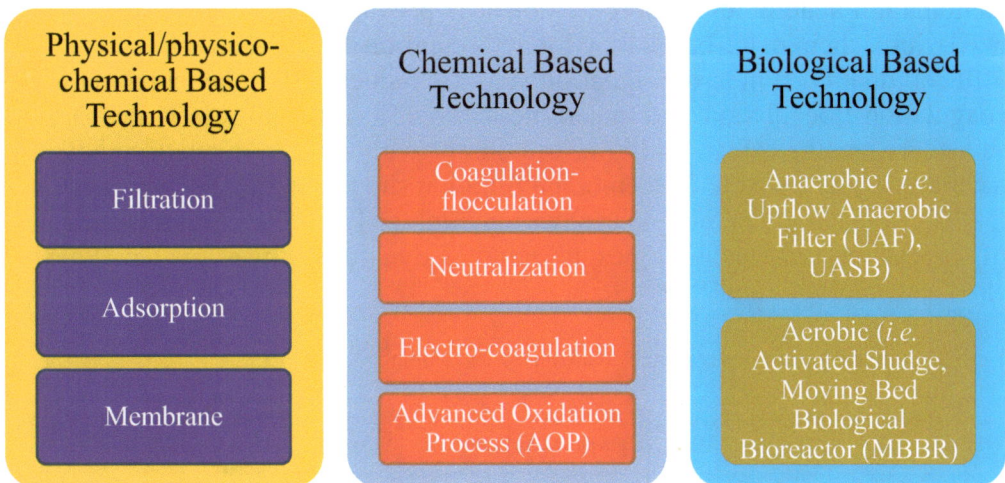

Physical/physico-chemical Based Technology	Chemical Based Technology	Biological Based Technology
Filtration	Coagulation-flocculation	Anaerobic (*i.e.* Upflow Anaerobic Filter (UAF), UASB)
Adsorption	Neutralization	
	Electro-coagulation	Aerobic (*i.e.* Activated Sludge, Moving Bed Biological Bioreactor (MBBR)
Membrane	Advanced Oxidation Process (AOP)	

Fig. (3). Wastewater technology according to the methods used for pollutant degradation.

Physical, chemical and biological technology can also have been categorized into two categories, according to the level of innovation, namely: conventional and advanced technology as shown in Fig. (**4**). For example, coagulation-flocculation could fall into conventional chemical based technology, while Up-flow Anaerobic Sludge Blanket (UASB) could fall into advanced biological based technology.

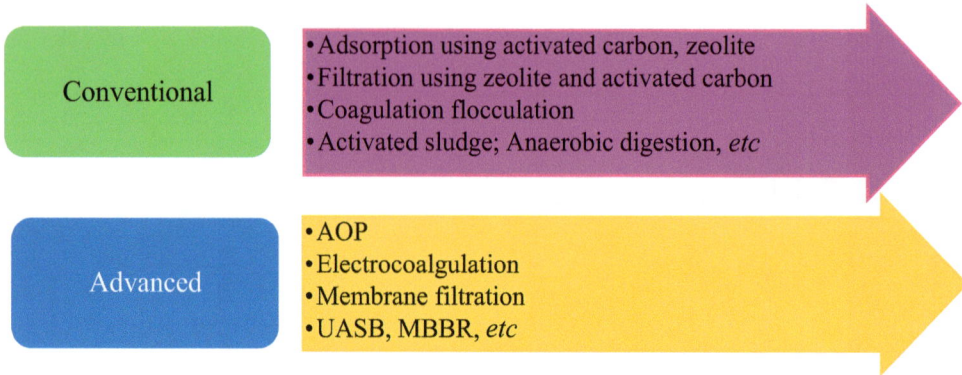

Fig. (4). Wastewater technology according to the level of innovation.

As already mentioned above, each of the methods have their own drawbacks for full scale application. For instance, coagulation-flocculation generates high chemical sludge volume despite having a short retention time. In the contrary, combination anaerobic- aerobic treatment is proven to generate less sludge and needs less chemicals, but has a longer retention time. Generally, to achieve effluent quality standard regulation and to reduce cost, the combination of coagulation-flocculation and anaerobic-aerobic is used (known as primary and secondary treatment). In addition, to comply with stricter regulation, advanced technology such as AOP or membrane filtration should be used. In a consequence, the investment and operational cost will go higher. Table **10** represents the review of several current technologies to remove dye in the wastewater (Physical, chemical, biology, and integrated).

Table 10. Current review of dyes removal advanced technology.

S. No.	Type of Technology	Type of Dyes	Initial Dye Concentration (mg/L as COD)	Type of Reactor	Reactor Volume (L)	Retention Time (mins)	% Removal (%)	Refs.
			CHEMICAL BASED TECHNOLOGY					
1	Electrochemical oxidation	1. Reactive dyes (reactive blue 19, reactive red 195 and yellow 145) 2. Mixed of real textile wastewater	75 - 750	batch reactor with magnetic stirring	0.1	60	Reactive dyes removal 100% Real mixed wastewater: 59% and 48%	[15]

(Table 10) cont.....

S. No.	Type of Technology	Type of Dyes	Initial Dye Concentration (mg/L as COD)	Type of Reactor	Reactor Volume (L)	Retention Time (mins)	% Removal (%)	Refs.
2	Advanced oxidation processes (AOPs)	A simulated mixture of Reactive Yellow 145, Reactive Red 195 and Reactive Blue 221	125	batch	1	10	90%	[53]
3	Electrocoagulation	reactive dyes (Optilan MF and Novacron NF)	478 – 800 (as COD)	batch	200	5; 10; 15	36 – 80%	[14]
PHYSICO-CHEMICAL BASED TECHNOLOGY								
1	Adsorption using Silica nanoparticlesbased technologies (Fe_3O_4@SiO_2-COOH nanoparticles)	malachite green (MG)	20	batch	0.04	120	97.5%	[54]
2	Adsorption using Fe_3O_4/□-Al_2O_3 hybrid composite	anionic azo dye Acid Black 1 (AB1)	500 (as COD)	Batch (in vials)	0.2	180 - 720	>90%	[55]
BIOLOGICAL BASED TECHNOLOGY								
1	Aerobic membrane bioreactor	reactive Blue	250	continue	10	nA	80-90% removal for total organic removal 91 -100% for decolourization	[17]
2	Biological degradation using single cell microorganisms enzyme Trichoderma asperellum laccase	malachite green	122.66	batch	5	98.58 (incubation time)	97.18% under	[56]
INTEGRATED/HYBRID BASED TECHNOLOGY								
1	UF-RO integrated membrane system combined with coagulation/flocculation	Real Hairwork Dyeing Effluent wastewater	2140 (as COD)	continue	0.2	72 min (2 min coagulation, 10 min flocculation, 60 min sedimentation)	90 -95%	[57]

(Table 10) cont.....

S. No.	Type of Technology	Type of Dyes	Initial Dye Concentration (mg/L as COD)	Type of Reactor	Reactor Volume (L)	Retention Time (mins)	% Removal (%)	Refs.
2	Coagulation/flocculation - ultrafiltration in TIO2-modified membranes	Reactive black 5 dye	10	batch	1	2 min coagulation, 20 min flocculation, 30 mins sedimentation and 90 min filtration	Almost 100%	[58]
3	Hybrid Sequencing Batch Reactor (SBR) anaerobic - Aerobic - nanofiltration (NF)	Textile wastewater containing Reactive Blue 21 and Sodium Dodecyl Sulfate (SDS)	1047 – 2750 (as COD)	Sequencing batch	2	72 hours	96% of dye and 97% of COD	[16]

BIODEGRADATION MECHANISMS OF DYES

Three principal mechanisms of biological dye removal processes are biosorption, bioaccumulation and biodegradation (Fig. **5**).

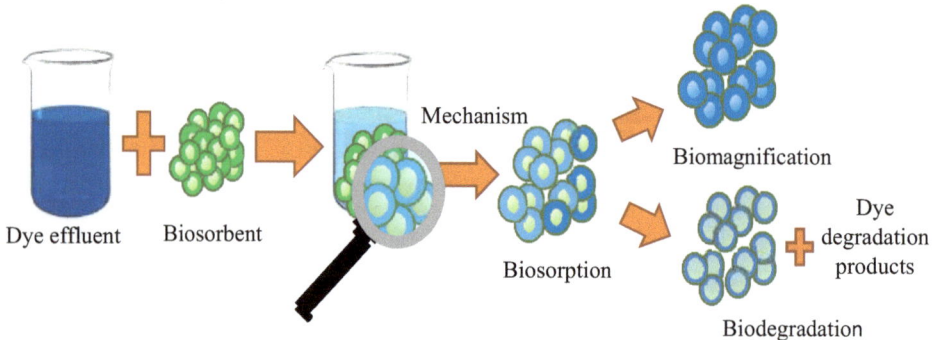

Fig. (5). Mechanism of biological dye removal.

Biosorption

Biosorption mechanisms are described as the process of substance/ sorbate (*i.e.* compounds, organic pollutants, metal ion, dyes) removal from solution by biological material (known as sorbent). Biological materials can be the form of living or dead organisms. In dyes biotreatment technology, full scale application, one of the primary challenges is to pick the most cost effective and highest-efficiency biosorbent that is readily available in the market [59]. Another challenge is the biosorbent that could also be restored with little care handling for quite some time. For that reason, dead biomass is preferable to living biomass. Microbial biomass (fungi, archaea, bacteria, yeast and microalga) shows better dye biosorption than macroscopic biomass (seaweeds, moss, *etc.*) [59]. Material of biological origin for biosorbent includes microbial biomass, archaea, cyanobacteria, filamentous fungi [60 - 63], yeast and microalgae [64 - 66], seaweeds (macroalgae), industrial wastes (food wastes), activated and anaerobic sludges [67, 68], *etc.*), agricultural wastes (fruit/vegetable wastes [69, 70], rice straw, wheat bran, sugar beet pulp, soybean hulls, cassava root husks [71] *etc.*), natural residues (plant residues [72], pine sawdust [73], tree barks, weeds [74], moss [75]) and other materials (eggshell [76], calcium alginate [77] chitosan, cellulose, *etc.*). The effectiveness of the biosorption processes can be evaluated by various models *viz.* equilibrium isotherm and kinetic [78].

Biosorption mechanisms could be understood empirically using isothermal equilibrium model and kinetic model (adsorption rate) presented in Fig. (**6**) and Table (**11**).

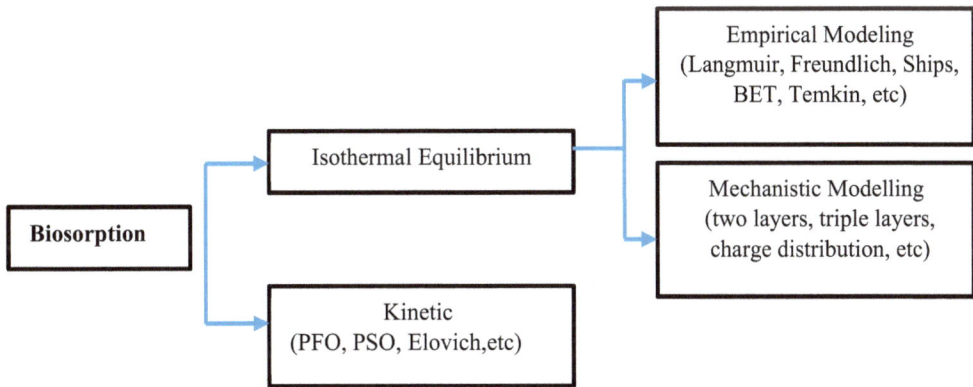

Fig. (6). Batch Model of Biosorption.

Table 11. Empirical Equilibrium Isotherm.

Model	Equation	Linear Equation	Plot
Langmuir	$q_e = \dfrac{q_m K_L C_e}{1 + K_L q_m}$	$\dfrac{C_e}{q_e} = \dfrac{C_e}{q_m} + \dfrac{1}{q_m K_L}$	$\dfrac{C_e}{q_e}$ vs Ce
Freundlich	$q_e = K_F C_e^{1/n}$	$\log q_e = \ln K_F + \dfrac{1}{n} \log C_e$	Log qe *vs* log Ce
Sips	$q_e = \dfrac{K_S C_e^\beta S}{1 + a_S C_e^\beta S}$	$\beta_s \ln(C_e) = -\ln\left(\dfrac{K_s}{q_e}\right) + \ln(a_s)$	$-\ln\left(\dfrac{K_s}{q_e}\right)$ vs $\ln(C_e)$

(Table 11) cont.....

Model	Equation	Linear Equation	Plot
Tempkin	$q_e = \dfrac{RT}{b_T} \ln A_T C_e$	$\dfrac{q_e}{} = \left(\dfrac{RT}{b_T}\right) \ln A_T + \left(\dfrac{RT}{b_T}\right) \ln C$	q_e VS C_e
Redlich-Peterson	$q_e = \dfrac{K_R C_e}{1 + a_R C_e^g}$	$\ln\left(K_R \dfrac{C_e}{q_e} - 1\right)$ $= g\ln(C_e) + \ln(a_R)$	$\ln\left(K_R \dfrac{C_e}{q_e} - 1\right)$ vs $\ln(C_e)$
BET	$\dfrac{q_e}{} = \dfrac{q_S C_{BET} C_e}{(C_s - C_e)\left[1 + (C_{BET-1}\right.}$	$\dfrac{C_e}{q_e(C_s - C_e)} = \dfrac{1}{q_S C_{BET}} + \dfrac{(C_{BET} - 1)}{q_S C_{BET}} \dfrac{C_e}{C_s}$	$\dfrac{C_e}{q_e(C_s - C_e)}$ vs $\dfrac{C_e}{C_s}$

Ref: Unuabonah, Omorogie, and Oladoja, 2018 [78]

Langmuir and Freundlich were classical isotherm sorption equations as the basic principle of derivation of modern isotherm equations [79]. Langmuir equation was originally planned to describe the adsorption of a gas on solids, thus far it is also used to measure and differentiate the adsorption capacities of numerous adsorbents including biosorption [80]. This model assumed a mono layer where all sorption sites are of the same types and they are all occupied at the end of the process. Meanwhile, Freundlich isotherm describes multilayer sorption, where sorption occurs on heterogeneous surfaces and there was an exponential distribution of active sites and energies [80]. Sip's isotherm describes heterogeneous sorption systems with advantages that can predict increasing adsorbate concentration.

Another sorption model that can be used to understand biosorption mechanisms is mechanistic models. Mechanistic models describe solute sorption onto the surfaces of biomass, which basically depend on the initial characterization of biomass. This model also is counted the certain solution chemistry of the solutes [79]. Mechanistic modeling of dye biosorption has been attempted in several investigations [69, 74, 76].

To describe the reaction order of biosorption system, the kinetic model can be used. Adsorption kinetics describe the time progress of the adsorption process. Kinetic models have described the reaction order of biosorption systems based on solution concentration [78]. Absorption kinetic models can be seen in Table **12**.

Table 12. Absorption kinetic models.

Kinetic	Linear Equation
Zero Order	$q_t = q_0 + k_0 t$
First order	$\ln(q_t) = \ln(q_o) + k_1 t$
Second Order	$\dfrac{1}{q_t} = \dfrac{1}{q_0} + \dfrac{1}{q_e} t$

(Table 12) cont.....

Kinetic	Linear Equation
Pseudo First Order	$$log(q_m - q_t) = log(q_m) - \frac{k_1}{2.303}t$$
Pseudo Second Order	$$\frac{1}{q_t} = \frac{1}{kq_e^2} + k_2 t$$
Elovich	$$q_t = q_0 + \frac{1}{\beta} ln(\alpha\beta) + \frac{1}{\beta} ln(t)$$

Ref: Unuabonah, Omorogie, and Oladoja, 2018) [78]

Dye biosorption studies generally reported Pseudo First Order (PFO) and Pseudo Second Order (PSO) as kinetic models. PFO rate equation of Lagergren may have been the first equation for the sorption in liquid/solid systems based on solid capacity. PFO corresponds to a diffusion-controlled process, while PSO corresponds to an adsorption reaction-controlled process which is known as chemisorption [81 - 83]. Azizian [84] reported that the sorption process follows PFO at high initial concentration of solute, while it follows PSO at a lower initial concentration of solute.

Based on the models that were previously described, a compilation of recent studies of dye removal by biosorption of various dyes were presented in Table **13**.

Table 13. Studies of modeling dye removal.

Dye	Sorbent	Operation condition	Results	Model	References
Methyl orange	*Oedogonium subplagiostomum* AP1	pH = 6.5, contact time = 5.5 days, C_{in} dye = 500 mg/L, algal dose = 400 mg/L	Dye removal 97%	Langmuir isotherm Pseudo-second order kinetic	[66]
Remazol Black B (RBB)	*Phormidium animale*	pH = 2, T = 45 °C, contact time = 1440 minutes, C_{in} dye = 93.16 mg/L, and biosorbent dosage = 4 g/L	Dye removal 99.66%	Langmuir isotherm model, Pseudo-second order kinetic	[64]
Direct Black Eco TFA (DBET)	Cassava root husks	pH = 2, T = 30°C, stirring speed = 150 rpm,	Adsorption capacity 10 mg/g	Langmuir isotherm with Pseudo-second order model	[71]

(Table 13) cont.....

Dye	Sorbent	Operation condition	Results	Model	References
Remazol Brilliant Violet-5R	Eggshell Waste	pH = 6.0 ± 0.2, T = 20 ± 2 °C, contact time = 30 min, C_{in} dye = 20 mg/L, 1.5 g biomass/100 mL, 160 µm, speed = 700 rpm,	95% sorption efficiency and sorption capacity of 1,3 mg/g	Langmuir isotherm, Pseudo-second order kinetic model.	[76]
Methyl red	*L. fusiformis* strain W1B6	pH = 5.5–8.5, T = 30 °C, contact time = 2 h, inoculum 10% (v/v), and C_{in} dye 100 mg/L, the specific growth rate = 0.273/h	Decolorization 96%,	Langmuir isotherm Pseudo first-order	[85]
Methylene Blue	*H. Rob* (CCLHR) and *Paspalum maritimum* (PMT)	pH = 7, constant agitation speed = 150 rpm	Biosorption capacity PMT 56.1798 mg/g and CCLHR 76.3359 mg/g	-	[74]
Thioflavin T (TT) and Methylene Blue (MB)	*Vesicularia dubyana*	pH = 6, T = 25°C, contact time = 1440 min, sorbent dose 0,5 g/L	Maximum sorption capacities were 119 ± 11 mg/g for TT and 229 ± 9 mg/g for MB.	Both TT and MB fitted Langmuir isotherm	[75]
methylene blue	White rot fungi dead	pH = 6, contact time 90 min, C_{in} dye = 50 mg/L, sorbent dose = 0.3 g	Max adsorption capacity is 23.69 mg/g.	Langmuir isotherm and Pseudo-second order	[60]
Rhodamine B	Microalgae *Chlorella pyrenoidosa*	pH = 8, T = 25 °C, time = 120 minutes, dye concentration = 100 mg/l, sorbent dose 0,4 g/L	Maximum capacities of 63.14, mg/g	Sips model Pseudo–second order model	[65]

(Table 13) cont.....

Dye	Sorbent	Operation condition	Results	Model	References
Methylene blue	*Raphanus raphanistrum*	pH = 6, T = 303 K, contact time = 75 min, C_{in} dye = 25 mg/L, bio sorbent dosage = 0.1g, and average bio sorbent size = 63 - 212 μm	98.8% of maximum biosorption efficiency	Langmuir isotherm Pseudo-second order kinetic model	[72]
Reactive Red 238 dye	*Corylus avellana* L	pH = 6.4, T = 30 °C, C_{in} dye = 200 mg/l, biosorbent mass = 0.1 g, shaking speed = 200 rpm	maximum dye biosorption capacity 74.527 mg/g	Langmuir Pseudo-second order model	[70]
Malachite green	Calcium alginate nanoparticles	pH = 10, T = 25 °C, contact time = 5 min and C_{in} dye = 150 mg/L, biomass dosage 0.08 mg/L	98.5% dye removal	Langmuir isotherm Pseudo-second order	[77]
Acid Blue 29	Pine sawdust	pH = 2.5, T = 28 °C, contact time = 2 h, C_{in} dye = 100 μg/mL,	Decolorization was up to 94.50%	Langmuir model and phytotoxicity test with L. sativa	[73]
Methylene Blue	Dead biomass of *Aspergillus fumigatus*	pH = 13, T = 30°C, contact time 120 min, C_{in} dye = 12 mg/L, ratio biosorbent/sol = 2 g/L	93,43% biosorption efficiency	Langmuir and Freundlich	[62]
Methyl orange	Fungus *Aspergillus flavus*	T = 30°C, pH = 5.5, contact time = 40 min, C biomass = 2g/L, C dye =1 ppm	Biosorption efficiency 53.62%	-	[63]
Acid Orange 7 Dye	*Ceratophylum demersum* (living and nonliving)	pH = 2, T = 25 °C, C_{in} dye = 90 mg/L, biomass dosage (living) = 9 g/L, (nonliving) 1 g/L	Maximum uptake capacities (living) 4 mg/g and (nonliving) 32 mg/g	Pseudo-second order kinetic model.	[86]

(Table 13) cont.....

Dye	Sorbent	Operation condition	Results	Model	References
Congo Red	Kola nut pod carbons	1.0 g of KPC from an aqueous solution of 100 mg/g at a stirring speed of 200 rpm and a contact time of 900 min.	Remove CR 40%	KPCs follows a first-order kinetics.	[87]
Basic red-12	Sewage sludge	T = 303 K, pH = 6, contact time = 2 h, biosorben dosage = 12.5 g/L, C dye =250 ppm	Adsorption capacity was found to be up to 295.85 mg/g	Langmuir isotherms pseudo first-order kinetics	[88]
Reactive Yellow 42 and Reactive Red 45	Citrus sinensis biomass	C reactive red = 150 mg/L, C reactive yellow 42 = 300 mg/L, biosorbent dose 0.05 g, pH = 2, time = 60 min.	Sorption capacity was 65.57 mg/g (reactive yellow 42) and 150 mg/g (reactive red 45)	Freundlich isotherm Pseudo-second-order rate	[69]
Burazol Blue ED	Dried anaerobic sludge	pH = 0.5, T = 25 °C, contact time = 75 min, C_{in} dye = 150 mg/L, biomass dosage = 0.4 g/L	maximum uptake capacities for BB dye were 127.5mg/g	The Langmuir isotherm Pseudo second-order	[67]
Rhodamine B	Anaerobic sludge	pH = 7.0, T = 38 °C, VSS of 3.0 g/l and C_{in} dye = 200 μmol/l	Biosorption capacity 15 mg/g	The Langmuir and Freundlich adsorption models Pseudo second-order	[68]

Bioaccumulation

Bioaccumulation is defined as an intracellular sorbate accumulation. This process consists of two steps, viz., biosorption followed by transport of sorbate into inside cells [89, 90]. Table **14** shows some studies in which dyes removal occurred *via* bioaccumulation.

Tabel 14. Review of Bioaccumulation studies.

Dye	Microorganism	Operation Condition	Results	References
Violet 90 metal-complex dye	Candida tropicalis	pH = 3, T = 30 ∘C, agitation speed = 140 rpm, contact time = 15 days, reducing sugar = 3.2 g/L and initial dye concentration = 711.1 mg/L	Maximum uptake was 56.28 mg/g	[91]
Acid red 18 and Reactive black 5	Schizophyllum commune and Trametes versicolor	pH = 2, T = 25 °C, 180 rpm and contact time = 72 hours	Uptake capacity was up to 82.1 mg/g for acid red 18, and 179.1 mg/g for reactive black 5 for *S. commune* and 76.1 mg/g (acid red 18) and 178.3 mg/g (reactive black 5) for *T. versicolor.*	[92]
Remazol Blue, Remazol Black B and Remazol Red RB,	Saccharomyces cerevisiae	pH = 3, T = 25 °C, rotary shaker = 100 rpm, contact time = 18 days, initial dye concentration 400 mg/L	S. cerevisiae cells bioaccumulated 62.0% of Remazol Black B, 11.6% of Remazol Blue dye and no Remazol Red RB	[93]

Biodegradation

Biodegradation of dye is a breakdown of complex dye molecules into simpler ones through the metabolism of a living organism. Complete biodegradation, known as mineralization, is a complete breakdown of organic molecules into carbon dioxide, water and/or any other inorganic end products [94].

Biodegradation of dyes may be mediated through different mechanisms, such as enzymes, chemical reduction by biogenic reductants like sulfide, low molecular weight redox mediators, or a combination of both [95]. Several enzymes that act in dye biodegradation include azo reductase, NADH-DCIP reductase, Lignin Peroxidase, Laccase, Polyphenol Oxidase (Tyrosinase), and Veratrol Alcohol Oxidase [95, 96].

Table **15** represents studies about removal of dye *via* biodegradation mechanisms.

Table 15. Review of removal of dye *via* biodegradation.

Dye	Microorganism	Operation Condition	Results	References
Malachite green	*Pseudomonas veronii*	pH = 7.1, T=32.4 °C inoculum = 2.5 ×107 cfu/mL, C_{in} dye 50 mg/L, contact time 7 days	Degrade 93.5% Degradation products: maleylacetate to produce CO_2 and water	[97]
Acid red 337	*Bacillus megaterium* KY848339.1 isolated from textile wastewater	pH = 7, T = 30°C C_{in} dye = 500 mg/ L contact time 24 h inoculum = 10% wt/v,	Remove 91% dye Degraded products: small aliphatic compounds and CO_2.	[98]
Novacron Super Black G	*Alcaligenes faecalis* AZ26 and Bacillus spp	T = 37 °C. pH = 8.0 C_{in} dye 200 mg/L, contact time 96 h	Decolorization 90%	[99]
Reactive Yellow F3R	*Aeromonas hydrophila* SK16	pH = 8, T = 37 °C, C_{in} dye = 100 mg/L, 5% inoculum, a combined aerobic–micro aerophilic process for 48 h (shaking at 120 rpm for initial 24 h followed by static condition until 48 h)	Degradation products: Cis 2 Butene, 1,3,5-Triazine, Aniline, Naphthalene	[21]
Drimaren Red CL-5B	*Aeromonas hydrophila* MTCC 1739 and *Lysinibacillus sphaericus* MTCC 9523	5% grown culture (24 h) in 100 mL nutrient broth at pH 8, T 37 °C, 100 mg/L dye, shaking (120 rpm for 24 h) cum static (microaerophilic) conditions for 72 h in 250 mL flask	• *Aeromonas hydrophila MTCC 1739* biodegraded: naphthalen-2-amine, 2-nitroso naphthalene, naphthalene, 1,3,5-triazine-2,4,6-triol, Sodium 2-amino-3-hydroxy benzene sulfonate, sodium (3E)-4-aminopenta- 1,3-diene-2-sulfonate, 2-aminophenol • *Lysinibacillus sphaericus MTCC 9523* biodegraded: 2-nitrosonaphthalene, Naphthalene, 1,3,5-triazinane-2,4-6-trione, 1,3,5-triazine-2,4(1H,3H)-dione.	[100]
Reactive Blue 160	Isolate *Microbacterium* sp. B12	T= 35 ± 2°C and pH = 5.0, time 24 h, C dye 250 mg/L (Microaerobic)	96% decolourization of RB160	[101]

Textile Dyes Removal   *Sustainable Materials, Vol. 1* **163**

(Table 15) cont.....

Dye	Microorganism	Operation Condition	Results	References
C.I. Reactive Blue 172	*Providencia rettgeri* Strain HSL1	C dye = 50 mg/L, time = 20 h, T = 30°C (microaerophilic)	Decolorize 100% Biodegradation of RB 172: 4-(ethenylsulfonyl) aniline and 1-amino-1-(4-aminophenyl) propan-2-one.	[102]
Reactive Black B	*Morganella* sp. HK-1	C dye = 20 g/L, T= 30°C, time = 24 h (static conditions)	Biodegradation products: Disodium 3,4,6-triamino-5-hydroxynaphthalene-2,7-disulfonate, 4-aminophenylsulfonylethyl hydrogen sulfate, naphthalene-1-ol, aniline and benzene	[103]
Reactive Black	*Aeromonas* spp	static conditions with 350 mL cultures of MMR (pH = 7) C dye = 100 mg/L (Incubated microaerophilic) T = 30°C, time = 168 h/until no colour.	Biodegradation products: α-ketoglutaric acid with transient accumulation of 4-aminobenzenesulphonic acid (sulphanilic acid), 4-amino, 3-hydronapthalenesulphonic acid and 4-amino,--hydronapthalene 2,7 disulphonic acid .	[104]
Congo red	*Shewanella xiamenensis* BC01	Incubated statically under anaerobic at T = 37°C, time = 8 h	Biodegradation products: 4,4′-diamino-1,1′- biphenyl and 1,2′-diamino naphthalene 4-sulfonic acid	[105]
Crystal violet	*Pseudomonas aeruginosa, Clostridium perfringens* and *Proteus vulgaris* in organic and inorganic media	-	Biodegradation products: N,N,N,N "-tetramethylpararosa niline, [N;N-dimethylamino phenyl][N-methylaminophenyl] benzophenone, 4-methyl amino phenol, N; N- dimethyl amino benzaldehyde, and phenol.	[106]
Congo Red	*Bacillus thuringiensis* RUN1	Biomass = 5mg/L, $C_{dye-broth}$ =100 mg/L at static anoxic condition.	Biodegradation products: sodium (4-amino-3-diazenyl naphthalene-1-sulfonate), phenylbenzene, 3-diazonium naphthalen-1-ol, sodium phenyl phenoxide 4-(4-oxocyclohe-a-1,5,dienyl)cyclohexa-4-ene-1,2,-dione, 2-(1-amino-2-diazenyl-2-formylvinyl) benzoic acid	[107]
Acid Orange 7	*Bacillus cereus* (MTCC 9777) RMLAU1	100 mg/L dye, pH = 8.0, T = 33 °C, 3.0% inoculum, incubation time = 96 h (static culture condition)	Degraded products: sulfanilic acid	[108]
Acid Orange 7	*Enterococcus faecalis* strain ZL	C_{in} dye = 100 mg/L, T =37 °C. (yeast extract=0.1%, glycerol = 0.1%, inoculum = 2.5%)	Decolorization efficiency 98% was achieved in only 5 h Degraded products: sulfanilic acid	[109]

(Table 15) cont.....

Dye	Microorganism	Operation Condition	Results	References
Remazol red	*Pseudomonas aeruginosa* BCH	T = 40 °C, pH = 7, contact time = 20 min, C dye = 50 mg/L	Degraded up to 97% Biodegradation products: 2-[(3-diazenylphenyl) Sulfonyl] ethanesulfonate, Naphthalene-1-o, aniline, 4-choloro-1,2,3-triazi-e-2-amine, 2-Chloro-1,2,3 triazine	[110]
Methyl red	isolated *Sphingomonas paucimobilis*	T = 30 °C, pH = 9, C_{in} dye = 750 ppm time contact = 10 h	99.63% decolorization	[21]
Methyl red	*Brevibacillus laterosporus* MTCC 2298	Time contact = 12 h, C dye, under static condition	93% degradation Degraded products: heterocyclic substituted aryl amine (m/z 281), p-(N,N di formyl)-substituted para-di amino benzene derivative (m/z 355) and p-di-amino benzene derivative (m/z 282)	[111]
Reactive Black-5	*Shewanella putrefaciens* strain AS96	Incubated T = 35°C (static conditions)	Biodegradation products: 1-amino-2-naphthol-4-sulfonate, sulfanilic acid, and nitroaniline	[112]
Reactive Black 5	*Rhodopseudomonas palustris* W1	-	Biodegradation products: such as TAHNDSDP1,TAHNDS, and Catechol.	[113]

FUTURE PROSPECTS FOR APPLICATION

Further research work should focus on: how to reduce limitation factors of microbial activities, a specific type of biodegradation agents that are responsible for dye degradation, biodegradation pathway to understand more about biodegradation mechanisms, and environmental factors. Moreover, due to the potential formation of more toxic by-products of dyes degradation (i.e: aromatic amines), it is important to determine the nature of the degradation by-products and to establish their (non) toxicity to the environment. The development of new species of microorganisms through genetic engineering should also be considered since many microbial activities are resisted by dyes toxicity. Complete degradation of textile dyes are still challenged, therefore, for successful application required extensive research studies. Table **16** presents the current studies on biodegradation of textile dyes by different biodegradation agents.

Table 16. Review of current biological treatment technology using single/mix culture microorganisms, algae and enzyme.

No.	System	Microbes/Enzyme	Dyes wastewater	Performance	Suggests/Prospect	Reference
1	Aerobic Membrane bioreactor (ABR) lab's scale Seed: mix cultures with Concentration 4 and 8 g/L. V reactor = 10 L, operated in 4 working periods (100 days)	Mix cultures	Synthetic wastewater containing The reactive Blue Bezactiv S-GLD 150 (BB150) (4-amino-5-hydroxy-3,6-bis ((4 - ((2 - (sulfonatooxy) ethyl) sulfonyl) phenyl) azo)-naphthalene 2,7-disulfonate)	COD removal 80-90% in 8g/L^{-1} MLVSS	Effluent was suitable for reuse in the dyeing process	[17]
2	Pure culture isolates (*Alcaligenes faecalis AZ26, Bacillus cereus AZ27 and Bacillus sp. AZ28*), HRT = 48 h.	Inoculum of *Alcaligenes faecalis AZ26, Bacillus cereus AZ27 and Bacillus sp. AZ28*	Textile reactive dyes:	decolorization was up to 95 and 85% for *B. cereus* AZ27 and *Bacillus sp.* AZ28, *B. cereus* AZ27 and *Bacillus sp.* AZ28 decolorized 65–85% and 55–75%, respectively *A. faecalis* AZ26 decolorized dyes up to (~78%)	M.O in this study have potential for practical application, but need to be further studied.	[99]
3	Pure culture (*Pseudomonas aeruginosa* (RS1), *Thiosphaera pantotropha* (ATCC 35512). Dye concentration = 50-250 mg/L. Time interval 18 h, 24 h, 48 h, 72 h and 96 h, incubated in rotary flash	*Pseudomonas aeruginosa* (RS1), *Thiosphaera pantotropha* (ATCC 35512	Dye reactive yellow 145 (RY145)	50 mg/L RY145 decolorization could be achieved within 96 h and 72 h, for *Pseudomonas aeruginosa* (RS1) and *Thiosphaera pantotropha*.	a good potential for dye decolorization and also promoted partial mineralization	[114]

(Table 16) cont.....

No.	System	Microbes/Enzyme	Dyes wastewater	Performance	Suggests/Prospect	Reference
4	Upflow anaerobic sludge bed (UASB) - aerobic treatment integrated system, Reactor Volume = 5 litter for both, HRT = 24, 12, 6, and 3 h for over 4 months.	UASB= mesophilic granular sludge activated sludge = domestic WWTP	2-Naphthol Red dye	COD removal in UASB = 85.6%, COD removal in Activated Sludge = 92.4%. OLR = 2.97 g COD /L.d 98.4% of decolourization	Suitable for full-scale applications, need further trial	[23]
5	Hybrid anaerobic-aerobic biological treatment technology. Real wastewater = COD 2200-2800 mg/L). HRT anaerobic SBR = 48 h, aerobic SBR = 6 h. MLSS 6000 mg/L-8000 mg/L.	Activated sludge seed adapted to the synthetic and real textile wastewater	Synthetic wastewater COD≈3000 mg/L, and real textile wastewater	COD (99.5%), TKN (99.3%) and color (78.4%). Sludge adaptation is a key role in improving the reactor performance	A hybrid biological system, constituting anaerobic SBR and aerobic SBR.	[115]
6	Aerobic Batch system, V 0.5 L, with mix cultures inoculated with basal medium (270 mg-VSS L-1).	A modified selective cultures.	Direct Black 22 (DB22),	nA	intermittent aeration is effective to improve color removal	[19]
7	Lab-scale sequential anaerobic–aerobic treatment system, V = 5 L, HRT was maintained at 9 h.	Activated sludge mix cultures were taken from return sludge (RAS)	Synthetic wastewater contained Acid Red 18 (AR18)	>85% of influent COD was removed at all times,	Prospective to full scale application	[116]
8	Lab scale experiment using 5% pre-grown culture (24 h) in 100mL nutrient broth at pH 8, 37 °C, Dye concentration was 100 mg/L, shaking for 120 rpm for 24 h,	*Aeromonas hydrophila* MTCC 1739 and *Lysinibacillus sphaericus* MTCC 9523 non-adapted bacteria	Textile azo dye Drimaren Red CL-5B	*A. hydrophila* could decolorized dye to maximum 91.96%, *L. sphaericus* remove ddye up to 88.35% at max HRT 72 h	*A. hydrophila* MTCC 1739 and *L. sphaericus* MTCC 9523 can be used for decolorization and biodegradation of azo dye containing textile wastewater.	[21]

(Table 16) cont.....

No.	System	Microbes/Enzyme	Dyes wastewater	Performance	Suggests/Prospect	Reference
9	Laboratory scale experiment with volume 1 L. Using sequencing anaerobic-aerobic in an alternating anaerobic 24 hours /aerobic 24 hours system, settling 1 min, decanting 1 min, 50% return ratio.1 min feeding, and 7 min idling.	Domestic WWTP activated sludge mix culture	Mordant Orange 1	dye mineralization of 61 ± 2%, dye removal anaerobically 88 ± 1%, and an aerobic aromatic amines removal of 70 ± 3%, retention time 48 hours	could be a possible alternative to the conventional activated sludge process	[117]
10	Tests were performed in sequencing batch reactors (SBR), V = 1 L Biomass 3.10-3.88 gVSS/L; dye load 0.005-0.01kgCOD/ m³d.	Mix culture from domestic WWTP Activated sludge	dyeing bath Wastewater contained Remazol Black 5, Remazol Yellow RR and Remazol Brilliant Red 21	color removal efficiency = 70-80%, and COD removal > 96%.	ready for application and can be easily implemented in the full-scale textile WWTP	[118]
11	The experiment was carried out in bubble-column SBRs with a working volume of 1.5 L. The volumetric organic loading rate (OLR) being 201 kgCOD m⁻³d⁻¹.	Aerobic granular sludge from Municipal WWTP	Synthetic wastewater containing Acid Red 14	Dye removal above 90%, COD removal 80-90%, with an average of 55%.	SRT flexibility in AGS creates various microbial population	[119]
12	1 L Schott Duran glass bottles inoculated with 0.68 g mixed liquor suspended solids (MLSS). AGS operated in SBR mode with 72 h cycle for 16 cycles.	Activated sludge	Synthetic wastewater with Yellow dye (YD) Reactive yellow 15	Azo dye, total organic carbon and ammoniacal nitrogen removal efficiencies of 89 - 100%, 79 - 95% and 92 - 100%, respectively. reactor operated for 80 days	This technology is considered to be promising	[120]

(Table 16) cont.....

No.	System	Microbes/Enzyme	Dyes wastewater	Performance	Suggests/Prospect	Reference
13	The SBR with an effective volume of 4 L, cycle time of 6 h 4 min filling, 60 min anaerobic, 286–290 min aerobic, 3–7 min settling and 3 min of effluent discharge, volumetric exchange ratio (VER) = 71%.	Activated sludge	Real azo dyeing wastewater	73% color removal and 68% COD removal with a cycle time of 24 h	Further study is needed to overcome the instability of the system	[121]
14	a sequential anaerobic-aerobic biotreatment reactor.	Activated sludge collected from sewage treatment plant	azo dye reactive red (RR2)	96%, 93%, and 82% maximum COD removals were observed with samples containing RR2 at the initial concentration of 10, 20, and 40 mg/L	-	[1]
15	Bioreactor working volume = 3.5 L,	Activated sludge from municipal wastewater treatment plant	Reactive Orange 16 (RO16)	COD removal 97% was achieved after 2 weeks	MBBR technology was a promising technology to treat dye containing wastewater	[20]
16	YPD (yeast extract, peptone, and dextrose medium contained 50 mg/l carmoisine) was inoculated with 5% (v/v) and incubated anaerobically at 30 °C under both static and shaking conditions.	Saccharomyces cerevisiae ATCC 9763	Azo dye contain carmoisine	carmoisine (50 mg/l) was eliminated after 7 h of incubation under anaerobic	Saccharomyces cerevisiae ATCC 9763 has great potential for use in the treatment of wastewater	[122]
17	Experiment was carried out in Erlenmeyer flasks under microaerobic conditions. inoculum was 10% (v/v), azo dye concentration was 600 mg L^{-1}, incubated at 55°C for 48 h.	thermophilic microflora	azo dye — Direct Black G (DBG)	decolorization ratio could reach 97% with 8 h of incubation at optimal conditions	Thermophilic microflora revealed a practical application potential in the treatment of azo-dyes wastewater	[123]

(Table 16) cont.....

No.	System	Microbes/Enzyme	Dyes wastewater	Performance	Suggests/Prospect	Reference
18	CW-MFC bioreactors with total working volume of 5 L HRT = 1 day.	mixed-culture sludge from wastewater treatment plant of glove manufacturing company	Acid Red 18 (AR18), Acid Orange 7 (AO7) and Congo Red (CR)	The highest decolourisation rate was achieved by Acid Red 18 (AR18), 96%, followed by Acid Orange 7 (AO7), 67% and Congo Red (CR), 60%.	wetland plants helped to reduce the internal resistance and improve power performance	[124]
19	Experiments were performed in 250 mL shaking flasks with 100 mL culture medium. ARB with the inoculation size of 6% (v/v), incubated at 30 °C and 160 rpm for 24 h.	halotolerant yeast Candida tropicalis SYF-1	Acid Red B (ARB)	more than 95.0% of 50 mg/L, within 12 h	-	[22]
20	10 g L-1 yeast extract (YE), 5 g L-1 NaCl and 20 mg L-1 AR14 were mixed. Samples were collected after 24 and 48 h of incubation.	SRB mix with sludge from activated sludge WWTP	Synthetic textile wastewater azo dye Acid Red 14 (AR14)	*Oerskovia paurometabola* could decolorize 91% after 24 h in static anaerobic culture	Promising technology but need process scale-up for treating azo dye containing textile wastewaters.	[125]
21	Purified laccase from Weissella viridescens LB37 was immobilized to the Fe_3O_4–chitosan NPs. Dyes Concentration was 50 mg/L	laccase enzyme from Weissella viridescens LB37.	Reactive Black 5, Direct blue 15, and Evans blue	efficiency of in the range of 94–96%, under pH 6, contact time of 60 min, and at a temperature of 30 °C	The IMB-LAC enzyme is effective for synthetic azo dyes wastewater removal	[25]
22	Volume reactor 250 ml, at 45 °C. dye concentration of 10-50mg/L and nanobiocatalyst dosage (60.0-300.0 mg/L)	Laccase from genetically modified *Aspergillus*	Anionic dyes (*i.e.* Direct Red 23 (DR23) and Acid Blue 92 (AB92))	decolorization effectiveness of the nanobiocatalyst was more than 75% for both dyes	Nanobiocatalyst is one technology that should be explored further	[126]

(Table 16) cont.....

No.	System	Microbes/Enzyme	Dyes wastewater	Performance	Suggests/Prospect	Reference
23	Batch Experiments with variable parameters such as loading amount of enzyme, solution pH, temperature, reaction time and dye concentration on the dye decolorization were conducted	Horseradish peroxidase (HRP)	Acid Blue 113 and Acid black 10 BX	Decolorized 95.4% Acid blue 113 and 90.3% Acid black 10 BX within 35 min at 30 °C.	The immobilized HRP was proved to be an effective biocatalyst in the decolorization of azo dyes.	[127]
24	Batch experiment with V = 50 mL, using reactive azo dyes = 500 mg /L, yeast extract, sucrose and dextrose. The pH was adjusted to 7.0± 0.2, autoclave at 121∘C for 15 minutes, inoculated with 5 mL of fungal inoculums	*Trichoderma viride,Trichoderma koningii, Trichoderma harzianum, Aspergillus niger, Aspergillus flavus,* and *Fusariumoxysporum*	five azo dyes: Red HE7B, Reactive Violet-5, Red Black-B, Light Navy Blue HEG, Dark NavyBlue H2GP)	*T. koningii* was recognized as the best decolorizer based on the average decolorization rate of five azo dyes.	Potential of *H. koningii* needs to be demonstrated for its application in the treatment of real dye bearing wastewater	[128]

*nA: not available

CONCLUSION

The occurrence of toxic and carcinogenic compounds in textile dyes creates aesthetic problems and affects the aquatic ecosystem. Dyes Biological treatment technology is a preferable technology to remove textile dyes from aqueous environment. Biological treatments are proven to be low cost, produce less sludge and need lesser amount of chemicals. However, the conversion of non-toxic dyes into toxic metabolites under an anaerobic environment has caused bottlenecks toward the development of the more efficient biological treatment technology. To date, there are three biological agents that are proven to be able to treat textile dyes such as microorganisms, algae and enzyme, and they form either as single or mixed cultures. To get a maximum effectivity and efficiency for full-scale application, in terms of pollutants removal performance, robustness and costs, biological treatments should be integrated with other technology, *i.e.*: physio-chemical, chemical or biology.

CONSENT FOR PUBLICATION

Not applicable.

CONFLICT OF INTEREST

The authors declare no conflict of interest, financial or otherwise.

ACKNOWLEDGEMENTS

This book chapter is supported by Research Centre for Environmental and Clean Technology, National Research and Innovation Agency Republic of Indonesia.

REFERENCES

[1]   Hameed BB, Ismail ZZ. Decolorization, biodegradation and detoxification of reactive red azo dye using non-adapted immobilized mixed cells. Biochem Eng J 2018; 137: 71-7.
[http://dx.doi.org/10.1016/j.bej.2018.05.018]

[2]   Przystaś W, Zabłocka-Godlewska E, Grabińska-Sota E. Efficiency of decolorization of different dyes using fungal biomass immobilized on different solid supports. Braz J Microbiol 2018; 49(2): 285-95.
[http://dx.doi.org/10.1016/j.bjm.2017.06.010] [PMID: 29129408]

[3]   Haji A, Naebe M. Cleaner dyeing of textiles using plasma treatment and natural dyes: A review. J Clean Prod 2020; 265: 121866.
[http://dx.doi.org/10.1016/j.jclepro.2020.121866]

[4]   Yurtsever A, Sahinkaya E, Çınar Ö. Performance and foulant characteristics of an anaerobic membrane bioreactor treating real textile wastewater. J Water Process Eng 2020; 33(November): 101088.
[http://dx.doi.org/10.1016/j.jwpe.2019.101088]

[5]   Kurade MB, Waghmode TR, Xiong JQ, Govindwar SP, Jeon BH. Decolorization of textile industry effluent using immobilized consortium cells in upflow fixed bed reactor. J Clean Prod 2019; 213: 884-91.
[http://dx.doi.org/10.1016/j.jclepro.2018.12.218]

[6]   Bilińska L, Blus K, Foszpańczyk M, Gmurek M, Ledakowicz S. Catalytic ozonation of textile wastewater as a polishing step after industrial scale electrocoagulation. J Environ Manage 2020; 265(April): 110502.
[http://dx.doi.org/10.1016/j.jenvman.2020.110502] [PMID: 32275237]

[7]   Ferreira ESB, Hulme AN, McNab H, Quye A. The natural constituents of historical textile dyes. Chem Soc Rev 2004; 33(6): 329-36.
[http://dx.doi.org/10.1039/b305697j] [PMID: 15280965]

[8]   Benkhaya S. S. M' rabet, and A. El Harfi, "A review on classifications, recent synthesis and applications of textile dyes,". Inorg Chem Commun 2020; 115: 107891.
[http://dx.doi.org/10.1016/j.inoche.2020.107891]

[9]   Berradi M, Hsissou R, Khudhair M, *et al.* Textile finishing dyes and their impact on aquatic environs. Heliyon 2019; 5(11): e02711.
[http://dx.doi.org/10.1016/j.heliyon.2019.e02711] [PMID: 31840123]

[10]   Cooper P. Colour in Dyehouse Effluent. Society of Dyers and Colourists, 1995: p. 200.

[11]   Hou M, Baughman GL, Perenich TA. Estimation of water solubility and octanol/water partition coefficient of hydrophobic dyes. Dyes Pigments 1991; 16(4): 291-7.
[http://dx.doi.org/10.1016/0143-7208(91)85018-4]

[12]   Brüschweiler BJ, Küng S, Bürgi D, Muralt L, Nyfeler E. Identification of non-regulated aromatic amines of toxicological concern which can be cleaved from azo dyes used in clothing textiles. Regul Toxicol Pharmacol 2014; 69(2): 263-72.

[http://dx.doi.org/10.1016/j.yrtph.2014.04.011] [PMID: 24793261]

[13] Samsami S, Mohamadizaniani M, Sarrafzadeh M-H, Rene ER, Firoozbahr M. Recent advances in the treatment of dye-containing wastewater from textile industries: Overview and perspectives. Process Saf Environ Prot 2020; 143: 138-63.
[http://dx.doi.org/10.1016/j.psep.2020.05.034]

[14] Núñez J, Yeber M, Cisternas N, Thibaut R, Medina P, Carrasco C. Application of electrocoagulation for the efficient pollutants removal to reuse the treated wastewater in the dyeing process of the textile industry. J Hazard Mater 2019; 371(November 2018): 705-11.
[http://dx.doi.org/10.1016/j.jhazmat.2019.03.030]

[15] Santos DHS, Duarte JLS, Tavares MGR, et al. Electrochemical Degradation and Toxicity Evaluation of Reactive Dyes Mixture and Real Textile Effluent Over DSA® Electrodes. Chemical Engineering and Processing - Process Intensification 2020; 153: 107940.

[16] Khosravi A, Karimi M, Ebrahimi H, Fallah N. Sequencing batch reactor/nanofiltration hybrid method for water recovery from textile wastewater contained phthalocyanine dye and anionic surfactant. J Environ Chem Eng 2020; 8(2): 103701.
[http://dx.doi.org/10.1016/j.jece.2020.103701]

[17] Khouni I, Louhichi G, Ghrabi A. Assessing the performances of an aerobic membrane bioreactor for textile wastewater treatment Influence of dye mass loading rate and biomass concentration. Process Safety and Environmental Protection 2020; 135: 364-82.

[18] Sathishkumar K, AlSalhi MS, Sanganyado E, Devanesan S, Arulprakash A, Rajasekar A. Sequential electrochemical oxidation and bio-treatment of the azo dye congo red and textile effluent. J Photochem Photobiol B 2019; 200(September): 111655.
[http://dx.doi.org/10.1016/j.jphotobiol.2019.111655] [PMID: 31655456]

[19] Oliveira JMS, de Lima e Silva MR, Issa CG, Corbi JJ, Damianovic MHRZ, Foresti E. Intermittent aeration strategy for azo dye biodegradation: A suitable alternative to conventional biological treatments? J Hazard Mater 2020; 385: 121558.
[http://dx.doi.org/10.1016/j.jhazmat.2019.121558] [PMID: 31732337]

[20] Ong C, Lee K, Chang Y. Biodegradation of mono azo dye-Reactive Orange 16 by acclimatizing biomass systems under an integrated anoxic-aerobic REACT sequencing batch moving bed biofilm reactor. J Water Process Eng 2020; 36(March): 101268.
[http://dx.doi.org/10.1016/j.jwpe.2020.101268]

[21] Srinivasan S, Sadasivam SK. Exploring docking and aerobic-microaerophilic biodegradation of textile azo dye by bacterial systems. J Water Process Eng 2018; 22(February): 180-91.
[http://dx.doi.org/10.1016/j.jwpe.2018.02.004]

[22] Tan L, Shao Y, Mu G, Ning S, Shi S. Enhanced azo dye biodegradation performance and halotolerance of Candida tropicalis SYF-1 by static magnetic field (SMF). Bioresour. Technol. J. 2020; Vol. 295.
[http://dx.doi.org/10.1016/j.biortech.2019.122283]

[23] Gadow SI, Li YY. Development of an integrated anaerobic/aerobic bioreactor for biodegradation of recalcitrant azo dye and bioenergy recovery: HRT effects and functional resilience. Bioresour Technol Rep 2020; 9(January): 100388.
[http://dx.doi.org/10.1016/j.biteb.2020.100388]

[24] Holkar CR, Jadhav AJ, Pinjari DV, Mahamuni NM, Pandit AB. A critical review on textile wastewater treatments: Possible approaches. J Environ Manage 2016; 182: 351-66.
[http://dx.doi.org/10.1016/j.jenvman.2016.07.090] [PMID: 27497312]

[25] Nadaroglu H, Mosber G, Gungor AA, Adıguzel G, Adiguzel A. Biodegradation of some azo dyes from wastewater with laccase from Weissella viridescens LB37 immobilized on magnetic chitosan nanoparticles. J Water Process Eng 2019; 31(December): 100866.
[http://dx.doi.org/10.1016/j.jwpe.2019.100866]

[26]   El-Sheekh MM, Gharieb MM, Abou-El-Souod GW. Biodegradation of dyes by some green algae and cyanobacteria. Int Biodeterior Biodegradation 2009; 63(6): 699-704.
[http://dx.doi.org/10.1016/j.ibiod.2009.04.010]

[27]   Kumar N, Sinha S, Mehrotra T, Singh R, Tandon S, Thakur IS. Biodecolorization of azo dye Acid Black 24 by *Bacillus pseudomycoides*: Process optimization using Box Behnken design model and toxicity assessment. Bioresour Technol Rep 2019; 8(July): 100311.
[http://dx.doi.org/10.1016/j.biteb.2019.100311]

[28]   Xiao H, Zhao T, Li CH, Li MY. Eco-friendly approaches for dyeing multiple type of fabrics with cationic reactive dyes. J Clean Prod 2017; 165: 1499-507.
[http://dx.doi.org/10.1016/j.jclepro.2017.07.174]

[29]   Jalandoni-Buan AC, Decena-Soliven ALA, Cao EP, Barraquio VL, Barraquio WL. Characterization and identification of Congo red decolorizing bacteria from monocultures and consortia. Philipp J Sci 2010; 139(1): 71-8.

[30]   Phalakornkule C, Polgumhang S, Tongdaung W, Karakat B, Nuyut T. Electrocoagulation of blue reactive, red disperse and mixed dyes, and application in treating textile effluent. J Environ Manage 2010; 91(4): 918-26.
[http://dx.doi.org/10.1016/j.jenvman.2009.11.008] [PMID: 20042267]

[31]   Fang S, Feng G, Guo Y, Chen W, Qian H. Synthesis and application of urethane-containing azo disperse dyes on polyamide fabrics. Dyes Pigments 2020; 176: 108225.
[http://dx.doi.org/10.1016/j.dyepig.2020.108225]

[32]   Qiu J, Tang B, Ju B, Zhang S, Jin X. Clean synthesis of disperse azo dyes based on peculiar stable 2,6-dibromo-4-nitrophenyl diazonium sulfate. Dye Pigment 2020; 173(September): 107920.
[http://dx.doi.org/10.1016/j.dyepig.2019.107920]

[33]   Aspland JR. A series on dyeing. Chapter 8: disperse dyes and their application to polyester. Text Chem Color 1992; 24(12): 8-13.

[34]   Sirianuntapiboon S, Chairattanawan K, Jungphungsukpanich S. Some properties of a sequencing batch reactor system for removal of vat dyes. Bioresour Technol 2006; 97(10): 1243-52.
[http://dx.doi.org/10.1016/j.biortech.2005.02.052] [PMID: 16023339]

[35]   Burkinshaw SM, Son YA. The dyeing of supermicrofibre nylon with acid and vat dyes. Dyes Pigments 2010; 87(2): 132-8.
[http://dx.doi.org/10.1016/j.dyepig.2010.03.009]

[36]   Silkstone K. The Influence of Polymer Morphology on the Dyeing Properties of Synthetic Fibres. Rev Prog Color Relat Top 1982; 12(1): 22-30.
[http://dx.doi.org/10.1111/j.1478-4408.1982.tb00221.x]

[37]   Al-Duri B, McKay G. Prediction of binary system for kinetics of batch adsorption using basic dyes onto activated carbon. Chemical Engineering Science 1991; 46(1): 193-204.

[38]   Wu J, Li Q, Li W, *et al.* Efficient removal of acid dyes using permanent magnetic resin and its preliminary investigation for advanced treatment of dyeing effluents. J Clean Prod 2020; 251: 119694.
[http://dx.doi.org/10.1016/j.jclepro.2019.119694]

[39]   Patil S, Renukdas S, Patel N. Removal of methylene blue, a basic dye from aqueous solutions by adsorption using teak tree (*Tectona grandis*) bark powder Int J Environ Sci 2011; 1: 711-26.

[40]   Nguyen TA, Juang RS. Treatment of waters and wastewaters containing sulfur dyes: A review. Chem Eng J 2013; 219: 109-17.
[http://dx.doi.org/10.1016/j.cej.2012.12.102]

[41]   Sarkar S, Banerjee A, Chakraborty N, Soren K, Chakraborty P, Bandopadhyay R. Structural-functional analyses of textile dye degrading azoreductase, laccase and peroxidase: A comparative in silico study. Electron J Biotechnol 2020; 43: 48-54.

[http://dx.doi.org/10.1016/j.ejbt.2019.12.004]

[42]   Benkhaya S, M'rabet S, El Harfi A. Classifications, properties, recent synthesis and applications of azo dyes. Heliyon 2020; 6(1): e03271.
[http://dx.doi.org/10.1016/j.heliyon.2020.e03271] [PMID: 32042981]

[43]   Brüschweiler BJ, Merlot C. Azo dyes in clothing textiles can be cleaved into a series of mutagenic aromatic amines which are not regulated yet. Regul Toxicol Pharmacol 2017; 88: 214-26.
[http://dx.doi.org/10.1016/j.yrtph.2017.06.012] [PMID: 28655654]

[44]   Zhang S, Ma W, Ju B, *et al.* Continuous dyeing of cationised cotton with reactive dyes. Color Technol 2005; 121(4): 183-6.
[http://dx.doi.org/10.1111/j.1478-4408.2005.tb00270.x]

[45]   Ujiie H. Fabric Finishing: Printing Textiles. In: Sinclair R. Textile and Fashion. Materials, Design and Technology. Elsevier Ltd 2015: 507-29.
[http://dx.doi.org/10.1016/B978-1-84569-931-4.00020-9]

[46]   O'Neill C, Hawkes FR, Hawkes DL, Lourenço ND, Pinheiro HM, Delée W. Colour in textile effluents - sources, measurement, discharge consents and simulation: a review. J Chem Technol Biotechnol 1999; 74(11): 1009-18.
[http://dx.doi.org/10.1002/(SICI)1097-4660(199911)74:11<1009::AID-JCTB153>3.0.CO;2-N]

[47]   Xin B, Chen G, Zheng W. Bioaccumulation of Cu-complex reactive dye by growing pellets of *Penicillium oxalicum* and its mechanism. Water Res 2010; 44(12): 3565-72.
[http://dx.doi.org/10.1016/j.watres.2010.04.004] [PMID: 20421123]

[48]   Sadettin S, Dönmez G. Bioaccumulation of reactive dyes by thermophilic cyanobacteria. Process Biochem 2006; 41(4): 836-41.
[http://dx.doi.org/10.1016/j.procbio.2005.10.031]

[49]   Clarke EA, Anliker R. Safety in Use of Organic Colorants: Health and Safety Aspects. Rev Prog Color Relat Top 1984; 14(1): 84-9.
[http://dx.doi.org/10.1111/j.1478-4408.1984.tb00048.x]

[50]   Rawat D, Mishra V, Sharma RS. Detoxification of azo dyes in the context of environmental processes. Chemosphere 2016; 155: 591-605.
[http://dx.doi.org/10.1016/j.chemosphere.2016.04.068] [PMID: 27155475]

[51]   Hitam C N C, Jalil A A. A review on exploration of Fe_2O_3 photocatalyst towards degradation of dyes and organic contaminants 2020.
[http://dx.doi.org/10.1016/j.jenvman.2019.110050]

[52]   Li W, Mu B, Yang Y. Feasibility of industrial-scale treatment of dye wastewater *via* bio-adsorption technology. Bioresour Technol 2019; 277(January): 157-70.
[http://dx.doi.org/10.1016/j.biortech.2019.01.002] [PMID: 30638884]

[53]   Bilińska L, Gmurek M, Ledakowicz S. Textile wastewater treatment by AOPs for brine reuse. Process Saf Environ Prot 2017; 109: 420-8.
[http://dx.doi.org/10.1016/j.psep.2017.04.019]

[54]   Mohammadi Galangash M, Mohaghegh Montazeri M, Ghavidast A, Shirzad-Siboni M. Synthesis of carboxyl-functionalized magnetic nanoparticles for adsorption of malachite green from water: Kinetics and thermodynamics studies. J Chin Chem Soc (Taipei) 2018; 65(8): 940-50.
[http://dx.doi.org/10.1002/jccs.201700361]

[55]   Jung KW, Choi BH, Ahn KH, Lee SH. Synthesis of a novel magnetic $Fe3O4/\gamma$-Al2O3 hybrid composite using electrode-alternation technique for the removal of an azo dye. Appl Surf Sci 2017; 423: 383-93.
[http://dx.doi.org/10.1016/j.apsusc.2017.06.172]

[56]   Shanmugam S, Ulaganathan P, Swaminathan K, Sadhasivam S, Wu YR. Enhanced biodegradation and detoxification of malachite green by Trichoderma asperellum laccase: Degradation pathway and

product analysis. Int Biodeterior Biodegradation 2017; 125: 258-68.
[http://dx.doi.org/10.1016/j.ibiod.2017.08.001]

[57]   Song Y, Hu Q, Sun Y, *et al.* The feasibility of UF-RO integrated membrane system combined with coagulation/flocculation for hairwork dyeing effluent reclamation. Sci Total Environ 2019; 691: 45-54.
[http://dx.doi.org/10.1016/j.scitotenv.2019.07.130] [PMID: 31306876]

[58]   N de C L Beluci et al, "Hybrid treatment of coagulation/flocculation process followed by ultrafiltration in TIO 2 -modified membranes to improve the removal of reactive black 5 dye. Sci. Total Environ 2019; Vol. 664: pp. 222-9.
[http://dx.doi.org/10.1016/j.scitotenv.2019.01.199]

[59]   Fomina M, Gadd GM. Biosorption: current perspectives on concept, definition and application. Bioresour Technol 2014; 160: 3-14.
[http://dx.doi.org/10.1016/j.biortech.2013.12.102] [PMID: 24468322]

[60]   Abbas SH, Mohammed AA, Ali WH. Biosorption of Methylene Blue from Aqueous Solution Using Wastes Micelium of Fungal Biomass Type White Rot Fungi. J Eng Sustain Dev 2018; 22(4): 1-11.
[http://dx.doi.org/10.31699/IJCPE.2018.4.1]

[61]   Saraf S, Vaidya VK. Comparative Study of Biosorption of Textile Dyes Using Fungal Biosorbents. Int J Curr Microbiol Appl Sci 2015; 2(2): 357-65.

[62]   Kabbout R, Taha S. Biodecolorization of textile dye effluent by biosorption on fungal biomass materials. Phys Procedia 2014; 55: 437-44.
[http://dx.doi.org/10.1016/j.phpro.2014.07.063]

[63]   Mahmooda Takey , Shaikh T, Mane N, Majumder DR. Bioremediation of Xenobiotics: Use of Dead Fungal Biomass As Biosorbent. Int J Res Eng Technol 2014; 3(1): 565-70.
[http://dx.doi.org/10.15623/ijret.2014.0301094]

[64]   Bayazıt G, Tastan BE, Gül D. Biosorption, isotherm and kinetic properties of common textile dye by phormidium animale. Glob NEST J 2019; 22(1): 1-7.
[http://dx.doi.org/10.30955/gnj.002984]

[65]   da Rosa ALD, Carissimi E, Dotto GL, Sander H, Feris LA. Biosorption of rhodamine B dye from dyeing stones effluents using the green microalgae *Chlorella pyrenoidosa*. J Clean Prod 2018; 198: 1302-10.
[http://dx.doi.org/10.1016/j.jclepro.2018.07.128]

[66]   Maruthanayagam A, Mani P, Kaliappan K, Chinnappan S. In vitro and In silico Studies on the Removal of Methyl Orange from Aqueous Solution Using Oedogonium subplagiostomum AP1. Water Air Soil Pollut 2020; 231(5): 232.
[http://dx.doi.org/10.1007/s11270-020-04585-z]

[67]   Caner N, Kiran I, Ilhan S, Iscen CF. Isotherm and kinetic studies of Burazol Blue ED dye biosorption by dried anaerobic sludge. J Hazard Mater 2009; 165(1-3): 279-84.
[http://dx.doi.org/10.1016/j.jhazmat.2008.09.108] [PMID: 19013018]

[68]   Wang Y, Mu Y, Zhao QB, Yu HQ. Isotherms, kinetics and thermodynamics of dye biosorption by anaerobic sludge. Separ Purif Tech 2006; 50(1): 1-7.
[http://dx.doi.org/10.1016/j.seppur.2005.10.012]

[69]   Asgher M, Bhatti HN. Mechanistic and kinetic evaluation of biosorption of reactive azo dyes by free, immobilized and chemically treated Citrus sinensis waste biomass. Ecol Eng 2010; 36(12): 1660-5.
[http://dx.doi.org/10.1016/j.ecoleng.2010.07.003]

[70]   Deniz F, Kepekci RA. A promising biosorbent for biosorption of a model hetero-bireactive dye from aqueous medium. Fibers Polym 2017; 18(3): 476-82.
[http://dx.doi.org/10.1007/s12221-017-6826-3]

[71]   Scheufele FB, Staudt J, Ueda MH, *et al.* Biosorption of direct black dye by cassava root husks:

Kinetics, equilibrium, thermodynamics and mechanism assessment. J Environ Chem Eng 2020; 8(2): 103533.
[http://dx.doi.org/10.1016/j.jece.2019.103533]

[72]   Pallavi P, King P, Prasanna Kumar Y. Experimental and statistical modeling on the biosorption of methylene blue from an aqueous solution using raphanus raphanistrum leaves. J Pharm Sci Res 2017; 9(12): 2319-28.

[73]   Guari EB, de Almeida ÉJR, de Jesus Sutta Martiarena M, Yamagami NS, Corso CR. Azo Dye Acid Blue 29: Biosorption and Phytotoxicity Test. Water Air Soil Pollut 2015; 226(11): 361.
[http://dx.doi.org/10.1007/s11270-015-2611-3]

[74]   Silva F, Nascimento L, Brito M, da Silva K, Paschoal W Jr, Fujiyama R. Biosorption of methylene blue dye using natural biosorbents made from weeds. Materials (Basel) 2019; 12(15): 2486.
[http://dx.doi.org/10.3390/ma12152486] [PMID: 31387319]

[75]   Pipíška M, Valica M, Partelová D, Horník M, Lesný J, Hostin S. Removal of Synthetic Dyes by Dried Biomass of Freshwater Moss Vesicularia Dubyana: A Batch Biosorption Study. Environments 2018; 5(1): 10.
[http://dx.doi.org/10.3390/environments5010010]

[76]   Rápó E, Aradi LE, Szabó Á, Posta K, Szép R, Tonk S. Adsorption of Remazol Brilliant Violet-5R Textile Dye from Aqueous Solutions by Using Eggshell Waste Biosorbent. Sci Rep 2020; 10(1): 8385.
[http://dx.doi.org/10.1038/s41598-020-65334-0] [PMID: 32433528]

[77]   Geetha P, Latha MS, Koshy M. Biosorption of malachite green dye from aqueous solution by calcium alginate nanoparticles: Equilibrium study. J Mol Liq 2015; 212: 723-30.
[http://dx.doi.org/10.1016/j.molliq.2015.10.035]

[78]   Unuabonah EI, Omorogie MO, Oladoja NA. Modeling in adsorption: Fundamentals and applications.Composite Nanoadsorbents. Elsevier Inc. 2018; pp. 85-118.

[79]   Park D, Yun YS, Park JM. The past, present, and future trends of biosorption. Biotechnol Bioprocess Eng; BBE 2010; 15(1): 86-102.
[http://dx.doi.org/10.1007/s12257-009-0199-4]

[80]   Ayawei N, Ebelegi AN, Wankasi D. Modelling and Interpretation of Adsorption Isotherms. J Chem 2017; 2017: 1-11.
[http://dx.doi.org/10.1155/2017/3039817]

[81]   Simonin JP. On the comparison of pseudo-first order and pseudo-second order rate laws in the modeling of adsorption kinetics. Chem Eng J 2016; 300: 254-63.
[http://dx.doi.org/10.1016/j.cej.2016.04.079]

[82]   Ho YS, McKay G. Pseudo-second order model for sorption processes. Process Biochem 1999; 34(5): 451-65.
[http://dx.doi.org/10.1016/S0032-9592(98)00112-5]

[83]   Ho YS, McKay G. A Comparison of chemisorption kinetic models applied to pollutant removal on various sorbents. Process Saf Environ Prot 1998; 76(4): 332-40.
[http://dx.doi.org/10.1205/095758298529696]

[84]   Azizian S. Kinetic models of sorption: a theoretical analysis. J Colloid Interface Sci 2004; 276(1): 47-52.
[http://dx.doi.org/10.1016/j.jcis.2004.03.048] [PMID: 15219428]

[85]   Sari IP, Simarani K. Decolorization of selected azo dye by *Lysinibacillus fusiformis* W1B6: Biodegradation optimization, isotherm, and kinetic study biosorption mechanism. Adsorpt Sci Technol 2019; 37(5-6): 492-508.
[http://dx.doi.org/10.1177/0263617419848897]

[86]   Daneshvar E, Kousha M, Koutahzadeh N, Sohrabi MS, Bhatnagar A. Biosorption and bioaccumulation studies of acid Orange 7 dye by *Ceratophylum demersum*. Environ Prog Sustain

Energy 2013; 32(2): 285-93.
[http://dx.doi.org/10.1002/ep.11623]

[87]   Ayanda OS, Adeyi O, Durijayive B, Olafisoye O. Adsorption kinetics and Intraparticulate diffusivities of congo red onto kola nut pod carbon. Pol J Environ Stud 2012; 21(5): 1147-52.

[88]   Dave PN, Kaur S, Khosla E. Removal of basic dye from aqueous solution by biosorption on to sewage sludge. Indian J Chem Technol 2011; 18(3): 220-6.

[89]   Lichtfouse E, Wilson LD, Morin-Crini N, *et al.* Adsorption-oriented processes using conventional and non-conventional adsorbents for wastewater treatment. In: Green Adsorbents for Pollutant Removal. Springer Nature 2019; Vol. 18: pp. 23-71.
[http://dx.doi.org/10.1007/978-3-319-92111-2_2]

[90]   Chojnacka K. Biosorption and bioaccumulation – the prospects for practical applications. Environ Int 2010; 36(3): 299-307.
[http://dx.doi.org/10.1016/j.envint.2009.12.001] [PMID: 20051290]

[91]   Okur M, Saraçoğlu N, Aksu Z. Use of response surface methodology for the bioaccumulation of Violet 90 metal-complex dye by Candida tropicalis. TURKISH JOURNAL OF ENGINEERING AND ENVIRONMENTAL SCIENCES 2014; 38(2): 217-30.
[http://dx.doi.org/10.3906/muh-1409-2]

[92]   Sahadevan R, Manikam V, Miranda LR, Pennathur G. A Comparative Study on the Accumulation of Acid Red 18 and Reactive Black 5 Dyes by Growing Schizophyllum commune and Trametes versicolor. Int J Chem Sci 2008; 6(2): 553-68.
[http://dx.doi.org/10.1002/apj]

[93]   Aksu Z. Reactive dye bioaccumulation by Saccharomyces cerevisiae. Process Biochem 2003; 38(10): 1437-44.
[http://dx.doi.org/10.1016/S0032-9592(03)00034-7]

[94]   Kaushik P, Malik A. Mycoremediation of Synthetic Dyes : An Insight into the Mechanism, Process, Optimization and Reactor Design.Microbial Degradation of Synthetic Dyes in Wastewaters. Springer International Publishing Switzerland 2015; pp. 1-25.
[http://dx.doi.org/10.1007/978-3-319-10942-8_1]

[95]   Telke AA, Kadam AA, Govindwar SP. Bacterial Enzymes and their Role in Decolorization of Azo Dyes.Microbial Degradation of Synthetic Dyes in Wastewaters. Springer International Publishing Switzerland 2015; pp. 149-68.
[http://dx.doi.org/10.1007/978-3-319-10942-8_7]

[96]   Dave SR, Patel TL, Tipre DR. Bacterial Degradation of Azo Dye Containing Wastes.Microbial Degradation of Synthetic Dyes in Wastewaters. Springer International Publishing Switzerland 2015; pp. 57-83.
[http://dx.doi.org/10.1007/978-3-319-10942-8_3]

[97]   Song J, Han G, Wang Y, *et al.* Pathway and kinetics of malachite green biodegradation by *Pseudomonas veronii.* Sci Rep 2020; 10(1): 4502.
[http://dx.doi.org/10.1038/s41598-020-61442-z] [PMID: 32161360]

[98]   Ewida AYI, El-Sesy ME, Abou Zeid A. Complete degradation of azo dye acid red 337 by *Bacillus megaterium* KY848339.1 isolated from textile wastewater. Water Science 2019; 33(1): 154-61.
[http://dx.doi.org/10.1080/11104929.2019.1688996]

[99]   Hossen MZ, Hussain ME, Hakim A, Islam K, Uddin MN, Azad AK. Biodegradation of reactive textile dye Novacron Super Black G by free cells of newly isolated *Alcaligenes faecalis* AZ26 and *Bacillus* spp obtained from textile effluents. Heliyon 2019; 5(7): e02068.
[http://dx.doi.org/10.1016/j.heliyon.2019.e02068] [PMID: 31338473]

[100] Srinivasan S, Shanmugam G, Surwase SV, Jadhav JP, Sadasivam SK. In silico analysis of bacterial systems for textile azo dye decolorization and affirmation with wetlab studies. Clean (Weinh) 2017;

45(9): 1600734.
[http://dx.doi.org/10.1002/clen.201600734]

[101] Roat C, Kadam A, Patel T, Dave S. Biodegradation of Diazo Dye, Reactive Blue 160 by Isolate Microbacterium sp. B12 Mutant: Identification of Intermediates by LC-MS. Int J Curr Microbiol Appl Sci 2016; 5(3): 534-47.
[http://dx.doi.org/10.20546/ijcmas.2016.503.063]

[102] Lade H, Govindwar S, Paul D. Low-cost biodegradation and detoxification of textile Azo Dye C.I. reactive blue 172 by providencia rettgeri strain HSL1. J Chem 2015; 2015: 1-10.
[http://dx.doi.org/10.1155/2015/894109]

[103] Pathak H, Soni D, Chauhan K. Evaluation of in vitro efficacy for decolorization and degradation of commercial azo dye RB-B by Morganella sp. HK-1 isolated from dye contaminated industrial landfill. Chemosphere 2014; 105: 126-32.
[http://dx.doi.org/10.1016/j.chemosphere.2014.01.004] [PMID: 24480425]

[104] Shah M. Evaluation of Aeromonas Spp. In Microbial Degradation and Decolorization of Reactive Black in Microaerophilic – Aerobic Condition. J Bioremediat Biodegrad 2014; 5(6)
[http://dx.doi.org/10.4172/2155-6199.1000246]

[105] Ng IS, Chen T, Lin R, Zhang X, Ni C, Sun D. Decolorization of textile azo dye and Congo red by an isolated strain of the dissimilatory manganese-reducing bacterium Shewanella xiamenensis BC01. Appl Microbiol Biotechnol 2014; 98(5): 2297-308.
[http://dx.doi.org/10.1007/s00253-013-5151-z] [PMID: 23974367]

[106] Ali SAM, Akthar N. a Study on Bacterial Decolorization of Crystal Violet Dye By *Clostridium perfringens, Pseudomonas aeruginosa* and *Proteus vulgaris*. Int J Pharm Biol Sci 2014; 4(2): 89-96.

[107] Olukanni OD, Osuntoki AA, Awotula AO, Kalyani DC, Gbenle GO, Govindwar SP. Decolorization of dyehouse effluent and biodegradation of Congo red by *Bacillus thuringiensis* RUN1. J Microbiol Biotechnol 2013; 23(6): 843-9.
[http://dx.doi.org/10.4014/jmb.1211.11077] [PMID: 23676913]

[108] Garg SK, Tripathi M. Process parameters for decolorization and biodegradation of orange II (Acid Orange 7) in dye-simulated minimal salt medium and subsequent textile effluent treatment by *Bacillus cereus* (MTCC 9777) RMLAU1. Environ Monit Assess 2013; 185(11): 8909-23.
[http://dx.doi.org/10.1007/s10661-013-3223-2] [PMID: 23636502]

[109] Lim CK, Bay HH, Aris A, Abdul Majid Z, Ibrahim Z. Biosorption and biodegradation of Acid Orange 7 by Enterococcus faecalis strain ZL: optimization by response surface methodological approach. Environ Sci Pollut Res Int 2013; 20(7): 5056-66.
[http://dx.doi.org/10.1007/s11356-013-1476-5] [PMID: 23334551]

[110] Jadhav SB, Phugare SS, Patil PS, Jadhav JP. Biochemical degradation pathway of textile dye Remazol red and subsequent toxicological evaluation by cytotoxicity, genotoxicity and oxidative stress studies. Int Biodeterior Biodegradation 2011; 65(6): 733-43.
[http://dx.doi.org/10.1016/j.ibiod.2011.04.003]

[111] Gomare SS, Govindwar SP. *Brevibacillus laterosporus* MTCC 2298: a potential azo dye degrader. J Appl Microbiol 2009; 106(3): 993-1004.
[http://dx.doi.org/10.1111/j.1365-2672.2008.04066.x] [PMID: 19187152]

[112] Khalid A, Arshad M, Crowley DE. Decolorization of azo dyes by Shewanella sp. under saline conditions. Appl Microbiol Biotechnol 2008; 79(6): 1053-9.
[http://dx.doi.org/10.1007/s00253-008-1498-y] [PMID: 18461315]

[113] Wang X, Cheng X, Sun D. Autocatalysis in Reactive Black 5 biodecolorization by *Rhodopseudomonas palustris* W1. Appl Microbiol Biotechnol 2008; 80(5): 907-15.
[http://dx.doi.org/10.1007/s00253-008-1657-1] [PMID: 18762937]

[114] Garg N, Garg A, Mukherji S. Eco-friendly decolorization and degradation of reactive yellow 145

textile dye by *Pseudomonas aeruginosa* and *Thiosphaera pantotropha*. J Environ Manag 2020; 263: 110383.
[http://dx.doi.org/10.1016/j.jenvman.2020.110383]

[115] Shoukat R, Khan S J, Jamal Y. Hybrid anaerobic-aerobic biological treatment for real textile wastewater. J Water Process Eng 2019; 29: 100804.
[http://dx.doi.org/10.1016/j.jwpe.2019.100804]

[116] Sadri Moghaddam S, Alavi Moghaddam MR. Aerobic Granular Sludge for Dye Biodegradation in a Sequencing Batch Reactor With Anaerobic/Aerobic Cycles. Clean (Weinh) 2016; 44(4): 438-43.
[http://dx.doi.org/10.1002/clen.201400855]

[117] Yan LKQ, Fung KY, Ng KM. Aerobic sludge granulation for simultaneous anaerobic decolorization and aerobic aromatic amines mineralization for azo dye wastewater treatment. Environ Technol 2018; 39(11): 1368-75.
[http://dx.doi.org/10.1080/09593330.2017.1329354] [PMID: 28488938]

[118] Tomei MC, Soria Pascual J, Mosca Angelucci D. Analysing performance of real textile wastewater bio-decolourization under different reaction environments. J Clean Prod 2016; 129: 468-77.
[http://dx.doi.org/10.1016/j.jclepro.2016.04.028]

[119] Franca RDG, Vieira A, Mata AMT, Carvalho GS, Pinheiro HM, Lourenço ND. Effect of an azo dye on the performance of an aerobic granular sludge sequencing batch reactor treating a simulated textile wastewater. Water Res 2015; 85: 327-36.
[http://dx.doi.org/10.1016/j.watres.2015.08.043] [PMID: 26343991]

[120] Sarvajith M, Reddy GKK, Nancharaiah YV. Textile dye biodecolourization and ammonium removal over nitrite in aerobic granular sludge sequencing batch reactors. J Hazard Mater 2018; 342: 536-43.
[http://dx.doi.org/10.1016/j.jhazmat.2017.08.064] [PMID: 28886566]

[121] Manavi N, Kazemi AS, Bonakdarpour B. The development of aerobic granules from conventional activated sludge under anaerobic-aerobic cycles and their adaptation for treatment of dyeing wastewater. Chem Eng J 2017; 312: 375-84.
[http://dx.doi.org/10.1016/j.cej.2016.11.155]

[122] Kiayi Z, Lotfabad TB, Heidarinasab A, Shahcheraghi F. Microbial degradation of azo dye carmoisine in aqueous medium using Saccharomyces cerevisiae ATCC 9763. J Hazard Mater 2019; 373(March): 608-19.
[http://dx.doi.org/10.1016/j.jhazmat.2019.03.111] [PMID: 30953978]

[123] Chen Y, Feng L, Li H, Wang Y, Chen G, Zhang Q. Biodegradation and detoxification of Direct Black G textile dye by a newly isolated thermophilic microflora. Bioresour Technol 2018; 250: 650-7.
[http://dx.doi.org/10.1016/j.biortech.2017.11.092] [PMID: 29220809]

[124] Oon YL, Ong SA, Ho LN, *et al.* Constructed wetland–microbial fuel cell for azo dyes degradation and energy recovery: Influence of molecular structure, kinetics, mechanisms and degradation pathways. Sci Total Environ 2020; 720: 137370.
[http://dx.doi.org/10.1016/j.scitotenv.2020.137370] [PMID: 32325554]

[125] Franca RDG, Vieira A, Carvalho G, *et al.* Oerskovia paurometabola can efficiently decolorize azo dye Acid Red 14 and remove its recalcitrant metabolite. Ecotoxicol Environ Saf 2020; 191: 110007.
[http://dx.doi.org/10.1016/j.ecoenv.2019.110007] [PMID: 31796253]

[126] Kashefi S, Borghei SM, Mahmoodi NM. Covalently immobilized laccase onto graphene oxide nanosheets: Preparation, characterization, and biodegradation of azo dyes in colored wastewater. Journal of Molecular Liquids 2019; 276.

[127] Sun H, Jin X, Long N, Zhang R. Improved biodegradation of synthetic azo dye by horseradish peroxidase cross-linked on nano-composite support. Int J Biol Macromol 2017; 95: 1049-55.
[http://dx.doi.org/10.1016/j.ijbiomac.2016.10.093] [PMID: 27984149]

[128] Gajera HP, Bambharolia RP, Hirpara DG, Patel SV, Golakiya BA. Molecular identification and characterization of novel Hypocrea koningii associated with azo dyes decolorization and biodegradation of textile dye effluents. Process Saf Environ Prot 2015; 98: 406-16.
[http://dx.doi.org/10.1016/j.psep.2015.10.005]

CHAPTER 7

Fungal Bioremediation of Pollutants

Evans C. Egwim[1,*], **Oluwafemi A. Oyewole**[2] and **Japhet G. Yakubu**[2]

[1] *Department of Biochemistry, Federal University of Technology, Minna, Nigeria*

[2] *Department of Microbiology, Federal University of Technology, Minna, Nigeria*

Abstract: Advancement in industrialization and urbanization has caused an influx of contaminants into the environment polluting the soil, water, and air. These contaminants come in various forms and structures, including heavy metals, petroleum hydrocarbons, industrial dyes, pharmaceutically active compounds, pesticides, and many other toxic chemicals. The presence of these pollutants in the environment poses a serious threat to living things, including humans. Various conventional methods have been developed to tackle this menace, though effective, are however not safe for the ecosystem. Interestingly, bioremediation has offered a cheap, effective, and environmentally safe method for the removal of recalcitrant pollutants from the environment. White-rot fungi (WRF), belonging to the basidiomycetes, have shown class and proven to be an excellent tool in the bioremediation of the most difficult organic pollutants in the form of lignin. White-rot fungi possess extracellular lignin modified enzymes (LMEs) made up of laccases (Lac), manganese peroxidase (MnP), lignin peroxidase (LiP), and versatile peroxidase (VP) that are not specific to a particular substrate, causes opening of aromatic rings and cleavage of bonds through oxidation and reduction among many other pathways. The physiology of WRF, non-specificity of LMEs coupled with varying intracellular enzymes such as cytochrome P450 removes pollutants through biodegradation, biosorption, bioaccumulation, biomineralization, and biotransformation, among many other mechanisms. The application of WRF on a laboratory and pilot scale has provided positive outcomes; however, there are a couple of limitations encountered when applied in the field, which can be overcome through improvement in the genome of promising strains.

Keywords: Basidiomycetes, Bioaccumulation, Biodegradation, Biomineralization, Bioremediation, Biosorption, Biotransformation, Fungi, Heavy metals, Industrial dyes, Laccases, Mycoremediation, Peroxidase, Pesticides, Petroleum hydrocarbons, Pharmaceutical products, Pollutants, Oxidation, Synthetic pesticides, White-rot fungi.

* **Corresponding author Evans C. Egwim:** Department of Biochemistry, Federal University of Technology Minna, Nigeria; E-mail: evanschidi@gmail.com

Inamuddin (Ed.)

INTRODUCTION

Man's effort to provide an easy and comfortable life for himself through industrialization and urbanization has become a threat to him and the ecosystem at large [1 - 4]. Infrastructural construction, mining, transformation of raw materials, electroplating, smelting, extraction of petroleum, and farming, among many other anthropogenic activities have caused and is still causing deleterious effects on the ecosystem [5]. Most of these anthropogenic activities, release harmful substances into the environment (air, soil and water), which if not properly cleaned or disposed of effect biotic activities in the ecosystem [6]. These harmful substances include polymers, cyanide compounds [7], papers and pulp [8], heavy metals [9 - 11], pesticides, industrial dyes, pharmaceutically active compounds (PhACs) [12], petroleum hydrocarbons [13, 14], chlorendic acids [15] among many others. Some of these pollutants such as heavy metals (*i.e.* mercury, cadmium, arsenic, chromium, copper, selenium, and lead), when present in the environment in large amounts impair the metabolism of living organisms including man [11, 16]. Some of these heavy metals cause cancer, skin inflammation, nausea, dizziness, and headache [17]. Other pollutants such as pesticides have caused chronic illnesses leading to the global loss of about 1 million lives annually [18]. This event of loss of lives and resources is no exception to petroleum pollutants, which arises as a result of oil spills during drilling of oil wells, leakage of underground tanks, vandalization of storage tanks among many more occurrences. This eventually affects the diversity of biological niches, deaths of aquatic organisms, affects the productivity of plants and causes starvation in man [14].

The debilitating effects of pollutants in the environment cause a man to respond rapidly through conventional methods in a bid to alleviate the damages it causes [19, 20]. Some of these conventional methods developed include; soil flushing, land filling, burning, vitrification, electroreclamation/electrokinetics, solidification and stabilization, removal and containment among many others, though rapid but do not eradicate the pollutants, rather they change their location and state from one form to another [4, 19, 21]. The disadvantage of conventional methods is that they are not environmentally friendly, they are expensive, require more labor and expose more surfaces to pollutants. The use of conventional chemicals to treat a polluted environment also has adverse effects in the long run [2]. Some of the methods applied in the treatment of soils affect the soil structure and quality, and the efficiency of some of the methods is limited to certain depths [2, 9, 19]. As a result of all these setbacks, man has searched for a means to alleviate the environmental problems posed by pollutants, out of which biological methods have proven to be environmentally friendly, with little or no adverse effects on the environment [2, 20]. A process where biological materials are used in the cleanup of pollutants from the environment is known as bioremediation [2, 4].

Bioremediation involves the transformation or degradation of hazardous pollutants from a toxic form to a less toxic absorbable form [2]. Among the natural bioremediation processes, microbial bioremediation has stood to be the most effective in the removal of pollutants from the environment [4]. Microorganisms remove pollutants either by biodegradation, bioaccumulation, biosorption, biotransformation or biomineralization [10]. Bioremediation strategy could either be *in-situ* (*i.e.* treatment of pollutants in the site of contamination) or *ex-situ* (*i.e.* taking the pollutants away from the contamination sites for treatment) [2, 4, 20, 22]. However, the former is safe and cheap unlike the latter, which is expensive and a potential threat to the health of laborers involved and exposes the pollutants to more surfaces [2, 4]. Among the bioremediation technologies, the use of fungi, particularly those belonging to the basidomycetes has distinguished themselves as an effective tool in the remediation of an environment polluted by recalcitrant xenobiotics, a technology known as mycoremediation [2, 23, 24]). Their metabolites and body structures coupled with the fact that they can withstand toxic compounds and still perform in an environment depleted of nutrients have made them cheap and effective for a safe and sustainable strategy in alleviating polluted environment [23, 25].

Pollutants and Their Classification

There are different types of pollutants that are generated by various industries. These pollutants could come in the form of solid, liquid, or gases. Their nature, chemical composition and structure could vary from simple monomers to a complex polymer of aromatic rings. They include petroleum hydrocarbons, chemical compounds (*i.e.* industrial dyes and pesticides), and heavy metals [26].

Petroleum Hydrocarbons

The use of petroleum based products is almost inevitable in the world we live today. Aside from being used as a form of energy to power machines, vehicles, trains, aeroplanes, generators, heating mantles, and cookers; some components of petroleum are also used in the production of various forms of plastics, chairs, rubbers, fibbers for reinforcement of concretes, fittings in electric appliances, cutleries, among many other uses [13, 27]. Right from the exploration of petroleum to its conversion (refining), down to its transportation and storage, the environment one way or the other gets polluted by harmful substances present in petroleum [13, 28, 29]. Petroleum is composed of a various complex mixture of hydrocarbons [28], which is distinguished mainly into aromatic hydrocarbon, saturated hydrocarbon and non-hydrocarbon compounds [30, 31]. Aromatic hydrocarbons are complex with high melting and boiling point due to the presence of benzene ring(s), which makes them difficult to degrade away from the environment [32]. These features are however absent in saturated hydrocarbon, which their molecular structure is made up of carbon-hydrogen bond and carbon-carbon bond. These bonds could either be straight, branched, or in circle, they lack complex molecular structures (*i.e.* benzene ring) with lower melting and boiling point. Saturated hydrocarbons are easily degraded as such do not persist in the environment [31].

Persistent organic pollutants (POPs) in the environment exist among the polycyclic aromatic hydrocarbons (PAHs), which are made up of complex benzene structures [33, 34]. The European Community (EC) and the United States Environmental Protection Agency (USEPA) have prioritized the control of 16 PAHs. These are: acenaphthene, anthracene, pyrene, fluorene, naphthalene, fluoranthene, phenanthrene, benzo(b)fluoranthene, benzo(a)anthracene, benzo[k]fluoranthene, benzo(a)pyrene, benzene(ghi)perylene, indene, diphenyl(a,h)anthracene, acenaphthylene and (1,2,3-cd) pyrene [30, 34 - 36]. PAHs get into the environment through the burning of fossil fuel (*i.e.* premium motor spirit (PMS), diesel, kerosene), incineration of petroleum products (*i.e.* synthetic plastics, tyres), oil spills, untreated effluents from chemical and petroleum industries, among many others release these harmful and toxic substances into the environment affecting the vitality of living things in the ecosystem [13, 28, 31, 33, 37, 38]. Sulphur compounds such as thiols, sulphides, disulphides, naphthobenzothiophene cyclic sulphides, benzothiophene, dibenzothiophene; oxygen compounds such as alcohols, ethers, esters, carboxylic acids, furans and ketones; and nitrogen compounds such as pyridine, indoline, carbazole, benzo(a)carbazole, benzo(f)quinolone, indole, and nitriles are all part of non-hydrocarbon compounds present in petroleum as well as some metals, which most times are contained in the nonvolatile part of petroleum hydrocarbon [26, 33, 35]. When compared to the three groups of petroleum hydrocarbon, maximum carbon numbers are seen among the non-hydrocarbon compounds (*i.e.* porphyrin). In water, non-hydrocarbon compounds are insoluble with high boiling and fusion point. Thus, they persist in the environment alongside PAHs and are difficult to remove causing deleterious effects to live things such as mutagenicity [28, 30].

Heavy Metals

Metals and metalloids that exhibit metallic features such as malleability, conductivity, ductility, ligand specificity, malleability and cation stability with a relatively high atomic weight, density and an atomic number ≥ 20 excluding alkaline earth lanthanides, alkaline metals, and actinides are regarded as heavy metal (HM) [11, 34, 39, 40]. Seiyaboh and Izah [41] simply define HM to be metals with a specific gravity greater than 5 cm^3. Some HM (*i.e.* copper, manganese, iron and zinc) are essential micronutrients required for metabolism in a biological system in minute quantity and become dangerous when their concentration increases above the safe level [11, 34]. The environment is constantly exposed to an influx of HM pollutants from various anthropogenic activities such as mining, metal forging, electroplating, burning of fossil fuel, smelting, sewage sludge, dyeing, agricultural activities, forest burning, biosolids, ore mining, electronic waste, batteries waste, coal combustion, wood preservatives, personal care products (PCPs) such as cosmetics, biosolids waste treatment plant, tannery and other untreated industrial effluents among geological activities such as weathering of rocks and volcanic eruptions all release varied concentration of HM to the environment (Table **1**) [9, 10, 16, 39 - 43]. Biological degradation of HM is not feasible as such, they remain persistent in the environment taking one form or the other [9, 40, 42].

Table 1. Some heavy metals and their various sources [16, 17].

Heavy Metals	Sources
Zn	Mining, refining, biosolids smelting, electroplating industry.
Pb	Combustion of leaded gasoline, municipal sewage, mining and smelting of metalliferous ores, glass manufacturing, paints, and industrial waste supplemented with Pb, lead batteries, X-ray shields, ammunition.
Cd	Sewage sludge, application of phosphate fertilizers, geogenic sources, metal refining and smelting, anthropogenic activities, burning of fossil fuel, production of batteries, welding, alloy pigment.
Hg	Wood and peat burning, volcano eruptions, forest fire, emissions from industries producing caustic soda, coal, mining, electrical industries, batteries, dentistry.
Ni	Automobile batteries, forest fire, kitchen appliances, land fill, industrial effluents, surgical instruments, steel alloys, volcanic eruptions, gas exchange in ocean and bubble bursting, weathering of geological materials and soils.
As	Mining and smelting, volcanoes, petroleum refining, wood preservatives, herbicides, animal feed additives, semiconductors, coal power plants.
Cr	Metal processing, chrome plating, solid waste, tanneries, sludge, metallurgy, leather tannings, electroplating industry, anti-corrosives, dyeing, cooking systems and boilers.
Cu	Smelting and refining, biosolids, electroplating industry, mining.

Uncontrolled release of HM into the environment has gained global attention with different reports of their presence in the food we eat causing varying degrees of health disorders in humans (Table **2**) [11]. HM can be accumulated by a biological systems such as plants and animals, they can be transferred from one trophic level to another since they can't be degraded [40]. Even at low concentrations, most HM (lead, selenium, mercury, silver, cadmium, arsenic among many others) cause neurotoxic and cytotoxic effects, which could be mutagenic and carcinogenic to the biological system [10, 34, 40, 41]. Several reports of HM pollution have been made in the past. An incidence of HM poisoning was recorded in Minamata Bay, Japan in 1963 where lots of lives were lost as a result of consuming shellfish that have accumulated a high concentration of mercury in their system [11]. HM pollutants pose a serious threat to all forms of life directly or indirectly with no method so far suitable for total elimination [16].

Table 2. Some heavy metals and their toxic effects on human health [10, 17, 44].

Heavy Metal	EPA Regulatory Limit (ppm)	Toxic Effects
Ba	2.0	Elevated blood pressure, respiratory failure, cardiac arrhythmias, gastrointestinal dysfunction, and muscle twitching.
Cr	0.1	Diarrhoea, liver diseases, reproductive toxicity, renal failure, lung cancer, hair loss, chronic bronchitis, bronchopneumonia, irritation of the skin and the respiratory tract, liver diseases, nausea, headache.
Ag	0.10	Exposure may cause dermal tissues to turn grey or blue-grey, irritation of the respiratory tract and the lungs, breathing problems, stomach pain.
Cu	1.3	Kidney, liver and brain damage, metabolic disorders, headache, chronic anaemia, vomiting, abdominal pain, nausea.
Ni	0.2	Dry cough, chest pain, itching of the skin, headache, cancer of the sinuses, nose, lungs; dermatitis, kidney diseases, neurotoxic, immunotoxic, genotoxic.
Hg	2.0	Brain damage, impaired vision, insomnia, sclerosis, memory loss, autoimmune diseases, gingivitis, depression, dysphasia, ataxia, drowsiness, deafness, kidney and lungs failure.
Cd	5.0	Headache, coughing, vomiting, lymphocytosis, hypochromic anaemia, endocrine disruptor, testicular atrophy, itai itai, kidney diseases, emphysema, prostrate and lung cancer.
Zn	0.5	Fatigue, impotence, dizziness, lethargy, depression, macular degeneration, icterus, hematuria, vomiting, liver and kidney failure, seizures, gastrointestinal irritation, ataxia, prostate cancer, macular degeneration.
Pb	15	Anorexia, hyperactivity, insomnia, reduced fertility, renal damage, reduced fertility, chronic nephropathy, elevated blood pressure, impaired neurons, learning deficits, shortened attention.
As	0.01	Brain damage, conjunctivitis, skin cancer, dermatitis, respiratory and cardiovascular impairment,
Se	50	Hepatotoxicity, dysfunction of natural killer cells, gastrointestinal disturbance.

Chemical Pollutants

Chemical compounds have been used in the production of various products that are widely used in homes, farmlands as well as industries. Most of these chemicals such as pesticides, and synthetic dyes are formulated by complex aromatic compounds and remain persistent and very harmful when they get into the environment.

Synthetic Pesticides

The presence of pest in the environment especially on farmlands have proven to be problematic causing loss of farm produce [45, 46]. The pest could be plants or animals breeding in an unwanted environment. Their presence in the environment has detrimental effects directly or indirectly on other organisms. They also affect the productivity of crops in farmlands, as such their eradication is paramount [47]. Chemical substances that kill pests are known as pesticides and are formulated to target specific organisms such as birds, rodents, insects, mollusks, and weeds among many others [48]. Pesticides get into the environment during preparation or application on farmlands. There are different classifications of pesticides. Some pesticides are classified based on the target pest such as fungicides, bactericides, nematicides, algicide, rodenticide, avicide, molluscicide, miticide, piscicide, insecticides, and herbicides [49, 50]. Another classification of pesticides is based on the parent chemical compound from which they are made, which includes; pyrethroids, organochlorines, organophosphate, thiocarbamates, carbamates, among others [49, 50]. Some of these pesticides have a long life span as such persist in the environment. Examples of persistent pesticides include organochlorine (*i.e.* Dichlorodi-phenyltrichloroethane (DDT)), whereas others such as carbamates (*i.e.* aldicarb and carbaryl), organophosphates (*i.e.* diazinon and chlorpyrifos), phenoxyacid derivatives (*i.e.* 2,4-Dichlorophenoxyacetic acid), pyrethroids (*i.e.* cypermethrin and cyfluthrin), and chloroacetanilides have a short life span and do not persist in the environment [47, 49, 51, 52]. Some of the chemical substances used to formulate some of these pesticides include atrazine, bisphenol A, thiamethoxam, azinphos methyl, glyphosate, 2,4-D (2,4-dichlorophenoxyacetic acid), and linuron among others [48, 53].

The ways in which synthetic pesticides are used in large amounts in the environment go on to affect other non-targeted organisms, some of the chemical substances accumulate in the soil and subsequently find their way to water bodies when eroded during rainfall, and others leach down where they contaminate the underground water and become harmful directly to aquatic lives and indirectly to humans when they feed on such organism [34, 47]. Some of these pesticides have other pathways aside oral route, which they use to affect human health. Other routes include; dermal (skin), eyes and respiratory (nose) routes, which can cause either acute or chronic effects [49, 51]. The mechanism used by most pesticides includes; disruption of enzyme activities and or the impairment of the nervous system of the target pest. These actions similarly occur in humans when exposed to these pesticides [50]. Some of the classes of pesticides and the harmful effects they cause on humans are summarized in Table 3. Despite the adverse effects recorded in pesticides, developing nations still import pesticides in large amounts to help crop production. In 2016, the statistical data base of FAO known as FAOSTAT reported that an estimate of $1,590,160.326 worth of pesticides were exported to Africa [54]. From January to November 2015, statistics on the export of pesticides from China's customs reported that 13.9% of the total pesticides produced were exported to 44 countries in Africa with Nigeria followed by South Africa and Ghana topping the chart [54].

Table 3. Types of pesticides and the toxic effects they have on humans.

Class of Pesticide	Example	Toxic Effects	Source
Organochlorines (chlorinated hydrocarbons)	Endolsulfan, chlordane, aldrin, endrin, and dieldrin	Cancer, seizures, tremors, dizziness, anorexia, headache, nausea, malaise, vomiting, dermatitis, apprehension, diarrhea, irritability, muscle weakness, altered reflexes, excessive sweating, spasms, anxiety, mental confusion, coma, endocrine disorders, death	[50 - 52, 55]
Pyrethrin and pyrethroids	Esfenvalerate, deltamethrin, cyhalothrin, permethrin, cypermethrin, allethrin	Seizures, salivation, temporary paraesthesias, irritability, tremor, ataxia, spasms, prostration, incoordination, drooling, clonic and tonic convulsions, skin irritation, hypersensitivity, miosis of eyes, neurotoxicity	[50 - 52, 55]
Organophosphates	Naled, diazinon, azinphos, trichlorfon, dimethoate, temephos, clorpyriphos, fenthion	Fasciculations, hypersecretion, vomiting, nausea, miosis, coma, impair functions of cholinesterase enzymes, spasms, hyperreflexia, tremors, paralysis, hyperglycemia, pallor, tachycardia, hypertension, memory disorders, headache, ataxia, respiratory depression, diarrhea, coma, bradycardia, miosis, cancer, Alzheimer's and Parkinson's disease	[50 - 52, 55]

(Table 3) cont.....

Class of Pesticide	Example	Toxic Effects	Source
Thiocarbamates and carbamates	EPTC (S-ethyl N, N-dipropylthiocarbamate), Propham, propoxur, carbofuran, methomyl, thiodicarb, barban, triallate, carbaryl, nabam	Tachycardia, hypertension, paralysis, lip tingling, incoordination, respiratory failure, bronchial secretions, pulmonary edema, hypoxemia, bradycardia, bronchospasm, dyspnea, urinary incontinence, abdominal pain, vomiting, tearing, sweating, necrosis and apoptosis of immune cells	[50 - 52, 55]

Industrial Dyes

Initially, dyes were obtained solely from natural means, but over the cause of time, they have been successfully produced in the laboratory on a large scale to meet their increasing demands in various sectors including textile, plastic, paper printing, pharmaceutical, cosmetics and the food industries [7, 56, 57]. Annually, about 800,000 tons of dyes are produced globally, and this value represents a small fraction of the overall production of chemicals in the chemical industries. About 10-15% of synthetic dyes are lost to the environment during application to various uses. Synthetic dyes used in varying industries are classified either as basic, acid, vat, direct, reactive, disperse, mordant, metal complex, or sulphur dyes [56, 58, 59]. The textile industry being the major consumer of synthetic dyes accounts for over 10,000 dyes and 70% of them are the azo dyes whose structure is complex and difficult to degrade from the environment. One of the major sources of dye entry into the environment is through incomplete exhaustion of the aqueous form of the dye during dyeing processes of textile fibers as well as untreated effluents from the textile industry being disposed into the environment [34, 57, 60]. This textile effluent contains a lot of toxic and hazardous substances such as azoic, sulphur, indigoids, nitrates, soaps, acids, heavy metals, complex compounds and other auxiliary chemicals, which are added during the dyeing process when in contact with the environment and impair the normalcy of the ecosystem [57]. Most of the route of disposing of textile effluents is through water bodies such as rivers, lakes and oceans due to their high solubility in water and can react with other chemical substances to further form other harmful and toxic substances, which are deleterious to live in the aquatic system and man at large through bioaccumulation and transfer to humans through the food chain [60, 61].

Pharmaceutical Products

Pharmaceutical products are not limited to only drugs used in the treatment of human and animal diseases, they also include other PCPs such as cosmetics, soaps, toothpaste, fragrances, lotions, synthetic musks, sunscreens UV filters, and preservatives among many others, which are used on a daily basis. These constitute a class of pollutants in the environment [62 - 66]. Pharmaceutical compounds known as pharmaceutically active compounds (PhACs), are biologically active and are utilized for therapeutic uses both for the treatment of human and animal diseases [12, 63, 66]. This PhACs include hormones, anti-inflammatory drugs, antiepileptic drugs, blood lipids, regulators, antineoplastics, antimicrobial agents, cytostatic drugs, β-blockers, and antibiotics among many more [12]. The presence of pharmaceutical pollutants in the environment has raised great concern among the general public particularly antibiotics and hormones for their wide spread application in livestock agriculture and human therapy [62]. The last two decades have seen a series of projects funded by the European Commission (EC) on the topic, of pharmaceutical in the environment. Findings from the project have helped to elucidate the emergence of pharmaceutical pollutants as a threat to the environment [67]. There are different classes of antibiotics such as sulfonamides (*e.g.* sulfadimethoxine, sulfamethoxazole), macrolides (*e.g.* roxithromycin, erythromycin), fluoroquinolones (*e.g.* ciprofloxacin, norfloxacin) among many others that have long half-life making them biologically active for a long period of time [65, 66].

Indiscriminate use and constant disposal of antibiotics into the environment create resistant strains of bacteria, which are now of public health concern [64, 68]. These strains of bacteria become resistant to multiple drugs making their eradication difficult, resulting in mobility and mortality among their infected hosts [65]. The presence of other PhACs pollutants in the environment such as hormones (*i.e.* steroid estrogens) has been related to the endocrine disrupting effects of biotas inhabiting polluted water bodies [12, 62]. Aside from indiscriminate use and disposal of pharmaceutical products in the environment, other sources of pharmaceutical pollution include treated and untreated effluents from hospitals and clinics, pharmaceutical factories and pharmacies, droppings from livestock treated with antibiotics, municipal waste water treatment plants and agricultural run offs [65, 68, 69]. Pharmaceutical pollutants have been reported to be found present in ground water, surface water and drinking water [62, 67]. Despite efforts put by government bodies to regulate the disposal of pharmaceutical products into the environment, still, over 600 pharmaceutical compounds have been detected in the environment on a global scale [64, 67]. This has raised concerns about the negative impacts this will have on our health system and life particularly in the aquatic environment as most pollutants end up in the water bodies [67, 69].

Effects of Pollutants in the Soil

Soil is a living system that supports myriad living organisms including humans. However, anthropogenic activities and other geological events are gradually destroying the health of soil. Healthy soil is productive soil [35, 70]. For a soil to be productive simply means, it has its own capacity to provide abundant micro and macro nutrients needed for the growth of plants and other macro and microorganisms. Thus, sustaining the biodiversity of the soil's ecosystem [70]. This biodiversity is today being destroyed by anthropogenic activities, which release an influx of varying pollutants from industrial dyes, petroleum hydrocarbons, heavy metals, pharmaceuticals, synthetic chemicals, paper and pulp industries among many more [41, 71]. The presence of these pollutants in the soil have serious detrimental effects on the soil biota including plants [70]. They destroy microorganisms through impairment of the metabolic pathways and inhibition of their enzymes [72], which in turn reduce or eliminate the biomass of microorganisms responsible for making nutrients available for plant use in the ecosystem through degradation of organic materials [70]. Some of these environmental pollutants are recalcitrant making their effects persist in the environment [17]. Large hectares of land have been made unfit for agricultural practices (Fig. **1**), crops destroyed, and livestock lost as a result of HM poisoning [39].

Fig. (1). Unsafe soil polluted by heavy metals [34].

Industrial wastes have lots of illicit effects on the biodiversity of the ecosystem [34]. All forms of life are affected directly or indirectly. If microorganisms are affected by pollutants the soil fertility is affected, likewise the soil biota. The soil becomes desolate and void of most life that keeps it healthy, this to a great extent affects the productivity of farmlands causing scarcity of food to man [35, 70]. Genotoxic pollutants affect plant's symbiosis, disintegrate cell membrane, inhibit the activities of cytoplasmic enzymes and disrupt organelle function. By doing so, physiological processes such as photosynthesis, cellular respiration, carbohydrate metabolism, and synthesis of protein are all impaired [9, 26]. In a bid to increase soil fertility, fertilizers, biosolids (*i.e.* manure), and municipal waste sludge are added to the soil. This however may introduce industrial pollutants such as HM, synthetic dyes, polyaromatic amines, polyvinyl chloride (PVC), polyethylene (PE), polyethylene terephthalate (PET), polystyrene (PS), sulphur among many other compounds into farm lands [34, 73 - 75]. The presence of recalcitrant pollutants often contaminates the food chain and undergoes biomagnification with serious cytotoxic and genotoxic effects capable of causing death to man [61]. The most conventional way of treating pollutants (*i.e.* incineration, land flushing, and landfills) in the environment causes destruction to the soil structure exposing it to erosion and depleting soil nutrients [30, 73].

Petroleum hydrocarbon has shown its potency in terms of the destruction of vegetation. Large hectares of land have been marred by oil spills either intentionally (vandalization of pipelines by opposition groups, oil thefts) or accidentally (*i.e.* metal corrosion of storage tanks, pipelines, accidents of transport vehicles among many others) [26, 33, 70]. Petroleum hydrocarbon affects soil porosity and permeability due to its low emulsifying ability, higher viscosity, and small density [26]. They easily get absorbed into the soil thus, reducing oxygen concentration. The nutrient balance of the soil is affected such as the carbon/nitrogen and carbon/phosphorus concentration, and the soil pH, salinity, and conductivity are affected due to the abundant carbon present in petroleum hydrocarbon [26, 30]. Such soils experience the destruction of normal biota as petroleum products contain lots of toxic substances that are genotoxic, cytotoxic and carcinogenic in nature [31]. Petroleum products disposed of in landfills over time experience thermal and chemical degradation releasing toxic recalcitrant into the soil, some are leached to underground water systems where it causes lots of damage to a living organism in the aquatic system [11, 28, 26, 33, 34].

Effects of Pollutants in the Aquatic Environment

The aquatic environment provides man with drinking water, food, recreation, occupation, and transportation [76]. All these could be lost with increased anthropogenic activities generating a large amount of pollutants on a daily basis, which are often discharged into the environment. Pollutants generated from the land often find themselves one way or the other in the aquatic environment. About 72% of the earth's crust is made up of water making it the largest collector of pollutants [11]. Waste effluents generated from various industries such as chemical, metal, fertilizer, pesticide, petroleum, pharmaceutical, textile, construction, food and paper industries are being discharged into the aquatic environment [76]. Sewage waste from homes is also collected in municipal tanks. Some are treated and reused, others are emptied into the open ocean introducing various forms of chemical pollutants from detergents, cosmetics, plastics among many others [37]. This waste may remain on the surfaces of water bodies others whose densities are heavier than water sinks forming part of the benthic zone of water bodies [37, 59]. Some of the chemical pollutants that remain on the surfaces of water such as dyes, plastics and other chemical pollutants form thin layers of film that affects light penetration from the sun to the primary producers, such as planktons [59, 75]. Thus, photosynthesis is hampered causing scarcity of food in water bodies, which could lead to starvation and death of aquatic organisms [59].

Pollutants such as plastic may undergo a diverse forms of degradation such as biodegradation, hydrolysis, mechanical abrasion, thermal oxidation and photodegradation releasing toxic chemical substances such as polyethylene, polyethylene terephthalate, polystyrene, polypropylene, polyvinylchloride and polyamide, which are cytotoxic and toxigenic to aquatic organisms [37, 75]. Microplastics may be ingested by marine animals, which could affect their digestion of food resulting in starvation and death [74, 77], death of marine animals have been recorded as a result of entanglement (Fig. **2**) others have been made immobile exposing them to dangers as a result of predation in the aquatic environment [77]. Run offs from farmland supplemented with organic or chemical fertilizers cause eutrophication resulting in algal bloom, which is dangerous to any aquatic environment [59]. Sewage discharged in water bodies could result in reduced or depleted oxygen concentration, causing a rise in biological oxygen demand (BOD) [20, 59]. HM contamination from industrial effluents impair metabolism of microorganisms, and higher organisms [77]. Bioaccumulation of HM occurs among aquatic organisms, which could greatly affect the health of man owing to its high toxicity [77]. Oil spills are very toxic to aquatic organisms. They are carcinogenic, neurotoxic, cytotoxic, and toxigenic among many forms of health disorders [41].

Humans who live in the coastal areas are often displaced as a result of oil spills; occupation is also lost as the aquatic system has experienced migration and deaths of fishes and other aquatic vertebrates [77]. The body of dead fishes, molluscs, crustaceans, sea turtles among other marine organisms causes fouling smells making the water unfit for consumption and other domestic use (Fig. **3**) [26, 77]. Some of the chemical substances present in the aquatic system get volatilized into the atmosphere affecting other forms of lives. The contaminants can cause various range of toxicological health problems to humans and animals including haemotoxicity (destruction of red blood cells), carcinogenicity (ability or tendency to induce cancer), genotoxicity (ability to induce non-transmissible DNA damage), mutagenicity (capacity to incite transmissible genetic mutations), teratogenicity (induction of malformation of embryo or foetus), cytotoxicity (ability of being toxic to cells), neurotoxicity (damage to the brain and nervous system), cardiotoxicity (capacity to cause damage to heart muscles, hepatoxicity (ability to elicit damage to the liver), and ocular toxicity (ability to induce eye disorders) [20, 26, 77, 78].

Fig. (2). Unopened stomach of northern fulmars (*Fulmarus glacialis*) shows presence of ingested plastic (left) and entanglement of Northern gannet on Texel beach; Netherlands (right) [77].

Fig. (3). 2010 Oil spills in Louisiana and Mississippi, Alabama, causes death of marine animals such as dolphin washed ashore (left) [79]; Shell oil spills in Niger Delta region of Nigeria displacing fishermen from their occupation (right) [80].

Effects of Pollutants in the Air

Various anthropogenic activities have caused the release of varying harmful substances into the atmosphere, which are in the form of gases such as carbon (*i.e.* CO, CO_2, CH_4), nitrogen (NO, NO2, N_2O_4, NH^{4+}, NH_3), particulate matter (PM2.5 and PM10), polycyclic aromatic hydrocarbon, ozone (O_3), heavy metal (HM) (*i.e.* Ni, Pb, As, Cd), sulphur compounds, volatile phenols [5, 20, 78 - 83]. Anthropogenic activities make these aforementioned compounds suspend in the atmosphere affecting its serenity and causing poor vision to drivers (Fig. 4) [84, 85]. Various chronic and acute respiratory disorders have been reported to be associated with air pollution in men. This includes asthma, atherosclerosis, lung cancer, high blood pressure, stroke, myocardial infarction, thrombosis, neurological damage, blindness, miscarriage, nausea, and coma, among many other health disorders [5, 78, 80, 86]. Combustion of fossil fuels emits CO, and sulfur compounds from the exhausts of vehicles, generators and other machines, which greatly affects the climate. So far, the climate has received a great deal of attention concerning the effects air pollutants are causing [78, 81, 87]. There has been an increase in global warming with an increase in the industrial activities of man both in developed and developing countries [5, 78, 87], causing the depletion of atmospheric oxygen concentration due to an increase in CO emission [84, 88, 89].

Fig. (4). Air pollution in Trabzon, Turkey, causing poor vision [92].

The stratospheric ozone layers protecting the earth from direct impacts of sun's ultraviolet (UV) rays is been destroyed on a daily basis raising great concern about the fate of man in the nearest future if this emission is remained unchecked [89, 90]. Stratospheric ozone protects the earth from varying ultraviolet rays from the sun. Spectral radiation from the sun consists of UV A (320-400 nm), UV B (290-320 nm), and UV C (200-290 nm). Most UV A (95%) with little of UV B (1-5%) reaches the earth's surfaces whereas UV C is being absorbed by O_2 and ozone layers in the atmosphere [93]. Depletion of these ozone layers by air pollutants results in the penetration of UV A and UV B, which are capable of causing cancer, photoaging and sunburns among humans [84, 88, 91]. Other effects of air pollutants result in the formation of acid rain, which causes lots of problems to the environment, such as acidification of soil, rivers and lakes, corrosion of metals and roofing sheets, and other building materials [5, 79, 89]. Acid rain occurs as a result of interaction between O_3, nitrogen oxide and sulfur oxide [88]. Air pollutants such as fluorides, ozone, carbon monoxide, sulfur dioxide, and nitrogen oxides suspended in the atmosphere over time get deposited on vegetation causing a collapse of leaves, the malfunction of plants organelles such as interference with stomata opening of plants, thus hampering photosynthesis [5, 81]. Other effects of air pollution on plants are seen in the yellowing of leaves, stunted growth and low yield of products [81, 88]. Smog from the reaction between nitrogen oxide and various forms of hydrocarbon results in the formation of yellow brown haze in the atmosphere, which interferes with visibility, particularly during winter [88].

Bioremediation

Bioremediation is simply the cleanup of an environment (*i.e.* air, soil or water) polluted by organic contaminants using biological means, which could either be plants, microorganisms or their metabolites [4, 92, 93]. The use of microorganisms in the remediation of polluted environment has gained interest from various research groups, due to their ubiquity, wider diversity, faster growth rate, ability to be genetically enhanced and manipulated to endure harsh conditions, and provision of diverse metabolic mechanisms [20]. These microorganisms could be indigenous to the polluted environment others are isolated elsewhere showing potential as an effective tool for remediation of a particular contaminant [21, 93]. The toxicity of pollutants is often neutralized or reduced to a less toxic form through biomineralization, biodegradation, biotransformation and biosorption [4, 10]. For bioremediation to be successful, disciplines such as chemistry, geology, ecology, and microbiology need to be interplayed [22]. In recent times, microbial remediations have received a great deal of attention in the detoxification of pollutants in the environment [4].

Bioremediation Techniques

There are two techniques involved in bioremediation based on transportation and removal of contaminants. They are *in situ* and *ex situ* bioremediation.

In situ in this technique of bioremediation, the pollutants are treated at the site of contamination. This technique is safe, cheap and does not pose any further risks associated with exposure to laborers and other environmental surfaces [4, 93]. This technique is however limited to some particular depth, especially in polluted soil and there is no sufficient level of monitoring and manipulation of environmental conditions [22]. Commonly utilized *in situ* processes are biosparging, bioventing, bioaugmentation and bioaccumulation.

Biosparging

In this process of *in situ* bioremediation, air and nutrients under pressure are injected below the underground water table so as to increase the concentration of oxygen, thus, enhancing the rate of microbial degradation of pollutants [4, 94]. Biosparging helps to increase the rate at which pollutants are degraded by ensuring contact and mixing between soil water in the saturated zone [96]. The design, construction and installation of small diameter air injectors allow control of specific amount of oxygen injected into the contaminated site. This help to regulate volatilization of pollutants into the atmosphere during degradation [21].

Bioventing

Bioventing simply involves the supply of low airflow to microorganisms in the unsaturated (vadose) zone above the ground water level, stimulating them to degrade pollutants at a faster rate [93]. In bioventing, only the amount of oxygen required is supplied. This process is suitable for petroleum hydrocarbon pollutants and reduces the risk of volatilization pollutants [4, 21, 94].

Bioaugmentation

For bioremediation to be effective there is usually a replacement of exhausted microbial cells. Addition of indigenous or exogenous microorganism to a polluted environment for bioremediation purposes is known as bioaugmentation [93]. Acclimatization of exogenous microorganism to a polluted environment is feasible but slow, and is quite difficult for them to compete for nutrients and other essential resources with indigenous ones [4, 21].

Biostimulation

This technique simply involves the addition of specific nutrients required for the proliferation of indigenous microorganisms. These nutrients could be in the form of fertilizers, trace minerals, and growth promoters, which are been injected into the contaminated site as well as conditioning environment to suit the temperature, pH, and oxygen requirements in other to enhance the rapid growth and metabolism of microorganisms [93].

Ex situ

Bioremediation that requires the treatment of pollutants away from the polluted site is known as *ex situ*. It requires the transportation of excavated soils or pumping of pollutants to the treatment site [4, 22, 93, 94]. *Ex situ* treatment is thorough and efficient in the treatment of pollutants. It is independent of the natural environmental conditions and could be easily monitored unlike *in situ* processes [93]. *Ex situ* process; however, exposes laborers and other environmental surfaces to the pollutants [4, 93]. Depending on the stage of the pollutants, *ex situ* process could be characterized either as a solid phase (*i.e.* land farming, composting and biopiles) or slurry phase (*i.e.* bioreactors) [22].

Solid Phase

This phase mostly involves the treatment of organic waste such as animal manures, leaves, agricultural waste as well as domestic and industrial waste, municipal solid waste, sewage sludge. Some techniques used in the treatment of pollutants in solid phase include land farming, composting and biopiles [4, 21, 22].

Land Farming

Land farming is simply a technique that involves the excavation of polluted soil spreading it over a prepared bed and tilled at time intervals until pollutants are completely degraded. The essence of tilling the soil at regular intervals is to provide the soil with sufficient oxygen required for aerobic degradation. This technique is cheap and simple, and does not require constant monitoring. This treatment process is, however, limited to a certain depth (10-35 cm) of the soil [22].

Composting

In this *ex situ* method of bioremediation, the polluted soil is covered in a compost heap mixed with organic waste from farmlands and manure. This organic waste helps in providing nutrients and elevating the temperature required for the growth and proliferation of microorganisms ultimately facilitating the degradation of pollutants [4, 22].

Biopiles

Biopile is a technology that is derived from composting and land farming. It involves utilization of engineered microbial cells to degrade pollutants arranged as piles above ground with apparatus that allow the supply of oxygen and a leachate system, which helps in reducing the concentration of pollutants and prevents volatilization of pollutants in the biopiles [22, 93].

Slurry Phase

In the slurry phase, bioremediation is relatively rapid when compared to other treatment processes. This is because the pollutants are placed in a bioreactor mixed with water, nutrients and microorganism. In the bioreactor, optimum conditions such as temperature, pH, oxygen and nutrient source can be attained and maintained for better microbial degradation of pollutants. When this process is completed, liquid in the bioreactor is separated, which could be treated before discarding [21, 22].

Fungi

Fungi are eukaryotes that were initially placed under the kingdom 'Plantae' [95]. With time, study and advancement in technology and knowledge on evolution made it possible for fungi to be placed under the kingdom 'Fungi'. They were separated from the plant kingdom because they possess chitosans, glucans, and chitin in their outer cell wall [25, 44, 95 - 98]. Fungi or funguses are the plural of fungus; they include unicellular yeasts multicellular molds, and macroscopic filamentous mushrooms [25, 99]. Planet earth is home to various groups of organism with fungi being second to arthropods in terms of biodiversity with an estimate of 1.5 million species. Despite this large value, this kingdom remains largely unexplored with only about 100,000 species currently identified and classified [97, 100, 101]. Literature report by Newbound *et al.* [102] classified fungi into five major groups; they are Zygomycota, Ascomycota, Glomeromycota, Chytridiomycota and Basidiomycota. A typical fungus body is made up of a collection of hyphae known as mycelium. A mycelium consists of long, branched filamentous hypha made up of one or more cells enclosed in a tube like cell wall, which extends as the fungus grow through the substrate. Hyphae are approximately 4-6 μm in diameter and can serve as the vegetative part of fungi. The hyphae can be separated into various cells by an internal cross wall known as septa and each of the septa contains organelles (*i.e.* mitochondria, ribosomes, nuclei) that flow within the cells [95].

Fungi like bacteria are ubiquitous and unique. Their genetic, physiological, morphological features make it possible for them to survive and thrive in varying matrices (water, soil, and air) where they play essential roles in maintaining the stability of the ecosystem [6]. Fungi are present in the air in the form of spores, conidia and hyphae and are most times transient. The aquatic environments (fresh water, estuarine and marine) are constantly colonized by fungi. However, the soil serves as the major habitat for fungi proliferation where they are often found in a symbiotic relationship (*i.e.* mutualism and parasitism) with other organisms [25, 99]. All fungi are heterotrophs and have been found to thrive in extreme environmental conditions, such as those contaminated by toxic pollutants [101, 102]. They have been found growing in industrial effluents, petroleum hydrocarbon and heavy metal polluted environments. They have also been seen growing on organic debris and are major degraders of agricultural waste and carried out saprophytic mode of feeding enabled by their mycelia and the digestive enzymes they secrete into the environment capable of degrading polymers such as cellulose and lignin [6, 25, 99, 101, 103]. Also, the rhizomorphs developed by varying fungal mycelia can grow and cover large area of contaminated environments far away from the parent mycelial and can withstand toxic substances owing to the component of their cell wall [23]. This unique feature has made fungi an important tool in bioremediation processes. The technology that involves the use of fungi and their metabolites in the remediation of environmental pollutants is known as Mycoremediation [16, 24, 95].

Mycoremediation

Mycoremediation is a technology that involves the use of fungi or their metabolites in the suppression, removal, detoxification and destruction of contaminants capable of causing harmful effects on the environment [2, 25, 104]. Mycoremediation helps in restoring balance to the ecosystem and ensuring stability. Both macrofungi (mushrooms) and filamentous fungi (mold) are well adapted to growing in adverse conditions than bacteria and yeast [24, 105]. Their pattern of growth (*i.e.* network of mycelia), body structure, production of extracellular enzymes and bioactive compounds, and ability to co-metabolize chemical pollutants and translocate nutrients and water from environmental pollutants make their usage for bioremediation purposes feasible [24]. Varying literature have reported the pervasive nature of fungi making them capable to survive in stressed environmental conditions such as pH, temperature, nutrients, moisture among others [23, 25]. Mycoremediation is a cheap, environmentally friendly, and efficient method of bioremediation. It helps in alleviating an environment polluted by different types of pollutants [16, 20].

Among the five groups of fungi mentioned earlier, the Basidiomycetes have distinguished themselves to be the most efficient group for bioremediation purposes [2, 24]. They have the potential to break down the most difficult groups of polymers from the environment. Basidiomycetes are grouped into two; white-rot fungi (WRF) and brown-rot fungi (BRF) [100]. BRF can degrade cellulose efficiently but cannot degrade recalcitrant lignin and lignin-like compounds. This is, however, different from WRF, as they can efficiently degrade cellulose, hemicellulose, lignin and lignin-like compounds from the environment [100]. They are the only group of fungi capable of degrading wood owing to the complex groups of enzymes such as lignin peroxidases (LiP), Lacs (Lac), phenoloxidases and manganese peroxidases (MnP) they secrete [23, 38]. The white-rot fungi (WRF) can degrade a wide range of environmental pollutants including pesticides, pentachlorophenol, industrial dyes, coal tars, heavy fuels, PAH, PCBs, monoaromatic hydrocarbon (Benzene, Toluene, Ethylbenzene, Xylene (BTEX)) among many more into CO_2, H_2O and basic elements [20, 24, 101]. They have been successful in the bioremediation of environments polluted by heavy metals and radioactive compounds [16, 100]. Most known WRF are basidiomycetes, although a few ascomycete genera within the Xylariaceae are also capable of white rot decay [1].

White Rot Fungi

White rot is a term commonly used to delineate the characteristic appearance of wood decay produced by a fungal attack. White rot fungi (WRF) breakdown lignin and lignin-like compounds in wood under nitrogen depleted conditions to utilize nutrients from the wood [100, 106 - 110]. Wood attacked by WRF makes them appear bleached, with a stringy, spongy or laminated structure (Fig. **5**), thus, the name white-rot fungi [106, 112, 111, 112]. White rot fungi (WRF) is not a taxonomic grouping of fungi but rather, a physiological one. As such, it is not limited only to basidiomycetes; it also includes some ascomycetes within the xylariaceae family [1, 112, 113]. It is interesting to note that, the enzymes secreted by the WRF break down lignin so as to utilize the cellulose present in the interior of the wood. The complete mineralization of lignin by WRF and other carbohydrate compounds to H_2O and CO_2 is attributed to their extracellular enzymes, which they secrete onto the wood [101, 107, 111, 112]. The first fungus to be discovered with ligninolytic potential was *Phanerochaete chrysosporiun*. Since its discovery in the 1970s, it has become the most studied fungus used in various bioremediation processes of harmful pollutants [113].

Fig. (5). WRF *Phanerochaete chrysosporiun* characteristic whitish, stringy and spongy appearance when degrading lignin and lignin-like compounds in wood [112].

Table **4** shows different types of pollutants remediated by *P. chrysosporiun* [113, 114]. Since then, various studies have been carried out on WRF leading to the discovery of other genera showing credible potential as tools for bioremediation processes. They include *Bjerkandera, Phlebia, Coriolopsis, Trametes,* and *Pleourotus* [114]. Remediation of pollutants by WRF is achieved through various mechanisms including; biosorption, biodegradation, biomineralization, adsorption, bioaccumulation (Fig. **6**). Table **5** carries information on the type of pollutants these genera have been able to successfully neutralize.

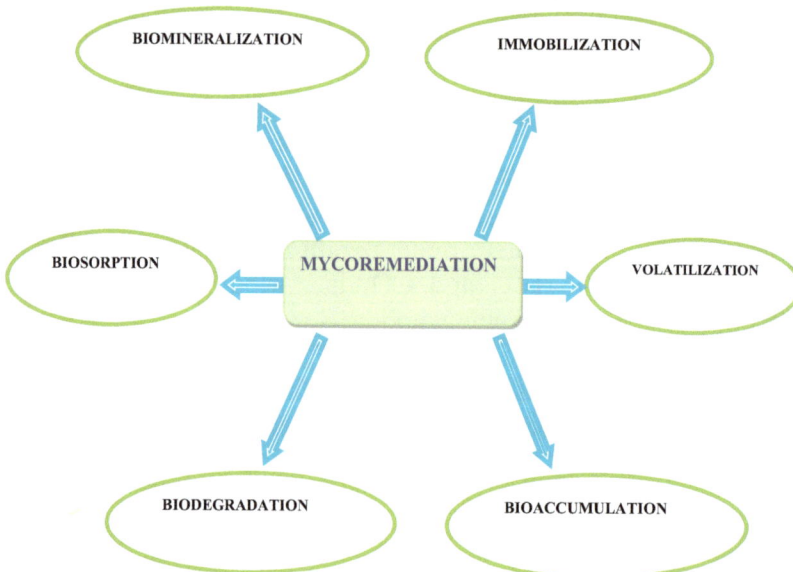

Fig. (6). Mechanisms used in mycoremediation of pollutants from the environment.

Table 4. Bioremediation of different pollutants by *Phanerochaete chrysosporiun.*

Types of pollutants	Pollutants	Sources
Heavy metals	Nickel, chromium, arsenic, boron, cadmium, mercury	[7, 115, 116]
Synthetic dyes	Remazol brilliant blue royal (RBBR), mordant blue 9, methylene blue, astrazon red FBL, Red-80, orange G, triphenylmethane dyes, naphthalenic dye amaranth dye, atrazine, indigo carmine, Anthraquinone dyes, azo dye	[1, 7, 113, 114, 116 - 118]
Drugs	Naproxen, diclofenac, ibuprofen	[1, 114]
Chlorophenols	Pentachlorophenol (PCP), 2,4,6-trichlorophenol, 2,4-dichlorophenol, 4-chlorophenol.	[1, 113, 114]
PAHs	Pyrene, phenanthrene, perylene, anthracene, benzo[a]anthracene, trinitrotoluene (TNT), pentachlorophenols (PCP), fluoranthene, fluorine, BTEX.	[1, 37, 113 - 115, 117 - 120]
Pesticides	Bentazon, trifluralin, MCPA, simazine, and dieldrin, 1,1,1-trichloro-2-2-bis(4-methoxyphenyl)ethane (DDT), dioxin 2,7-dichlorodibenzo-p-dioxin	[1, 114, 116]

Table 5. Bioremediation of pollutants by different white-rot fungi.

Fungus	Pollutants	Sources
Pleurotus ostreatus	Benzo[a]pyrene, carbamazepine, Polychlorinated biphenyls (PCBs), anthracene, benzo[a]anthracene, Naphthalen sulphonic acid, anthracene; Acid Orange 7, Mordant Violet 5, Acid Orange 8, Chromium, Nickel, Cadmium, Zinc,	[1, 113, 114, 117, 118, 120]
Trametes versicolor	Fluoranthene, chrysene, trichlorobenzenes, naproxen, Sulfonamide, phenanthrene, benzo[k]fluoranthene, benzo[a]anthracene, benzo[b]fluoranthene, pyrene, perylene, 1,1,1-trichloro-2,2-bi-(4-methoxyphenyl)ethane (DDT), polybrominated flame retardant, triphenylmethane dyes, trichlorobenzenes (TCBs), naphthalenic dye amaranth, atrazine. arsenic, nickel, aluminium, chromium, lead, zinc, cadmium, mercury	[1, 113, 114, 116, 117, 120]
Coriolus versicolor	PCBs, chrysene, perylene, phenanthrene, Acid black 210, reactive black B(S), Crystal violet, Acid blue 193, reactive black BL/LPR	[1, 113, 118]
Irpex lacteus	Phenanthrene, benzo[a]anthracene, benzo[a]pyrene-1,6-quinone,) (BP1,6-quinone), 3-hydroxybenzo[a]pyrene (3-OHBP), Dark blue 2SGL-01 dye, PCBs, Levafix blue E-RA dye, chromium, lead, zinc, cadmium, mercury	[7, 113, 114, 116, 117]
Bjerkandera adusta	PCBs, Fluoxetine, citalopram, sulfametoxazole, diazepem, naproxen, chlorobenzoic acids, ibuprofen, carbazepine, daunomycin, hexachlorocyclohexane, pentachlorophenol, nickel, lead, chromium	[1, 113, 114, 116 - 118, 120]
Anthracophyllum discolor	Pyrene, pentachlorophenol, anthracene, 2,4-dichlorophenol	[114]

(Table 5) cont.....

Fungus	Pollutants	Sources
Trametes modesta	Trinitrotoluene (TNT), dyes	[1, 113]
Lentinus tigrinus	Fluoranthene, pyrene, perylene, 1,1,1-trichloro-2,2-bi-(4-methoxyphenyl)ethane (DDT)	
Phlebia acerina	Dichloroaniline isomers o-hydroxybiphenyl, TNT, 1,1'-binaphthalene and naphthalene, PCBs	[7, 113]

Enzyme System of WRF

The WRF possesses a series of extracellular ligninolyttic enzymes collectively called lignin modifying enzymes (LMEs). The LMEs lack substrate specificity, aside lignin and lignin-like compounds; they attack wide arrays of both synthetic and organopollutants including the discoloration of synthetic dyes [121 - 123]. Their extracellular nature allows them to act on insoluble and non-polar compounds unlike the intracellular enzyme cytochrome P450 [123]. The three major components of LMEs are; lignin peroxidase (LiP), manganese peroxidase (MnP) and lacasses, which are produced as a response to a nutrient depleted environment [25, 95, 123 - 127]. Another type of enzyme has been reported in *Pleurotus* sp., and some other fungi known as versatile peroxidase (VP) [128]. This enzyme has catalytic properties of MnP, LiP, and other peroxidases of plants and microorganisms [120, 128].

Lignin Peroxidase

Lignin peroxidase (LiP, EC 1.11.1.14) is a glycosylated heme protein that catalyzes H_2O_2 dependent oxidation of aromatic compounds having lignin-like structures [93, 129]. They can oxidize diverse groups of chemical compounds including non-aromatic phenolic compounds due to their high redox potential, which is higher than most peroxidases [93, 130]. Lignin peroxidase is activated by oxidation of H_2O_2 to give Fe^{4+} with a porphyrin cation radical, which mediates the oxidation of phenolic substances [129]. Oxidation of xenobiotics by LiP is achieved through free radicals generated from the secreted metabolites such as veratryl alcohol, which is secondary metabolite and have been found to have an essential role in LiP catalysis and also serve as an ideal stimulant produced by WRF. The H_2O_2 oxidizes LiP to produce two electron intermediates known as compound I. One electron from compound I cause oxidation of substrate leading to the formation of compound II, which is a more reduced enzyme intermediate. The enzyme is returned to the ferric state when compound II oxidizes the substrate by the other electron. In the presence of H_2O_2, compound II is very reactive. However, in the presence of excess H_2O_2 and a poor substrate, it causes the formation of compound III an inactive form of the enzyme when compound II reacts with the poor substrate. The presence of veratryl alcohol is suitable for compound II and favors its conversion to the resting enzyme, hence the catalytic cycle is completed (Fig. **7**) [131].

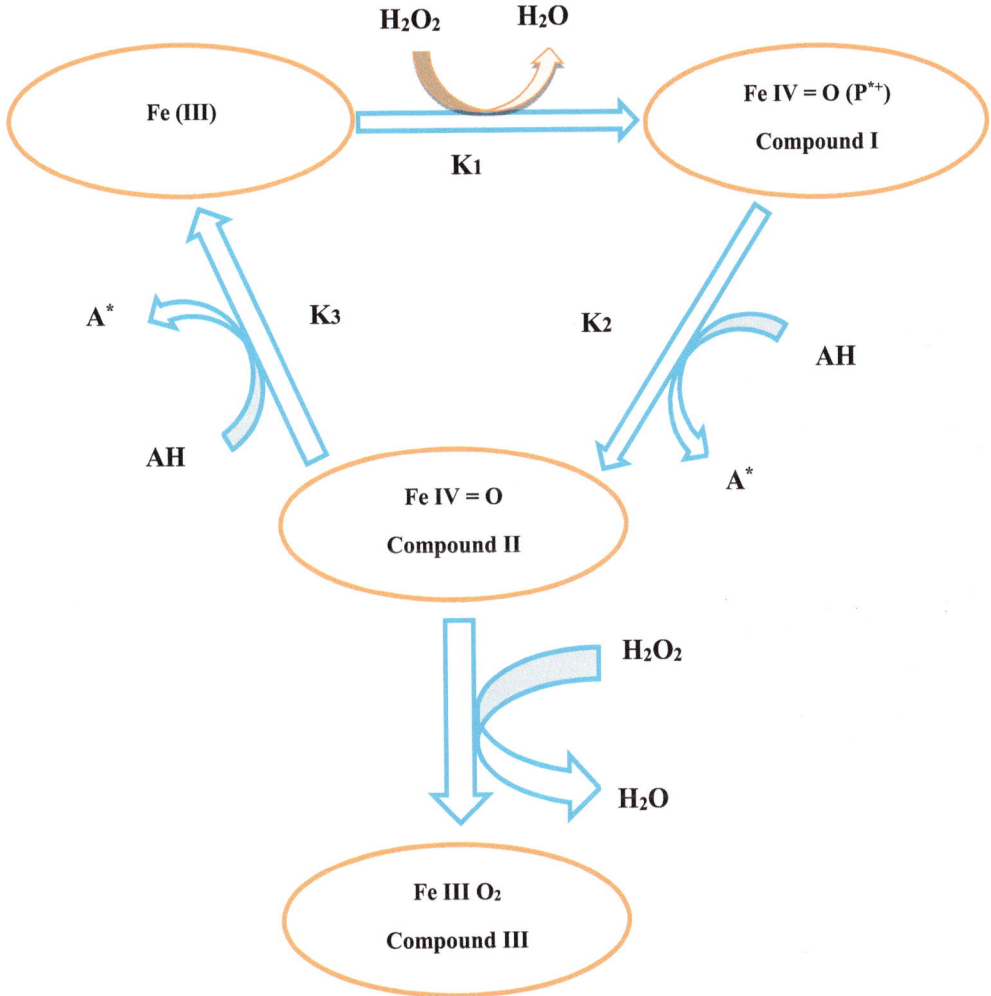

H_2O_2 H_2O

Fe (III)

Fe IV = O (P^{*+})

Compound I

K_1

A^* K_3 K_2

AH

AH

Fe IV = O

Compound II

A^*

H_2O_2

H_2O

Fe III O_2

Compound III

Fig. (7). Catalytic cycle of LnP.

Manganese Peroxidase

Manganese peroxidase (MnP, EC 1.11.1.13) like LiP, is a glycosylated heme protein that was first discovered in *P. chrysosporium* [93, 110, 128]. Manganese peroxidase also uses H_2O_2 to oxidize Mn^{2+} to Mn^{3+} a highly reactive chelate oxalate that serves as an oxidant, but is not specific to a particular substrate (Fig. **8**) [128, 130, 131]. Mn^{3+} is generally accepted as an oxidant that can diffuse away from its point of production and oxidize varying monomers, phenolic compounds, phenolic lignin dimers, synthetic lignin and dyes [128 - 130].

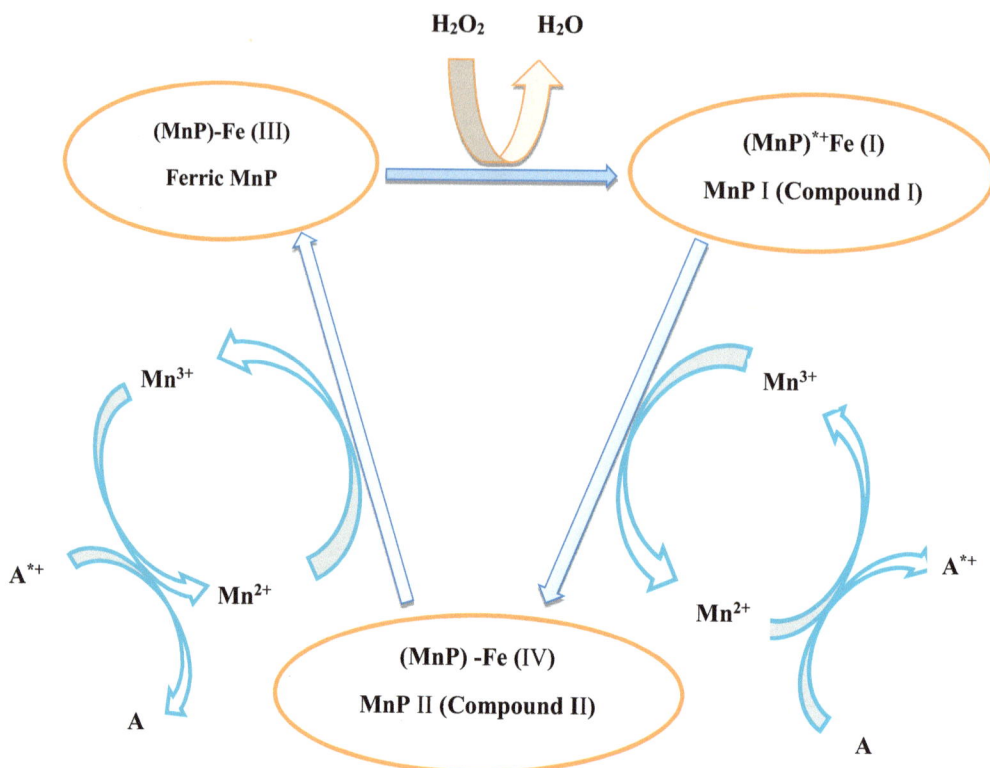

H_2O_2 H_2O

(MnP)-Fe (III)

Ferric MnP

$(MnP)^{*+}Fe$ (I)

MnP I (Compound I)

Mn^{3+}

Mn^{3+}

A^{*+}

A^{*+}

Mn^{2+}

Mn^{2+}

(MnP) -Fe (IV)

MnP II (Compound II)

A

A

Fig. (8). Catalytic cycle of MnP peroxidase [134].

Versatile Peroxidase

Versatile peroxidase (VP, EC 1.11.1.16) is the hybrid of LiP and MnP [110]. The catalytic potentials of LiP and MnP is exhibited by VP making it active against a wide range of substrate [93, 130]. Versatile peroxidase like MnP and LiP produces oxidants mediated by H_2O_2. The oxidant produced oxidizes both phenolic and non-phenolic aromatic compounds and many other compounds with lignin-like structures, making their application in the environmental degradation of pollutants highly desirable [129, 130].

Laccase

Laccase (Lac, benzenediol, oxygen oxidoreductases, EC 1.10.3.2) is a polyphenol oxidase enzyme that contains copper in its active site [37, 110, 129, 131 - 135]. Unlike the peroxidases (LiP, MnP and VP) that use H_2O_2 as the final electron acceptor; Lac uses molecular oxygen (O_2) as the final electron acceptor to release free radicals (Fig. **9**) [120, 125]. The free radicals (*i.e.* reactive oxygen species (ROS) H_2O_2) oxidized by Lac cause cleavage of bonds from complex chemical structures [93, 101, 125, 130, 135]. Lac was discovered first in 1883, which includes it among the earliest enzymes described [120, 135]. In the presence of O_2, Lac converts phenolic compounds to radicals of quinones and further to quinones [129, 134]. Though the redox potentials of Lac are low compared to most of the ligninolytic enzymes but are sufficient enough to cause oxidation of various recalcitrant compounds in the environment including; chlorinated phenol, pesticides, synthetic dyes, PAHs among many others [123, 129, 131, 136, 135].

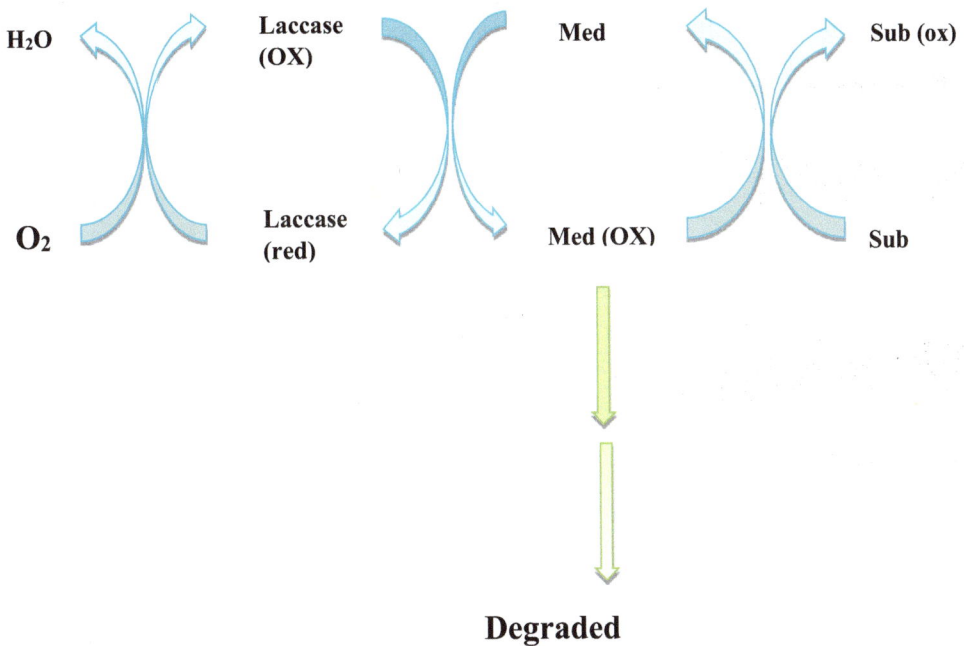

Degraded

Fig. (9). Catalytic cycle of laccase [134].

Cytochrome P450s Monooxygenase

Cytochrome P450s (CYP450s) is a heme-thiolate, oxidoreductase intracellular enzyme used by WRF for intracellular degradation of pollutants [12, 24, 114]. The WRF can absorb pollutants into their hyphae, once inside, CYP450s act on it with other intracellular enzymes, causing the oxidation of its substrate through the incorporation of a molecule of oxygen [125]. Epoxidation, dealkylation, sulfoxidation and hydroxylation can occur with NAD(P)H acting as the electron donor [24]. CYP450s have been reported in the degradation and mineralization of some pharmaceutical products including; antianalgesic drugs, lipid regulator drugs, anti-inflammatory drugs, antiepileptic drugs and diphenyl ether herbicides

[127]. Various other intracellular enzymes aside CYP450s have been involved in the breakdown of pollutants in WRF, including; phenol 2-monooxygenases, nitroreductases, reductive dehalogenases, and miscellaneous transferases [127].

Mycoremediation of Pollutants

Various types of pollutants and the effects they imposed on the ecosystem have been previously discussed. The metabolites (*i.e.* enzymes) and the physiology of WRF have made them an ideal tool for bioremediation purposes. There are different mechanisms used by WRF for the bioremediation of pollutants. They either biodegrade pollutants through actions of enzymes to break bonds of complex structures into simpler non-toxic forms or adsorb or bioaccumulate the pollutants in the case of heavy metals [116].

Mycoremediation of Petroleum Hydrocarbons

Extracellular LMEs secreted by WRF has been found to be effective in the bioremediation of recalcitrant PAH, a group of petroleum hydrocarbon that is often found polluting the environment. The LMEs are stimulated by a nutrient depleted environment, thereby making this technology cost effective [116]. The mechanism of degradation of PAH by WRF is similar to that used in the degradation of lignin and lignin-like compounds, which utilizes the free radical mechanism [101].

For degradation of recalcitrant xenobiotics, the agents involved (*e.g.* LMEs) must find a way of breaking the bonds holding the rings together by reactive oxygen species (ROS) and other oxidants. This is exactly what LMEs do. They may act individually or synergistically to cause opening of rings and cleavage of bonds [95]. Since they are nonspecific to the substrate and can diffuse into compounds (*i.e.* M^{3+} chelates), they can cause oxidation of aromatic compounds either directly or indirectly with the formation of ether peroxide [95]. Once this is established, the rings are opened, and the long chains cleaved such that the compound is changed to a non-toxic, biodegradable form.

Studies have reported the indirect oxidation of phenolic compounds mediated by radicals of LMEs causing the spontaneous opening of rings to produce derivatives of muconic acid and the decarboxylation of the carboxyl groups formed to mineralize CO_2 among various other intermediates produced in the process (Fig. 10) [95]. In the case of intracellular degradation of PAH by WRF, CYP450s and other intracellular enzymes such as transferases play a key role in achieving it, through the oxidation of the xenobiotic with further metabolism of the intermediate products in the tricarboxylic acid (TCA) cycle to produce CO_2 [6,

116, 120]. Various studies have reported the potential of WRF in the biodegradation of PAHs. A study from Tortella *et al.* [114] reported the Lac producing potentials of *Trametes versicolor*, which it employs in the degradation of PAHs such as pyrene, phenanthrene, fluoranthene, benzo(k)fluoranthene, benzo(b)fluoranthene, chrysene, perylene, benzo(a)anthracene.

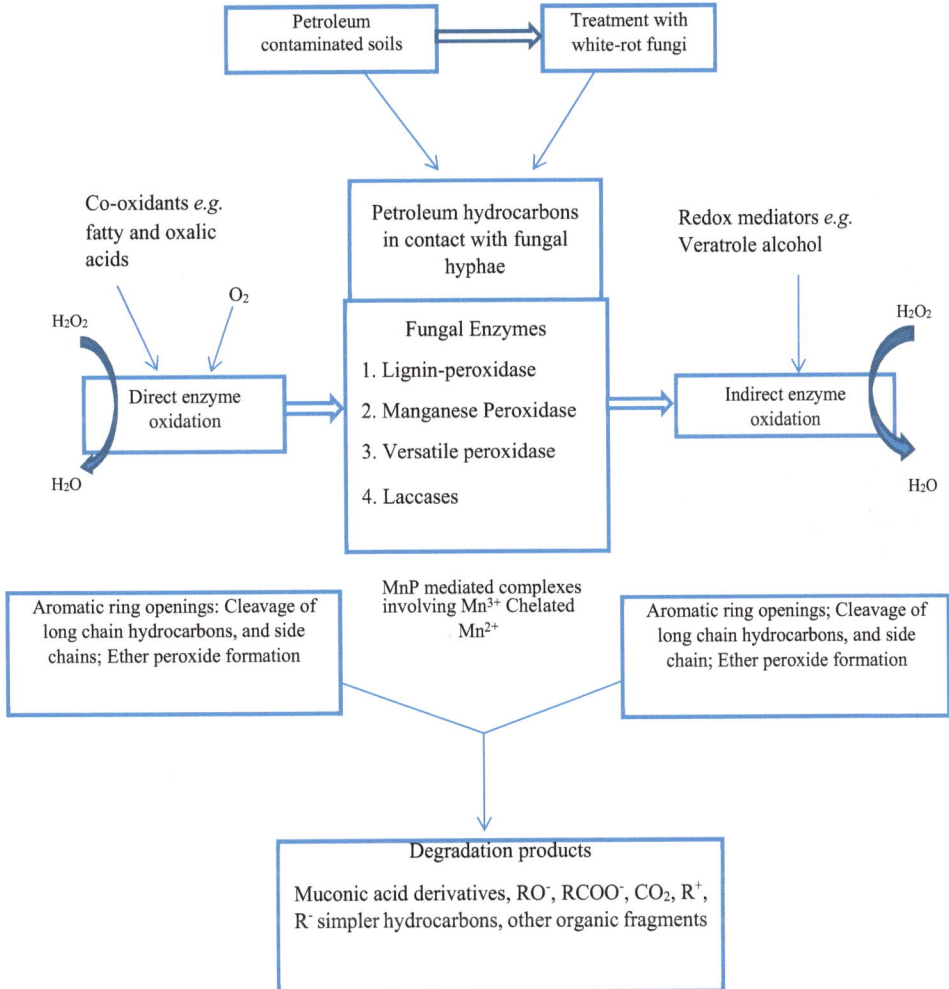

Fig. (10). Mechanism of mycoremediation of petroleum-contaminated soil by white rot fungi [97].

Though PAHs are known to exhibit high hydrophobicity, an obstacle frequently encountered in liquid medium was, however, overcome by the inclusion of organic solvent facilitating the degradation of four representatives PAHs (anthracene, phenanthrene, pyrene and fluoranthene) by *I. lacteus in vitro*. Thanks to the cascades of LMEs produced [120]. Likewise, the WRF fungi *Agrocybe* sp., which was isolated in Thailand, was able to degrade both low and high molecular weight PAHs. At a concentration of 100 ppm, *Agrocybe* sp. was able to degrade 99% of fluorene and phenanthrene and 92% of anthracene after six days of incubation [120]. Better removal of pyrene and phenanthrene has been associated with high Lac production of 3000 U/L in *T. versicolor* and 2700 U/L in *P. ostreatus* as reported by Singh *et al.* [127]. Zhao *et al.* [136] studied manganese peroxidase, MnP-Tra-48424, isolated, purified and characterized from *Trametes* sp. 48424 on the effects of pH, temperature, and presence of metal ions and solvent on the degradation of PAHs (*i.e.* fluorene, pyrene, fluoranthene, anthracene, and phenanthrene). In their study, it was observed that pH and temperature had greater effects on MnP-Tra-48424 activities, whereas contaminants such as metal ions and solvents had little or no effects on MnP-Tr--48424 activity. However, it was noted that total inhibition of MnP-Tra-48424 was recorded in the presence of Al^{3+}. Total degradation of 10 mg/l of studied PAHs was obtained after 24 h of incubation at a temperature of 68°C and a pH of 5. Deviation from these pH and temperature values recorded a declination of MnP-Tra-48424 activity [136]. Wang *et al.* [137] investigated the ability of spent mushroom (*Lentinula edodes*) the removal of 200 mg/kg of PAHs (*i.e.* fluoranthene, naphthalene, anthracene, pyrene, and phenanthrene) for 42 days in greenhouse conditions at pH of 4.3 ± 0.11 and temperature was maintained between 20-30°C. At the end of 42 days of the study, it was observed that above 74% of PAHs were removed with remarkable activities of MnP and Lac detected.

The addition of adsorbent such as pulp and kapok have been related to an increased (93%) rate of degradation of 3,000 mg of crude oil by *Polyporus* sp. S133 after 60 days of incubation at a temperature of 25°C. It was noted that an increase in the flow of oxygen led to an increase (73%) rate of degradation by *Polyporus* sp. S133 [138]. Hadibarata *et al.* [139] delineated that *Polyporus* sp. S133 utilizes dioxygenases and phenoloxidases as the pathways in the oxidation of phenanthrene. These degradation pathways are used by *Polyporus* sp. S133 was determined as a result of the detection of phenanthrene metabolites produced after the degradation process. They include 2,2'-diphenic acid, 9,10-phenanthrenequinone, salicylic acid, and catechol. Ring cleavage of phenanthrene as a result of 9,10-oxidation gave rise to 9,10-phenanthrenequinone as the major fate of *Polyporus* sp. S133 degradation of phenanthrene. *Trametes polyzona* RYNF13 a thermotolerant WRF was used by Teerapatsakul *et al.* [140] in the degradation of three PAHs (*i.e.* fluorene, phenanthrene and pyrene). In the study, *T. polyzona* RYNF13 was observed to be highly effective in the complete breakdown of phenanthrene even at a concentration of 100 mg/L for 18 days at 30°C including pyrene (52%) and fluorene (90%). An increase in temperature from an initial 30°C to 42°C reduced the effectiveness of the *T. polyzona* RYNF13 limiting degradation of pyrene to 30%, fluorene 48% and phenanthrene 68%. It is important to note that MnP was abundantly produced during the degradation of PAHs at all temperatures, however, at 37°C and 42°C MnP and Lac were observed whereas LnP was only detected at the initial incubation temperature of 30°C.

Mycoremediation of Dyes

Discoloration and detoxification of synthetic dyes produced by textile, paper, printing, pharmaceutical, food and chemical industries have been feasible using WRF [114, 120, 127]. The mechanism of discoloration of these dyes by WRF is either through degradation by extracellular LMEs, adsorption by fungal mycelia, or through biosorption followed by intracellular mineralization caused by metabolic processes [56, 121]. The LMEs are broad in action and have been able to cause oxidation of the chromophores responsible for coloration [116]. Diverse WRF has been reported to decolorize textile effluents through the LMEs system, they include; *Coriolus versicolor, Pleurotus ostreatus, Lentinus tigrinus, Phlebia tremellosa, Bjerkandera adusta, Phlebia acerina, Inonotus hispidus, P. chrysosporium, Tramtes versicolor* among others [7, 111, 114, 116, 118, 127]. Their efficiency depends on the fungal species and the type of dye. Pinedo-Rivilla *et al.* [7] reported different degradation efficiency between *Trametes* sp., which was able to decolorize 76.07% of the textile effluent and *Phanerochaete* sp. having the highest discoloration rate 82.01%. It has also been noted that activities of members of LMEs differ with different fungal species and substrates *T. versicolor* was able to degrade different types of azo anthraquinone dyes while *B. adusta* is able to degrade various reactive and acid dyes at a faster rate. More secretion of MnP and Lac was observed in *T. versicolor* while *B. adusta* produced more LnP as reported by Tišma *et al.* [118] and Zhang *et al.* [141].

The efficiency of purified MnP-Tra-48424 from *Trametes* sp. 48424 in the decolorization of 100 mg/L of Remazol brilliant blue royal (RBBR), indigo dye, methyl green and remazol brilliant violet 5R resulted in 85.0%, 94.6%, 93.1% and 88.4% decolorization respectively within 4 h. Jebapriya and Gnanadoss [121] stated that adsorption as a type of mechanism in the discoloration of dye effluents is inefficient and generally contributes less than 50% of the total discoloration as *C. versicolor* was examined for discoloration of dye effluents through adsorption. It was discovered that only 5-10% of discoloration occurred through adsorption. This is however contrary to the findings of Pinedo-Rivilla *et al.* [7]. They reported that dead cells of *P. ostreatus* BP were able to decolorize 82.35% of RBBR. As mentioned earlier, the rate of discoloration by WRF regardless of the mechanism is dependent on the type of fungal species and the type of dye. Bhattacharya *et al.*

[142] also reported the decolorization potentials of *P. ostreatus* against Congo red. It was observed that various environmental conditions such as oxygen, pH, nutrient and temperature all played crucial roles in the decolorization of Congo red. Agitation and biostimulation of the fermentation medium with NPK fertilizer and adjustment of pH to neutral provided and impressive decolorization rate above 85% within 72 h of incubation at 30°C. *Daedalea flavida* a WRF was reported by Rani *et al.* [143] showcasing its potential in the decolorization of four azo dyes (metanil yellow, trypan blue, amaranth and chlorazole black) in the presence or absence of external carbon source. *D. flavida* was observed to decolorize 99% of Amaranth after 5 days; 82.5% of metanil yellow after 10 days; 72% of chlorazole black E and 99% of trypan blue after 10 days of incubation at 36°C with high Lac activity detected. Vasdev *et al.* [144] reported the decolorization of 0.3g/L of triphenyl methane dyes (malachite green, crystal violet and bromophenol blue) was successful by six strains of WRF.

All the dyes were decolorized although at a different pace. Malachite green was decolorized the fastest (88%) followed by crystal violet (79%) and bromophenol blue (74%) within 72 h with the detection of varying concentrations of Lac. Yuan *et al.* [145] similarly reported the activity of Lac in the decolorization of alizarin red by *Antrodiella albocinnamomea*. *A. albocinnamomea* was demonstrated to achieve 93.37% decolorization rate within 8 days at optimum pH of 5.5 and temperature of 28°C. Chairin *et al.* [146] demonstrated how a redox mediator such as 1-hydroxybenzotriazole (HBT) was able to enhance the decolorization of dyes (RBBR, bromophenol blue, methyl orange, Congo red acridine orange, and relative black 5) by Lac of *T. polyzona* when added to the reaction mixture. It was observed that all reaction mixtures with HBT had an improved performance against those without HBT. The decolorization rate varies between 58 – 100% of reaction with HBT against 27- 100% without HBT after 24 h. Senthilkumar *et al.* [178] studied the pH range at which *P. chrysosporium* will be able to efficiently remove Amido black 10B from synthetic effluent. It was reported that *P. chrysosporium* was able to grow across a wide range of pH 2-8. However, the best pH for removal of Amido black 10B from the effluent was recorded at pH 7 (95%) at 37°C. Sumandono *et al.* [147] screened for Lac producing WRF, of the 24 species collected only 9 were able to oxidize guaiacol at a concentration of 500 ppm indicating positive production of Lac. However, only one strain identified as KRUS-G was able to decolorize RBBR right from the first day of inoculation to the 5th day (Fig. **11**). When *P. crysosporium* ATCC 34541C and *Ceriporiopsis subvermispora* ATCC 90467 were compared to the newly isolated strain KRUS-G. It was observed that strain KRUS-G outperformed them in the decolorization of RBBR across all concentrations (100, 500, 1000, 1500 ppm). The highest decolorization (89%) of RBBR at 100 ppm was obtained by KRUS-G against *C. subvermispora* ATCC 90467 (70%) and *P. chrysosporium* ATCC 34541C (20%)

at pH of 4 and temperature of 28°C. It is interesting to note that KRUS-G was still able to maintain high decolonization of RBBR from pH 4 (89%), pH 5 (80%) and pH 6 (82%) still maintaining a temperature of 28°C for 6 days.

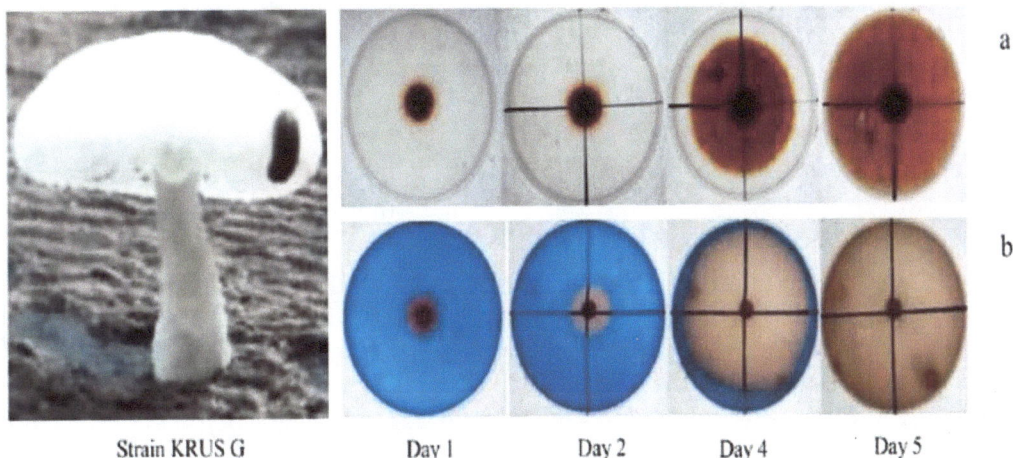

Fig. (11). (a) Oxidation of guaiacol (500 ppm) by KRUS-G and **(b)** Rapid decolorization of RBBR (500 ppm) on PDA medium [150].

Involvement of LMEs particularly Lac in decolonization of industrial dye was also demonstrated by Kunjadia *et al.* [179], whereby Lac from three *Pleurotus* species (*P. sapidus, P. ostreatus*, and *P. florida*) enhanced the decolorization of three azo dyes (coralene navy blue, coralene dark red and coralene golden yellow). *P. florida* was reported to have the highest (98%) decolorization rate followed by *P. sapidus* (92%) whereas *P. ostreatus* had the least decolorization rate (88%) across all the azo dyes within 10 days at a temperature of 25°C. Various studies [148 - 153] have all highlighted the essential roles of LMEs system and other intra and extracellular components of various WRF in the decolonization of different dyes.

Mycoremediation of Pesticides

Most chemicals such as paraquat, glyphosate, aldrin, endosulfan, fipronil, pendimenthaline, tribromophenol, DDT, lindane, dieldrin, PCBs among others used in the formulation of pesticides have deleterious effects on living organisms and the ecosystem [116, 154, 155]. Some of these chemicals have been banned and the use of others has been limited [116]. Various studies have reported the efficiency of WRF in the remediation of these persistent pollutants including; *P. ostreatus, Talaromyces helices, Agaricus augustus, Trametes versicolor*, Bjerkandera adusta, *Lentinus edodes, P. aurea, Trametes hirsutus, P. brevispora, P. acanthocystis, Phanerochaete chrysosporium* among others [1, 7, 120, 156].

The mechanism used in the mycoremediation of pesticides by WRF is based on chemical reactions and modifications, which could happen intracellularly with the help of intracellular enzymes to cause a reduction, oxidation, hydroxylation, dehydrogenation, oxidation and reduction of alkyl groups as well as the destruction of aromatic rings. They can also cause biotransformation and mineralization of pesticides through the help of free radicals catalyzed by LMEs and other extracellular enzymes [1, 116, 157].

Akhtar and Mannan [116] reported the biotransformation of aldrin and dieldrin by *P. ostreatus*. They proposed that the biotransformation occurred as a result of hydroxylation and epoxidation reactions. Other reactions such as dechlorination, deoxygenation, esterification, demethylation, depolymerisation, decarboxylation and dehydrogenation mediated by LMEs have also been reported to be associated with biotransformation of pesticides [1, 116]. The free radical mechanisms used by LMEs in the degradation of pollutants make it possible for the addition of oxidants and other free radicals in the aromatic ring causing opening of rings and cleavage of bonds. It is important to note that the removal or addition of an electron to a chemical compound in its ground state makes it very reactive as such it is prone to give or accept electrons from other elements. This has formed the basis of actions of LMEs system enabling them to degrade and detoxify varying pesticides they have never had contact with [1]. Pinedo-Rivilla *et al.* [7] reported the biotransformation of tribromophenol (TBP) (a wood preservative) to tribromoanisole when treated with *Agaricus augustus* and *T. versicolor* [7]. A study by [1] informed us of the active role played by purified LnP and MnP isolated from *P. chrysosporium* in the biotransformation of 2,7-dichlorodibenz--p-dioxin. The enzymes were able to carry out mineralization of 2,7-dichlorodibenzo-p-dioxin through the cleavage of rings due to the two Cl atoms removed as a result of stepwise oxidation, reduction and methylation reactions [1].

Xiao *et al.* [156] reported a study on the efficiency of different species of *Phlebia* in the degradation of aldrin and dieldrin. After 42 days of incubation, it was discovered that *P. acanthocystis* had the highest (56%) capacity to degrade aldrin followed by *P. aurea* (54%) and *P. brevispora* (52%). This shows that different species from the same genus can still differ in the rate at which they breakdown xenobiotics from the environment, which is totally dependent on the metabolites they produced as well as the environmental conditions. Some WRF can still break down pesticides independent of the LMEs system since they have other pathways aside from extracellular enzymes such as co-metabolism of absorbed pesticide in fungal mycelia.

Reports of Prakash [120] support the versatility of WRF, which utilizes all available mechanisms in the breakdown of pesticides. The degradation of lindane by *P. chrysosporiun* was found to be independent of ligninolytic enzymes frequently used by WRF. The toxicity of xenobiotics on the microorganisms used for bioremediation is common. However, filamentous fungi have proven to be effective even in the presence of substrate with high concentration, although the level of tolerance differs among species. Fulekar *et al.* [1] demonstrated tolerance of up to 1,800 mg/L of PCBs by *T. versicolor* and were still able to modify 70% of the PCBs whereas *P. chrysosporium* modified 73% of the PCBs but at a concentration of 600 mg/L [1]. Degradation of various organochlorines (Aldrin, HCH (hexachlorocyclohexane), DDT, dieldrin, endosulfan, endrin and heptachlor) by six *Pleourotus* sp., was feasible but the level of degradation of the organochlorines varies from one species to another [158]. A study by Adelaja *et al.* [159] also reported *P. ostreatus* in the degradation of soil contaminated by 2,2-dichlorovinyl dimethyl phosphate (DDVP). They reported that a higher concentration of DDVP reduced the removal rate of *P. ostreatus* from 57.74% (10% v/w) to 31.64% (60% v/w) after 90 days of incubation. The degradation of lindane by *Phlebia lindtneri* and *Phlebia brevispora* were reported by Xiao and Kondo [160] to be highly efficient in a nitrogen depleted environment with a degradation rates of 73.3% and 87.2%, respectively after 25 days of incubation compared to 64.9% (*P. lindtneri*) and 75.8% (*P. brevispora*) in a nitrogen rich medium. They also suggested that CYP450s played a pivotal role in the degradation of lindane.

Mycoremediation of Pharmaceutical Products

The degradation of PhACs has been a major problem due to their toxic nature and complex structure. This has made them recalcitrant to degradation [161 - 163]. However, WRF fungi on the other hand have an edge over recalcitrant compounds owing to the cascade of LMEs, which are nonspecific to a particular substrate [162]. The major mechanism of degradation and transformation of PhACs by WRF is through extracellular secretion of LMEs that causes the opening of aromatic ring structures and cleavage of bonds by oxidation and other ROS [161 - 163]. The WRF can also absorb the PhACs and co-metabolize them in their mycelia through the action of various intracellular enzymes such as CYP450s [116]. Various WRFs have been reported for the degradation and transformation of PhACs, although their efficiency varies from one species to another and on the type of chemical structure of the PhACs. Reports of Asif *et al.* [162] documented the use of various WRF including; *T. versicolor, P. chrysosporiun, Dichomitus squalens* and *B. adusta*, in the degradation and removal of diverse PhACs such as naproxen, carbamazepine, diazepam, ibuprofen, propranolol, tetracyclin, caffeine, benzophenone, triclosan among many others in a bioreactor. Actions of LMEs are almost inevitable in the degradation of PhACs [162]. The complete removal of diclofenac, citalopram, naproxen, sulfamethoxazole, carbamazepine, and ibuprofen by *B. adusta* and *Bjerkandera* sp. R1, was achieved, whereas only 23-57% of diazepem was degraded [114]. Also, *P. ostreatus* was able to completely degrade carbamazepine through intracellular action of CYP450s and MnP [114].

Various literature have reported that *T. versicolor* is the most efficient in the removal of the most difficult PhACs in treatment plants [163]. Rodarte-Morales *et al.* [161] studied the degradability of *T. versicolor, I. lacteus, Ganoderma lucidum* and *P. chrysosporiun* when exposed to PhACs. After 7 days of incubation, it was observed that ibuprofen was completely degraded by all the test isolates. However, clofibric acid and carbamazepine were recalcitrant to all the test isolates except *T. versicolor* showing significant degradation. Reports of the capacity of *T. versicolor* inoculated in a non-sterile bioslurry system of a membrane biological reactor (MBR) to degrade spiked hydrochlorothiazide (HZT) compounds present on the sludge have been reported [164]. The fungus was able to achieve a total degradation of 66.9% of 12 drugs out of the initial 28 drugs that were detected before the spike.

Migliore *et al.* [165] established in their study that purified Lac alone was not able to degrade oxytetracycline (OTC) by *P. ostreatus*. However, the mycelia growth of *P. ostreatus* in the reaction mixture containing 50 and 100 µg/mL of OTC was reduced to 0.6 µg/mL within the first 7 days, and a complete removal after the following 7 days at 25°C. This claim is similar to the results obtained by Marco-Urrea *et al.* [166]. They reported that Lac and MnP of *I. lacteus, T. versicolor, P. chrysosporium* and *Ganoderma lucidum* were not sufficient in the removal of 10 mg/L of carbamazepine, ibuprofen, and clofibric acid *in vitro*. However, they reported that the fungal mycelia couple with intracellular enzymes such as CYP450s successfully removed the PhACs compounds after 7 days. The roles played by mycelia and LMEs of two species of *Pleurotus* sp. P1 and another basidiomycete identified as BNI in the remediation of carbamazepine (CBZ) and 17α-ethinylestradiol (EE2) have been demonstrated by Santos *et al.* [167]. All three strains completely removed EE2 after 7 days of incubation with *Pleurotus* sp. P1 has the highest performance. However, only BNI was able to remove 47% of CBZ after 28 days of incubation. This performance by BNI was not attributed to the mycelia but LMEs system as Lac was detected in a large amount (282 U/L) with little presence of LiP (6 U/L) whereas MnP was not detected after the first 7 days. Lac production was significantly detected (1512 U/L) at the end of the 28 days incubation period at 30°C above any other LMEs. *P. chrysosporium* was demonstrated by Xia-li *et al.* [168] to be efficient in degrading 53% of 10 mg/L of sulfamethoxazole (SMX) after 24 h and 74% after 10 days. High indulgence of Lac (6237 U/L) in the degradation of SMX was detected after 7 days of incubation at pH 4.5 and a temperature of 30°C compared to LiP (5.3 U/L) and

MnP (17.9). They went further to establish that fungal mycelia were excluded in biodegradation of SMX since the inactivated biomass did not produce a significant difference in SMX concentration. Singh *et al.* [169] made reference to MnP, LiP and Lac in the degradation of ciprofloxacin (CIP) by *P. ostreatus*. Interestingly, other extracellular enzymes such as hemicellulase and endo-β-D-1,4-glucanase were recorded in their study. Imperatively, all the aforementioned enzymes increased activities with an increase in CIP concentration. At 100 ppm of CIP, 82.3% were degraded. This value increased to 95.07% when the concentration was increased to 500 ppm after 14 days of incubation at 25°C ± 2.

Aracagök *et al.* [170] also highlighted the removal of 97% diclofenac (50 mg/L) by crude laccase (2.1 U/mL) of *Trametes trogii* within 48 h at pH 5 and a temperature of 30°C. They went further to establish that diclofenac degradation was not a result of mycelia biosorption but rather a product of degradation by enzymes such as Lac and CYP450s when the inactivated cells (0.05 g) only removed 5% of diclofenac at the end of the experiment. Data collected from various studies by Marcelina, and Wioletta [171], reported biodegradation as the major mechanism used by *T. versicolor* and *P. ostreatus* in the degradation of various PhACs with active involvement of Lac, MnP and CYP450s. Although, there was no information on the actual mechanism utilized by *Ceriporia lacerata* and *T. versicolor* in the degradation of streptomycin [172], however, the efficiency of the isolates cannot be overlooked since it was reported that *C. lacerata* was able to degrade 71.1% (100 ppm) and 50.05% (400 ppm) of streptomycin whereas *T. versicolor* degraded 88.8% and 60.65% of 100 and 400 ppm of streptomycin, respectively within 7 days at pH 5 and temperature of 25 ± 2°C. Alharbi *et al.* [173] mentioned how combining two or more PhACs (carbamazepine (CBZ), diclofenac (DCF), sulfamethoxazole (SMX), and trimethoprim (TMP)) affects the biodegradation potentials of commercially available Lac of *T. versicolor*. The result obtained from their study strongly highlighted that synergistic toxicity from the PhACs affected the efficiency of the enzyme. Lac was able to degrade DCF completely but when combined with other PhACs its efficiency reduced after 48 h, likewise TMP (95% - 35%), CBZ (82% - 34%) and SMX (56% - 49%) under the same conditions and exposure time.

Mycoremediation of Heavy Metal

Heavy metals are recalcitrant to biodegradation as such, are difficult to remove from the environment. However, biovolatilization, biosorption and bioaccumulation are some of the mechanisms WRF uses to remove HM and metalloids from the environment [116]. Biovolatilization involves intracellular reactions of enzymes on organic or inorganic compounds leading to the conversion of these compounds to their volatile derivatives through reduction, oxidation, dealkylation, or methylation [11]. While bioaccumulation involves live microbial cells and depends on the cell's metabolism, biosorption does not and can occur on both living and dead microbial cells [11, 116, 174]. The cell wall of WRF is the first cellular component that comes in contact with HM, they play important role in protecting the cell as well as regulating what goes in and out of the cell [11]. The cell wall of WRF is composed of various polymers including chitosans, chitin, polyphosphates, glucans, proteins, polysaccharides and lipids, which contain various functional groups such as hydroxyl, sulfate, carboxyl, imino, sulfhydryl, carbonyl, imidazole, phosphate, sulfonate, thioether among other moieties that interact with various HM [11, 72, 96]. This interaction could either be in the form of precipitation, complexation, ion exchange, adsorption, chelation, and crystallization among many more, and is influenced by the chemical nature of the metals as well as fungal biomass [11]. *P. ostreatus* was reported to efficiently remove Mn, Cr, Cu, Co, Ni and Zn from coal and waste water effluents through biosorption and bioaccumulation. However, the efficiency of *P. ostreatus* was enhanced by the addition of surfactant to the water effluents, which increase the metal binding site and surface area on the fungal hypha. It was also noted that the activity of antioxidants and metallothioneins, which helps in regulating the cytotoxic effects of HM was high [116].

The WRF like *Phlebia radiata, B. fumosa,* and *T. versicolor* have all been reported [96] to cause solubilization and leaching of HM such as Fe, Al, Li from contaminated soil through the production of phosphatases, organic acids, protons, oxalic acid, which causes the formation of metal oxalate complexes. The removal of Cu and Pb by *P. chrysosporium* was reported by Vaishaly *et al.* [174]. However, the efficiency of LMEs secreted by *P. chrysosporiun* was reduced over time, which could be associated with distortion of LMEs active sites [177]. The observations of Zhao *et al.* [136] is in relation to that of Vaishaly *et al.* [174], which reported reduced activities of LMEs present in *P. chrysosporiun* when it was exposed to Cd and Pb for a long period of time. Kapahi and Sachdeva [16] reported the bioaccumulation of Cd, Cr, Pb and Zn in the fruiting body of *A. bisporus, P. chrysosporiun, Termitomyces clypeatu* and *P. ostreatus.* Likewise, fungi belonging to *Boletus, Suillus, Agaricus, Cortinarius, Leccinum, Amanita* and *Phellinus* have all been applied in the complexation of various HM [24]. Pal and Vimala [175] demonstrated different forms (*i.e.* immobilized, biosorbent or viable cells) of *P. chrysosporiun* MTCC 787 in the removal of Cr in the laboratory using malt extract liquid medium and basal salt liquid medium. In their study, they were able to establish the fact that, the removal rate of HM varies from one form of the cell to another and they all have different optimal temperatures and pH utilized in the removal of Cr. The highest removal rate (97.9%) in basal salt liquid medium at 37°C was recorded in biosorbent form at pH 2 whereas, in malt extract liquid medium, immobilized cells recorded the highest rate (85.7%) at pH 9. However, an increase in temperature to 40°C resulted in the biosorbent having the highest removal rate (80%) at pH 9 after 72 h of incubation. This shows that pH and temperature have a great influence on mycoremediation of HM. Stanley and Immanuel [176] reported the potential of *Lentinus subnudus* in alleviating soil pollution by HM (*i.e.* Zn, Cu, Mn, Cd, Fe, V, Pb, Cr, and Ni). In the study, *L. subnudus* was sufficient in reducing the levels of Cr, Cd, V, Ni and Pb than Cu, Mn, Zn and Fe. The removal rate was least (33.5%) for Mn and highest (98.16%) for V over 180 days of incubation at room temperature and pH range of 7.24-7.85. Li *et al.* [177] explore other mechanism of WRF aside the LMEs mostly used in HM remediation. They studied the effect of extracellular polymeric substances (EPS), which are mostly compost of nucleic acid, protein, polysaccharides and other cellular components in the immobilization of Pb^{3+}. It was reported that mycelia of *P. chrysosporiun* with EPS was able to immobilize Pb^{3+} above 40% than mycelia without EPS (30%) at all concentration (10, 50, and 100 mg/L) and their efficiency was dependent on pH concentration, as it was observed that increase in pH enhanced mycelia of *P. chrysosporiun* with EPS to immobilize Pb^{3+} after seven days of incubation.

Advantages of Mycoremediation

Mycoremediation of xenobiotics by WRF has attracted a lot of attention from around the world. The impacts of anthropogenic activities on the ecosystem are no longer new as different methods have been devised to curb the harmful effects these pollutants pose on the environment. Conventional methods of xenobiotics removal are rapid but expensive, not safe, require lots of manpower and technicality and the pollutants are often not destroyed [93]. On the contrary, mycoremediation provides a safe method for the removal of pollutants from the environment. They are cheap and do not require lots of labor. The treated environment is restored without any defects to their nature and at the end of the process, the pollutants are often destroyed completely or transformed into a less toxic form [2]. The use of biological sorbent is not sufficient enough to remove pollutants since the technique they utilize is limited to biosorption and immobilization, which is not the case with WRF which utilizes biodegradation, biomineralization, solubilization, among others. The WRF has emerged as a novel tool for the bioremediation of pollutants. Their fruiting body provides them with lots of intracellular spaces for bioaccumulation and co-metabolism of pollutants. The cell wall of WRF is provided with various polymers, which contain various functional groups that interact with various xenobiotics they come in contact with [11, 72, 96]. Their enzyme system is equipped with different arsenal used in attacking a wide range of recalcitrant pollutants and is diversified in action.

Bacteria need to be precondition so as to induce essential enzymes in the degradation of pollutants, which is not the case with WRF. They need no preconditioning and can release wide arrays of extracellular enzymes to degrade, solubilize, volatilize and immobilize recalcitrant pollutants. For bacteria, enzymes responsible for bioremediation to be expressed require quite a large concentration of pollutants, which is not the case with WRF [106]. Although bacteria grow faster than most WRF, they are however greatly affected by the toxicity of pollutants. The physiology of WRF provides them with the capacity to generate antioxidants used to overcome toxicity posed by pollutants. The free radicals catalyzed by LMEs are readily available to attack pollutants in the environment [180]. The rate at which WRF mycelia can grow and the distance they can cover provides them with an incredible mechanical support and a tool used to penetrate through pollutants and exude enzymes that cause the destruction of bonds [181]. The growth conditions required by WRF are cheap (*i.e.* sawdust, wheat hays, rice husks) and simple, unlike other bioremediation technique that requires a lot of attention. The WRF can also carry out bioremediation in a nutrient depleted environment specifically nitrogen [131]. Mycoremediation with WRF is not limited to the viable cells. Their crude enzymes can also be used [4, 93].

Limitations of Mycoremediation

Just like every other bioremediation technology, mycoremediation is also faced with challenges. Most studies on the use of WRF for bioremediation of pollutants have been performed on pilot scale in the laboratory where conditions can be optimized for effective outcomes [24, 117]. However, few studies have been performed under field conditions. This is due to the fact that WRF grows slowly and compete less with bacteria that are fast growers giving them little chance to survive against the myriad of organisms in the environment [20]. The regulation of physiochemical and climatic conditions is quite not feasible and monitoring the process is difficult in field application [20]. Some of the conditions that generally influence the success of WRF in bioremediation technology include; nutrient, bioavailability and toxicity of pollutants, pH, temperature, moisture, humidity, and oxygen [24, 94, 182].

Nutrients

Every living thing requires nutrients for growth, metabolism and reproduction. This is no different with WRF as their success is tied to their ability to grow at a faster rate, which is encouraged by the presence of necessary nutrients such as carbon, nitrogen, phosphorus among many others [2]. Failure to meet their growth requirements will make them retarded and unable to carry out their metabolic functions, which in turn affects the purpose for which they were made [24].

Introducing WRFs that are foreign to an environment can affect their ability to obtain nutrients, since they are new to the environment they need to adjust and compete for nutrients with other microorganisms such as bacteria that grow faster than them [95]. Bhattacharya *et al.* [142] reported the effective removal of azo dye by *P. ostreatus* when NPK fertilizer was added to the fermentation medium from an initial removal rate of 20% without NPK fertilizer to 78% when NPK was added. This shows how important nutrients are to the growth and mycoremediation potentials of WRF. Various studies have stressed how nutrients influenced the rate of degradation of various xenobiotics [143, 146, 178].

Bioavailability of Pollutants

Regardless of the mycoremediation mechanism utilized by WRF, if the molecules of the pollutants are inaccessible by fungal mycelia or their metabolites then removal of pollutants won't be feasible. Pollutants such as PAH are not readily available because of their characteristic high hydrophobic nature. However, the addition of surfactants can help reduce their hydrophobicity thus, increasing their surface area in contact with the WRF [95, 115, 119, 154]. Pozdnyakova *et al.* [119] highlighted how the bioavailability of pollutants can affect WRF degradation of PAHs pollutants. In their study, it was observed that the molecules of anthracene were reported to be poorly available for degradation by the tested WRF due to its low ionization and solubility potentials of 7.43 eV and 0.07 mg/L respectively thus, making anthracene the least degraded. However, phenanthrene having high ionization potentials (8.03 eV) and a high solubility (1.8 mg/L) makes its molecules much more available for degradation. In the case of fluorene, there has been a successful combination of high solubility (1.98 mg/L) with low ionization potentials (7.88 eV) making its molecules very much available and accessible by mycelia of WRF and their metabolites.

Temperature

Temperature is very important for the growth of microorganisms, though the temperature requirements may vary from one species of WRF to another. The growth of WRF and their metabolic activities are greatly affected by temperature. Likewise, the removal of pollutants from the environment [115]. Temperature influences the activities of LMEs and other intracellular enzymes in WRF and when it is below the minimum or above the maximum requirement will make them inactive, whereas they work best at optimum temperature. A study by Zhang *et al.* [141] recorded manganese peroxidase, MnP-Tra-48424 purified from *Trametes* sp.48424 was slightly inactive at a temperature of 20-30°C and 80°C, but optimal activity was observed at a temperature range of 70-75°C. Oyewole *et al.* [183] reported 28°C as optimum for biodegradation of *Manihot esculentus* effluent by *Pleurotus ostreatus* and *Gloeophyllum sepiarium*. Most WRF can function within the temperature range of 20-40°C and their optimum temperature varies [95]. An increase in temperature from 30°C to 37°C and 42°C has been demonstrated [140] to affect the activities of LMEs produced by *T. polyzona* RYNF13, since they recorded a decrease in the breakdown of PAHs from an initial degradation rate of 52% to 30% in pyrene, 90% to 48% in fluorene and 100% to 68% in phenanthrene. Similarly, Bhattacharya *et al.* [142] reported a decrease in the efficiency of *P. ostreatus* in the decolorization of Congo red when the temperature was set above 35°C.

Effects of pH

The pH of an environment is crucial for the growth of WRF. Just like temperature, the optimum pH also varies among species of WRF [94]. However, most WRF function well at a pH above 6.0. When the pH of the environment drops below the minimum requirement or above the concentration they can tolerate, it will make WRF inactive and unable to remove pollutants effectively [95]. But when the pH is optimum, the isolate will be very active [2]. This pH range could be different in terms of purified enzymes as demonstrated in the study carried out by Zhang *et al.* [141], where purified MnP-Tra-48424 from *Trametes* sp.48424 was 100% active at a pH of 5.0, and the enzyme activity reduced when the pH moves further away from the optimum. Li *et al.* [177] also reported the importance of pH during mycoremediation processes, as an increase in pH of the medium enhanced EPS of *P. chrysosporiun* in the immobilization of Pb^{3+}. At pH 3 only about 18% of Pb^{3+} was removed. However, an increase of pH to 7 caused an incredible removal of Pb^{3+} above 65% within 7 days. Likewise, the study of Bhattacharya *et al.* [142] observed a significant decolorization of Congo red by *P. ostreatus* when the pH of the medium was neutral. Yuan *et al.* [146] demonstrated the effects of pH on *A. albocinnamomea*. Maximum decolorization was achieved at a pH of 5.5. However, an increase or decrease from this pH point exhibited a reduction in the decolonization of Alizarin red by *A. albocinnamomea*. Oyewole *et al.* [183] reported pH 8 as optimum for biodegradation of *Manihot esculentus* effluent by *Pleurotus ostreatus* and *Gloeophyllum sepiarium*.

Relative Humidity

Relative humidity is important in the growth of WRF particularly in a field application. A relative humidity range of 70-80% generally favors the growth and fruiting of WRF whereas, below 40%, their growth is quite discouraging [95]. The mycoremediation of PAHs by *Lentinula edodes* in the study demonstrated by Wang *et al.* [137] shows how full potentials of *L. edodes* were achieved at a relative humidity of $59.6 \pm 0.6\%$.

Toxicity of Pollutants

The concentration of pollutants greatly affects the performance of WRF. Their growth, metabolism, development, and reproduction are greatly affected by the toxicity of pollutants such as azo dyes, PAH, HM among others, by impairing enzyme function, blockage of metabolic pathways, producing oxidative stress as well as causing distortion of functional groups present on mycelia of WRF, which makes it difficult for them to remove pollutants from the environment [8, 95]. A study by Kristanti *et al.* [138] demonstrated how an increase in crude oil concentration affected degradation by *Polyporus* sp. S133 as lowest degradation (19%) was recorded at 15,000 ppm followed by 10,000 ppm (25%), 7,500 ppm (37%), and 5,000 ppm (46%). Co-contamination with phenanthrene and manganese has been reported to inhibit the growth of *Pleurotus eryngii* when their concentration was increased [184]. Similar observations were reported on the effects of pollutants on mycoremediation potentials of WRF such as *P. chrysosporiun* [185]. They reported that *P. chrysosporiun* removal rate of Cd was highest (100%) at lower concentrations (*i.e.* 10-50 µM). However, the removal rate was greatly affected (22%) when the concentration was increased to 500 µM within 2 days of exposure.

Oxygen

Degradation of pollutants by WRF is mostly carried out by cascade of LMEs, which causes oxidation of aromatic rings and cleavage of bonds by the addition of O_2 [94, 95]. The presence of O_2 in the environment is of great importance to the action of WRF to remove pollutants and goes a long way to affect the rate and extent of degradation of pollutants. Gheorghe and Ion [8] reported total removal of lignin by *P. chrysosporium* when the supply of O_2 was maintained at 100%. Similarly, Kristanti *et al.* [138] reported a significant increase (73%) in the rate at which *Polyporus* sp. S133 degrades crude oil. Bhattacharya *et al.* [142] reported an increase in *P. ostreatus* decolorization of Congo red in an incubator with a shaker. At 50 rpm, decolorization by *P. ostreatus* was recorded between 70.62% to 99.81%, an increase to 150 rpm caused an increase (82.10% to 100%) in the decolorization of Congo red. Agitation causes the production of oxygen in the fermentor, thus, providing oxygen for aerobic processes. Senthilkumar *et al.* [178] also reported the significance of aerating mycoremediation processes as it was shown in their study how *P. chrysosporiun* was enhanced in the decolorization of Amido black 10 in the presence of oxygen.

Moisture Content

Absence of water in an environment makes it difficult for the metabolic activities of living things. The water activity of fungi is generally different from bacteria.

The WRF can grow in an environment with little moisture content. In fact, it has been stated that the secretion of LMEs by WRF is stimulated under stressed conditions such as limited access to water [127]. When the moisture content is too much, especially in the soil it displaces O_2 present in soil pores. Thus, WRF mycelia are devoid of O_2, which makes it difficult for their LMEs to cause oxidation of pollutants since the process requires O_2 [2]. Njoku *et al.* [46] delineated how moisture content decreased the activity of *P. pulmonarius* in the degradation of DDVP pesticide.

Presence of Contaminants

Contaminants such as solvents, HM, and salts can interfere with the mycoremediation of a particular pollutant. In the case of mycoremediation of PAHs, the presence of HM can hinder the effectiveness of WRF by binding to the active sites of enzymes or functional groups on the cell wall of WRF. Zhang *et al.* [141] reported the inhibition of the enzyme MnP-Tra-48424 from *Trametes* sp. 48424 in the degradation of fluorene, pyrene, fluoranthene, anthracene, and phenanthrene. The presence of other contaminants such as microorganisms is often a challenge in mycoremediation. A study by Kristanti *et al.* [138] reported poor degradation (10%) of crude oil in a non-sterile soil using *Polyporus* sp. S133 against an incredible 93% degradation in an autoclaved soil. The removal of manganese by *Pleurotus eryngii* was reported to be inhibited by the presence of saponin due to the antimicrobial properties it possesses [184].

CONCLUSION AND FUTURE PERSPECTIVE

The application of WRF in the degradation and mineralization of pollutants to a less or nontoxic form is a cheap, effective, environmentally safe process that requires less labor when compared to conventional methods. The WRF is the only group of microorganism that can degrade lignin and lignin-like compounds from the environment through their versatile LMEs system, which confer to them the ability to degrade a wide range of recalcitrant compounds that poses a serious threat to man and the ecosystem. In doing so, knowledge of the physiology and biochemistry of the particular WRF needs to be put into consideration, likewise the nature of the pollutants, since some WRF can go on to metabolize substances that could be more toxic than the initial recalcitrant compound.

Environmental conditions (*i.e.* pH, temperature, moisture, nutrient) need to be considered as well. So far, there have been successful reports on laboratory and pilot scale application of WRF in bioremediation processes, however, much work needs to be done since its application under field conditions have not yielded the expected result. Exploring more genera will help in screening the right strains that could overcome challenges encountered by the already known species.

Advancement in biotechnology and genetics has provided the platform for understanding the genes responsible for the secretion of various LMEs. Expanding knowledge in fields of ecology, engineering, enzymology, biochemistry, molecular biology, genetics and fungal physiology will give environmentalists the opportunity to explore other options such as up-regulation of enzymes responsible for catalytic potentials of WRF. Genes responsible for growth can also be enhanced so as to overcome the slow growth rate and still maintain its catalytic features. Modifications to the genome of promising WRF should be done to meet the physiochemical and climatic conditions, which have been a constraint to the full working capacity of WRF under field conditions. If this is achieved, environmental pollution currently observed will no longer be a problem to the ecosystem owing to the catalytic prowess of WRF.

CONSENT FOR PUBLICATION

Not applicable.

CONFLICT OF INTEREST

The authors declare no conflict of interest, financial or otherwise.

ACKNOWLEDGEMENT

Declared none.

REFERENCES

[1] Fulekar MH, Pathak B, Fulekar J, Godambe T. "Bioremediation of Organic Pollutants Using Phanerochaete chrysosporium" in Fungi as Bioremediators, Soil Biology. Berlin, Heidelberg: Springer 2013; Vol. 32: pp. 135-57.

[2] Gnanasalomi DVV, Jebapriya RG, Gnanadoss JJ. Bioremediation of Hazardous Pollutants Using Fungi. Intern J Comp Alg Int Intel Res 2013; 2: 273-8.

[3] Ugya AY, Imam TS. Efficiency of the Decomposition Process of *Agaricus bisporus* in the Mycoremediation of Refinery Wastewater: Romi Stream Case Study. World J Pharm Res 2017; 6(2): 200-11.

[4] Pande V, Pandey SC, Sati D, Pande V, Samant M. Bioremediation: an emerging effective approach towards environment restoration. Environmental Sustainability 2020; 3(1): 91-103.
[http://dx.doi.org/10.1007/s42398-020-00099-w]

[5] Manisalidis I, Stavropoulou E, Stavropoulos A, Bezirtzoglou E. Environmental and health impacts of air pollution: a review. Front Pub Heal 2020; 8: 14.
[http://dx.doi.org/10.3389/fpubh.2020.00014]

[6] Hussein AN. Role of Fungi in Bioremediation. Adv Biotechnol Microbiol 2019; 12(4): 77-81.

[7] Pinedo-Rivilla C, Aleu J, Collado I. Pollutants Biodegradation by Fungi. Curr Org Chem 2009; 13(12): 1194-214.
[http://dx.doi.org/10.2174/138527209788921774]

[8] Kulshreshtha S, Mathur N, Bhatnagar P. Mycoremediation of Paper, Pulp and Cardboard Industrial

Wastes and Pollutants.Fungi as Bioremediators, Soil Biology. Berlin, Heidelberg: Springer 2013; Vol. 32: pp. 77-116.
[http://dx.doi.org/10.1007/978-3-642-33811-3_4]

[9]   Sepehri M, Khodaverdiloo H, Zarei M. Fungi and Their Role in Phytoremediation of Heavy Metal-Contaminated Soils.Fungi as Bioremediators, Soil Biology. Berlin, Heidelberg: Springer 2013; Vol. 32: pp. 13-345.
[http://dx.doi.org/10.1007/978-3-642-33811-3_14]

[10] Dixit R, Wasiullah D. Malaviya, K. Pandiyan, U.B. Singh, A. Sahu, R. Shukla, and D. Paul, "Bioremediation of Heavy Metals from Soil and Aquatic Environment: An Overview of Principles and Criteria of Fundamental,". Sustain 2015; 7: 2189-212.
[http://dx.doi.org/10.3390/su7022189]

[11] Siddiquee S, Rovina K, Azad SA, Naher L, Suryani S, Chaikaew P. Heavy Metal Contaminants Removal from Wastewater Using the Potential Filamentous Fungi Biomass: A Review. J Microb Biochem Technol 2015; 7(6).
[http://dx.doi.org/10.4172/1948-5948.1000243]

[12] Olicón-Hernández DR, González-López J, Aranda E. Overview on the Biochemical Potential of Filamentous Fungi to Degrade Pharmaceutical Compounds Front Microb 2017; 8.
[http://dx.doi.org/10.3389/fmicb.2017.01792]

[13] Das N, Chandran P. Microbial degradation of petroleum hydrocarbon contaminants: an overview. Biotechnol Res Int 2011; 2011: 1-13.
[http://dx.doi.org/10.4061/2011/941810] [PMID: 21350672]

[14] Yuniati MD. Bioremediation of petroleum-contaminated soil: A Review. IOP Conf Ser Earth Environ Sci 2018; 118(012063): 012063.
[http://dx.doi.org/10.1088/1755-1315/118/1/012063]

[15] Jambon I, Thijs S, Torres-Farradá G, *et al.* Fenton-Mediated Biodegradation of Chlorendic Acid A Highly Chlorinated Organic Pollutant By Fungi Isolated From a Polluted Site Front Microb 2019; 10
[http://dx.doi.org/10.3389/fmicb.2019.01892]

[16] Kapahi M, Sachdeva S. Bioremediation Options for Heavy Metal Pollution. J Health Pollut 2019; 9(24): 191203.
[http://dx.doi.org/10.5696/2156-9614-9.24.191203] [PMID: 31893164]

[17] Li C, Zhou K, Qin W, *et al.* A Review on Heavy Metals Contamination in Soil: Effects, Sources, and Remediation Techniques. Soil Sediment Contam: Int J 2019; 28(4): 380-94.
[http://dx.doi.org/10.1080/15320383.2019.159210]

[18] Satish GP, Ashokrao DM, Arun SK. Microbial degradation of pesticide: A review. Afr J Microbiol Res 2017; 11(24): 992-1012.
[http://dx.doi.org/10.5897/AJMR2016.8402]

[19] Dada EO, Njoku KI, Osuntoki AA, Akinola MO. A review of current techniques of <in situ> Physico-chemical and biological remediation of heavy metals polluted soil. Ethiop J Environ Stud Manag 2015; 8(5): 606-15.
[http://dx.doi.org/10.4314/ejesm.v8i5.13]

[20] Gupta S, Wali A, Gupta M, Annepu SK. Fungi: An Effective Tool for Bioremediation.Plant-Microbe Interactions in Agro-Ecological Perspectives. Singapore: Springer 2017; pp. 593-606.

[21] Kumar A, Bisht BS, Joshi VD, Dhewa T. Review on Bioremediation of Polluted Environment: A Management Tool. Int J Environ Sci 2011; 1(6): 1079-93.

[22] Luka Y, Highina BK, Zubairu A. Bioremediation: A Solution to Environmental Pollution-A Review. Am J Eng Res 2018; 7(2): 101-9.

[23] Treu R, Falandysz J. Mycoremediation of hydrocarbons with basidiomycetes—a review. J Environ Sci Health B 2017; 52(3): 148-55.

[http://dx.doi.org/10.1080/03601234.2017.1261536] [PMID: 28121269]

[24]   Bosco B, Mollea C. Mycoremediation in Soil; Environmental Chemistry and Recent Pollution Control Approaches. IntechOpen 2019.
        [http://dx.doi.org/10.5772/intechopen.84777]

[25]   Chaurasia PK, Bharati SL, Mani A. "Significances of Fungi in Bioremediation of Contaminated Soil," in New and Future Developments in Microbial Biotechnology and Bioengineering. Amsterdam: Elsevier 2016; pp. 281-94.

[26]   Ossai IC, Ahmed A, Hassan A, Hamid FS. Remediation of soil and water contaminated with petroleum hydrocarbon: A review. Environmental Technology & Innovation 2020; 17: 100526.
        [http://dx.doi.org/10.1016/j.eti.2019.100526]

[27]   Oyewole OA, Leh-Togi ZSS, Oladoja OE, Terhemba IT. Isolation of Bacteria from Diesel Contaminated Soil for Diesel Remediation. J Biosci 2019; 28: 33-41.

[28]   Oyewole OA, Raji RO, Musa OI, Enemanna CE, Abdulsalam ON, Yakubu JG. Enhanced degradation of crude oil with *Alcaligenes faecalis* ADY25 and iron oxide nanoparticle. Int J Appl Biol Res 2019; 10(2): 62-72.

[29]   Srivastava M, Srivastava A, Yadav A, Rawat V. Source and Control of Hydrocarbon Pollution. IntechOpen 2019; pp. 1-21.
        [http://dx.doi.org/10.5772/intechopen.86487]

[30]   Wang S, Xu Y, Lin Z, Zhang J, Norbu N, Liu W. The Harm of Petroleum-Polluted Soil and its Remediation Research. AIP Conf Proc 2017; 1864: 020222-8.
        [http://dx.doi.org/10.1063/1.4993039]

[31]   Davoodi SM, Miri S, Taheran M, Brar SK, Galvez-Cloutier R, Martel R. Bioremediation of Unconventional Oil Contaminated Ecosystems under Natural and Assisted Conditions: A Review. Environ Sci Technol 2020; 54(4): 2054-67.
        [http://dx.doi.org/10.1021/acs.est.9b00906] [PMID: 31904944]

[32]   Singh K, Chandra S. Treatment of petroleum hydrocarbon polluted environment through bioremediation: a review. Pak J Biol Sci 2013; 17(1): 1-8.
        [http://dx.doi.org/10.3923/pjbs.2014.1.8] [PMID: 24783772]

[33]   Ahmed F, Fakhruddin A. A. "A Review on Environmental Contamination of Petroleum Hydrocarbons and its Biodegradation,". Int J Environ Sci Nat Res 2018; 11(3): 555811.
        [http://dx.doi.org/10.19080/IJESNR.2018.11.555811]

[34]   Rodríguez-Eugenio N, McLaughlin M, Pennock D. "Soil Pollution: a hidden reality," Food and Agriculture Organization of the United Nations. FAO 2018; pp. 1-142.
        https://books.google.com.ng/books/about/Soil_Pollution.html?id=LxpeDwAAQBAJ&source=kp_boo k_description&redir_esc=y

[35]   Abdel-Shafy HI, Mansour MSM. A review on polycyclic aromatic hydrocarbons: Source, environmental impact, effect on human health and remediation. Egyptian Journal of Petroleum 2016; 25(1): 107-23.
        [http://dx.doi.org/10.1016/j.ejpe.2015.03.011]

[36]   Njoku KL, Yussuf A, Akinola MO, Adesuyi AA, Jolaoso AO, Adedokun AH. Mycoremediation of petroleum hydrocarbon polluted soil by <i>Pleurotus pulmonarius</i>. Ethiop J Environ Stud Manag 2017; 9(1): 865-75.
        [http://dx.doi.org/10.4314/ejesm.v9i1.6s]

[37]   Auta HS, Emenike CU, Fauziah SH. Distribution and importance of microplastics in the marine environment: A review of the sources, fate, effects, and potential solutions. Environ Int 2017; 102: 165-76.
        [http://dx.doi.org/10.1016/j.envint.2017.02.013] [PMID: 28284818]

[38] Lee AH, Lee H, Heo YM, *et al.* A proposed stepwise screening framework for the selection of polycyclic aromatic hydrocarbon (PAH) -degrading white rot fungi. Biopr Biosyst Eng 2020; 43, 767–83.
[http://dx.doi.org/10.1007/s00449-019-02272-w]

[39] Chibuike UG, Obiora SC. Heavy metal polluted soils: effect on plants and bioremediation methods. Appl Environ Soi Sci 2014; 1: 1-12.
[http://dx.doi.org/10.1155/2014/752708]

[40] Oyewole OA, Zobeashia SSL, Oladoja EO, Raji RO, Odiniya EE, Musa AM. Biosorption of heavy metal polluted soil using bacteria and fungi isolated from soil. Spri Nat App Sci 2019; 1: 857.

[41] Seiyaboh EI, Izah SC. Impacts of Soil Pollution on Air Quality under Nigerian Setting. J Soi Wat Sci 2018; 3(1): 45-53.

[42] Lenart-Boroń A, Boroń P. The Effect of Industrial Heavy Metal Pollution on Microbial Abundance and Diversity in Soils - A Review. IntechOpen 2014. [Online]
[http://dx.doi.org/10.5772/57406]

[43] Abioye OP, Oyewole OA, Oyeleke SB, Adeyemi MO, Orukotan AA. Biosorption of lead, chromium and cadmium in tannery effluent using indigenous microorganisms. Braz J Biol Sci 2018; 5(9): 25-32.
[http://dx.doi.org/10.21472/bjbs.050903]

[44] Varma A, Choudhary D, Chandra P, Eds. . Mycoremediation of Environmental Pollutants from Contaminated Soil. Mycorrhizosphere and Pedogenesis. Singapore: Springer 2019.

[45] Özkara A, Akyıl D, Konuk M. Pesticides, Environmental Pollution, and Health. IntechOpen 2016.
[http://dx.doi.org/10.5772/63094]

[46] Njoku KL, Ulu Z, Adesuyi AA, Jolaoso AO, Akinola MO. Mycoremediation of Dichlorvos Pesticide Contaminated Soil by *Pleurotus pulmonarius* (Fries) Quelet. Poll 2018; 4(4): 605-15.

[47] Anjum MM, Ali N, Iqbal S. Pesticides and Environmental Health; A Review Agr Res Tech Ope Acc J 2017; 5
[http://dx.doi.org/10.19080/ARTOAJ.2017.05.555671]

[48] Amoatey P, Baawain MS. Effects of pollution on freshwater aquatic organisms. Water Environ Res 2019; 91(10): 1272-87.
[http://dx.doi.org/10.1002/wer.1221] [PMID: 31486195]

[49] Kim KH, Kabir E, Jahan SA. Exposure to pesticides and the associated human health effects. Sci Total Environ 2017; 575: 525-35.
[http://dx.doi.org/10.1016/j.scitotenv.2016.09.009] [PMID: 27614863]

[50] Adewunmi AA, Fapohunda SO. Pesticides and food safety in Africa. Eur J Biol Res 2018; 8(2): 70-83.

[51] Bernardes MFF, Pazin M, Pereira LC, Dorta DJ. Impact of Pesticides on Environmental and Human Health. IntechOpen 2015. [Online]
[http://dx.doi.org/10.5772/59710]

[52] Nicolopoulou-Stamati P, Maipas S, Kotampasi C, Stamatis P, Hens L. Chemical Pesticides and Human Health: The Urgent Need for a New Concept in Agriculture. Front. Pub. Heal 2016; Vol. 4: p. 148. [Online]
[http://dx.doi.org/10.3389/fpubh.2016.00148]

[53] Rajmohan KS, Chandrasekaran R, Varjani S. A Review on Occurrence of Pesticides in Environment and Current Technologies for their Remediation and Management. Ind J Microb 2020; 60: 125-38.
[http://dx.doi.org/10.1007/s12088-019-00841-x]

[54] Bertrand PG. Uses and Misuses of Agricultural Pesticides in Africa: Neglected Public Health Threats for Workers and Population. IntechOpen 2019. [Online]
[http://dx.doi.org/10.5772/intechopen.84566]

[55]   Pesticides; Children's Health and the Environment 2008.www.who.int/ceh

[56]   Ramachandran R, Gnanadoss JJ. Mycoremediation for the Treatment of Dye Containing Effluents. Int J Comp Algor 2013; 2: 286-93.

[57]   Hassaan MA, El-Nemr A. Health and Environmental Impacts of Dyes: Mini Review. Amer J Environ Sci Eng 2017; 1(3): 64-7.

[58]   Islam MR, Mostafa MG. Textile Dyeing Effluents and Environment Concerns - A Review. J Environ Sci Nat Resour 2018; 11(1&2): 131-44.

[59]   Lellis B, Fávaro-Polonio CZ, Pamphile JA, Polonio JC. Effects of textile dyes on health and the environment and bioremediation potential of living organisms. Biotechnology Research and Innovation 2019; 3(2): 275-90.
[http://dx.doi.org/10.1016/j.biori.2019.09.001]

[60]   Omotosho AO, Oyeleke SB, Daniyan SY, Egwim EC. Isolation of Ligninolytic Enzymes Producing Microbes from Textile Effluent Contaminated Soil. UMYU Journal of Microbiology Research (UJMR) 2020; 4(2): 36-41.
[http://dx.doi.org/10.47430/ujmr.1942.007]

[61]   Mia R, Selim MD, Mojnun A-M. Shamim, M. Chowdhury, S. Sultana, H. Naznin, "Review on various types of pollution problem in textile dyeing & printing industries of Bangladesh and recommendation for mitigation,". J Tex Eng Fash Tech 2019; 5(4): 220-6.

[62]   Liu JL, Wong MH. Pharmaceuticals and personal care products (PPCPs): A review on environmental contamination in China. Environ Int 2013; 59: 208-24.
[http://dx.doi.org/10.1016/j.envint.2013.06.012] [PMID: 23838081]

[63]   De Sotto RB, Kim KI, Kim S, Song KG, Park Y. Identification of metabolites produced by *Phanerochaete chrysosporium* in the presence of amlodipine orotate using metabolomics. Water Sci Technol 2015; 72(7): 1140-6.
[http://dx.doi.org/10.2166/wst.2015.317] [PMID: 26398029]

[64]   Kapoor D. Impact of Pharmaceutical Industries on Environment, Health and Safety. J Cri Rev 2015; 2(4): 25-30.

[65]   Kalyva M. Fate of pharmaceuticals in the environment; A reviewUMEA Universitet, Sweden: Master's Reports, Department of Ecology and Environmental Science (EMG) 2017.

[66]   Costa F, Lago A, Rocha V, *et al.* A Review on Biological Processes for Pharmaceuticals Wastes Abatement—A Growing Threat to Modern Society. Environ Sci Technol 2019; 53(13): 7185-202.
[http://dx.doi.org/10.1021/acs.est.8b06977] [PMID: 31244068]

[67]   Küster A, Adler N. Pharmaceuticals in the environment: scientific evidence of risks and its regulation. Philos Trans R Soc Lond B Biol Sci 2014; 369(1656): 20130587.
[http://dx.doi.org/10.1098/rstb.2013.0587] [PMID: 25405974]

[68]   Tarfiei A, Eslami H, Ebrahimi AA. Pharmaceutical Pollution in the Environment and Health Hazards. J Environ Heal Sust Develop 2018; 3(2): 492-5.

[69]   Sayadi MH, Trivedy RK, Pathak RK. Pollution of Pharmaceuticals in Environment. J Indus Pollut Contr 2010; 26(1): 89-94.

[70]   Adipah S. Introduction of Petroleum Hydrocarbons Contaminants and its Human Effects. J Environ Sci Pub Heal 2019; 3(1): 1-9.

[71]   Bünemann EK, Bongiorno G, Bai Z, *et al.* Soil quality – A critical review. Soil Biol Biochem 2018; 120: 105-25.
[http://dx.doi.org/10.1016/j.soilbio.2018.01.030]

[72]   Igiri BE, Okoduwa SIR, Idoko GO, Akabuogu EP, Adeyi AO, Ejiogu IK. Toxicity and Bioremediation of Heavy Metals Contaminated Ecosystem from Tannery Wastewater: A Review. J Toxicol 2018;

2018: 2568038.
[http://dx.doi.org/10.1155/2018/2568038]

[73]   Verma R, Vinoda KS, Papireddy M, Gowda ANS. Toxic Pollutants from Plastic Waste- A Review.
       Procedia Environ Sci 2016; 35: 701-8.
       [http://dx.doi.org/10.1016/j.proenv.2016.07.069]

[74]   Chae Y, An YJ. Current research trends on plastic pollution and ecological impacts on the soil
       ecosystem: A review. Environ Pollut 2018; 240: 387-95.
       [http://dx.doi.org/10.1016/j.envpol.2018.05.008] [PMID: 29753246]

[75]   Li P, Wang X, Su M, Zou X, Duan L, Zhang H. Characteristics of Plastic Pollution in the
       Environment: A Review. Bullet Environ Conta Tox 20201; 107: 577–84.
       [http://dx.doi.org/10.1007/s00128-020-02820-1]

[76]   Häder D-P, Banaszak AT, Villafañe VE, Narvarte MA, González RA, Helbling EW. Anthropogenic
       pollution of aquatic ecosystems: Emerging problems with global implications. Sci Tot Environ 2020;
       713: 136586.
       [http://dx.doi.org/10.1016/j.scitotenv.2020.136586] [PMID: 31955090]

[77]   Kühn S, Rebolledo BEL, van Franeker JA. Deleterious Effects of Litter on Marine Life.Marine
       Anthropogenic Litter, M Bergmann, L Gutow, M Klages. Texel, Netherlands: Springer 2015; pp. 75-
       116.
       [http://dx.doi.org/10.1007/978-3-319-16510-3_4]

[78]   Maduna K, Tomašić V. Air pollution engineering. Physical Sciences Reviews 2017; 2(12): 1-29.
       [http://dx.doi.org/10.1515/psr-2016-0122]

[79]   Gheorghe IF, Ion B. The Effects of Air Pollutants on Vegetation and the Role of Vegetation in
       Reducing Atmospheric Pollution.The Impact of Air Pollution on Health, Economy, Environment and
       Agricultural Sources. Rijeka, Croatia: IntechOpen 2011; pp. 242-80.

[80]   Olmo NRS, do Nascimento Saldiva PH, Braga ALF, Lin CA, de Paula Santos U, Pereira LAA. A
       review of low-level air pollution and adverse effects on human health: implications for
       epidemiological studies and public policy. Clinics (São Paulo) 2011; 66(4): 681-90.
       [http://dx.doi.org/10.1590/S1807-59322011000400025]

[81]   Anderson JO, Thundiyil JG, Stolbach A. Clearing the air: a review of the effects of particulate matter
       air pollution on human health. J Med Toxicol 2012; 8(2): 166-75.
       [http://dx.doi.org/10.1007/s13181-011-0203-1] [PMID: 22194192]

[82]   Abioye OP, Oyewole OA, Aransiola SA, Usman AU. Application of Box-Behnken model to study
       biosorption of lead by *Saccharomyces cerevisiae* and *Candida tropicalis* isolated from electrical and
       electronic waste dumpsite. Global NEST International Journal 2020; 22(1): 95-101.

[83]   Ariste AF, Batista-García RA, Vaidyanathan VK, *et al.* Mycoremediation of phenols and polycyclic
       aromatic hydrocarbons from a biorefinery wastewater and concomitant production of lignin modifying
       enzymes. J Clean Prod 2020; 253: 119810.
       [http://dx.doi.org/10.1016/j.jclepro.2019.119810]

[84]   Ghorani-Azam A, Riahi-Zanjani B, Balali-Mood MM. Effects of air pollution on human health and
       practical measures for prevention in Iran J Res Med Sci 2016; 21
       [http://dx.doi.org/10.4103/1735-1995.189646]

[85]   Zuhara S, Isaifan R. The Impact of criteria air pollutants on soil and water; A review. Journal of
       Environmental Science and Pollution Research 2018; 4(2): 278-84.
       [http://dx.doi.org/10.30799/jespr.133.18040205]

[86]   Buteau S, Geng X, Labelle R, Smargiassi A. Review of the effect of air pollution exposure from
       industrial point sources on asthma-related effects in childhood. Environ Epidemiol 2019; 3(6): e077.
       [http://dx.doi.org/10.1097/EE9.0000000000000077] [PMID: 33778345]

[87]   Khafaie MA, Yajnik CS, Salvi SS, Ojha A. Critical review of air pollution health effects with special

concern on respiratory health. J Ar Pollut Heal 2016; 1(2): 123-36.

[88]   Boamah PO, Onumah J, Takase M, Bonsu PO, Salifu T. "Air Pollution Control Techniques," *Glo.* J Biosci Biotechnol 2012; 1(2): 124-31.

[89]   Dean A, Green D. Climate change, air pollution and human health in Sydney, Australia: A review of the literature. Environ Res Lett 2018; 13(5): 053003.
[http://dx.doi.org/10.1088/1748-9326/aac02a]

[90]   Türk YA, Kavraz M. Air Pollutants and Its Effects on Human Healthy: The Case of the City of Trabzon.Advanced Topics in Environmental Health and Air Pollution Case Studies. Shanghai, China: IntechOpen 2011; pp. 252-68.

[91]   Kathuria S, Puri P, Nandar SK, Ramesh V. Effects of air pollution on the skin: A review. Indian J Dermatol Venereol Leprol 2017; 83(4): 415-23.
[http://dx.doi.org/10.4103/0378-6323.199579] [PMID: 28195077]

[92]   Fouda A, Khalil A, El-Sheikh H, Abdel-Rhaman E, Hashem A. Biodegradation and Detoxification of Bisphenol-A by Filamentous Fungi Screened from Nature. J Adv Biol Biotechnol 2015; 2(2): 123-32.
[http://dx.doi.org/10.9734/JABB/2015/13959]

[93]   Vishnoi N, Dixit S. Bioremediation: New Prospects for Environmental Cleaning by Fungal Enzymes.Recent Advancement in White Biotechnology Through Fungi. Switzerland: Springer 2019; Vol. 3: pp. 17-52.
[http://dx.doi.org/10.1007/978-3-030-25506-0_2]

[94]   Hara E, Uchiyama H. Degradation of Petroleum Pollutant Materials by Fungi.Fungi as Bioremediators, Soil Biology. Berlin, Heidelberg: Springer 2013; Vol. 32.
[http://dx.doi.org/10.1007/978-3-642-33811-3_5]

[95]   Dickson UJ, Coffey M, Mortimer R, Di Bonito M, Ray N. Mycoremediation of Petroleum Contaminated Soils: Progress, Prospects and Perspectives. Environ Sci: Processes Impacts 2019; 21: 1446-58.
[http://dx.doi.org/10.1039/C9EM00101H]

[96]   Danesh YR, Tajbakhsh M, Goltapeh EM, Varma A. Mycoremediation of Heavy Metals.Fungi as Bioremediators, Soil Biology. Berlin, Heidelberg: Springer 2013; Vol. 32: pp. 245-67.
[http://dx.doi.org/10.1007/978-3-642-33811-3_11]

[97]   Spiteller P. Chemical Ecology of Fungi. Nat Prod Rep 2015; 32: 971-93.
[http://dx.doi.org/10.1039/C4NP00166D]

[98]   Singh SK, Singh PN, Gaikwad SB, Maurya DK. Conservation of Fungi: A Review on Conventional Approaches.Microbial Resource Conservation, Soil Biology. Berlin, Heidelberg: Springer 2018; Vol. 54: pp. 223-37.
[http://dx.doi.org/10.1007/978-3-319-96971-8_8]

[99]   Deshmukh R, Khardenavis AA, Purohit HJ. Diverse Metabolic Capacities of Fungi for Bioremediation. Indian J Microbiol 2016; 56(3): 247-64.
[http://dx.doi.org/10.1007/s12088-016-0584-6] [PMID: 27407289]

[100]   Carris LM, Little CR, Stiles CM. Introduction to Fungi. Introduction to Fungi. The Plant Health Instructor. American Phytopathological Society, 2012.

[101]   Anastasi A, Tigini V, Varese GC. The Bioremediation Potential of Different Ecophysiological Groups of Fungi.Fungi as Bioremediators, Soil Biology. Berlin, Heidelberg: Springer 2013; Vol. 32: pp. 29-49.
[http://dx.doi.org/10.1007/978-3-642-33811-3_2]

[102]   Newbound M, Mccarthy MA, Lebel T. Fungi and the urban environment: A review. Landsc Urban Plan 2010; 96(3): 138-45.
[http://dx.doi.org/10.1016/j.landurbplan.2010.04.005]

[103] Wisecaver JH, Slot JC, Rokas A. The evolution of fungal metabolic pathways. PLoS Genet 2014; 10(12): e1004816.
[http://dx.doi.org/10.1371/journal.pgen.1004816] [PMID: 25474404]

[104] Barrech D, Ali I, Tareen M. A Review on Mycoremediation—the fungal bioremediation. Pure Appl Biol 2018; 7(1): 343-8.
[http://dx.doi.org/10.19045/bspab.2018.70042]

[105] Chatterjee S, Sarma MK, Deb U, Steinhauser G, Walther C, Gupta DK. Mushrooms: from nutrition to mycoremediation. Environ Sci Pollut Res 2017; 24: 19480–93.
[http://dx.doi.org/10.1007/s11356-017-9826-3]

[106] Rhodes CJ. Mycoremediation (bioremediation with fungi) – growing mushrooms to clean the earth. Chem Spec Bioavail 2014; 26(3): 196-8.
[http://dx.doi.org/10.3184/095422914X14047407349335]

[107] Young D, Rice J, Martin R, *et al.* Degradation of Bunker C Fuel Oil by White-Rot Fungi in Sawdust Cultures Suggests Potential Applications in Bioremediation. PLoS One 2015; 10(6): e0130381.
[http://dx.doi.org/10.1371/journal.pone.0130381] [PMID: 26111162]

[108] Ntwampe S, Chowdhury F, Sheldon M, Volschenk H. Overview of parameters influencing biomass and bioreactor performance used for extracellular ligninase production from Phanerochaete chrysosporium. Braz Arch Biol Technol 2010; 53(5): 1057-66.
[http://dx.doi.org/10.1590/S1516-89132010000500008]

[109] Adewole M, Olanrewaju O. Enhancing the Performance of Three White-rot Fungi in the Mycoremediation of Crude Oil Contaminated Soil. Biotechnology Journal International 2017; 18(4): 1-10.
[http://dx.doi.org/10.9734/BJI/2017/34267]

[110] Rodríguez-Couto S. Industrial and environmental applications of white-rot fungi. Mycosphere 2017; 8(3): 456-66.
[http://dx.doi.org/10.5943/mycosphere/8/3/7]

[111] Korcan SF, Cig˜erci IH, Konuk M. White-Rot Fungi in Bioremediation.Fungi as Bioremediators, Soil Biology. Berlin, Heidelberg: Springer 2013; Vol. 32: pp. 371-90.
[http://dx.doi.org/10.1007/978-3-642-33811-3_16]

[112] Singh N, Singh J. Secretomics of Wood-Degrading Fungi and Anaerobic Rumen Fungi Associated with Biodegradation of Recalcitrant Plant Biomass.Recent Advancement in White Biotechnology Through Fungi. Switzerland: Springer 2019; Vol. 3: pp. 1-16.
[http://dx.doi.org/10.1007/978-3-030-25506-0_1]

[113] Tomasini A, León-Santiesteban HH. The Role of the Filamentous Fungi in Bioremediation.Boca Raton, London, New York: CRC Press, Taylor and Francis Group 2018; pp. 3-21.

[114] Tortella G, Dura´n N, Rubilar O, Parada M, Diez MC. Are white-rot fungi a real biotechnological option for the improvement of environmental health? Crit Rev Biotech 2013; 1-8.
[http://dx.doi.org/10.3109/07388551.2013.823597]

[115] Liu SH, Zeng GM, Niu QY, *et al.* Bioremediation mechanisms of combined pollution of PAHs and heavy metals by bacteria and fungi: A mini review. Bioresour Technol 2017; 224: 25-33.
[http://dx.doi.org/10.1016/j.biortech.2016.11.095] [PMID: 27916498]

[116] Akhtar N, Mannan MA. Mycoremediation: Expunging environmental pollutants Biotech Rep 2020; 26: 00452.
[http://dx.doi.org/10.1016/j.btre.2020.e00452]

[117] Gao D, Du L, Yang J, Wu WM, Liang H. A critical review of the application of white rot fungus to environmental pollution control. Crit Rev Biotechnol 2010; 30(1): 70-7.
[http://dx.doi.org/10.3109/07388550903427272] [PMID: 20099998]

[118] Tišma M, Zelic B, Vasic-Racki D. White-rot fungi in phenols, dyes and other xenobiotics treatment a brief review. Crot J F Sci Tech 2010; 2(2): 34-47.

[119] Pozdnyakova NN, Balandina SA, Dubrovskaya EV, Golubev CN, Turkovskaya OV. Ligninolytic basidiomycetes as promising organisms for the mycoremediation of PAH-contaminated Environments. IOP Conf Ser Earth Environ Sci 2018; 107: 012071.
[http://dx.doi.org/10.1088/1755-1315/107/1/012071]

[120] Prakash V. "Mycoremediation of Environmental Pollutants," *Int. J. ChemTec.* Res 2017; 10(3): 149-55.

[121] Jebapriya GR, Gnanadoss JJ. Bioremediation of Textile Dye Using White Rot Fungi: A Review. Int J Curr Res Rev 2013; 5(3): 1-11.

[122] Cupul WC, Vázquez RR. Mycoremediation of Atrazine in a Contaminated Clay-Loam Soil and its Adsorption-Desorption Kinetic Parameters.Soil Contamination - Current Consequences and Further Solutions. Shanghai, China: IntechOpen 2016; pp. 193-207.
[http://dx.doi.org/10.5772/64743]

[123] Hou L, Ji D, Dong W, *et al.* The Synergistic Action of Electro-Fenton and White-Rot Fungi in the Degradation of Lignin. Front Bioeng Biotechnol 2020; 8: 99.
[http://dx.doi.org/10.3389/fbioe.2020.00099] [PMID: 32226782]

[124] Adenipekun OC, Lawal R. Uses of mushrooms in bioremediation: A review. Biotechnol Mol Biol Rev 2012; 7(3): 62-8.

[125] Anderson C, Juday G. Mycoremediation of Petroleum: A Literature Review. J Environ Sci Eng 2016; A5: 397-405.

[126] Ellouze M, Sayadi S. White-Rot Fungi and their Enzymes as a Biotechnological Tool for Xenobiotic Bioremediation.Management of Hazardous Wastes. Shanghai, China: IntechOpen 2016; pp. 104-20.
[http://dx.doi.org/10.5772/64145]

[127] Singh RK, Tripathi R, Ranjan A, Srivastava AK. Fungi as potential candidates for bioremediation.Abatement of Environmental Pollutants. Edinburgh, London: Elsevier 2020.
[http://dx.doi.org/10.1016/B978-0-12-818095-2.00009-6]

[128] Bugg TDH, Ahmad M, Hardiman EM, Rahmanpour R. Pathways for degradation of lignin in bacteria and fungi. Nat Prod Rep 2011; 28(12): 1883-96.
[http://dx.doi.org/10.1039/c1np00042j] [PMID: 21918777]

[129] Young DMA. https://commons.clarku.edu/mosakowskiinstitute/ Bioremediation with White-Rot Fungi at Fisherville Mill: Analyses of Gene Expression and Number 6 Fuel Oil degradation. Mosakowski Institute for Public Enterprise 2012: 12.

[130] Karigar CS, Rao SS. Role of Microbial Enzymes in the Bioremediation of Pollutants: a review. Enzyme Res 2011; 2011: 805187.
[http://dx.doi.org/10.4061/2011/805187]

[131] Periasamy D, Mani S, Ambikapathi R. White Rot Fungi and Their Enzymes for the Treatment of Industrial Dye Effluents.Recent Advancement in White Biotechnology Through Fungi. Switzerland: Springer 2019; Vol. 3: pp. 73-100.
[http://dx.doi.org/10.1007/978-3-030-25506-0_4]

[132] Brijwani K, Rigdon A, Vadlani PV. Fungal laccases: production, function, and applications in food processing. Enzyme Res 2010; 2010: 1-10.
[http://dx.doi.org/10.4061/2010/149748] [PMID: 21048859]

[133] Arregui L, Ayala M, Gómez-Gil X, *et al.* Laccase: structure, function, and potential application in water bioremediation. Microb. Cel. Fact 2019; Vol. 18: p. 200. [Online]
[http://dx.doi.org/10.1186/s12934-019-1248-0]

[134] Shanmugapriya S, Manivannan G, Gopal S, Natesan S. Extracellular Fungal Peroxidases and Lacs for

Waste Treatment: Recent Improvement.Recent Advancement in White Biotechnology Through Fungi. Switzerland: Springer 2019; Vol. 3.
[http://dx.doi.org/10.1007/978-3-030-25506-0_6]

[135]   Janusz G, Pawlik A, Swiderska-Burek U, *et al.* Laccase Properties, Physiological Functions, and Evolution Int J Mol Sci 2020; 21: 966.
[http://dx.doi.org/10.3390/ijms21030966]

[136]   Zhao M, Zhang C, Zeng G, Huang D, Cheng M. Toxicity and bioaccumulation of heavy metals in *Phanerochaete chrysosporium.* Trans Nonferrous Met Soc China 2016; 26(5): 1410-8.
[http://dx.doi.org/10.1016/S1003-6326(16)64245-0]

[137]   Wang C, Yu D, Shi W, Jiao K, Wu B, Xu H. Application of spent mushroom (*Lentinula edodes*) substrate and acclimated sewage sludge on the bioremediation of polycyclic aromatic hydrocarbon polluted soil. RSC Advances 2016; 6(43): 37274-85.
[http://dx.doi.org/10.1039/C6RA05457A]

[138]   Kristanti RA, Hadibarata T, Toyama T, Tanaka Y, Mori K. Bioremediation of crude oil by white rot fungi *Polyporus* sp. S133. J Microbiol Biotechnol 2011; 21(9): 995-1000.
[http://dx.doi.org/10.4014/jmb.1105.05047] [PMID: 21952378]

[139]   Hadibarata T, Tachibana S, Askari M. Identification of metabolites from phenanthrene oxidation by phenoloxidases and dioxygenases of *Polyporus* sp. S133. J Microbiol Biotechnol 2011; 21(3): 299-304.
[http://dx.doi.org/10.4014/jmb.1011.11009] [PMID: 21464602]

[140]   Teerapatsakul C, Pothiratana C, Chitradon L, Thachepan S. Biodegradation of polycyclic aromatic hydrocarbons by a thermotolerant white rot fungus <i>Trametes polyzona</i> RYNF13. J Gen Appl Microbiol 2016; 62(6): 303-12.
[http://dx.doi.org/10.2323/jgam.2016.06.001] [PMID: 27885193]

[141]   Zhang H, Zhang S, He F, Qin X, Zhang X, Yang Y. Characterization of a manganese peroxidase from white-rot fungus *Trametes* sp.48424 with strong ability of degrading different types of dyes and polycyclic aromatic hydrocarbons. J Hazard Mater 2016; 320(Jul): 265-77.
[http://dx.doi.org/10.1016/j.jhazmat.2016.07.065] [PMID: 27551986]

[142]   Bhattacharya S, Das A, G M, K V, J S. Mycoremediation of Congo red dye by filamentous fungi. Braz J Microbiol 2011; 42(4): 1526-36.
[http://dx.doi.org/10.1590/S1517-83822011000400040] [PMID: 24031787]

[143]   Rani C, Jana AK, Bansal A. Studies on the Biodegradation of Azo Dyes by White rot Fungi *Daedalea flavida* in the Absence of External Carbon Source 2nd International Conference on Environmental Science and Technology. 147-50.

[144]   Vasdev K. Decolorization of Triphenylmethane Dyes by Six White-Rot Fungi Isolated from Nature J Biorem Biodegrad 2011; 2.
[http://dx.doi.org/10.4172/2155-6199.1000128]

[145]   Yuan H-S, Dai Y-C, Steffen K. Screening and evaluation of white rot fungi to decolourise synthetic dyes, with particular reference to *Antrodiella albocinnamomea.* Mycol 2012; 3(2): 100-8.

[146]   Chairin T, Nitheranont T, Watanabe A, Asada Y, Khanongnuch C, Lumyong S. Biodegradation of bisphenol A and decolorization of synthetic dyes by laccase from white-rot fungus, *Trametes polyzona.* Appl Biochem Biotechnol 2013; 169(2): 539-45.
[http://dx.doi.org/10.1007/s12010-012-9990-3] [PMID: 23239411]

[147]   Sumandono T, Saragih H, Migirin T, Watanabe T, Amirta R. Watanabe, and R. Amirta, Decolorization of Remazol Brilliant Blue R by new isolated white rot fungus collected from tropical rain forest in East Kalimantan and its ligninolytic enzymes activity. Procedia Environ Sci 2015; 28: 45-51.
[http://dx.doi.org/10.1016/j.proenv.2015.07.007]

[148] Zahmatkesh M, Spanjers H, Toran MJ, Blánquez P, van Lier JB. Bioremoval of humic acid from water by white rot fungi: exploring the removal mechanisms. AMB Exp 2016; Vol. 6: p. 118. [Online]
[http://dx.doi.org/10.1186/s13568-016-0293-x]

[149] Zuleta-Correa A, Merino-Restrepo A, Hormaza-Anaguano A, Cardona Gallo SA, Cardona-Gallo SA. Use of white rot fungi in the degradation of an azo dye from the textile industry1. Dyna (Medellin) 2016; 83(198): 128-35.
[http://dx.doi.org/10.15446/dyna.v83n198.52923]

[150] Elshafei A, Elsayed M, Hassan M, Haroun B, Othman A, Farrag A. Biodecolorization of Six Synthetic Dyes by Pleurotus ostreatus ARC280 Laccase in Presence and Absence of Hydroxybenzotriazole (HBT). Annu Res Rev Biol 2017; 15(4): 1-16.
[http://dx.doi.org/10.9734/ARRB/2017/35644]

[151] Chaturvedi S. Azo Dyes Decolorization Using White Rot Fungi. Res Rev J Microb Biotech 2019; 8(1): 9-19.

[152] Kiran S, Huma T, Jalal F, *et al.* Lignin Degrading System of *Phanerochaete chrysosporium* and its Exploitation for Degradation of Synthetic Dyes Wastewater. Pol J Environ Stud 2019; 28(3): 1749-57.
[http://dx.doi.org/10.15244/pjoes/89575]

[153] Levin LN, Hernández-Luna CE, Niño-Medina G, *et al.* Decolorization and Detoxification of Synthetic Dyes by Mexican Strains of *Trametes* sp. Int J Environ Res Public Health 2019; 16(23): 4610.
[http://dx.doi.org/10.3390/ijerph16234610] [PMID: 31757086]

[154] Oyeleke SB, Oyewole OA, Dagunduro JN. Effect of Pendimethalin on physicochemical properties of soil. ISABB J F Agr Sci 2019; 1(2): 36-9.

[155] Gouma S, Papadaki A, Markakis G, Magan N, Goumas D. Studies on Pesticides Mixture Degradation by White Rot Fungi. J Ecol Eng 2019; 20(2): 16-26.
[http://dx.doi.org/10.12911/22998993/94918]

[156] Xiao P, Mori T, Kamei I, Kiyota H, Takagi K, Kondo R. Novel metabolic pathways of organochlorine pesticides dieldrin and aldrin by the white rot fungi of the genus *Phlebia*. Chemosphere 2011; 85(2): 218-24.
[http://dx.doi.org/10.1016/j.chemosphere.2011.06.028] [PMID: 21724225]

[157] Bhadouria R, Das S, Kumar A, Singh R, Singh VK. Mycoremediation of agrochemicals.Agrochemicals Detection, Treatment and Remediation, (na). Edinburgh, London: Elsevier 2020.
[http://dx.doi.org/10.1016/B978-0-08-103017-2.00022-2]

[158] Sadiq S, Haq IM, Ahmad I, Ahad K, Rashid A, Rafiq N. Bioremediation Potential of White Rot Fungi, *Pleurotus* Sp., against Organochlorines. J Bioremediat Biodegrad 2015; 6: 308.

[159] Adelaja OD, Njoku KL, Akinola MO. Mycoremediation of Pesticide Contaminated Soil Using Mushroom Pleurotus ostreatus. Int J Healt Eco Develop 2017; 3(2): 20-7.

[160] Xiao P, Kondo R. Potency of Phlebia species of white rot fungi for the aerobic degradation, transformation and mineralization of lindane J Microb 2020; 58
[http://dx.doi.org/10.1007/s12275-020-9492-x]

[161] Rodarte-Morales AI, Feijoo G, Moreira MT, Lema JM. Degradation of selected pharmaceutical and personal care products (PPCPs) by white-rot fungi. World J Microbiol Biotechnol 2011; 27(8): 1839-46.
[http://dx.doi.org/10.1007/s11274-010-0642-x]

[162] Asif MB, Hai FI, Singh L, Price WE, Nghiem LD. Degradation of Pharmaceuticals and Personal Care Products by White-Rot Fungi—a Critical Review. Curr Pollut Rep 2017; 3(2): 88-103.
[http://dx.doi.org/10.1007/s40726-017-0049-5]

[163] Silva A, Delerue-Matos C, Figueiredo S, Freitas O. The Use of Algae and Fungi for Removal of

Pharmaceuticals by Bioremediation and Biosorption Processes: A Review. Water 2019; 11(8): 1555.
[http://dx.doi.org/10.3390/w11081555]

[164] Llorens-Blanch G, Badia-Fabregat M, Lucas M, *et al.* Degradation of pharmaceuticals from Membrane Biological Reactor sludge with *Trametes versicolor.* Environ Sci Processes Impacts 2015; 17: 429-40.
[http://dx.doi.org/10.1039/C4EM00579A]

[165] Migliore L, Fiori M, Spadoni A, Galli E. Biodegradation of oxytetracycline by Pleurotus ostreatus mycelium: a mycoremediation technique. J Hazard Mater 2012; 215-216: 227-32.
[http://dx.doi.org/10.1016/j.jhazmat.2012.02.056] [PMID: 22436341]

[166] Marco-Urrea E, Pérez-Trujillo M, Vicent T, Caminal G. Ability of white-rot fungi to remove selected pharmaceuticals and identification of degradation products of ibuprofen by *Trametes versicolor.* Chemosphere 2009; 74(6): 765-72.
[http://dx.doi.org/10.1016/j.chemosphere.2008.10.040] [PMID: 19062071]

[167] Santos IKS, Grossman MJ, Sartoratto A, Ponezi AN, Durrant LR. Degradation of the Recalcitrant Pharmaceuticals Carbamazepine and 17Į-Ethinylestradiol by Ligninolytic Fungi. Chem Eng Trans 2012; 27: 169-75.

[168] Xia-li G, Zheng-wei Z, Hong-li L. Biodegradation of sulfamethoxazole by *Phanerochaete chrysosporium.* J Mol Liq 2014; 198: 169-72.
[http://dx.doi.org/10.1016/j.molliq.2014.06.017]

[169] Singh SK, Khajuria R, Kaur L. Biodegradation of ciprofloxacin by white rot fungus Pleurotus ostreatus. 3 Biotech 2017; 7(1): 69.
[http://dx.doi.org/10.1007/s13205-017-0684-y] [PMID: 28452015]

[170] Aracagök YD, Göker H, Cihangir N. N. "Biodegradation of diclofenac with fungal strains,". Arch Environ Prot 2018; 44(1): 55-62.

[171] Jureczko M, Wioletta P. *Pleurotus ostreatus* and *Trametes versicolor,* Fungal Strains as Remedy for Recalcitrant Pharmaceuticals Removal Current Knowledge and Future Perspectives. Biomed J Sci Tech Res 2018; 3(3).
[http://dx.doi.org/10.26717/BJSTR.2018.03.000903]

[172] Afolabi K. Bioremediation of Aminoglycoside Antibiotic (Streptomycin) in water by White Rot Fungi (*Ceriporia lacerata* and *Trametes versicolor*). 2019.

[173] Alharbi SK, Nghiem LD, van de Merwe JP, *et al.* Degradation of diclofenac, trimethoprim, carbamazepine, and sulfamethoxazole by laccase from *Trametes versicolor* : Transformation products and toxicity of treated effluent. Biocatal Biotransform 2019; 37(6): 399-408.
[http://dx.doi.org/10.1080/10242422.2019.1580268]

[174] Vaishaly AG, Mathew BB, Krishnamurthy NB, Krishnamurthy TP. Bioaccumulation Of Heavy Metals By Fungi. Int J Environ Chem Chromatogr 2015; 1(1): 15-21.

[175] Pal S, Vimala Y. Bioremediation of Chromium from Fortified Solutions by Phanerochaete Chrysosporium (MTCC 787). J Bioremed Biodegrad 2011; 2(5).
[http://dx.doi.org/10.4172/2155-6199.1000127]

[176] Stanley HO, Immanuel OM. Bioremediation potential of <i>Lentinus subnudus</i> in decontaminating crude oil polluted soil. Niger J Biotechnol 2015; 29(1): 21-6.
[http://dx.doi.org/10.4314/njb.v29i1.3]

[177] Li N, Zhang X, Wang D, Cheng Y, Wu L, Fu L. Contribution characteristics of the *in situ* extracellular polymeric substances (EPS) in *Phanerochaete chrysosporium* to Pb immobilization. Bioproc Biosys Eng 2017; 40: 1447–52.
[http://dx.doi.org/10.1007/s00449-017-1802-2]

[178] Senthilkumar S, Perumalsamy M, Janardhana Prabhu H. Decolourization potential of white-rot fungus Phanerochaete chrysosporium on synthetic dye bath effluent containing Amido black 10B. J Saudi

Chem Soc 2014; 18(6): 845-53.
[http://dx.doi.org/10.1016/j.jscs.2011.10.010]

[179] Kunjadia PD, Sanghvi GV, Kunjadia AP, Mukhopadhyay PN, Dave GS. Role of ligninolytic enzymes of white rot fungi (Pleurotus spp.) grown with azo dyes. Springerplus 2016; 5(1): 1487.
[http://dx.doi.org/10.1186/s40064-016-3156-7] [PMID: 27652061]

[180] Ali A, Guo D, Mahar A, *et al.* Mycoremediation of Potentially Toxic Trace Elements—a Biological Tool for Soil Cleanup: A Review. Pedosphere 2017; 27(2): 205-22.
[http://dx.doi.org/10.1016/S1002-0160(17)60311-4]

[181] Singh A, Gauba P. Mycoremediation: A Treatment for Heavy Metal Pollution of Soil. J Civ Eng Environ Techn 2014; 1(4): 59-61.

[182] Barh A, Kumari B, Sharma S, *et al.* Mushroom mycoremediation: kinetics and mechanism.Smart Bioremediation Technologies: Microbial Enzymes. Edinburgh, London: Elsevier 2020; pp. 1-24.

[183] Oyewole, S.B. Oyeleke, S.S.D. Muhammed, and R.U. Hamzah Biodegradation of cassava (*Manihot esculentus*) effluent using white rot fungus (*Pleurotus ostreatus*) and brown rot fungus (*Gloeophyllum sepiarium*). Continental Journal of Microbiology 2011; 5(1): 37-45.

[184] Wu M, Xu Y, Ding W, Li Y, Xu H. Mycoremediation of manganese and phenanthrene by *Pleurotus eryngii* mycelium enhanced by Tween 80 and saponin. Appl Microbiol Biotechnol 2016; 100(16): 7249-61.
[http://dx.doi.org/10.1007/s00253-016-7551-3] [PMID: 27102128]

[185] Zhang X, Shao J, Chen A, *et al.* Effects of cadmium on calcium homeostasis in the white-rot fungus Phanerochaete chrysosporium. Ecotoxicol Environ Saf 2018; 157: 95-101.
[http://dx.doi.org/10.1016/j.ecoenv.2018.03.071] [PMID: 29609109]

CHAPTER 8

Antifouling Nano Filtration Membrane

Sonalee Das[1,*] and **Lakshmi Unnikrishnan[1]**

[1] *CIPET : School for Advanced Research in Polymers (SARP) - APDDRL - Bengaluru 7P, Hi-Tech Defence and Aerospace Park (IT Sector), Jalahobli, Bengaluru North, Near Shell R&D Centre, Devanahalli, Bengaluru - 562 149, India*

Abstract: In the recent decade, membrane technology has gained immense interest in water purification, wastewater treatment, and water desalination. However, the major drawback which destroys the efficiency of membrane technology is fouling. Membrane fouling arises due to the non-specific interaction between fouling species and membrane surface. This major drawback can be overcome by preparation of antifouling membranes. Although there are various techniques involved in water filtration *i.e.* microfiltration, ultrafiltration, and nanofiltration. However, in this book chapter, we shall emphasize antifouling nanofiltration membranes, recent developments and future prospects. Further, we shall discuss the various fouling types, its consequences, mechanisms affecting fouling, challenges, and modification approaches in the antifouling membrane technology.

Keywords: Antifouling, Membrane fouling, Membrane technology, Nanofiltration, Water treatment.

INTRODUCTION

The major global challenge faced in this 21st century is safe drinking water and its scarcity which arises due to water pollution. The various factors which account for such crisis include urbanization, growing population, the imbalance between demand & supply, inefficient water treatment process and unsustainable economic practices [1]. It has been estimated that in 2050, the worldwide water demand would go up to 40% 2050. As a consequence, 1.8 billion people will be facing problems in water-deficient regions [1]. India too is facing a huge crisis in the sector of water pollution. As per the reports by World Economic Forum (2019) it has been reported that in India about 70% of surface water is unhealthy for intake [2]. Further, World Bank Report suggests that water pollution can lower economic

* **Corresponding author Sonalee Das:** CIPET: School for Advanced Research in Polymers (SARP) - APDDRL - Bengaluru7P, Hi-Tech Defence and Aerospace Park (IT Sector), Jalahobli, Bengaluru North, Near Shell R&D Centre, Devanahalli, Bengaluru - 562 149, India; E-mail: das.sonalee31@gmail.com

Inamuddin (Ed.)

growth as well as GDP growth in the downstream areas [2]. In India, the impact of water pollution would lead to a reduction of GDP growth by almost half a factor [2]. In addition, to economic loss, lack of water sanitation would also lead to the loss of more than 4 million lives in India per year [2]. Hence, to meet the demand for clean drinking water there needs to be a technological transition approach from a traditional to a modern aspect.

Membrane technology for the treatment of waste water has gained immense research interest over the last 2 decades. As per the reports the global membrane filtration market is estimated to grow at a CAGR of 6.4% and reach USD 19.6 billion by 2025 [3]. For the water treatment process membranes are being used as filters for various applications which includes water filtration (desalination), water purification as well as in industries (food and beverage) [4]. The major advantages of this technology are low energy consumption, continuous separation and simple up-scaling [5]. This technology is capable to treat various water streams with different salinity and feed compositions [6, 7]. However, different membrane technology processes *i.e.* microfiltration (MF), ultrafiltration (UF), nanofiltration (NF) and reverse osmosis (RO) with varying pore sizes of the membranes are involved in treating the fluid stream. Fig. (**1**) depicts comparative effectiveness of different filtration processes [8]. Fig. (**2**) shows the broad spectrum of different filtration technologies [9]. Although RO technology is the most accepted one amongst the various membrane technology for absolute purified water, however, it requires more energy to produce it and also removes some of the minerals from water that are considered to be beneficial for human consumption [8, 10]. On the other hand, NF technology allows these minerals to pass through it and also requires less energy to produce higher fluxes at low pressure *i.e.* (6-15 bar/ 600-1500 kPa) [8, 10]. In addition, it works best for applications that do not require a feed stream completely free of dissolved solids [10].

The major drawback, however, in the membrane technology is membrane fouling which arises due to different kinds of foulants in fluid stream resulting in blockage of pore, gel formation, adsorption of organic substances and inorganic precipitates, *etc.* resulting in temporary/ permanent decline of flux [1]. As a consequence, water permeation is reduced thereby restricting the process efficiency, deteriorating water quality, and increasing the energy consumption [11 - 14]. Various strategies have been adapted to mitigate membrane fouling like feed water pre-treatment [15], development of foul resistant membranes [16, 17] and membrane cleaning [18]. However, the design and development of intrinsic antifouling membrane surfaces can be a candid approach towards addressing membrane fouling since it reduces the post and pre-treatment frequency, cleaning cycles and replacement of membrane [19]. Hence, the preceding sections will discuss the fouling mechanism, a brief description regarding nanofiltration, antifouling nanofiltration membranes and their applications.

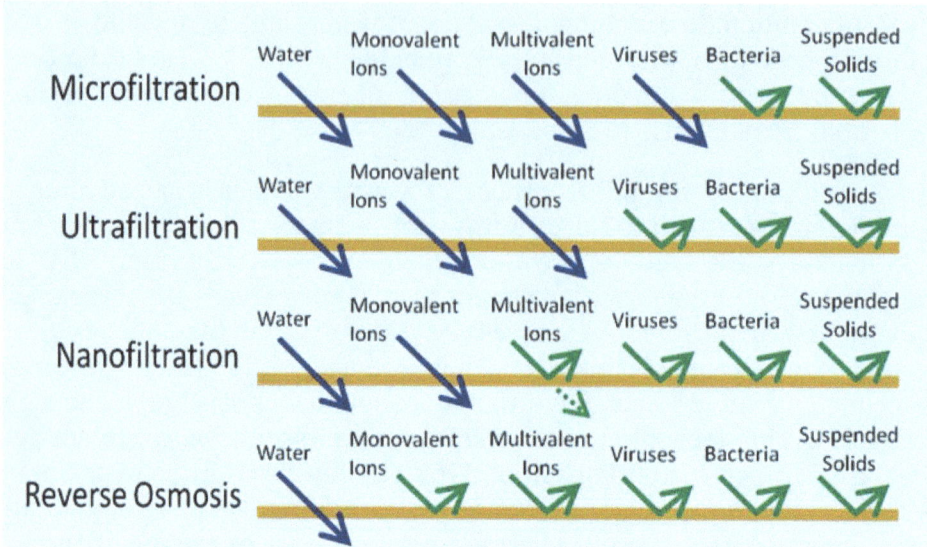

Fig. (1). Comparative effectiveness of different filtration processes [8].

Fig. (2). Broad spectrum of different filtration technologies [9].

MEMBRANE FOULING

It is a physicochemical occurrence, wherein, the membranes material composition dictates their interactions between feed water and foulants [19].

Classification of Membrane Fouling

Depending upon the attachment strength of foulants onto the membrane surface and on the degree of foulant removal, fouling can be divided into 04 types *i.e.* reversible, non-reversible/irreversible, backwashable and non-backwashable [20 - 23]. Reversible fouling occurs due to the loose deposition of foulants/contaminates on the membrane surface [21]. In reversible fouling, the foulants can be eliminated by physical methods such as hydraulic cleaning, pneumatic cleaning and mechanical/ultrasonication [21]. The most common technique is the hydraulic cleaning process which involves flushing and backwashing [22]. Fig. (**3**) depicts the conventional backwashing process [24]. On the other hand, pneumatic cleaning process involves air scouring, sparging, lifting and bubbling [22]. Mechanical/Ultrasonication physical cleaning process involves the use of sponge ball for backwashing/intermittent membrane operation under the influence of cross-flow filtration [22]. Non-reversible/irreversible fouling occurs due to the blockage of membrane pores and strong adhesion of contaminants on the membrane surface which involves chemical cleaning *i.e.* hydrolysis, solubilisation, and chelation [21, 22]. In general, cleaning in place (CIP) is the most frequently used chemical cleaning process in water treatment plants which involves the use of sodium hypochlorite and citric acid for removing organic matter and inorganic substances, respectively [25, 26]. The chemically enhanced backwash (CEB) technique conducted frequently involves the use of oxidants such as chlorine and ozone [25, 26]. Although chemical cleaning has its cons, however, the major pros of this process are the life span reduction of the membranes resulting in membrane replacement which can be expensive [27]. In case of backwashable fouling, the foulants can be removed finally after completion of each filtration cycle by changing permeate flow direction through the membrane pores [20]. On the other hand, non-backwashable fouling involves chemical cleaning of the membranes [20].

Depending upon the type of fouling material it can be classified into 04 categories *i.e.* inorganic/ scaling, partial/colloidal, microbial/biological and organic [20]. Inorganic fouling arises due to the aggregation of particles when their concentration exceeds the saturation limit [20]. Organic fouling is seen on the internal surface which arises due to the membrane clogging by organic substances [20]. On the other hand, biological fouling arises due to the deposition and growth of microorganisms *i.e.* bacteria, fungi, *etc.* on the surface of the membrane surface resulting in the formation of biofilms [23, 28]. Colloidal/partial fouling the most serious form of fouling in the membrane process arises due to the accumulation of particles such as clay, silica, aluminium, and iron on the surface of the membrane resulting in cake formation [22, 28]. The formation of this cake layer imparts hydraulic resistance to water flow thereby reducing product water flux [29].

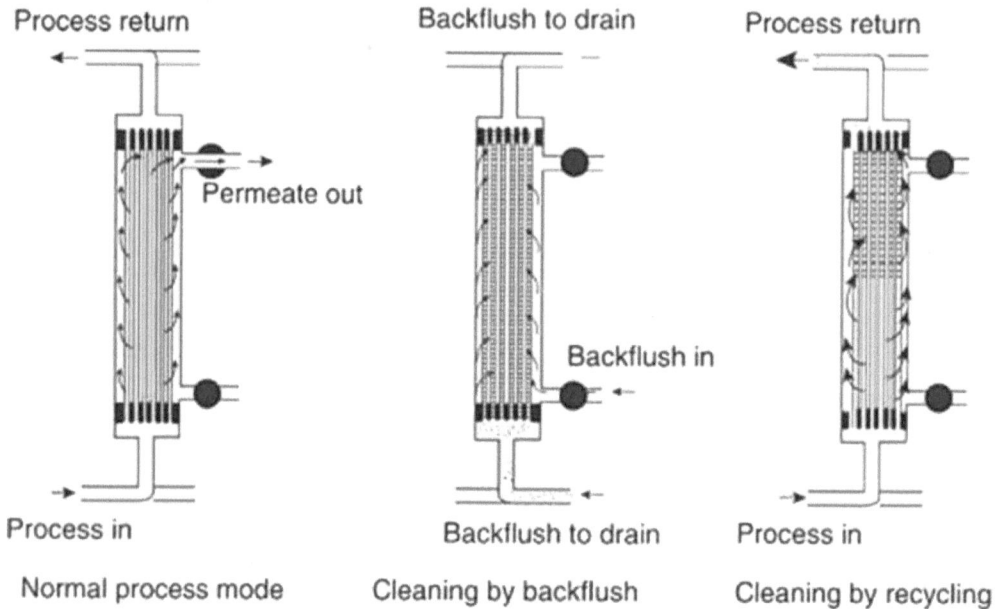

Fig. (3). Conventional backwashing process [24].

Depending upon the deposition of the foulants membrane fouling can be classified into three categories *i.e.* internal, external and concentration polarization. Internal fouling also referred to as pore blocking arises due to the deposition and adsorption of solutes and colloidal particles on the interior part of the pore membrane surface [30]. External fouling involves the deposition of particles, colloids and macromolecules on the surface of the membrane surface fouling layer is termed as gel/cake layer. In the case of the gel layer, the deposition on the surface of the membrane arises due to the pressure difference between the feed and permeate sides of the membrane. The deposition of solids on the membrane surface results in the formation of a cake layer [31]. On the other hand, the deposition of ions and solutes in the thin liquid layer beside the membrane surface results in the development of concentration polarization an inherent process in the filtration process [32]. It improves the resistance of the flow resistance and reduces the membrane flux [21].

Mechanism of Membrane Fouling

Membrane acts as a permselective barrier within two phases wherein, the desired species is separated using driving forces, such as pressure gradient, vapour partial pressure gradient, concentration gradient, or electrical potential gradient as illustrated in Fig. (**4**) [33, 34].

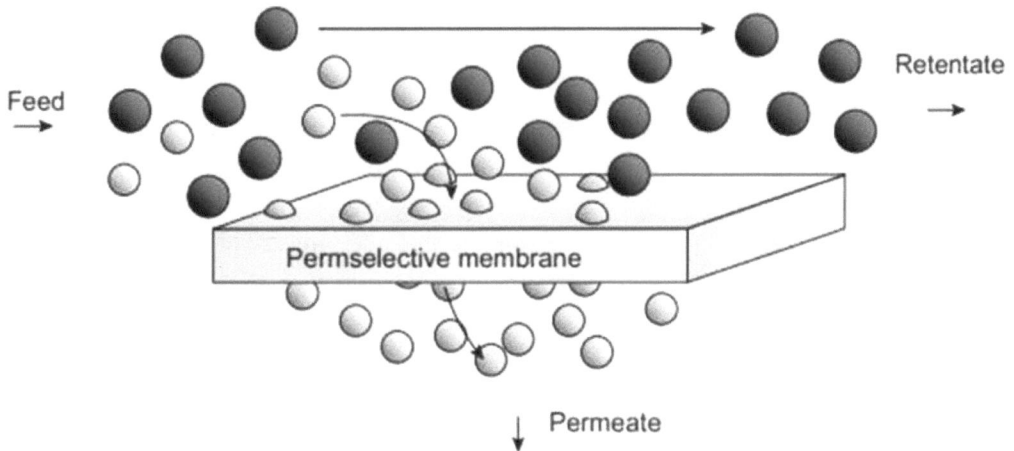

Fig. (4). Schematic diagram representing the membrane surface [34].

Flux and selectivity are two parameters that are used for determining membrane performance [33]. Flux is expressed in terms of volume, mass, or number of moles of a substance which can permeate through the membrane per unit membrane area and time [33]. On the other hand, the membrane selectivity is associated with the retention/separation factor. The rate of permeation is influenced by various factors *i.e.* electrical flow, pressure, concentration, or partial pressure [33].

Basically, there are 02 types of filtration technique involved in the membrane process *i.e.* dead end filtration and cross-flow filtration [28]. This filtration technique is based upon the feed stream direction relative to the orientation of the membrane surface, as depicted in Fig. (**5**) [35]. Generally, the dead end filtration technique is for small lab scale separation application wherein, the feed stream and permeate are perpendicular to the surface of the membrane [36]. This filtration technique is susceptible to fouling wherein, the membrane permeability can be improved *via* . backwashing [28]. In the case of cross-flow filtration process, the feed stream flows parallel to the surface of the membrane and the permeate passes through the membrane [28]. In cross-flow filtration process, the deposits can be removed by shear force of the flowing feed stream [28].

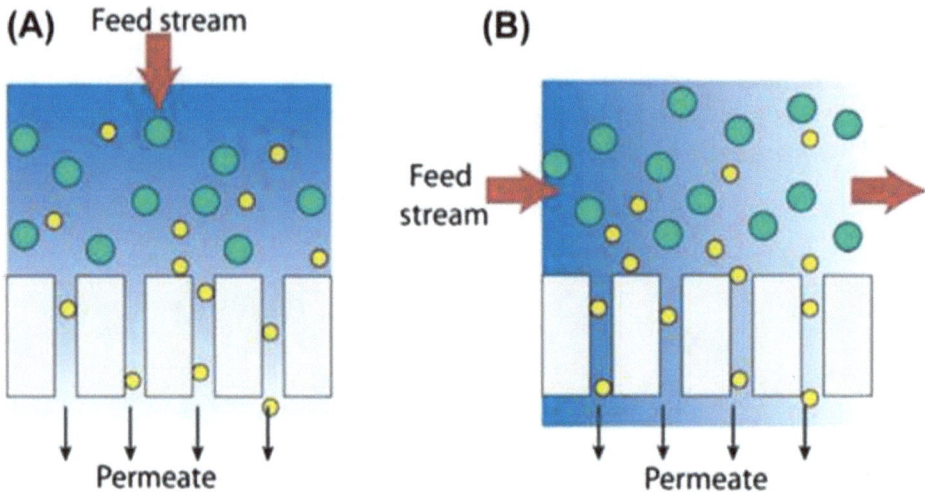

Fig. (5). Membrane process (a) Dead end filtration and (b) Cross-flow filtration [35].

Fig. (**6**) depicts the three stages of flux decline which start initially with a swift drop of the permeate flux (I) followed by a decrease in flux over a prolonged period (II) and finally the steady state flux (III).

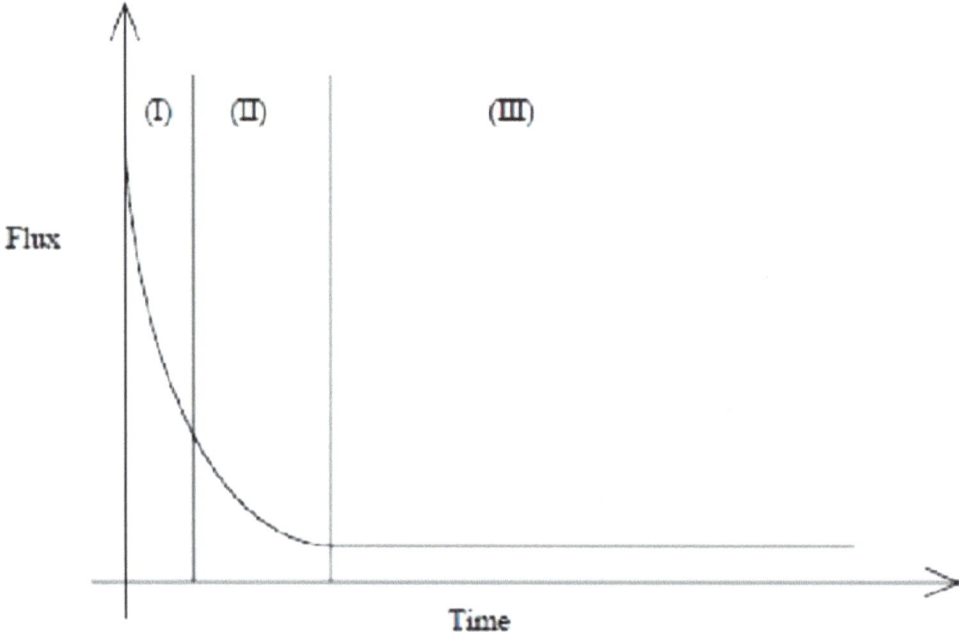

Fig. (6). Schematic diagram representing the different stages of flux decline [37].

Flux decline in filtration process arises due to 02 vital mechanisms *i.e.* (a) blockage of the membrane pore and (b) formation of cake/gel layer on the surface of the membrane. The blockage of the pore membrane increases the membrane resistance. On the other hand, cake/gel formation results in creating an additional resistance layer to the feed flow stream [20].

The rapid drop of the permeate flux (I) arises due to the rapid blocking of the membrane pores. The degree of membrane pore blockage is governed by two important factors *i.e.* the shape and relative size of particles and pores [20]. The degree of pore blockage is more prominent with particles and pores of similar shape and size [38 - 40]. As compared to cake/gel formation the process of pore blockage is a faster process since less than one layer of foulants is sufficient enough to cause full blockage [39, 41]. Further, a decrease in flux (II) arises due to the formation of cake/gel on the surface of the membrane surface with an increase in the amount of foulants. As a result, the flux permeation through the membrane surface declines w.r.t time [20].

Factors Affecting Membrane Fouling

The various factors which affect the membrane fouling are listed below [20, 42].

a. Membrane properties which includes its morphological features, membrane material, surface properties and texture, hydrophilicity/hydrophobicity, pore size and its distribution
b. Operational conditions include temperature, cross flow velocity, flow rate, transmembrane pressure, pH
c. Solution properties include particle size, nature of components, particle/foulants concentration, feed composition and its concentration and the interaction in between the membrane and foulants

a. Membrane Properties

In general, the industrial membranes used are based on polymeric materials which include polyvinyldiene fluoride (PVDF), polysulfone (PS), cellulose, polyethersulfone (PES), polycarbonate and inorganic membranes *i.e.* ceramic and sintered metal [42]. The selection of the membrane material should be on the basis of the lowest fouling tendency with appropriate mechanical and chemical tolerances. The roughness of the membrane material dictates the fouling mechanism by serving as a structural template for the foulant initial layers [42]. A smooth membrane surface leads to the formation of a thin and dense fouling layer whereas, a rough membrane surface with a ridge valley structure and high surface area leads to the accumulation of more foulants [43 - 45]. The hydrophobicity/hydrophilicity of the polymer membrane also influences the fouling phenomenon by determining the interaction of the membrane with water. Hydrophilic membranes with active groups have the tendency to form hydrogen bonds with water and thus exhibit higher wettability [20]. On the other hand, hydrophobic membranes being chemically and thermally stable as compared to hydrophilic ones exhibit water repelling properties [20]. It has been observed that the foulants in aqueous media, tend to be hydrophobic in nature [20]. The foulants tend to cluster on the hydrophobic surface owing to lower interfacial free energy [20]. To prevent attachment of foulants the membrane surface should exhibit greater charge density, larger membrane flux with preferential binding of water molecules [19 - 21]. Hence, higher hydrophilicity of the membrane surface would favour less accumulation of foulants due to lower adsorption on the membrane [19 - 21]. The pore size, geometry and distribution on the membrane surface also dictate fouling since they serve as a site for adsorption and deposition of foulants [46]. The larger pore size of the membranes facilitates the higher accumulation of the foulants, pore blocking and cake/gel formation and faster flux decay [20, 21]. Further, the membrane module structure also influences the membrane fouling *i.e.* vertical mounted membrane module facilitates better flow state and shearing effect as compared to horizontal ones [21] as depicted in Fig. (**7**). The electrostatic surface charge of the membrane also plays a pivotal role in influencing the membrane fouling [19]. Upon contact with aqueous media, the membranes acquire surface charge *i.e.* +ve/-ve due to the ionization of the functional group on its surface. The generation of such a strong electrostatic force of attraction due to the oppositely charged membrane surface and foulants promotes the membrane fouling phenomenon [19]. Hence, proper fabrication of membrane surfaces with electrostatic repulsive forces can help in mitigating the fouling process [19].

b. Operational Conditions

It is observed that an increment in temperature leads to an increase in the permeate flux while, decreasing the rate of fouling. Salahi *et al.* [47] reported that with an increase in feed temperature from 20 °C to 40 °C led to an increase in permeate flux by up to 60% which might be due to an increase in the diffusion rate of permeate through the membrane [20]. The cross flow velocity directly influences the permeate flux [20]. Higher cross flow velocity results in higher permeation flux and lesser accumulation of foulants thereby reducing the concentration polarization [20, 47].

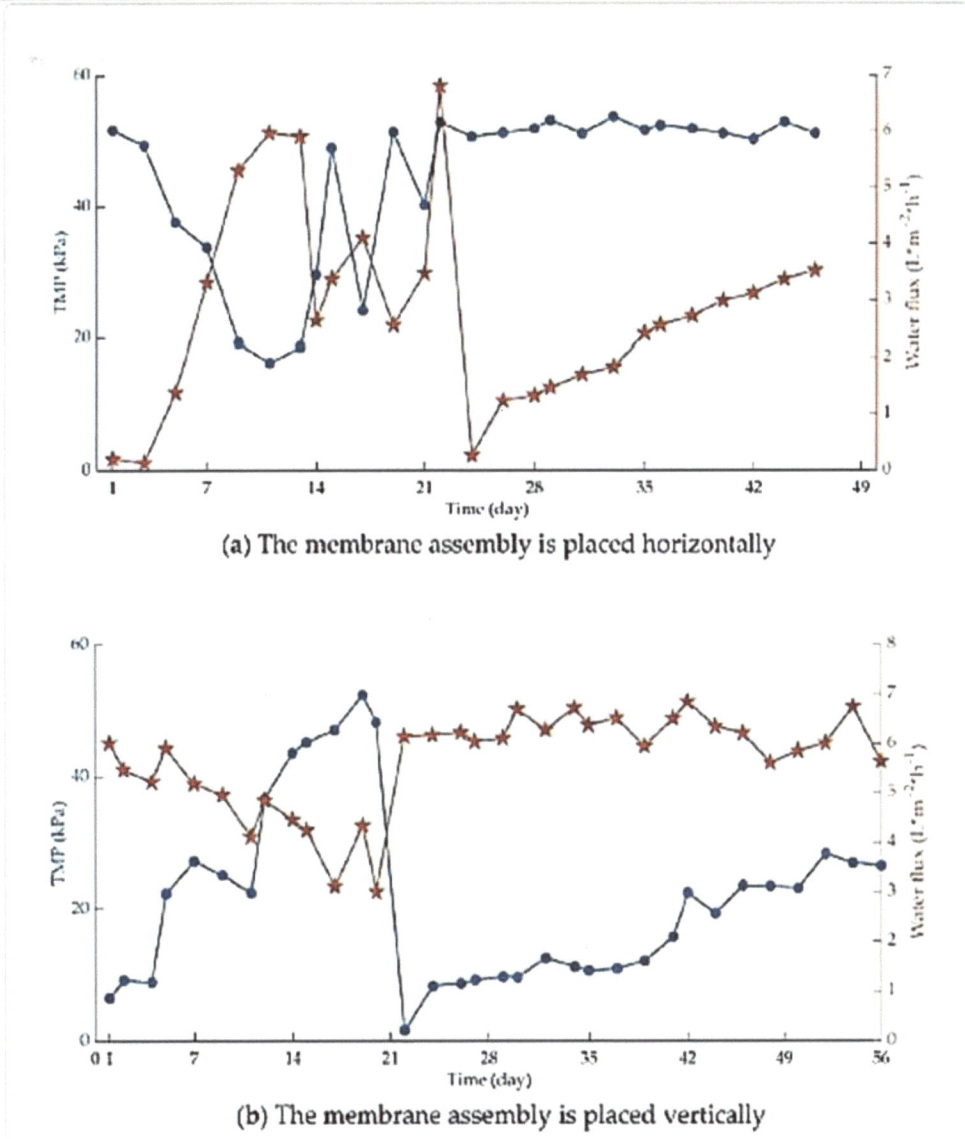

(a) The membrane assembly is placed horizontally

(b) The membrane assembly is placed vertically

Fig. (7). Diagram representing the influence of membrane module structure on fouling [21].

The transmembrane pressure (TMP) *i.e.* the pressure difference between the permeate stream and the feed also influences the permeation rate of the feed stream [21]. The higher the TMP, the higher the fluid force towards the membrane surface resulting in higher flux permeation as depicted in Fig. (**8**) [48].

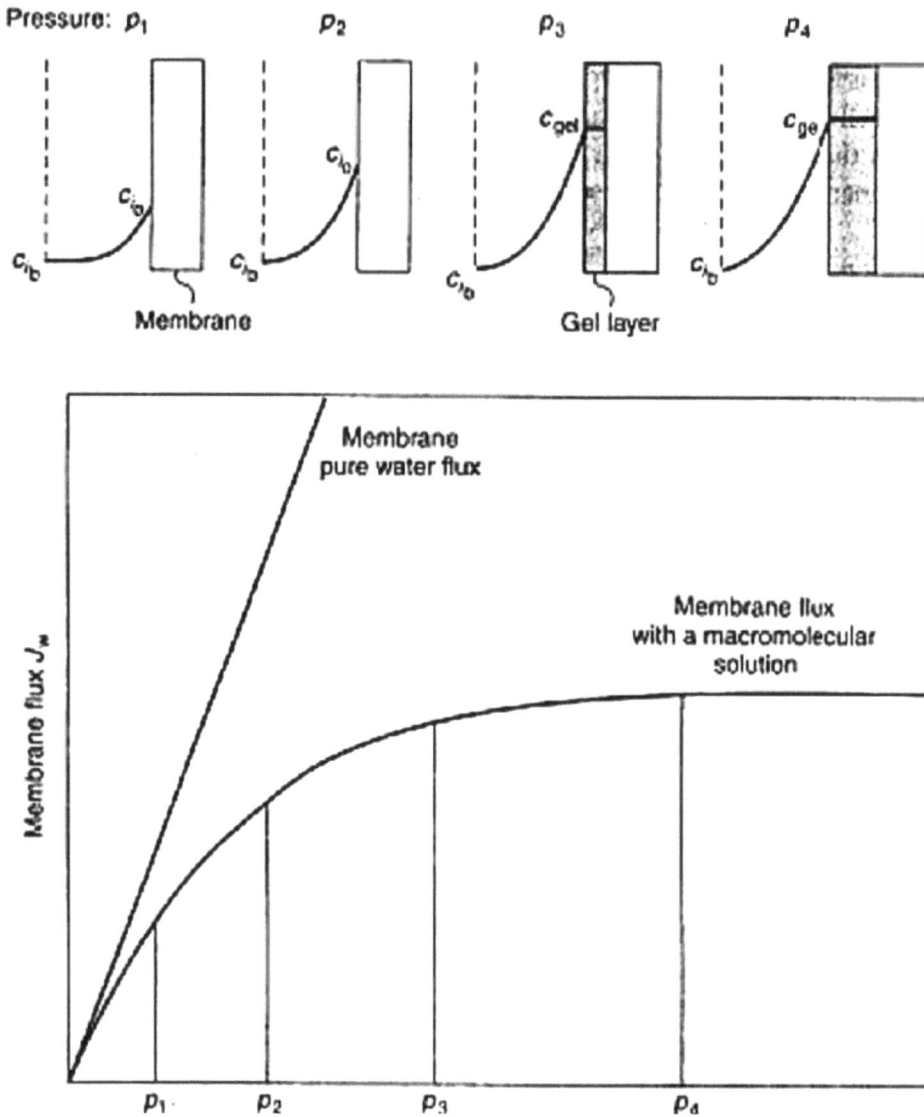

Fig. (8). Diagram representing the effect of pressure on the membrane flux [48].

The flux pressure is close to pure water flux pressure at low pressure p_1. With an increase in applied pressure p_2 the resultant higher flux increases the concentration polarization of the retained material. With a further increase in pressure to p_3 i.e. the critical pressure, the concentration polarization becomes high enough to reach the gel concentration c_{gel}. The formation of the gel layer marks the steady state level of flux permeation [20]. Experimentally it has been observed that no fouling occurs at the critical pressure state [48].

c. Solution Properties

Feed concentration and its composition directly influence membrane fouling. Higher feed concentration leads to fouling by increasing the adsorption of feed particles towards the membrane surface thereby lowering the permeate flux [20, 42]. The feed particles cause fouling by blocking the pores resulting in cake/gel formation. The larger the particle size, the higher the permeate flux and cake thickness [20]. Fine and coarser particles can lower the cake porosity since the particle can easily slide in between the large particles thereby filling the interstices [20]. The shape of the feed particles also affects the cake porosity *i.e.* lower the sphericity of the particle higher is the porosity [49]. Further, the pH and ionic strength of the feed also affect the cake size, particle charge, adhesion strength of particle and membrane surface [20]. It has been reported that lowering the pH decreases the molecular size of the feed, thereby enhancing its adsorption onto the membrane surface resulting in fouling [20].

NANOFILTRATION MEMBRANES

Nanofiltration (NF) membranes have come a long way since it was first documented in the late 80s. Nanofiltration (NF) membranes with a pore size of 1 nm corresponding to molecular weight cut-off (MWCO) of 300–500 Da have properties between ultrafiltration (UF) and reverse osmosis (RO). Dissociation of surface functional groups when in contact with aqueous solutions, render them slightly charged. For *e.g.*, when polymeric NF membranes contain ionizable groups (carboxylic groups and sulfonic acid groups) which result in the charged surface in the presence of a feed solution.

A pressure-driven separation technique like RO with semi-permeable membranes, NF comprises of the diffusion process. The removal of contaminants is by reverse osmosis process in case of NF also but differs only in terms of lower removal efficiencies. The slack membrane layer leads to easy passage of monovalent ions by NF (removal efficiency of 30 – 80%), while divalent ions are removed with high efficiency. This is because of stronger destabilization of divalent ions within the membranes. Further, NF has poor efficiency (<30%) in rejecting smaller organic molecules like methane, ethane, propane, acetone, *etc.* Thus, the major factors distinguishing NF with RO are a low rejection of monovalent ions and high rejection of divalent ions, apart from higher flux. This has led to the emergence of NF as a potential candidate for water / waste water treatment, pharmaceutical, biotechnology and food engineering. The properties and composition of a few commercial membranes are given in Table **1** and **2**.

Table 1. Commercial nanofiltration membranes *vis-a-vis* manufacturers specifications [50].

Membrane	Manufacturer	MWCO (Da)	Maximum temperature (°C)	pH range	Stabilized salt rejection (%)	Composition on top layer
NF270	Dow Filmtec[a]	200–400	45	2–11	>97%	Polyamide thin-film composite
NF200	Dow Filmtec[a]	200–400	45	3–10	50–65% CaCl$_2$ 3% MgSO$_4$ 5% Altrazine	Polyamide thin-film composite
NF90	Dow Filmtec[a]	200–400	45	3–10	85–95% NaCl >97% CaCl$_2$	Polyamide thin-film composite
TS80	TriSep[b]	150	45	2–11	99%	Polyamide
TS40	TriSep[b]	200	50	3–10	99%	Polypiperazineamide
XN45	TriSep[b]	500	45	2–11	95%	Polyamide
UTC20	Toray[c]	180	35	3–10	60%	Polypiperazineamide
TR60	Toray[c]	400	35	3–8	55%	Cross-linked polyamide composite
CK	GE Osmonics[d]	2000	30	5–6.5	94% MgSO$_4$	Cellulose acetate
DK	GE Osmonics[d]	200	50	3–9	98% MgSO$_4$	Polyamide
DL	GE Osmonics[d]	150–300	90	1–11	96% MgSO$_4$	Cross-linked aromatic polyamide
HL	GE Osmonics[d]	150–300	50	3–9	98% MgSO$_4$	Cross-linked aromatic polyamide
NFX	Synder[e]	150–300	50	3–10.5	99% MgSO$_4$ 40% NaCl	Proprietary polyamide thin-film composite
NFW	Synder[e]	300–500	50	3–10.5	97% MgSO$_4$ 40% 20% NaCl	Proprietary polyamide thin-film composite
NFG	Synder[e]	600–800	50	4–10	50% MgSO$_4$ 10% NaCl	Proprietary polyamide thin-film composite
TFC SR100	Koch[f]	200	50	4–10	>99%	Proprietary thin-film composite polyamide
SR3D	Koch[f]	200	50	4–10	>99%	Proprietary thin-film composite polyamide
SPIRAPRO	Koch[f]	200	50	3–10	99%	Proprietary thin-film composite polyamide
ESNA1	Nitto-Denko[g]	100–300	45	2–10	89%	Composite polyamide
NTR7450	Nitto-Denko[g]	600–800	40	2–14	50%	Sulfonated polyethersulfone

[a] Midland, Michigan, USA.
[b] Goleta, CA, USA.
[c] Tokyo, Japan.
[d] Le Mee sur Seine.
[e] Vacaville, CA, USA.
[f] Wilmington, Massachusetts, USA.
[g] Somicon AG, Basel, Switzerland.

Table 2. Commercial polymeric reverse osmosis and nanofiltration membranes for water purification application [51].

Membrane	Manufacturer	Selective Layer	Maximum Temperature (°C)	pH Range	Salt Rejection (%)
SW30HRLE-400	Dow Filmtec, USA	PA TFC	45	2–11	99.8 NaCl
NF270-400/34i	Dow Filmtec, USA	PA TFC	45	3–10	>97 NaCl
SWC4+	Hydranautics, USA	PA TFC	45	3–10	>99.7 NaCl
TM820C-370	Toray, USA	PA TFC	45	2–11	>99.5 NaCl
HB10255	Toyobo, Japan	CTA hollow fiber	40	3–8	>99.4 NaCl
TS40	Microdyn-Nadir, USA	Polypiperazineamide	45	1–12	40 NaCl >98.5 MgSO$_4$
TS80	Microdyn-Nadir, USA	PA TFC	45	1–12	80 NaCl >98.5 MgSO$_4$
AD-90	GE-Osmonics, USA	TFC	50	4–11	>99.5 NaCl 95% Boron
AG4040C	GE-Osmonics, USA	TFC	50	4–11	>99 NaCl
HL2540FM	GE-Osmonics, USA	TFC	50	3–9	>96 MgSO$_4$
CK4040FM	GE-Osmonics, USA	CA	30	5–6.5	>94 MgSO$_4$
8040-SW-400-34	Koch, USA	Proprietary PA TFC	45	4–11	>99.5 NaCl
4040-HR	Koch, USA	Proprietary PA TFC	45	4–11	>99.2 NaCl
MPS-34 2540 A2X	Koch, USA	Proprietary composite NF	50	0–14	35 NaCl 95 Glucose 97 Sucrose
NFX	Synder, USA	Proprietary PA TFC	50	2–11	40 NaCl >99 MgSO$_4$ >99 Lactose
NFW	Synder, USA	Proprietary PA TFC	50	2–11	20 NaCl >97 MgSO$_4$ >98.5 Lactose

Mechanism of Action

Nanofiltration is an extremely complex process dependent on the microhydrodynamic and interfacial events at the surface and within the pores. The rejection may be explained in terms of combination of steric, donnan dielectric and transport effects. Passage through solute is comparatively difficult through NF membranes, making it more complex than RO and UF. There are three modes of transport that solutes undergo through the membrane (Fig. **9**) [52],

- Diffusion wherein the molecules travel because of concentration gradient, as in the case of RO membranes
- Convection wherein the molecules just travel with flow through the larger pore as in the case of microfiltration
- Electromigration wherein molecular transport is because of attraction or repulsion from charges within and on the membrane surface

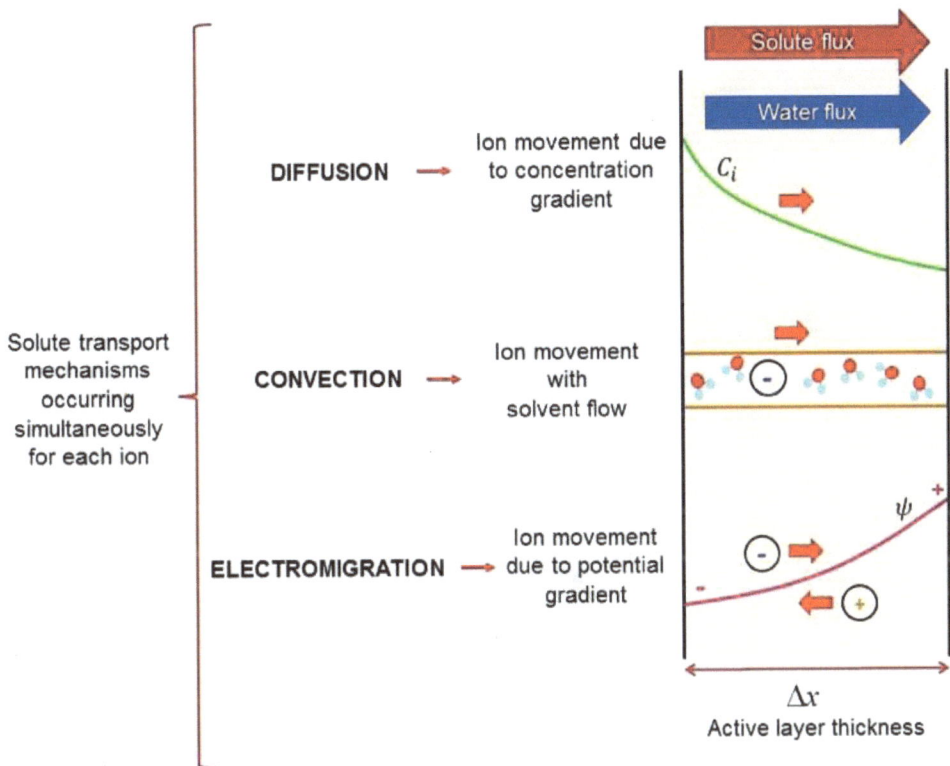

Fig. (9). Solution transport mechanism in NF membranes [52].

The exclusion mechanism is also unique in the case of nanofiltration, which considers surface charge impacts on small solute particles, and hydration impacts, wherein the molecules in the solution have a solvation shell of water molecules, apart from the steric/ size exclusion mechanism (Fig. **10**) [52]. Studies [53] have established the fact that the transport of neutral solutes is through a steric mechanism (size based exclusion). Donnan effect describes the membrane potential and equilibria between the charged species and its interface with the charged membranes. Literature [54 - 56] suggests that the charges on the membrane originates from the dissociation of ionizable groups on the membrane surface and within the pores. The dissociation of surface groups depends on the base membrane material, pH of the contacting solution and when the membrane nature is amphoteric, they may exhibit an isoelectric point at a specific pH [57]. The weak ion exchange capacity of the NF membrane may lead to the absorption of ions from the contacting solution on the membrane, leading to a change in the membrane charge [58, 59]. This may further affect the electrostatic repulsion or attraction taking place based on the ion valence and fixed charge of the membrane.

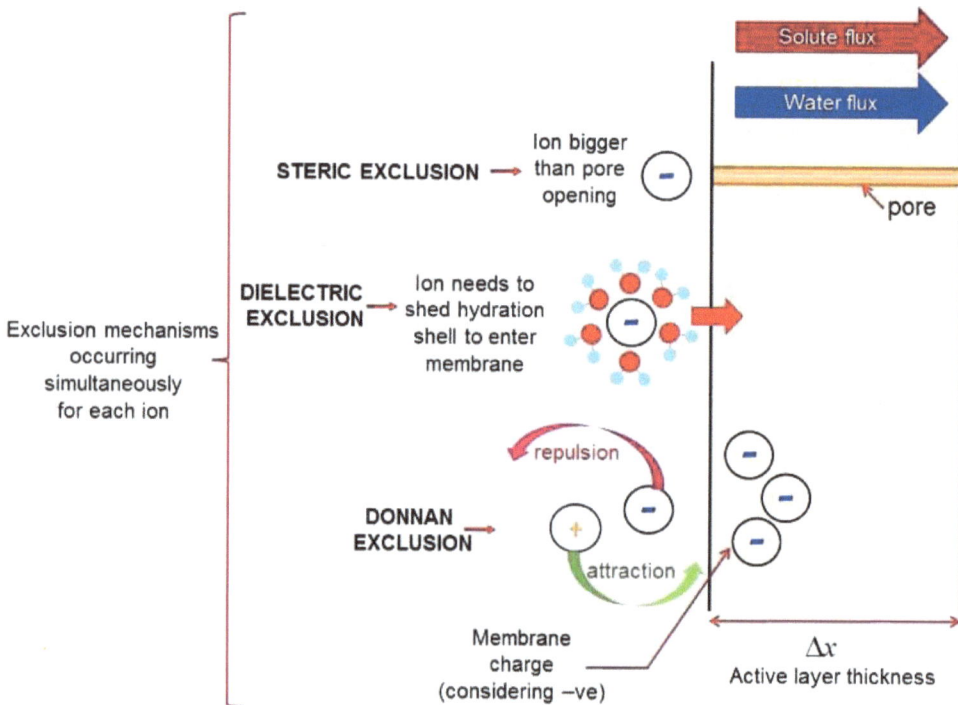

Fig. (10). Exclusion mechanism in NF membranes [52].

In the current technology, the dimensional parameters of NF active layer are very much close to the atomic length scales and this limits the characterization and understanding of the transport mechanism in NF. It is understood that the solute moving through a confined pore structure experiences drag forces and the movement is considered to be hindered. However, the real separation mechanism is still debated [60] and the role of dielectric exclusion is a contester in the debate [61]. This deals with the electrostatic repulsion or attraction taking place between the valence ions and fixed charge on the membrane surface, which depends greatly on the surrounding environment.

As discussed above, the transport and exclusion mechanisms are affected by size of membrane pore, solvent viscosity, thickness of membrane thickness, solute diffusivity, solution temperature, pH of solution, material dielectric constant [52] and pore size distribution.

Characterization of NF Membranes

It has been understood from the studies that the key factors driving the NF performance are pore size and its distribution. Hence, it can be easily concluded that in most NF a porous active layer is present [61]. The NF membranes may be characterized by

- Gas adsorption–desorption technique: The technique otherwise known as the Brunauer–Emmett–Teller (BET) method directly measures the pore size distribution [62 - 64].
- Atomic force microscopy (AFM): This microscopic technique also measures the pore size and distribution directly apart from topography, surface roughness and interactive forces between the membrane and colloids [65 - 68]
- Neutral solute rejection studies and models: Modeling technique helps in the indirect measurement of pore size and distribution, subjected to its coupling with other characterization techniques [69 - 71], that measure pore size directly.
- Transmission Electron Microscopy (TEM): Electron microscopy coupled with reverse surface impregnation also facilitates direct measurement of pore size and distribution [72].

Although each of the aforementioned methodologies can be useful in predicting the performance of NF membranes, a combination is sought-after, owing to the length scale of pores and limitations of these techniques. The charge properties of the NF membranes need to be calculated with utmost care, after considering factors like the nature of the contact solution, concentration and pH. They may be characterized across a range of process conditions using,

- Streaming potential, which assists in determining membrane surface zeta (ζ) potential, is a pseudo measurement of Donnan potential [73 - 76].
- Streaming current, an analogous measurement of streaming potential [77 - 79]
- Electro-osmosis, which again helps in determining the zeta (ζ) potential perpendicular to the membrane surface [80], *i.e.*, through the membrane pores
- Modeling and studies for charged solute rejection, which further aid in measuring the effective membrane charge density indirectly [73, 81 - 83]. This requires correlation with more authentic experimental techniques.

There are other constructive characterization measurements, like scanning electron microscopy (SEM) used for imaging the membrane surface, membrane cross-section and fouling layers [84 - 86]. Simple characterization studies like contact angle for hydrophilicity / hydrophobicity [87 - 89], which make use of comparatively inexpensive equipments to more advanced and expensive and sophisticated techniques such as spectroscopy methods [90 - 93] for studying membrane morphology and structural analysis are also utilized. To summarise, it may be stated that there is a surfeit of potential characterization techniques available for NF membranes. However, the key to successful analysis and interpretation lies in choosing the appropriate technique(s), with right resolution for the desired end purpose

Industrial Applications

Membrane filtration is a common process used widely in industrial applications wherein chemical processes are involved. Major areas are water filtration, pharmaceutical, biotechnological, chemical sugar and dairy sectors.

Food Industry: The membrane based technique covers an infinite number of areas like concentration of egg whites, clarification & extraction of ashes of porcine, bovine or bone gelation, clarification of meat brine, the concentration of vegetables & plants (Soy, canola oats) *etc.*

Dairy Industry: NF is an important part in dairy ingredient manufacturing application may be divided into milk, whey, and clarified brine.

Starch and sweetener Industry: NF offers improvement in the performance of products, viz., clarification of dextrose/fructose, concentration of rinse water from starch and concentration of maceration of water.

Sugar Industry: In this industry, NF can be used to clarify unprocessed juice w/o primary clarifiers thereby improving the efficiency of traditional methodologies.

The NF membranes can also clarify, divide and concentrate various sugar solutions during the production process.

Chemical Industry: NF membranes are commonly used to desalicate and purify dyes, pigments and optical brightness. They are also used to clean waste water and rinse water currents for the concentration of minerals like kaolin clay, titanium dioxide, calcium carbonate, clarification caustic agents, polymer production, or recuperation of metals.

Pharmaceutical Industry: NF finds a very significant role in improving the fermentation process through the recuperation of biomass for the manufacturing of antibiotics. The membrane filtration process improves productivity and reduces labour workload apart from maintenance costs. They are commonly used in enzyme production lines for concentrating enzymes.

Challenges in NF Membranes

Although NF membranes find application in a wide range of fields, there are reported impediments or challenges that used to be addressed while using NF membranes [94].

Membrane Fouling

The foremost challenge faced by a membrane filtration technique is the fouling of the membrane over time. In the case of NF, the fouling could be more complex owing to the involvement of nanoscale pores wherein fouling takes place and is difficult to understand. Due to fouling, the membranes face problems like the need for pre-treatment, membrane cleaning, limited recoveries, feed water loss and obviously reduced service span of the membranes. This would further affect the membrane stability & lifetime as well as concentration treatment. Thus, controlling fouling would reduce the need for cleaning frequency of the system, thereby enhancing the permeate yield.

Separation Between the Solutes

NFs are commonly applied for fractionation, specifically for salts, since the rejection of monovalent salts is. lesser than that of divalent or multivalent salts. The charge-induced separation can lead to negative rejections of monovalent ions or polyelectrolytes. As explained in the earlier sections, the rejection of divalent ions is >95% while monovalent ions of the same charge are somewhere between 20-80%. Thus, NF membranes are potential candidates for ion fractionation, a major factor behind the commercial growth of the NF process.

Post-treatment of Concentrates

Most of the membrane filtration techniques using pressure driven mechanism face a common problem of generating a retentate. They are often waste by-products of purification process, which need to be discharged or further treated. This problem remains unresolved because of the limitation of membrane technology to remove the unwanted fraction in the feed, due to the fact that the composition of feed and retentate are similar with only a difference in the concentration.

Chemical Resistance

This factor is somehow related to fouling, the need for cleaning, specifically in application involving solvent resistant NF. This is a common problem in membrane filtration, which need to be studied from a pragmatic angle. This can be addressed by controlling the membrane matrix, operating conditions, energy consumption, cleaning chemicals, permeate yields and overall environmental impact.

Insufficient Rejection in Water Treatment

NF was regarded as a low pressure RO process, consuming lesser amount of energy due to increased flux. This was the result of a loose membrane structure and polymers with high free volume. NF had gained immense interest in applications wherein complete removal of ions weren't necessary, as a low cost option. However, considering water treatment, there has been an increasing demand to ensure complete absence of all possible pollutants, even at lowest possible trace limits. Even though this criterion does not clash with the risk or toxicity levels, the debate is based on nitrate compounds (monovalent ions), a harmful ion for infant health. Hence, in certain cases, requirement may so arise to have a combination of techniques like RO & NF to ensure complete removal of all toxic ions, including pesticides.

Need for Modelling & Simulation Tools

The performance evaluation and lab-scale understanding of NF membranes through modelling comprises of two aspects - flux and rejection prediction. However, there are limits for such models to analyse the scaling of membranes to large-scale installations wherein permeate yield has to be taken care of.

ANTIFOULING NANOFILTRATION (AF-NF) MEMBRANES

It is imperative for the above discussion that intensive R&D is required for the fabrication of robust AF membranes for controlling fouling. As discussed the surface characteristics membrane can affect the fouling mechanism. Hence, proper strategy and mechanisms should be tailored to fabricate the AF-NF membranes for mitigating the fouling issues. As reported the major strategies used for the fabrication of AF-NF membranes includes passive strategy and active strategy [1]. In case of passive anti-fouling strategy the initial adsorption of foulants is prevented on the membrane surface without disturbing the intrinsic features of the foulants [1]. The passive anti-fouling strategy involves the tailoring of physicochemical and/or topographical features of the membrane surface in order to reduce the interactions in between the foulants and membrane surface thereby preventing the adsorption of foulants [1]. On the other hand, in case of active anti-fouling strategy proliferative fouling and suppression of microbial colonies is eradicated by destroying and inactivating the chemical structure of the microbial cells [1]. The active anti-fouling strategy involves the fabrication of anti-biofouling membranes with the help of antimicrobial agents which can engulf the bacterial cells by destroying their chemical pathways [1]. Based on the types of antimicrobial agents the active anti-fouling strategy can be classified into two types *i.e.* off-surface antifouling strategy and on-surface antifouling strategy with releasable antimicrobial agents and unreleasable antimicrobial agents respectively [1].

AF-NF membranes can be fabricated *via* . various advanced methods which includes surface modification *via* . grafting, coating and blending, lowering the surface roughness, inducing hydrophilicity, enhancing the electrostatic surface charge and inclusion of antibacterial layers [1, 19]. Fig. (**11**) depicts the various mechanisms and modification techniques to fabricate AF membranes for purification of water [1].

Recent Progress in the Fabrication of Anti-Fouling Nanofiltration (NF) Membranes

Qian *et al* [95]. in a recent literature described about the fabrication of anti-fouling nanofiltration membranes using polyethyleneimine (PEI) functionalized with perfluorobutylsulfonyl groups (FPEI) *via* . interfacial polymerization process. The covalent bonding and the presence of perfluoro groups led to decrease the surface energy 39.9 mJ m^{-2} to 31.1 mJ m^{-2} resulting in reduction of interaction in between the pollutants and the membrane surface. The developed NF membranes exhibited improved fouling resistance against bovine serum albumin (BSA) and humic acid (HA), indicating a decline in total flux of 9.9%

and 4.9%, respectively. Further, the flux recovery ratio (FRR) attained 99.1% and 98.0% for BSA and HA respectivley.

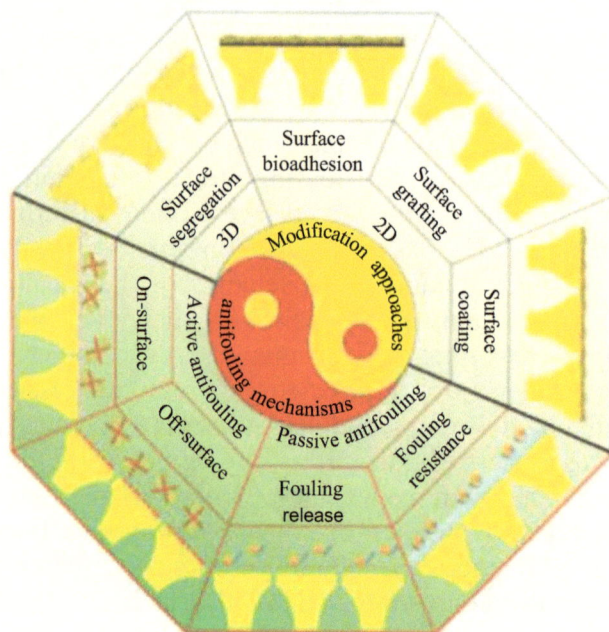

Fig. (11). Various approaches for the fabrication of antifouling membrane [1].

Qi *et.al* [96]. in a recent literature reported about the fabrication of novel nano-filtration membrane modified with polydopamine and hydroxyl propyl trimethyl ammonium chloride chitosan blended with chitosan and chelated silver nanoparticles. Scanning electron microscopy (SEM), contact angle, atomic force microscopy (AFM) were used to analyse the chemical composition, morphology, wettability, surface features antibacterial and antifouling properties and of the fabricated nanofiltartion membranes. The authors used volumetric flux and protein rejection parameters for evaluating the membrane performance. Further, the antifouling and antibacterial properties of the membranes were also investigated. The authors observed that the fabricated membranes exhibited superior antifouling property owing to their three-layer architecture which inhibited the bacterial growth on the membrane surface along with a flux recovery rate of over 96%. The higher hydrophilicity exhibited by the modified NF membranes due to the presence of amino groups and hydroxyl groups on the polydopamine surface resulted in mitigating the fouling issues.

Bagheripour *et al* [97]. developed polyether sulphone based hydrophilic antifouling nanofiltration membranes using water dispersible activated carbon/chitosan (ACh) nanoparticles. The authors used hydrophilic chitosan particles to modify the hydrophobic carbon nanoparticles to enhance dispersion and impart adsorptive sites for removal of solutes from water resulting in higher rejection rates. The loading of activated carbon/chitosan nanoparticles was varied as 0.05, 0.1, 0.5 and 1 wt.%. It was observed that the salt rejection rate increased to 97% for the modified nanofiltration membranes as compared to the polyether sulphone ones. It was observed that with the incorporation of nanoparticles the wettability of the composite membranes decreased and water flux increased from 21 to 30 (L/m^{2}h) at a loading of 0.5 wt %. AFM results indicated that the inclusion of nanoparticles also led to the decrease in roughness from 19.2 nm to 5.9 nm which led to increase the permeation flux and decline the fouling properties of the fabricated nanofiltration membrane. The SEM images indicated reduction in pore size with uniform distribution of nanoparticles thereby inhibiting the cake formation as depicted in Fig. (**12**).

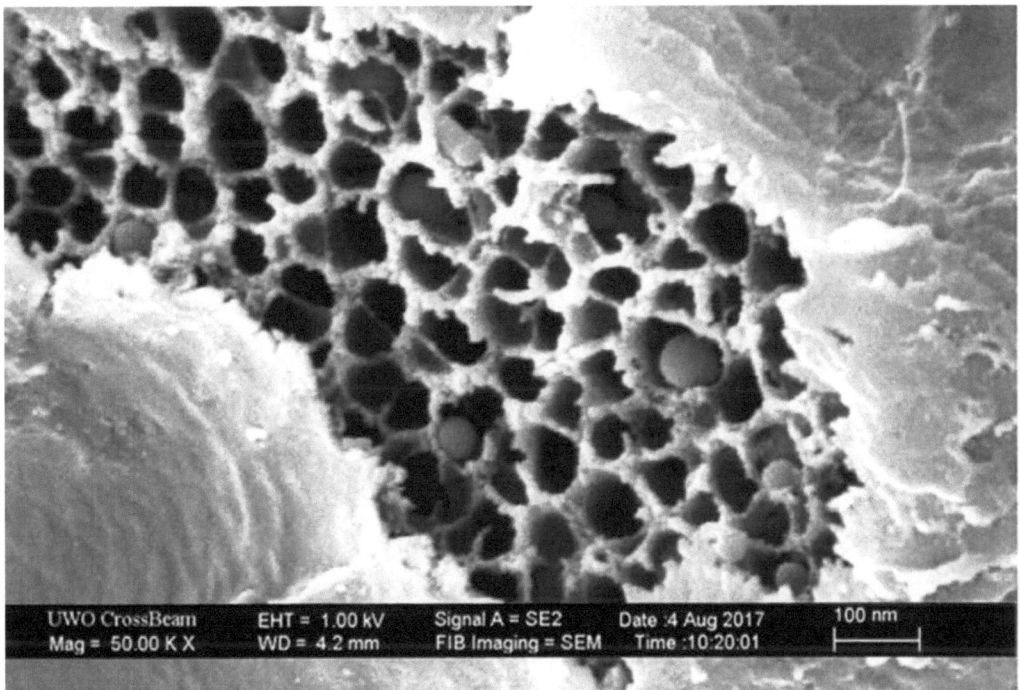

Fig. (12). SEM image of 0.5 wt.% loaded NF filtration membrane (Reproduced with permission from Elsevier Publications, [97]).

The antifouling performance result indicated an enhancement of fouling resistance rate of 80.9% at 0.5 wt.% loading of nanoparticles which was due to higher hydrophilicity, lower pore size, uniform dispersion of the nanoparticles [See Fig. (**13**)].

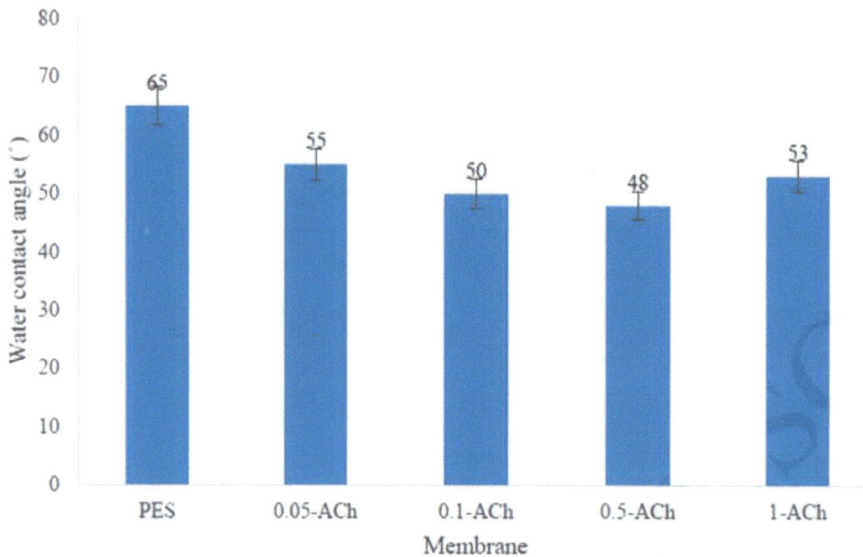

Fig. (13). Hydrophilicity of membrane w.r.t loading of nanoparticles (Reproduced with permission from Elsevier Publications, [97]).

Song *et al* [98]. developed a nanofiltration composite membrane with excellent antifouling ability based on graphene oxide. The authors selected graphene oxide owing to its chemical stability, anti-fouling ability and higher mechanical strength. Graphene oxide was deposited onto dopamine-modified polysulfone membrane by self-assembly method to eradicate low water flux, low ion retention capacity and poor stability of the membranes. Graphene oxide was also cross-linked with 1,3,5-benzenetricarbonyl trichloride (TMC) for the fabrication of graphene oxide composite nanofiltartion (NF) membrane. The fabricated graphene oxide composite nanofiltartion membrane were characterized using FTIR, XPS, AFM and SEM techniques to evaluate the morphological, chemical and structural features. The authors observed that the fabricated composite NF membranes exhibited superior anti-fouling performance with a rejection rate and permeation flux towards methyl blue up to 98% and 70 $kgm^{-2}h^{-1}$, respectively. The rejection rate and permeation flux towards inorganic ion *i.e.* PO_4^{-3} was found to be 92% and 120 $kgm^{-2}h^{-1}$, respectively. This findings can be attributed to the hydrophilic groups on the surface of Graphene oxide resulting in easy permeation of water with high flux without any hindrance. The higher negative zeta potential (-183.4 mV) owing to the carboxyl groups on graphene oxide surface enhanced the anti-fouling ability and the stability of the fabricated NF membranes with flux recovery ratio of up to 90%.

Fan *et al* [99]. reported about the inclusion of hydrophilic arginine into polyamide (piperazine, PIP) layer and trimesoyl chloride (TMC) *via* . interfacial polymerization technique to investigate the antifouling ability of the fabricated nanofiltration membrane. Different characterization techniques such as Fourier transform infrared spectroscopy (FTIR), SEM, AFM, water contact angle and zeta potential were carried out to evaluate the performance of the nanofiltration membrane. Arginine with abundant amine and carboxyl groups introduced hydrophilicity into the piperazine layer. Further, the zeta potential of the NF membrane also indicated higher negative value with the introduction of Arginine due to the presence of carboxyl groups as depicted in Fig. (**14**) The authors observed that with the increase in the loading of Arginine into polyamide layer from 10 to 40% the flux rate increased by 2 times from 18.56 L/m²h to 35.12 L/m²h. This phenomenon was due to the increase in hydrophilicity which resulted in higher permeation of water. The fabricated NF membranes also exhibited dye rejection rates (*i.e.* Congo red, methyl orange and orange GII) of over 90% which was related to the size of the separation pore layer. Further, the fabricated NF membranes also showed higher salt rejection rate for SO_4^{-2} due to higher strong electrostatic repulsion between the negatively charged membrane surface and SO_4^{-2} ions. Further, with the incorporation of arginine the flux decreased to 15% and fouling release rate increased to 89% indicating higher antifouling performance due to higher hydrophilicity and negative charges.

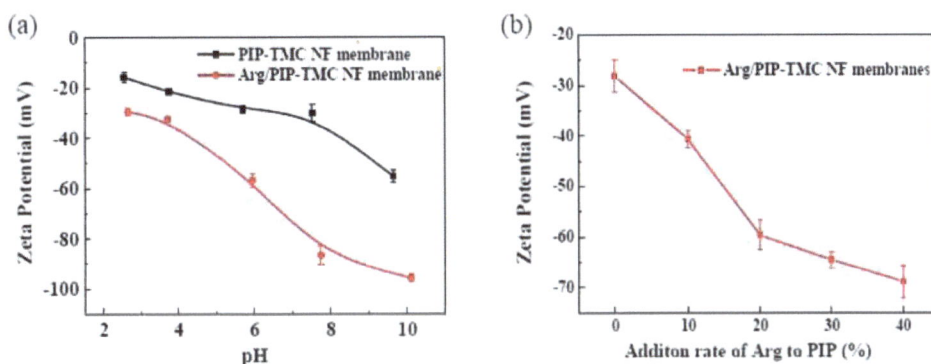

Fig. (14). Surface zeta potential of **a)** unmodified membrane and Arginine modified membrane, **(b)** Arginine modified membrane with varying Arginine % (Reproduced with permission from ACS Publications, [99]).

Zinadani *et al.* [100] developed antifouling polyethersulfone (PES) nanofiltration membranes by mixing with O-carboxymethylchitosan coated Fe_3O_4 nanoparticles *via* . phase inversion process. Fe_3O_4 has good adsorptive and O-carboxymethylchitosan has hydrophillic properties, hence both were used to exhibit synergistic effects in the fabricated membrane material. As depicted in Fig. **(15)** the carboxylate anions act as bidentated ligands and exhibit strong co-ordination with the Fe atoms. The polar head group of carboxylate anions gets anchored onto the magnetite surface and the polar tail group extends into the solution resulting in hydrophilic magnetite surface with higher dispersibility in an aqueous solution. FTIR, SEM, AFM and water contact angle techniques were used for the characterization of the NF membranes. The authors used Direct red 16 dye to examine the nanofiltration efficiency of the developed NF membranes. It was observed that with the increase in the loading of 0 to 0.5 wt%. of nanoparticles the water flux permeation of the membranes increased from 9.2 to 36 kg/m^{2}h due to the increase in hydrophilicity. The increase in hydrophilicity of the membrane from 64.4° to 52.5° with the addition of core-shell nanoparticles led to the attraction of water molecules towards the membrane matrix thereby enhancing the membrane flux. Further, the fabricated membranes exhibited higher flux recovery rate of 91.7% at a loading of 0.5 wt. % of O-carboxymethylchitosan/Fe$_3$O$_4$ nanoparticles. Also, the NF membranes exhibited higher rejection against Direct Red 16 dye due to the higher negative repulsive forces introduced by the nanoparticles.

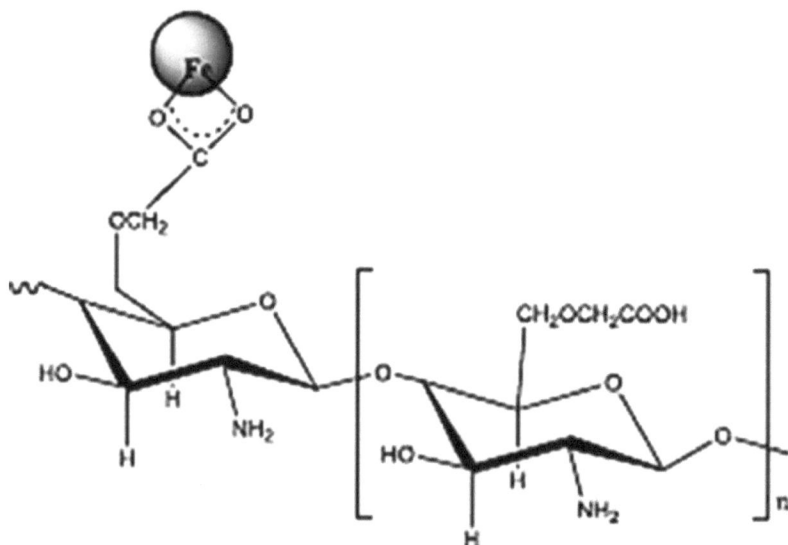

Fig. (15). Fe$_3$O$_4$ (magnetite) nanoparticle interaction with O-carboxymethyl Chitosan (Reproduced with permission from ACS Publications, [100]).

CONCLUSION

Nanofiltration (NF) membranes are widely preferred across industries due to their wider range of separation possibilities that are difficult to achieve with expensive counterparts. However, due to the drawbacks discussed in the chapter, the potential of NF is underdeveloped. This majorly involves the issue of membrane fouling and the corresponding need for cleaning, which may further affect the membrane's lifetime. Process control can be achieved by translation of transport concepts into models and simulation tools, which may trigger a wider commercial application for NF membranes. Feasible solution needs to be achieved to ensure the complete removal of contaminants by using novel process configurations and selecting appropriate membrane materials.

The chapter throws light on the significance of anti-fouling properties of the NF membrane and ways to obtain such structural integrity. The approach is majorly to prevent the adhesion of biofouling matter on the membrane surface. Methods involving the development of hydrophilic membranes, surface modification, polymer blending have also been discussed. The chapter also discusses the novel concept of nanocomposite membranes like metal oxides, graphene-based materials, carbon nanotubes, clays, zeolites, metal-organic frameworks, for improving the fouling resistance.

Embedding nanomaterials within the polymer matrix also alter the intrinsic properties of nanocomposite membranes like composition, surface area, pore size, hydrophilic/hydrophobic nature, *etc.* Although many studies have been undertaken in this direction, there is still a big room left out for designing commercially feasible techniques for addressing the major challenges faced by NF membranes. It is also recommended that studies may combine real complex situations involving waste wasters or industrial by-products with modeling and simulation tools to derive a lucid view of the potential of NF membranes for a variety of industrial applications.

CONSENT FOR PUBLICATION

Not applicable.

CONFLICT OF INTEREST

The authors declare no conflict of interest, financial or otherwise.

ACKNOWLEDGEMENT

Declared none.

REFERENCES

[1]   Zhang R, Liu Y, He M, *et al.* Antifouling membranes for sustainable water purification: strategies and mechanisms. Chem Soc Rev 2016; 45(21): 5888-924.
[http://dx.doi.org/10.1039/C5CS00579E] [PMID: 27494001]

[2]   https://www.weforum.org/agenda/2019/10/water-pollution-in-india-data-tech solu- tion/

[3]   https://www.marketsandmarkets.com/Market-Reports/membrane-filtration-mar-et-68840418.html?gclid=EAIaIQobChMIjcC5zsXH6wIVi38rCh22jgBmEAAYASAAEgI6b_D_BwE

[4]   Kurt E, Koseoglu-Imer DY, Dizge N, Chellam S, Koyuncu I. Pilot-scale evaluation of nanofiltration and reverse osmosis for process reuse of segregated textile dyewash wastewater. Desalination 2012; 302: 24-32.
[http://dx.doi.org/10.1016/j.desal.2012.05.019]

[5]   Baker WR. Membrane technology and applications. 2nd ed. John Wiley and Sons, Ltd. 2004; pp. 1-521.
[http://dx.doi.org/10.1002/0470020393]

[6]   Pendergast MM, Hoek EMV. A review of water treatment membrane nanotechnologies. Energy Environ Sci 2011; 4(6): 1946-71.
[http://dx.doi.org/10.1039/c0ee00541j]

[7]   Buonomenna MG. Membrane processes for a sustainable industrial growth. RSC Advances 2013; 3(17): 5694-40.
[http://dx.doi.org/10.1039/c2ra22580h]

[8]   https://aquaclearllc.com/technical-info/reverse-osmosis-vs-nanofiltration-and-other-filtration-tech-nologies/

[9]   https://www.freepurity.com/blogs/resources/micro-ultra-nano-filtration-vs-reverse-osmosis-whats-the-difference

[10]   https://www.pureaqua.com/reverse-osmosis-vs-nanofiltrationsystems/#:~:text=Nano-filtration%20requires%20less%20energy%20and

[11]   Shannon MA, Bohn PW, Elimelech M, Georgiadis JG, Mariñas BJ, Mayes AM. Science and technology for water purification in the coming decades. Nature 2008; 452(7185): 301-10.
[http://dx.doi.org/10.1038/nature06599] [PMID: 18354474]

[12]   Le-Clech P, Chen V, Fane TAG. Fouling in membrane bioreactors used in wastewater treatment. J Membr Sci 2006; 284(1-2): 17-53.
[http://dx.doi.org/10.1016/j.memsci.2006.08.019]

[13]   Fritzmann C, Löwenberg J, Wintgens T, Melin T. State-of-the-art of reverse osmosis desalination. Desalination 2007; 216(1-3): 1-76.
[http://dx.doi.org/10.1016/j.desal.2006.12.009]

[14]   Lee S, Cho J, Elimelech M. Influence of colloidal fouling and feed water recovery on salt rejection of RO and NF membranes. Desalination 2004; 160(1): 1-12.
[http://dx.doi.org/10.1016/S0011-9164(04)90013-6]

[15]   Valavala R, Sohn J, Han J, Her N, Yoon Y. Pretreatment in Reverse Osmosis Seawater Desalination: A Short Review. Environ Eng Sci 2011; 16: 205-12.

[16]   Madaeni SS, Samieirad S. Chemical cleaning of reverse osmosis membrane fouled by wastewater. Desalination 2010; 257(1-3): 80-6.
[http://dx.doi.org/10.1016/j.desal.2010.03.002]

[17]   Zhang W, Luo J, Ding L, Jaffrin MY. A Review on Flux Decline Control Strategies in Pressure-Driven Membrane Processes. Ind Eng Chem Res 2015; 54(11): 2843-61.
[http://dx.doi.org/10.1021/ie504848m]

[18]   Lin JCT, Lee D-J, Huang C. Membrane Fouling Mitigation: Membrane Cleaning. Sep Sci Technol 2010; 45(7): 858-72.
[http://dx.doi.org/10.1080/01496391003666940]

[19]   Choudhury RR, Gohil JM, Mohanty S, Nayak SK. Antifouling, fouling release and antimicrobial materials for surface modification of reverse osmosis and nanofiltration membranes. J Mater Chem A Mater Energy Sustain 2018; 6(2): 313-33.
[http://dx.doi.org/10.1039/C7TA08627J]

[20]   Abdelrasoul A, Doan H, Lohi A. Fouling in Membrane Filtration and Remediation Methods.Intech Open Acess 2013; 195-218.
[http://dx.doi.org/10.5772/52370]

[21]   Du X, Shi Y, Jegatheesan V, Haq IU. A Review on the Mechanism, Impacts and Control Methods of Membrane Fouling in MBR System. Membranes (Basel) 2020; 10(2): 24-35.
[http://dx.doi.org/10.3390/membranes10020024] [PMID: 32033001]

[22]   Membrane Fouling – A Mini Review. GSI Environmental 2016.

[23]   Khan SJ. Submerged and Attached Growth Membrane Bioreactors and Forward Osmosis Membrane Bioreactors for Wastewater Treatment, n, Emerging Membrane Technology for Sustainable Water Treatment.Emerging Membrane Technology for Sustainable Water Treatment. In Elsevier Publications 2016; pp. 277-96.

[24]   Singh R. Water and membrane treatment. Hybrid Membr Syst Water Purif 2005; 1: 57-130.
[http://dx.doi.org/10.1016/B978-185617442-8/50003-8]

[25]   Woo YC, Lee JJ, Oh JS, Jang HJ, Kim HS. Effect of chemical cleaning conditions on the flux recovery of fouled membrane. Desalination Water Treat 2013; 51(25-27): 5268-74.
[http://dx.doi.org/10.1080/19443994.2013.768754]

[26]   Porcelli N, Judd S. Chemical cleaning of potable water membranes: A review. Separ Purif Tech 2010; 71(2): 137-43.
[http://dx.doi.org/10.1016/j.seppur.2009.12.007]

[27]   Regula C, Carretier E, Wyart Y, *et al.* Ageing of ultrafiltration membranes in contact with sodium hypochlorite and commercial oxidant: Experimental designs as a new ageing protocol. Separ Purif Tech 2013; 103: 119-38.
[http://dx.doi.org/10.1016/j.seppur.2012.10.010]

[28]   Abdullah N, Rahman MA, Othman MHD, Jaafar J, Ismail AF. Membranes and Membrane Processes: Fundamentals. In: Basile A, Mozia S, Molinari R. (Ed.) Current Trends and Future Developments on (Bio-) Membranes, Photocatalytic Membranes and Photocatalytic Membrane Reactors. Elsevier: Amsterdam, The Netherlands 2018; pp. 45-70.

[29]   Ismail F, Khulbe KC, Matsuura T. Reverse Osmosis. Elsevier 2019; pp. 1-290.

[30]   Chen YQ, Li F, Qiao T, Ying H, Li W, Zhang J. Research on ultrafiltration membrane fouling based on chemical cleaning. Zhongguo Jishui Paishui 2013; 29(17): 51-4.

[31]   Blandin G, Verliefde A, Comas J, Rodriguez-Roda I, Le-Clech P. Efficiently combining water reuse and desalination through forward osmosis-reverse osmosis (FO-RO) hybrids: A critical review. Membranes (Basel) 2016; 6(3): 37.
[http://dx.doi.org/10.3390/membranes6030037] [PMID: 27376337]

[32]   Bhattacharjee S, Kim AS, Elimelech M. Concentration polarization of interacting solute particles in cross-flow membrane filtration. J Colloid Interface Sci 1999; 212(1): 81-99.
[http://dx.doi.org/10.1006/jcis.1998.6045] [PMID: 10072278]

[33]   Mulder M. Basic Principles of Membrane Technology. Springer Publications 2016; pp. 1-520.

[34]   Lu GQ, Diniz da Costa JC, Duke M, *et al.* Inorganic membranes for hydrogen production and purification: A critical review and perspective. J Colloid Interface Sci 2007; 314(2): 589-603.

[http://dx.doi.org/10.1016/j.jcis.2007.05.067] [PMID: 17588594]

[35]   Lee A, Elam JW, Darling SB. Membrane materials for water purification: design, development, and application. Environ Sci Water Res Technol 2016; 2(1): 17-42.
[http://dx.doi.org/10.1039/C5EW00159E]

[36]   Hsieh HP. Inorganic membrane reactor. Catal Rev, Sci Eng 1991; 33(1-2): 1-70.
[http://dx.doi.org/10.1080/01614949108020296]

[37]   Li NN, Fane AG, Ho WSW, Matsuura T. Advanced membrane technology and applications. John Wiley & sons Inc 2008; p. 1016.

[38]   Belfort G, Davis RH, Zydney AL. The behavior of suspensions and macromolecular solutions in crossflow microfiltration. J Membr Sci 1994; 96(1-2): 1-58.
[http://dx.doi.org/10.1016/0376-7388(94)00119-7]

[39]   Hlavacek M, Bouchet F. Constant flowrate blocking laws and an example of their application to dead-end microfiltration of protein solutions. J Membr Sci 1993; 82(3): 285-95.
[http://dx.doi.org/10.1016/0376-7388(93)85193-Z]

[40]   Hermia J. Constant pressure blocking filtration laws, application to power-law non-Newtonian fluids. Trans Am Inst Chem Eng 1982; 60: 183-7.

[41]   Granger J, Leclerc D. Filtration of dilute suspensions of latexes. Filtr Sep 1985; 22: 58-60.

[42]   Li H, Chen V. Membrane Fouling and Cleaning in Food and Bioprocessing, Membrane Technology. Elsevier Publications 2010; pp. 213-23.

[43]   Riedl K, Girard B, Lencki RW. Influence of membrane structure on fouling layer morphology during apple juice clarification. J Membr Sci 1998; 139(2): 155-66.
[http://dx.doi.org/10.1016/S0376-7388(97)00239-1]

[44]   Rana D, Matsuura T. Surface modifications for antifouling membranes. Chem Rev 2010; 110(4): 2448-71.
[http://dx.doi.org/10.1021/cr800208y] [PMID: 20095575]

[45]   Kochkodan V, Johnson DJ, Hilal N. Polymeric membranes: Surface modification for minimizing (bio)colloidal fouling. Adv Colloid Interface Sci 2014; 206: 116-40.
[http://dx.doi.org/10.1016/j.cis.2013.05.005] [PMID: 23777923]

[46]   Qiu HY, Xiao TH, Hu NE. The microporous membrane with different pore sizes was used tostudy the treatment of micro-polluted water in MBR. Technol Water Treat 2018; 44: 94-8.

[47]   Salahi A, Abbasi M, Mohammadi T. Permeate flux decline during UF of oily wastewater: Experimental and modeling. Desalination 2010; 251(1-3): 153-60.
[http://dx.doi.org/10.1016/j.desal.2009.08.006]

[48]   Li NN, Fane AG, Winston Ho WS, Matsuura T. Advanced membrane technology and applications. In John Wiley & sons Inc 2008.
[http://dx.doi.org/10.1002/9780470276280]

[49]   Vyas HK, Bennett RJ, Marshall AD. Influence of feed properties on membrane fouling in crossflow microfiltration of particulate suspensions. Int Dairy J 2000; 10(12): 855-61.
[http://dx.doi.org/10.1016/S0958-6946(01)00030-9]

[50]   Mohammad AW, Teow YH, Ang WL, Chung YT, Oatley-Radcliffe DL, Hilal N. Nanofiltration membranes review: Recent advances and future prospects. Desalination 2015; 356: 226-54.
[http://dx.doi.org/10.1016/j.desal.2014.10.043]

[51]   Yang Z, Zhou Y, Feng Z, Rui X, Zhang T, Zhang Z. A Review on Reverse Osmosis and Nanofiltration Membranes for Water Purification. Polymers (Basel) 2019; 11(8): 1252.
[http://dx.doi.org/10.3390/polym11081252] [PMID: 31362430]

[52]   Roy Y, Warsinger DM, Lienhard JH. Effect of temperature on ion transport in nanofiltration

membranes: Diffusion, convection and electromigration. Desalination 2017; 420: 241-57.
[http://dx.doi.org/10.1016/j.desal.2017.07.020]

[53]   Deen WM. Hindered transport of large molecules in liquid-filled pores. AIChE J 1987; 33(9): 1409-25.
[http://dx.doi.org/10.1002/aic.690330902]

[54]   Ernst M, Bismarck A, Springer J, Jekel M. Zeta-potential and rejection rates of a polyethersulfone nanofiltration membrane in single salt solutions. J Membr Sci 2000; 165(2): 251-9.
[http://dx.doi.org/10.1016/S0376-7388(99)00238-0]

[55]   Hagmeyer G, Gimbel R. Modelling the salt rejection of nanofiltration membranes for ternary ion mixtures and for single salts at different pH values. Desalination 1998; 117(1-3): 247-56.
[http://dx.doi.org/10.1016/S0011-9164(98)00109-X]

[56]   Hall MS, Lloyd DR, Starov VM. Reverse osmosis of multicomponent electrolyte solutions Part II. Experimental verification. J Membr Sci 1997; 128(1): 39-53.
[http://dx.doi.org/10.1016/S0376-7388(96)00301-8]

[57]   Childress AE, Elimelech M. Effect of solution chemistry on the surface charge of polymeric reverse osmosis and nanofiltration membranes. J Membr Sci 1996; 119(2): 253-68.
[http://dx.doi.org/10.1016/0376-7388(96)00127-5]

[58]   Afonso M, Hagmeyer G, *et al.* Streaming potential measurements to assess the variation of nanofiltration membranes surface charge with the concentration of salt solutions. Separ Purif Tech 2001; 22-23(1-2): 529-41.
[http://dx.doi.org/10.1016/S1383-5866(00)00135-0]

[59]   Schaep J, Vandecasteele C. Evaluating the charge of nanofiltration membranes. J Membr Sci 2001; 188(1): 129-36.
[http://dx.doi.org/10.1016/S0376-7388(01)00368-4]

[60]   Schäfer AI, Fane AG. Nanofiltration-principles and Applications. 1st ed., Elsevier Advanced Technology Oxford, 2005.

[61]   Oatley DL, Llenas L, Pérez R, Williams PM, Martínez-Lladó X, Rovira M. Review of the dielectric properties of nanofiltration membranes and verification of the single oriented layer approximation. Adv Colloid Interface Sci 2012; 173: 1-11.
[http://dx.doi.org/10.1016/j.cis.2012.02.001] [PMID: 22405540]

[62]   Fang Y, Bian L, Bi Q, Li Q, Wang X. Evaluation of the pore size distribution of a forward osmosis membrane in three different ways. J Membr Sci 2014; 454: 390-7.
[http://dx.doi.org/10.1016/j.memsci.2013.12.046]

[63]   Qian H, Zheng J, Zhang S. Preparation of microporous polyamide networks for carbon dioxide capture and nanofiltration. Polymer (Guildf) 2013; 54(2): 557-64.
[http://dx.doi.org/10.1016/j.polymer.2012.12.005]

[64]   Wang T, Yang Y, Zheng J, Zhang Q, Zhang S. A novel highly permeable positively charged nanofiltration membrane based on a nanoporous hyper-crosslinked polyamide barrier layer. J Membr Sci 2013; 448: 180-9.
[http://dx.doi.org/10.1016/j.memsci.2013.08.012]

[65]   Carvalho AL, Maugeri F, Silva V, Hernandez A, Palacio L, Pradanos P. AFM analysis of the surface of nanoporous membranes: application to the nanofiltration of potassium clavulanate. J Mater Sci 2011; 46(10): 3356-69.

[66]   Johnson DJ, Al Malek SA, Al-Rashdi BAM, Hilal N. Atomic force microscopy of nanofiltration membranes: Effect of imaging mode and environment. J Membr Sci 2012; 389: 486-98.
[http://dx.doi.org/10.1016/j.memsci.2011.11.023]

[67]   Misdan N, Lau WJ, Ismail AF, Matsuura T, Rana D. Study on the thin film composite poly(piperazine-amide) nanofiltration membrane: Impacts of physicochemical properties of substrate

[http://dx.doi.org/10.1016/j.desal.2014.03.036]

[68] Stawikowska J, Livingston AG. Assessment of atomic force microscopy for characterisation of nanofiltration membranes. J Membr Sci 2013; 425-426: 58-70.
[http://dx.doi.org/10.1016/j.memsci.2012.08.006]

[69] García-Martín N, Silva V, Carmona FJ, Palacio L, Hernández A, Prádanos P. Pore size analysis from retention of neutral solutes through nanofiltration membranes. The contribution of concentration–polarization. Desalination 2014; 344: 1-11.
[http://dx.doi.org/10.1016/j.desal.2014.02.038]

[70] Kiso Y, Muroshige K, Oguchi T, Hirose M, Ohara T, Shintani T. Pore radius estimation based on organic solute molecular shape and effects of pressure on pore radius for a reverse osmosis membrane. J Membr Sci 2011; 369(1-2): 290-8.
[http://dx.doi.org/10.1016/j.memsci.2010.12.005]

[71] Oatley DL, Llenas L, Aljohani NHM, *et al.* Investigation of the dielectric properties of nanofiltration membranes. Desalination 2013; 315: 100-6.
[http://dx.doi.org/10.1016/j.desal.2012.09.013]

[72] Stawikowska J, Livingston AG. Nanoprobe imaging molecular scale pores in polymeric membranes. J Membr Sci 2012; 413-414: 1-16.
[http://dx.doi.org/10.1016/j.memsci.2012.02.033]

[73] Cheng S, Oatley DL, Williams PM, Wright CJ. Positively charged nanofiltration membranes: Review of current fabrication methods and introduction of a novel approach. Adv Colloid Interface Sci 2011; 164(1-2): 12-20.
[http://dx.doi.org/10.1016/j.cis.2010.12.010] [PMID: 21396619]

[74] Déon S, Fievet P, Osman Doubad C. Tangential streaming potential/current measurements for the characterization of composite membranes. J Membr Sci 2012; 423-424: 413-21.
[http://dx.doi.org/10.1016/j.memsci.2012.08.038]

[75] Bauman M, Košak A, Lobnik A, Petrinić I, Luxbacher T. Nanofiltration membranes modified with alkoxysilanes: Surface characterization using zeta-potential. Colloids Surf A Physicochem Eng Asp 2013; 422: 110-7.
[http://dx.doi.org/10.1016/j.colsurfa.2013.01.005]

[76] Rice G, Barber AR, O'Connor AJ, *et al.* The influence of dairy salts on nanofiltration membrane charge. J Food Eng 2011; 107(2): 164-72.
[http://dx.doi.org/10.1016/j.jfoodeng.2011.06.028]

[77] Lee S, Lee E, Elimelech M, Hong S. Membrane characterization by dynamic hysteresis: Measurements, mechanisms, and implications for membrane fouling. J Membr Sci 2011; 366(1-2): 17-24.
[http://dx.doi.org/10.1016/j.memsci.2010.09.024]

[78] Luxbacher T. Electrokinetic characterization of flat sheet membranes by streaming current measurement. Desalination 2006; 199(1-3): 376-7.
[http://dx.doi.org/10.1016/j.desal.2006.03.085]

[79] Xie H, Saito T, Hickner MA. Zeta potential of ion-conductive membranes by streaming current measurements. Langmuir 2011; 27(8): 4721-7.
[http://dx.doi.org/10.1021/la105120f] [PMID: 21443169]

[80] Teixeira M, Rosa M, Nystrom M. The role of membrane charge on nanofiltration performance. J Membr Sci 2005; 265(1-2): 160-6.
[http://dx.doi.org/10.1016/j.memsci.2005.04.046]

[81] Kotrappanavar NS, Hussain AA, Abashar MEE, Al-Mutaz IS, Aminabhavi TM, Nadagouda MN. Prediction of physical properties of nanofiltration membranes for neutral and charged solutes. Desalination 2011; 280(1-3): 174-82.

[http://dx.doi.org/10.1016/j.desal.2011.07.007]

[82]   Kumar VS, Hariharan KS, Mayya KS, Han S. Volume averaged reduced order Donnan Steric Pore Model for nanofiltration membranes. Desalination 2013; 322: 21-8.
[http://dx.doi.org/10.1016/j.desal.2013.04.030]

[83]   Yang G, Shi H, Liu W, Xing W, Xu N. Investigation of Mg^{2+}/Li^+ separation by nanofiltration. Chin J Chem Eng 2011; 19(4): 586-91.
[http://dx.doi.org/10.1016/S1004-9541(11)60026-8]

[84]   Espinasse BP, Chae SR, Marconnet C, *et al.* Comparison of chemical cleaning reagents and characterization of foulants of nanofiltration membranes used in surface water treatment. Desalination 2012; 296: 1-6.
[http://dx.doi.org/10.1016/j.desal.2012.03.016]

[85]   Martínez MB, Jullok N, Negrin ZR, Van der Bruggen B, Luis P. Effect of impurities in the recovery of 1-(5-bromo-fur-2-il)-2-bromo-2-nitroethane using nanofiltration. Chem Eng Process 2013; 70: 241-9.
[http://dx.doi.org/10.1016/j.cep.2013.03.015]

[86]   Panda SR, De S. Preparation, characterization and performance of $ZnCl_2$ incorporated polysulfone (PSF)/polyethylene glycol (PEG) blend low pressure nanofiltration membranes. Desalination 2014; 347: 52-65.
[http://dx.doi.org/10.1016/j.desal.2014.05.030]

[87]   Baek Y, Kang J, Theato P, Yoon J. Measuring hydrophilicity of RO membranes by contact angles *via* sessile drop and captive bubble method: A comparative study. Desalination 2012; 303: 23-8.
[http://dx.doi.org/10.1016/j.desal.2012.07.006]

[88]   Do VT, Tang CY, Reinhard M, Leckie JO. Effects of hypochlorous acid exposure on the rejection of salt, polyethylene glycols, boron and arsenic(V) by nanofiltration and reverse osmosis membranes. Water Res 2012; 46(16): 5217-23.
[http://dx.doi.org/10.1016/j.watres.2012.06.044] [PMID: 22818949]

[89]   Hurwitz G, Guillen GR, Hoek EMV. Probing polyamide membrane surface charge, zeta potential, wettability, and hydrophilicity with contact angle measurements. J Membr Sci 2010; 349(1-2): 349-57.
[http://dx.doi.org/10.1016/j.memsci.2009.11.063]

[90]   Do VT, Tang CY, Reinhard M, Leckie JO. Degradation of polyamide nanofiltration and reverse osmosis membranes by hypochlorite. Environ Sci Technol 2012; 46(2): 852-9.
[http://dx.doi.org/10.1021/es203090y] [PMID: 22221176]

[91]   Lamsal R, Harroun SG, Brosseau CL, Gagnon GA. Use of surface enhanced Raman spectroscopy for studying fouling on nanofiltration membrane. Separ Purif Tech 2012; 96: 7-11.
[http://dx.doi.org/10.1016/j.seppur.2012.05.019]

[92]   Montalvillo M, Silva V, Palacio L, *et al.* Charge and dielectric characterization of nanofiltration membranes by impedance spectroscopy. J Membr Sci 2014; 454: 163-73.
[http://dx.doi.org/10.1016/j.memsci.2013.12.017]

[93]   Nanda D, Tung KL, Hung W-S, *et al.* Characterization of fouled nanofiltration membranes using positron annihilation spectroscopy. J Membr Sci 2011; 382(1-2): 124-34.
[http://dx.doi.org/10.1016/j.memsci.2011.08.026]

[94]   Van der Bruggen B, Mänttäri M, Nyström M. Drawbacks of applying nanofiltration and how to avoid them: A review. Separ Purif Tech 2008; 63(2): 251-63.
[http://dx.doi.org/10.1016/j.seppur.2008.05.010]

[95]   Qian Y, Wu H, Sun S-P, Xing W. Perfluoro-functionalized polyethyleneimine that enhances antifouling property of nanofiltration membranes. J Membr Sci 2020; 611: 118286.
[http://dx.doi.org/10.1016/j.memsci.2020.118286]

[96]   Qi Y, Zhu L, Gao C, Shen J. A novel nanofiltration membrane with simultaneously enhanced antifouling and antibacterial properties. RSC Advances 2019; 9(11): 6107-17.

antifouling and antibacterial properties. RSC Advances 2019; 9(11): 6107-17.
[http://dx.doi.org/10.1039/C8RA09875A] [PMID: 35517273]

[97]   Bagheripour E, Moghadassi AR, Hosseini SM, Ray MB, Parvizian F, Van der Bruggen B. Highly hydrophilic and antifouling nanofiltration membrane incorporated with water-dispersible composite activated carbon/chitosan nanoparticles. Chem Eng Res Des 2018; 132: 812-21.
[http://dx.doi.org/10.1016/j.cherd.2018.02.027]

[98]   Song X, Li Y, Zhao G, Lu Y, Li C, Meng H. A Novel Graphene Oxide Composite Nanofiltration Membrane with Excellent Performance and Antifouling Ability. J Membr Sci Technol 2018; 8(2): 2.
[http://dx.doi.org/10.4172/2155-9589.100018]

[99]   Fan L, Zhang Q, Yang Z, *et al.* Improving permeation and antifouling performance of the polyamide nanofiltration membrane through incorporating arginine. ACS Appl Mater Interfaces 2017; 9(15): 13577-86.

[100]   Zinadini S, Zinatizadeh AA, Rahimi M, Vatanpour V, Zangeneh H, Beygzadeh M. Novel high flux antifouling nanofiltration membranes for dye removal containing carboxymethyl chitosan coated Fe3O4 nanoparticles. Desalination 2014; 349: 145-54.
[http://dx.doi.org/10.1016/j.desal.2014.07.007]

CHAPTER 9

Microbes and their Genes involved in Bioremediation of Petroleum Hydrocarbon

Bhaskarjyoti Gogoi[1], Indukalpa Das[1], Shamima Begum[1], Gargi Dutta[1], Rupesh Kumar[1] and Debajit Borah[1,*]

[1] Department of Biotechnology, Assam Royal Global University, Guwahati-781035, Assam, India

Abstract: The catastrophic effect of petroleum contamination on the environment is a severe problem of global concern. Bioremediation is probably the easiest and most cost-effective way to treat the contaminants. Several microorganisms ranging from bacteria, fungi, yeast, algae, *etc.*, are known for their ability to biodegrade different hydrocarbons. Hydrocarbon degrading microorganisms are largely known for the release of biosurfactants and other surface-active biopolymers, which decrease the surface tension of oil particles into smaller entities for their easy degradation throughout the respective metabolic cycle. Such biopolymers are encoded by several genes and operon systems which are discussed briefly in this chapter. Information on such genes help in better understanding the molecular events involved in the microbial bioremediation of petroleum hydrocarbon.

Keywords: Bioremediation, Biosurfactant, Enzymes, Genes.

INTRODUCTION

Due to the massive use of fossil fuel and the derivatives of various petroleum-based products, environmental pollution has become a global issue of public concern that requires immediate attention from the scientific community to develop ways to solve the problem. Deliberate or accidental release of petroleum oil into soil and water largely affects the health of aquatic as well as terrestrial life forms, including human health in the form of chiefly respiratory diseases, kidney ailments, and even cancer [1]. A vast majority of the constituents of petroleum oil are polycyclic aromatic hydrocarbons (PAHs) which are mostly recalcitrant in nature and remain for a prolonged duration of time in the environment [1]. Petroleum contaminants are mostly treated by chemical methods, which involve surfactants, and physical methods involving burning, skimming, *etc.* [2]. However, skimming of petroleum is easier but the major problem with skimming

* **Corresponding author Debajit Borah:** Department of Biotechnology, Assam Royal Global University, Guwahati-781035, Assam, India; Tel: +91-9706394994; E-mail: dborah89@gmail.com

Inamuddin (Ed.)

is the clogging of the suction pipe due to clogging by debris and also solidification of waxy petroleum under ice-cold environment makes it even more difficult [3]. On the other hand, the open burning of petroleum products emits a huge amount of toxic gases into the environment which is more harmful [4]. Even though chemical surfactants are also used as a bioremediation agent but they release toxic non-biodegradable components into the environment [2].

Therefore, bioremediation proves itself as a future scope for solving such issues that involve the use of microorganisms such as various bacteria, fungi, algae, plants, *etc.* without releasing harmful residues to the environment [5, 6]. However, microorganisms including bacteria and fungi are chiefly used for their least requirements for maintenance and implementation nature. They release many surface active molecules which reduce the surface tension of oily substances for the initiation of hydrocarbon degradation and are governed by several independent or cascades of genes.

TYPES OF BIOREMEDIATION STRATEGIES

The pollution of soil, sediments, and both surface, as well as groundwater by petroleum hydrocarbon, possesses a risk to human health. The contaminants can be spread to a broad area from the site of their origin and affect the flora and fauna of the polluted area. Depending upon the various factors like contamination site, nature of contaminants and environmental conditions of the contaminated site, the remediation treatment methods are different. Also, the mechanism of the treatment procedure, regulatory requirements along with the time constraints, costs and the remediation process are chosen. The success of the use of any remediation methods depends on the design and adjustment of the system operation. The integration of two or more remediation processes in combination or in a sequence gives better efficacy in the removal of contaminants from the site of pollution [1].

PHYSICAL METHOD FOR BIOREMEDIATION OF PETROLEUM HYDROCARBON

The physical method is the most perceptible source of contamination that is removed from the oil spill in soil and the water body may involve mechanical removal methods such as skimming, manual removal (wiping), water gushing, *etc* [2]. A great component of the oil pollution problem results from the fact that the major oil-manufacturing countries are not the main oil consumers. It follows that huge transfers of petroleum must be made from areas of high production to those of high consumption [7]. Although physical removal of oil spillage is a primary response of treatment, it rarely achieves complete clean-up due to the usual blockage of the transporting pipes with debris.

CHEMICAL METHOD FOR BIOREMEDIATION OF PETROLEUM HYDROCARBON

To convert the toxic form of the hydrocarbon pollutants into less toxic forms, several chemical agents can be used. The chemicals change the physical and chemical properties of the pollutants. Dispersants and solidifiers are the two different classes of chemicals that are used to treat oil spills. Dispersants break the oil slicks by the activities of surface-active agents and converted them into smaller droplets. Then, the smaller droplets are transferred into the water column and finally undergo degradation. Solidifiers interact with oil and change the physical state of the oil which in turn reduces the release of hydrocarbons into the environment [8]. Several chemical remediation processes such as chemical leaching, chemical fixation, chemical oxidation, *etc.* are available to reduce the load of toxic pollutants in the surroundings of the spillage area [9].

BIOLOGICAL METHODS

Bioremediation is the most effective and proven remediation method for restoring polluted sites by removing the pollutants. In the bioremediation process, micro-organisms play a crucial role. The diverse enzymatic profile, micro-organisms are widely used to degrade petroleum hydrocarbons into less toxic forms. In comparison to the other conventional remediation processes, bioremediation is a more efficient process with relatively low cost. The use of bioremediation causes minimal site disruption with no additional nutrient requirements. Due to these advantages, bioremediation is a widely accepted popular remediation technique [10, 11].

The process of bioremediation can be achieved by any of the two treatments, *ex-situ* and *in-situ* treatment. The remediation strategies are considered based on the treatment cost, type of the pollutants, the severity of the pollution, and site of the pollution. In *ex-situ* bioremediation method, the pollutants are excavated from the polluted site and transferred to another site where the treatment is done. Whereas, the *in-situ* method of remediation involves no excavation of the pollutants; instead, the treatments are done on the site of the pollution. Several treatment methods are being used for the process of bioremediation. Fig. (**1**) shows the schematic representation of the different strategies for bioremediation techniques. Based on the age and the degree of a spill along with the physicochemical properties of the spillage matrix, the treatment procedures may be different [12].

EX-SITU BIOREMEDIATION

As shown in Fig. (**1**), the different *ex-situ* bioremediation methods are biopiling, windrows, bioreactor, and land farming.

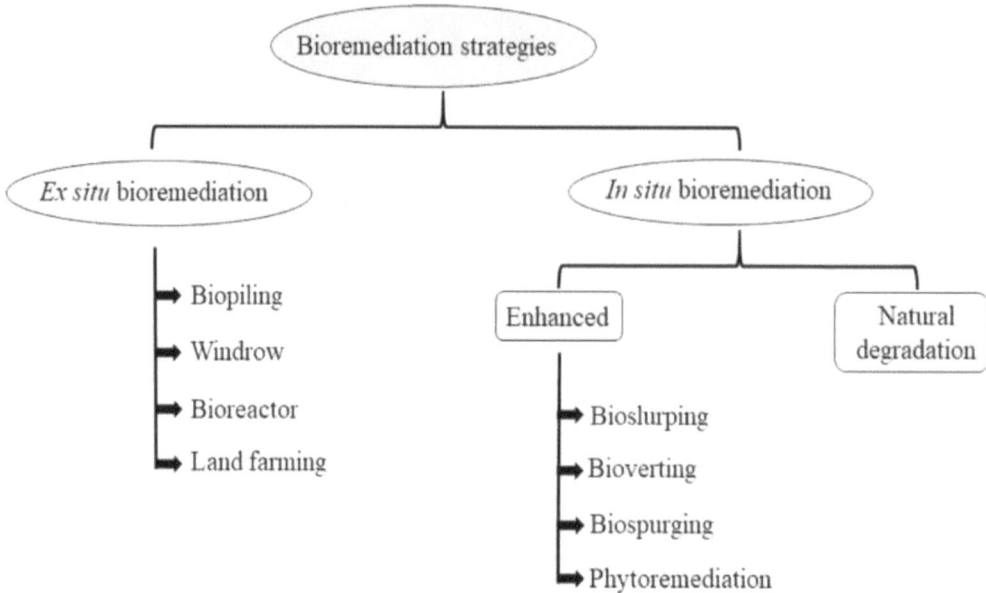

Fig. (1). Different strategic techniques for bioremediation of petroleum hydrocarbons.

Biopiling: It involves the piling through the topsoil of the excavated polluted soil matrix followed by augmentation of the nutrient content of the soil for better microbial growth. In this process, some components like aeration, nutrient, and leachate collection system, irrigation, and the treatment bed are also considered. In this technique of bioremediation, conditions like nutrients, temperature, and aeration can be controlled. This process is also helpful in limiting the volatilization of pollutants of lower molecular weight. In some biopiling systems, a heating system is also added to maintain the optimum temperature for microbial growth to reduce the bioremediation time [13]. Another advantage of using biopile method is to treat a large volume of the contaminated soil sample in a limited space and also the setup can be scaled up when requires.

Windrows: It involves the periodic turning of the piled polluted soil. This helps in enhancing the bioremediation process with the help of increased degradation abilities of microbial flora present in the soil. The addition of water induces aeration and uniform distribution of nutrients along with the distribution of the pollutants in the treatment site. As a result of it, microbial degradation activity increases at a faster rate in windrows as compared to conventional biopiling technique [14, 15].

Bioreactor: It is an engineered device to convert the raw materials into the product with the involvement of many enzymatic reactions under optimum conditions for efficient bioremediation of contaminants. The polluted samples can be used either as slurry or as dry matter. The device is engineered in such a way so that different growth parameters such as pH, temperature, substrate concentrations, the concentration of bacterium inoculum, *etc.* can be easily controlled. By reducing the limiting factors like bioaugmentation, the addition of nutrients, *etc.* makes the bioreactor an efficient device for bioremediation. Also, it is suitable for the treatment of soil and water contaminated with various volatile organic compounds. The changes in dynamics in the microbial population can be treated in bioreactor during the period of short as well as long term operation to treat petroleum hydrocarbon polluted slurry [16, 17].

Land Farming: It is the simplest bioremediation technique. Most of the time, it is considered as *ex-situ* bioremediation technique but sometimes it is also considered as an *in-situ* bioremediation process. In this process, the depth of the pollutants in the pile determines whether it is considered *ex-situ* or *in-situ*. If the excavated pollutants containing soil are considered for treatment on-site, it is considered as *in-situ* instead of *ex-situ* even though the polluted soil is excavated [18]. Generally, hydrocarbon polluted sites are treated by this land farming method, where two remediation mechanisms *viz.*, volatilization and remediation are involved. In this land farming procedure, a large volume of polluted soil can be treated with a low cost of capital and energy [19]. Land farming becomes less laborious and less time-consuming due to the lack of preliminary assessment of polluted sites [12].

In Situ Bioremediation

In this technique of bioremediation, the polluted soil is not excavated for transportation to the new site; instead, the treatment is done at the site of pollution itself. Due to this, *in-situ* bioremediation is less expensive in comparison to the *ex-situ* bioremediation. But the major concern is the on-site installation of sophisticated equipment. The natural *in-situ* bioremediation techniques are enhanced by using other techniques like bioventing, biosparging, bioslurping, and phytoremediation. The sites polluted with chlorinated solvents, heavy metals, different dyes, and hydrocarbons can be treated with these *in-situ* bioremediation techniques but soil porosity largely influences its application to the polluted sites [20 - 23].

Bioventing: In this enhanced *in-situ* bioremediation method, oxygen is delivered to unsaturated zones in a controlled manner with increased microbial activities. Although the airflow rate is one of the major parameters in bioventing the success rate of bioremediation depends on the uniform distribution of air, which is determined by the number of injection points for airflow [24 - 26]. In this process, additional moisture and nutrients can be added for the enhancement of bioremediation to transform pollutants into a harmless state [27, 28]. Since the bioventing remediation method is used for the aeration of the unsaturated zone, the process is useful for anaerobic bioremediation of unsaturated zone containing the contaminants [27, 28]. Since the bioventing remediation method is used for the aeration of the unsaturated zone, the process is useful for anaerobic bioremediation of unsaturated zone containing the contaminants [24 - 26].

Bioslurping: This method is the combination of bioventing, vacuum enhanced pumping, and soil vapor extraction. This method is used to remediate soil and groundwater with an indirect supply of oxygen [29]. Bioslurping helps in the remediation of capillary, saturated, and unsaturated zones with the recovery of free products like light non-aqueous phase liquids (LNAPLS). The soil contaminated with semi-volatile and volatile organic compounds can be mediated by using this *in-situ* technique [20]. However, this method is not widely used to remediate soil with low permeability. The advantage of this technique is basically lowering the cost of storage, treatment, and disposal [28].

Biosparging: Biosparging is a similar technique to bioventing. The only difference of biosparging from bioventing is the injection of air into the subsurface of the soil to enhance microbial activities. Due to the supply or injection of air to the saturated zone, the volatile organic compounds (pollutants) move upward to an unsaturated zone which induces the degradation of the pollutants. However, soil permeability and pollutant biodegradability are the two major factors influencing the effectiveness of biosparging [28]. The aquifers contaminated with diesel and kerosene are treated with biosparging method. It is also used for the bioremediation of benzene, toluene, ethylbenzene, and xylene (BTEX) from ground water [12, 30].

Phytoremediation: This technique is used to mitigate the effects of toxic materials by using plants. Based on the type of pollutants, several phytoremediation mechanisms are used. They are accumulation, degradation, filtration, stabilization, and volatilization, *etc.* Extraction, transformation, and sequestration are mostly used for the removal of elemental pollutants such as toxic heavy metals and radionuclides. Degradation, rhizoremediation, stabilization, volatilization, and mineralization are used to remove the organic pollutants [31, 32]. During the process of selection of plant species for phytoremediation, some important factors to be considered such as the root system of the plant, above-ground biomass, toxicity of the pollutants to the plants, adaptability and growth rate of the plant, and most importantly the time required by the plant to achieve the desired cleanliness. The resistance capacity of the plants to disease and pests are also important factors to be considered [33]. The plant rhizobacteria play an important role in the success of phytoremediation. It is suspected that the plant growth promoting rhizobacteria induces biomass production. They also enhance the tolerance capacity of the plant against unfavorable soil conditions [34, 35]. Nowadays, some specific genes of the plants are genetically modified to increase their tolerance capacity to heavy metals along with the increased metabolic rate [36]. It is important to the pretreatment of heavily polluted sites with increased microbial diversity and activity before the implementation of phytoremediation. This in turn reduces the toxic effects of the pollutants on the plants [12].

Microbial Bioremediation Method

The complexity of the degradation process is varied depending on the nature of the hydrocarbons and their concentration in the environment. Several microbes which include various fungi, bacteria, and yeasts involved in the degradation of petroleum hydrocarbon have been identified by various research groups [37]. Microorganisms such as *Arthrobacter, Mycobacterium, Sphingomonas, Burkholderia, Pseudomonas, Rhodococcus, Bacillus, Alkaligens, Acinetobacter, Flavobacterium, Micrococcus, Corynebacterium, etc.* are largely explored for the degradation of alkyl aromatic components in marine sediments [6]. It is observed that microbes with a broad enzymatic capacity are required for the degradation of hydrocarbon mixture [38 - 41].

A handful of reports also show the bacterial population belonging to the genus *Gordonia, Brevibacterium, Acromicrobium, Dietzia, Spingomonas, Burkholderia, Mycobacteria, etc.* involved in the bioremediation of soil contaminated with petroleum hydrocarbon [5, 42].

Among fungi, *Amorphotheca, Neosartorya, Talaromyces, Graphium, Aspergillus, Cephalosporium, Penicillium, etc.* have been reported to have petroleum degrading ability [5, 43]. Similarly, in the case of yeast *Candida, Yarrowia, Pichia, Rhodotorula mucilaginosa, Trichosporon mucoides, Candida lipolytica, Geotrichum* sp, *etc.* have also been reported for their activities in degrading petroleum hydrocarbon [5, 44].

Although protozoa and algae have been reported for their ability to degrade petroleum hydrocarbon, only a very few reports are available to support their involvement in hydrocarbon degradation. Algae known as *Prototheca zopfii* was reported to degrade *n*-alkanes, isoalkanes, and aromatic hydrocarbons [45]. Another group of algae including cyanobacteria, green algae, red algae, brown algae, and diatoms was also reported for their capabilities in degrading naphthalene [46].

To degrade the aromatic hydrocarbons by different microbes and plants, the hydrocarbon rings of aromatic hydrocarbons are attacked by enzymes in the presence of molecular oxygen. In an aerobic bacterial system, arenes are oxidized to form an early bi-products vicinal *cis*-dihydrodiols by dioxygenase enzyme. This early intermediate product is then converted into central intermediates protocatechuates and catechols by the enzyme intradiol or extradiol ring-cleaving dioxygenases. To convert the early dihydroxylated intermediates to central intermediate protocatechuates and catechols, either an ortho-cleavage pathway or a meta-cleavage pathway is followed. The central intermediates are then converted into the product which is the intermediates of Kreb's cycle and involve in ATP production [47 - 49].

ROLE OF BIOSURFACTANTS IN PETROLEUM HYDROCARBON DEGRADATION

Microbes releases various surface-active compounds which include enzymes, polysaccharides, lipids, and sugar moieties which reduce the surface tension of oily substances and make them available for the microbes for further degradation into less or no-toxic substances which are collectively known as biosurfactant in a broad sense [50]. Biosurfactants are one of the significant microbial products from an industrial and commercial perspective because of their wide use in several fields including agriculture, crude-oil recovery, cosmetics, pharmaceuticals, food processing, detergents, textiles, biomedicine, petroleum degradation, *etc* [3].

The capability to produce biosurfactants by microbes is the most essential feature for hydrocarbon degradation [27]. Biosurfactants possess the potential to decrease the surface tension of oily substances and also the interfacial tension with the formation foam which ultimately increases the solubility and the bio-availability of Non-Aqueous Phase Liquids (NAPLS). Based on the ionic properties, biosurfactants are generally classified into four classes anionic, non-ionic, cationic, and amphoteric. On the contrary, based on their chemical nature, they may also be classified into five different classes *viz*., glycolipids, phospholipids, polymeric biosurfactants, and lipopeptides [51]. Various classes of biosurfactants produced by diverse types of microbes are shown in Table 1.

ROLE OF MICROBIAL ENZYMES AND RESPONSIBLE GENES IN HYDROCARBON DEGRADATION

Microorganisms produce enzymes in the presence of carbon sources that are responsible for attacking the hydrocarbon compounds [7]. Many different enzymes and metabolic pathways are involved in the degradation of hydrocarbon contaminants. Among various enzymes, Cytochrome P450 plays a crucial role in the microbial degradation of hydrocarbon, polychlorinated biphenyls, fuel additives, and many other compounds [70]. Higher eukaryotes usually include several various types of Cytochrome-P450 (CYPs-P450) families that consist of a large number of specific CYPs-P450 that might participate as an assemblage of isoforms in the metabolic conversion of a given substrate [70, 71]. Cytochrome P450 enzyme can also be found in a few species of yeasts such as *Candida maltosa*, *Candida tropicalis*, *Candida apicola, etc.* which are involved in the biodegradation of petroleum hydrocarbon [72]. The wide range of mono and di-oxygenases and also catalases found in microorganisms such as *Enterobacter* sp. strain ODB01, *Bacillus subtilis, Pseudomonas aeruginosa, Bacillus cereus* also plays a great role in petroleum based hydrocarbon degradation [73, 74].

Table 1. Main classification of biosurfactants and producing microorganisms.

Metabolite	Sub-type	Source(s)	References
Glycolipids	Rhamnolipids	*Pseudomonas aeruginosa*	[52]
	Sophorolipids	*Torulopsis bombicola, Candida apicola*	[53, 54]
	Trehalolipids	*Rhodococcus erythropolis, Mycobacterium sp.*	[55]
Lipopeptides and lipoprotein	Peptide-lipid	*Bacillus licheniformis*	[56]
	Viscosin	*Pseudomonas fluorescens*	[57]
	Serrawettin	*Serratia marcenscens*	[58]
	Surfactin		[59, 60]
	Subtilisin	*Bacillus subtilis*	
	Polymyxin	*Bacillus polymyxa*	[61]
Fatty acids, neutral lipids and phospholipids	Fatty acid	*Corynebacterium lepus*	[62]
	Neutral lipids	*Nocardia erythropolis*	[63]
	Phospholipids	*Thiobacillus thiooxidans*	[64]
Polymeric surfactants	Emulsan	*Acinetobacter calcoaceticus*	[65]
	Biodispersan	*Acinetobacter calcoaceticus*	[66]
	Liposan	*Candida lipolytica*	[67]
	Carbohydrate-lipid-protein	*Pseudomonas fluorescens*	[68]
Particulate surfactant	Vesicles	*Acinetobacter calcoaceticus*	[69]

Microorganisms possess multiple systems to degrade alkanes with different chain lengths which induce oxidative effect to the methyl group at the end of the chain forming aldehyde fatty alcohol and fatty acid. Then, the further processing of hydrocarbon compounds occurs before entry into the TCA cycle. In the case of n-alkanes having a shorter chain length of range C1–C4, bacterial methane monooxygenase enzyme (MMO) initiates the degradation process. The family of MMO enzymes consists of two isoforms *viz.*, soluble di-iron methane monooxygenase (sMMO) and membrane-bound particulate Cu-containing methane monooxygenase (pMMO). But the substrates for both these enzymes are different. For sMMO the substrates hydrocarbon are saturated, unsaturated, linear, branched, and cyclic. On the other hand, pMMO are only functional against short length (up to 5C) alkanes and alkenes. Both the isoforms of MMO contain α-subunit. The α-subunit of sMMO is synthesized from the gene mmoX whereas, α-subunit of pMMO is encoded by pmoA genes [50]. Related to the MMOs, other monooxygenases namely propane or butane monooxygenase (BMO) is responsible for metabolizing gaseous alkanes. Kotani et al, 2003 sequenced a complete operon of PnA of Gordoniasp TY-5. The operon codes for an NADH dependent reductase enzyme along with a regulator protein. Deletion of one subunit from this reduces the growth of the bacteria in propane rich environment [75]. The subunit of propane monooxygenase encoding hydroxylase enzyme has higher sequence similarity with *Pseudomonas butanovora* butane monooxygenase (sBMO). *P. butanovora* is known to oxidize butane/1-butanol [76]. To oxidize the alkanes of different lengths, *P. aeruginosa* RR1 and *P. aeruginosa* PAO1 possess two enzymes alkane hydroxylases AlkB1 and AlkB2. AlkB1 is responsible for the oxidation of alkane of chain length C16-C24 whereas AlkB2 oxidizes the alkanes of length C10-C20. The enzymes alkane hydroxylases AlkB1 and AlkB2 are encoded by the genes alkB1 and alkB2, respectively. Both the enzymes act together to degrade alkane of length C10–C22but the expression level of alkB1 gene is more than the expression of gene alkB2. The variation in the expression pattern of these genes in two strains of *P. aeruginosa viz.*, RR1, and PAO1 were observed with the differential growth phase of the bacterium. Generally, induced expression of alkB2 occurs at the beginning of the exponential phase of the bacterial growth whereas expression of alkB1 is induced at the late exponential phase of the growth of the bacterium [77]. Another bacterium *Alcanivorax borkumensis* presents a complex system containing genes for encoding two alkanes hydroxylase AlkB1 and AlkB2 and for encoding 3 cytochrome P450 enzymes *viz.*, P450-1, P450-2, P450-3. AlkB1 enzyme is active for the oxidation of alkane of length C5-C12 whereas AlkB2 is active against alkanes of length C8-C16. In the presence of C10-C16 alkanes, both the genes alkB1 and alkB2 encoding AlkB1 and AlkB2, respectively are inducing their expression [78 - 80]. The expression of alkB1 is activated by an activator AlkS in the presence of alkanes [78, 79]. A list of enzymes involved in petroleum hydrocarbon degradation and their respective microbial sources are shown in Table **2**.

Table 2. List of microbial enzymes involved in hydrocarbon degradation and their sources.

Name of the Microbial Enzyme	Source Organism	References
sMMO	*Methylosinus trichosporium* OB3b	[81, 82]
sMMO	*Methylococcus capsulatus* (Bath)	
pMMO		
P450-1	*Alcanivorax borkumensis* SK2	[80]
P450-2		
P450-3		
AlmA	*Acinetobacter strain* DSM 17874	[83]
LadA	*Geobacillus thermodenitrificans* NG80-2	[84]
sMMO	*Gordonia* sp. TY-5	[75]
sBMO	*Pseudomonas butanovora*	[76]
CYP153	*Dietzia* sp. Strain DQ12-45-1b	[85]
AlkB	*Gordonia* strain SoCg	[86]
CYP153	*Acinetobacter* sp. EB104	[87]
P450	*Alcanivorax dieselolei* B-5	[88]
P450	*Rhodococcus erythropolis* strain PR4	[89]
2-Carboxybenzaldehyde dehydrogenase	*Nocardioides* sp. strain KP7	[90]
α-Subunit of the polycyclic aromatic hydrocarbon ring-hydroxylating dioxygenases (PAH-RHDα)	*Pseudomonas, Polaromonas, Sphingomonas, Acidovorax, Burkholderia, Mycobacterium, Gordonia, Terrabacter, Nocardioides*, and *Bacillus*	[91]
Gentisate 1,2-dioxygenase	*Polaromonas naphthalenivorans* CJ2	[92]
Catechol 2,3-dioxygenase	*Burkholderia* sp. AA1	[93]
Catechol 1,2-dioxygenase and catechol 2,3-dioxygenase	*Gordonia polyisoprenivorans*	[94]
Catechol dioxygenases	*Pseudomonas* sp., *Ochrobactrum* sp., *Rhodococcus*sp.	[95]
1,2-Dihydroxynaphthalene oxygenase	*Rhodococcus* sp. Strain b4, *Pseudomonas aeruginosa* PAO1	[96]

Genetic factors play crucial roles in conferring biodegradation potentials on microorganisms [97]. Plasmid DNA is likely to play a leading role in genetic modification as it is a highly movable form of DNA that can be transferred *via* conjugation or transformation and can report novel phenotypes, including hydrocarbon-oxidizing ability to beneficiary organisms [98]. For instance, the genes involved in the degradation of hydrocarbon, salicylate, camphor, octane, xylene, *etc.* in *Pseudomonas* sp. are reported to be present in their plasmids [98].

P. oleovoransis oxidises alkane with the help of enzymes synthesized *alkB* gene of OCT plasmid [99]. Some researchers reported the presence of naphthalene catabolic gene (*nah*) in the NAH7 plasmid of *P. putida* strain G7. Bacteria containing this gene degrade naphthalene by the formation of salicylate [100]. The naphthalene degrading bactium *Pseudomonas* sp. strain MC1 possesses pYIC1 plasmid. The genes responsible for naphthalene catabolism are clustered in*nahAa-Ab-Ac-Ad-B-F-C-Q-E-D* and *nahG-T-H-I-N-L-O-M-K-J* encoding enzymes for the conversion of naphthaleneto salicylate salicylate to pyruvate and acetyl CoA, respectively [101]. Otenio *et al.* 2005, reported the degradation of benzene (B), toluene (T) and xylene (X) and their mixtire by *Pseudomonas putida* CCMI 852 containing a TOL plasmid [102].

In *Acenetobacter* sp., two alkane hydroxylases are involved *viz.*, AlkMa and AlkMb. *alkMa* gene is induced in the presence in *n*-alkanes of length C_{22} by AlkRa whereas AlkRb induces the expression of *alkMb* in the presence of C_{16}-$C_{22}n$-alkanes [103]. *Pseudomonas putida* strain GPo1 containing OCT plasmid was characterized to explore the alkane degrading pathway by researchers [104]. The operon *alkBFGHJKL* present in the system was found responsible for encoding the enzymes which convert alkanes into acetyl CoA. This pathway is regulated by the *alkST* gene. This gene encodes for AlkT (rubredoxin reductase) and AlkS (positive regulator of *alkBFGHJKL* operon). AlkS activates the expression of *alkST* gene in the presence of alkanes [105]. Another enzyme known as cytochrome O ubiquinol oxidase (Cyo) of *Alcanivorax dieselolei* was reported for its ability to degrade long chain alkanes [106]. This Cyo enzymes is encoded by gene *cyo* [107, 108]. The overproduction of Cyo in the presence of long chain alkanes and pristane suppresses the action of AlmR in negative regulation of *almA* gene expression. Expression of *almA* induces its product AlmA hydrolase to degrade both branched as well as linear long chain alkane hydrocarbon [106].

Aromatic hydrocarbon degradation initiated by a ring hydroxylating dioxygenase (RHD) of microorganisms involving deoxygenation reaction which is found in *Mycobacterium vanbaalenii* PYR-1, *Mycobacterium* sp. strain SNP11, *etc* [109, 110, 111]. On the other hand, in some of the strains of *Pseudomonas putida*, the process starts with methyl monooxygenation [112, 113].

Microorganisms possess both aerobic and anaerobic pathways to degrade toluene, a major component of crude oil. There are 5 aerobic and 1 anaerobic pathways reported so far [114]. For example, some strains of *P. putida* uses xylene monooxygenase to oxidize toluene to benzaldehyde whereas, some other strains oxidize toluene into *cis*-toluene dihydrodiol by the action of toluene dioxygenase TodC1C2BA. The genes *xyl* present in pWWO plasmid of *P. putida* strain mt-2 encodes the enzymes involved in toluene degradation [115]. In this process, the methyl group of toluene is oxidized into (alkyl) benzoate followed by deoxygenation to produce (methyl) catechol. The (methyl) catechol upon meta cleavage enters into Kreb's cycle in which *meta*TOL gene gets involved which is a part of *xylXYZLTEGFJQKIH* gene cluster. On the other hand, the gene xylIE is responsible for the degradation of toluene or xylene in some of the strains of *Acinetobacter* sp., *Pseudomonas* sp., and *Kocuria* sp [116]. Some of the strains of *Ralstoniapickettii*, *Burkholderiacepacia* and *Pseudomonas mendocina* involve different monooxygenases to oxidize toluene m-cresol, o-cresol and p-cresol. It may also be noted that some strains of *Thauera aromatica* and *Azoarcus* sp. follow anaerobic pathways to degrade naphthalene/toluene by involving benzylsuccinate synthatase enzyme [117]. In this process, the product of upper catabolic operon *nahAaAbAcAdBFCED* converts naphthalene into salicylate. The resulting salicylate is converted into pyruvate and acetaldehyde by the product of lower operon *nahGTHINLOMKJ*. Similarly, *nah* gene products of NAH 7 like plasmid pKA1 in *P. flourescens* 5R oxidizes anthracene and phenanthrene to form hydroxynaphthoic acid [111]. In *Mycobacterium vanbaalenii* PYR-1, the involvement of NidAB and NidA3B3 enzymes for mono and dioxygenation of toluene has been reported [113]. The bph gene products of *Pseudomonas* sp., *Sphingomonas* sp., *Burkholderia* sp., *Rhodococcus* sp., and *Achromobacter* sp., are involved in the degradation of biphenyl and polychlorobiphenyls (PCBs) [118 - 120]. A list of various genes responsible for hydrocarbon degradation present in a wide range of microbes is shown in Table **3**.

Table 3. List of microbial genes involved in hydrocarbon degradation and their sources.

Name of the Genes	Source Organism	References
alkB1	*Pseudomonas aeruginosa* PAO1	[77]
alkB2		
alkB1	*Pseudomonas aeruginosa* RR1	
alkB2		
alkB1	*Alcanivorax borkumensis* AP1	[78]
alkB2		
alkB1	*Alcanivorax hongdengensis* A-11-3	[121]
alkB2		
p450-1		
p450-2		
p450-3		

(Table 3) cont.....

Name of the Genes	Source Organism	References
AlkMa	*Acinetobacter* sp. M-1	[122]
AlkMb		
almA	*Alcanivorax dieselolei* B5	[106]
alkB	*Pseudomonas putida* GPo1	[104]
nah	*Mycobacterium* sp. strain PYR-1	[123]
nod	*Rhodococcus* sp. strain NCIMB12038	[124]
phn	*Nocardioides* sp. strain KP7	[125]
nidA	*Rhodococcus wratislaviensis* IFP 2016	[126]
nah	*Pseudomonas stutzeri* AN10	[127]
nid	*Mycobacterium* sp	[128]
narB	*Rhodococcus* sp. NCIMB12038	[129]
phn	*Sphingomonas* sp. strain LH128	[130]
phn genes	*Acidovorax* sp	[131]
nidA, bphA3A4C	*Novosphingobium* sp. PCY, *Microbacterium* sp. BPW, *Ralstonia* sp. BPH, *Alcaligenes* sp. SSK1B, Achromobacter sp. SSK4	[132]
nah	*Pseudomonas aeruginosa* PAO1	[48]

FACTORS AFFECTING BIOREMEDIATION OF PETROLEUM HYDROCARBONS

Several successful applications of microbes in bioremediation have been reported so far. Despite all the methods used for the bioremediation of petroleum hydrocarbons, the use of a suitable microbial consortium is suggested to be the best method. The rate of degradation of the petroleum hydrocarbon varies depending on the catabolic activities of the microbes. However, several biotic and abiotic factors influence microbial bioremediation of petroleum hydrocarbon which includes temperature, oxygen level, availability of the nutrients, salinity, pH, humidity, nature of the petroleum hydrocarbons, microbial flora, bioavailability, *etc.* (Fig. **2**). The catabolic activity of the microbes is high inducing the rate of degradation when these factors are present at an optimum level. The effect of some of the biotic and abiotic factors are discussed below:

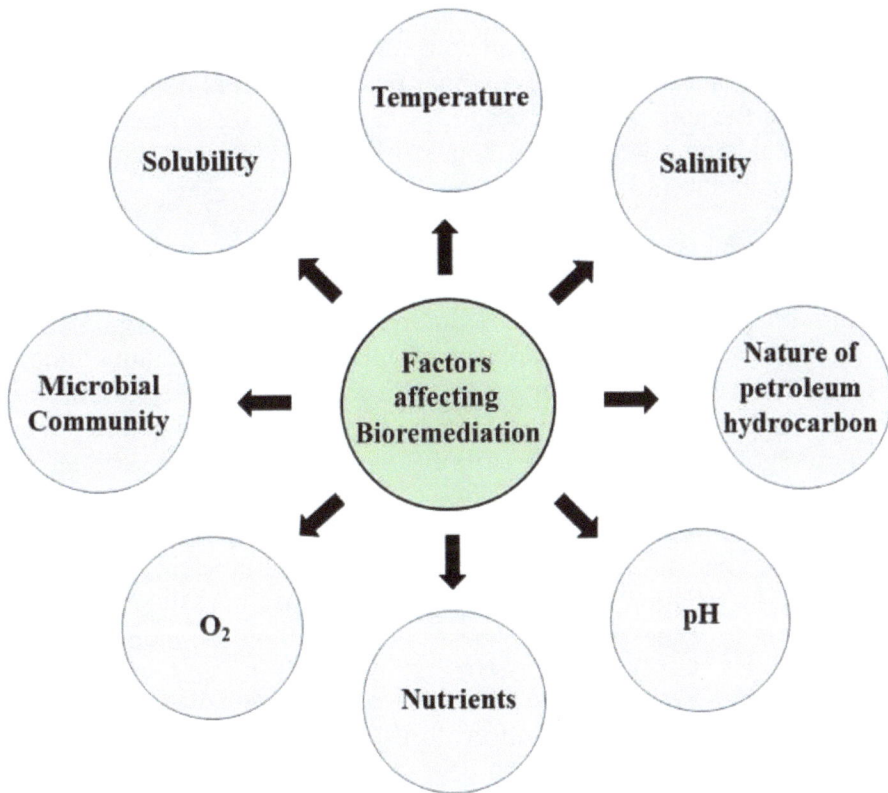

Fig. (2). Different factors affecting biodegradation of petroleum hydrocarbon.

Temperature: Temperature is one of the important factors influencing the degradation rate of petroleum hydrocarbon [133]. The rate of degradation of petroleum hydrocarbon at higher temperatures increases as compared to the rate of degradation at a lower temperature. For instance, the degradation rate of petroleum hydrocarbon was reported to maximum at 30-40 °C in soil, and the rate declines with a decrease in temperature [133]. The same pattern of degradation has been observed in marine water as well as in freshwater. In marine water, the maximum degradation occurs at 20-30 °C [39]. Whereas, the maximum degradation of petroleum hydrocarbon occurs at 15-20 °C in freshwater but it decreases below 15 °C [133]. At higher temperatures, the degradation is more because of the optimum enzymatic activity; whereas, the enzymatic activity is found to be less at reduced temperature [134].

Oxygen: Oxygen is considered the rate limiting variable in the biodegradation of petroleum hydrocarbons in the environment [135]. Previous studies have reported that the degradation of petroleum hydrocarbon increases in the presence of oxygen than in its absence [136]. The concentration of oxygen in soil depends on various factors such as its consumption rate by the microbial population present in the soil, type of the soil, porosity, *etc* [137]. Oxidation of substrate by oxygenase enzyme is considered a key step of catabolism in the biodegradation process of petroleum components [138].

Nutrients: Nutrients are the key important factor in the process of biodegradation. The level of nutrients sometimes works as limiting factor in the process of biodegradation. To degrade the hydrocarbons, microorganisms require carbon (C), nitrogen (N) and phosphorus (P). Hence, the optimum concentration of C, N, and P at the contamination site influences the bioremediation process of petroleum hydrocarbon [139]. Petroleum hydrocarbons are used by microorganisms as the source of carbon and water provides oxygen (O) and hydrogen (H) whereas other nutrients N and P are less in concentration. Therefore, in such a situation nutrients are supplied externally to increase the biodegradation process. On the other hand, excessive nutrients also reduce the activity of biodegradation as the excessive concentration of pollutants causes selective pressure on the petroleum degrading microbial community [39, 139].

Effect of pH: pH plays an important role in any biological as well as chemical reactions including enzymatic reactions [140]. pH is taken into consideration in the process of petroleum hydrocarbon degradation. Microbes grow at a maximum rate at optimum pH level depending upon the nature of the bacteria. For example, the growth of heterotrophic bacteria is maximum at neutral and alkaline pH but their growth rate is slow at higher pH. Although fungi can tolerate acidic conditions, fungi prefer neutral pH for their optimal growth. Previous studies reported that mineralization of octadecane and naphthalene by microbes occurs at pH 6.5. It was reported that an increase in pH from 6.5 to 8.0 increases the mineralization of octadecane whereas, no changes occur in the case of mineralization of naphthalene with the change in pH [141].

Salinity: In bioremediation, salinity plays an important role. Salinity also influences the growth of the microbes and affects their diversity [142]. It has been observed that estuarine sediments have a positive relationship between petroleum hydrocarbon mineralization and salinity [143]. Due to the increase in salinity, the metabolic rate of hydrocarbon degrading microbes decreases [144]. Increased salinity affects adversely the enzymatic activity of some major enzymes which ultimately reduces the rate of hydrocarbon biodegradation [145]. Besides, higher salt concentration also increases the osmotic pressure on bacterial cells which also affects the rate of hydrocarbon degradation.

Solubility: Besides, higher salt concentration also increases the osmotic pressure on bacterial cells which also affects the rate of hydrocarbon degradation [146].

Nature of the Petroleum Hydrocarbon: The rate of hydrocarbon degradation varies depending upon the nature of the petroleum. Petroleum hydrocarbons can be classified broadly in the following classes: aromatic, saturates, asphaltenes, and resins. Microbial activity on these hydrocarbons depends on their composition. The susceptibility of the hydrocarbon to the microbial attack determined the rate of degradation of the petroleum hydrocarbon, which depends on the nature of the components. The linear alkanes of petroleum hydrocarbons are more susceptible followed by branched alkanes, small aromatics, and lastly cyclic alkanes. There are some compounds present in petroleum hydrocarbons. They are very heavy in nature in comparison to the other hydrocarbon components [147]. In general, hydrocarbon susceptibility for microbial attack decreases in the order of n-alkenes > branched alkenes > aromatic (low molecular weight) > cyclic- alkanes [148].

Microbial Community: Among all the factors influencing the degradation process of petroleum hydrocarbons, the presence or availability of the microorganisms is one of the important factors [149]. This is considered a major and ultimate natural process to clean the petroleum pollutants from the environmental sites [150, 151]. Several species of bacteria, fungi, yeast, along with some algae possess the ability to degrade petroleum hydrocarbons. The microbial communities are imparting in catabolizing the pollutant hydrocarbons. They use the hydrocarbons as their food source and hence these types of microorganisms are readily available near the dumping areas or in the exposed areas of petroleum hydrocarbons, basically, in soil, bacteria and fungi play important role in petroleum hydrocarbon degradation [39].

CONCLUSION

Microbes play an invincible role in the biodegradation of petroleum hydrocarbon in nature. Even though chemical surfactants are also readily available at the commercial level but they are not environment friendly. Hence, bioremediation remains the sole eco-friendly method of treatment. Even though numerous microbes were found to have such potential but they efficiently vary from species to species. Hence, this is the need of the hour to search for better microbial species that can perform in a wide range of ecological and geographical conditions. In this connection, the exploration of essential genes plays a great role in governing the necessary characteristics in microbes eventually helping in the degradation process.

CONSENT FOR PUBLICATION

Not applicable.

CONFLICT OF INTEREST

The authors declare no conflict of interest, financial or otherwise.

ACKNOWLEDGEMENT

Declared none.

REFERENCES

[1]   Ossai IC, Ahmed A, Hassan A, Hamid FS. Remediation of soil and water contaminated with petroleum hydrocarbon: A review. Environ. Technol. Innov 2020; Vol. 17.

[2]   Borah D. Microbial bioremediation of petroleum hydrocarbon: An overview.Microbial Action on Hydrocarbons. Singapore: Springer 2018; pp. 321-41.
        [http://dx.doi.org/10.1007/978-981-13-1840-5_13]

[3]   Webb CC. Bioremediation of Marine Oil Spills 2005. http://home.eng.iastate.edu/ ~tge/ce421-521/CaseyWebb.pdf

[4]   Perera F. Pollution from fossil-fuel combustion is the leading environmental threat to global pediatric health and equity: Solutions exist. Int J Environ Res Public Health 2017; 15(1): 16.
        [http://dx.doi.org/10.3390/ijerph15010016] [PMID: 29295510]

[5]   Chaillan F, Le Flèche A, Bury E, *et al.* Identification and biodegradation potential of tropical aerobic hydrocarbon-degrading microorganisms. Res Microbiol 2004; 155(7): 587-95.
        [http://dx.doi.org/10.1016/j.resmic.2004.04.006] [PMID: 15313261]

[6]   Adebusoye SA, Ilori MO, Amund OO, Teniola OD, Olatope SO. Microbial degradation of petroleum hydrocarbons in a polluted tropical stream. World J Microbiol Biotechnol 2007; 23(8): 1149-59.
        [http://dx.doi.org/10.1007/s11274-007-9345-3]

[7]   Thapa B, Kc AK, Ghimire A. A Review On Bioremediation Of Petroleum Hydrocarbon Contaminants In Soil. J Sci Eng Technol 1970; 8(1): 164-70.
        [http://dx.doi.org/10.3126/kuset.v8i1.6056]

[8]   Speight JG. Crude oil in water systems. In: Speight JG, Ed. Natural Water Remediation Chemistry and Technology. Elsevier 2020; pp. 199-232.
        [http://dx.doi.org/10.1016/B978-0-12-803810-9.00006-1]

[9]   Motta F L, Stoyanov S R, Soares J B P. Application of solidifiers for oil spill containment: A review. Chemosphere 2018; 194: 837-46.
        [http://dx.doi.org/10.1016/j.chemosphere.2017.11.103]

[10]   Aktaş F. Bioremediation techniques and strategies on removal of polluted environment. J Eng Res Appl Sci 2013; 2(June): 107-15.

[11]   Peixoto RS, Vermelho AB, Rosado AS. Petroleum-degrading enzymes: bioremediation and new prospects. Enzyme Res 2011; 2011(1): 475193.
        [PMID: 21811673]

[12]   Azubuike CC, Chikere CB, Okpokwasili GC. Bioremediation techniques–classification based on site of application: principles, advantages, limitations and prospects. World J Microbiol Biotechnol 2016; 32(11): 180.
        [http://dx.doi.org/10.1007/s11274-016-2137-x] [PMID: 27638318]

[13]   Aislabie J, Saul DJ, Foght JM. Bioremediation of hydrocarbon-contaminated polar soils. Extremophiles 2006; 10(3): 171-9.
[http://dx.doi.org/10.1007/s00792-005-0498-4] [PMID: 16514512]

[14]   Coulon F, Al Awadi M, Cowie W, *et al.* When is a soil remediated? Comparison of biopiled and windrowed soils contaminated with bunker-fuel in a full-scale trial. Environ Pollut 2010; 158(10): 3032-40.
[http://dx.doi.org/10.1016/j.envpol.2010.06.001] [PMID: 20656385]

[15]   Bar CN, D. JR Finnamore, Bardos RP, Weeks JN. Biological Methods fo Assessment and Remediation of Contaminated Land: Case Studies. 2002; 79-109.

[16]   Chikere CB, Chikere BO, Okpokwasili GC. Bioreactor-based bioremediation of hydrocarbon-polluted Niger Delta marine sediment, Nigeri. 3 Biotech 2012; 2(1): 53-66.
[http://dx.doi.org/10.1007/s13205-011-0030-8]

[17]   Zangi-Kotler M, Ben-Dov E, Tiehm A, Kushmaro A. Microbial community structure and dynamics in a membrane bioreactor supplemented with the flame retardant dibromoneopentyl glycol. Environ Sci Pollut Res Int 2015; 22(22): 17615-24.
[http://dx.doi.org/10.1007/s11356-015-4975-8] [PMID: 26146373]

[18]   Nikolopoulou M, Pasadakis N, Norf H, Kalogerakis N. Enhanced ex situ bioremediation of crude oil contaminated beach sand by supplementation with nutrients and rhamnolipids. Mar Pollut Bull 2013; 77(1-2): 37-44.
[http://dx.doi.org/10.1016/j.marpolbul.2013.10.038] [PMID: 24229785]

[19]   Maila MP, Cloete TE. Bioremediation of petroleum hydrocarbons through landfarming: Are simplicity and cost-effectiveness the only advantages? Rev Environ Sci Biotechnol 2004; 3(4): 349-60.
[http://dx.doi.org/10.1007/s11157-004-6653-z]

[20]   Kim S, Krajmalnik-Brown R, Kim JO, Chung J. Remediation of petroleum hydrocarbon-contaminated sites by DNA diagnosis-based bioslurping technology. Sci Total Environ 2014; 497-498: 250-9.
[http://dx.doi.org/10.1016/j.scitotenv.2014.08.002] [PMID: 25129160]

[21]   Folch A, Vilaplana M, Amado L, Vicent T, Caminal G. Fungal permeable reactive barrier to remediate groundwater in an artificial aquifer. J Hazard Mater 2013; 262: 554-60.
[http://dx.doi.org/10.1016/j.jhazmat.2013.09.004] [PMID: 24095995]

[22]   Roy M, Giri AK, Dutta S, Mukherjee P. Integrated phytobial remediation for sustainable management of arsenic in soil and water. Environ Int 2015; 75: 180-98.
[http://dx.doi.org/10.1016/j.envint.2014.11.010] [PMID: 25481297]

[23]   Frascari D, Zanaroli G, Danko AS. In situ aerobic cometabolism of chlorinated solvents: A review. J Hazard Mater 2015; 283: 382-99.
[http://dx.doi.org/10.1016/j.jhazmat.2014.09.041] [PMID: 25306537]

[24]   Shah JK, Sayles GD, Suidan MT, Mihopoulos P, Kaskassian S. Anaerobic bioventing of unsaturated zone contaminated with DDT and DNT. Water Sci Technol 2001; 43(2): 35-42.
[http://dx.doi.org/10.2166/wst.2001.0070] [PMID: 11380202]

[25]   Mihopoulos PG, Suidan MT, Sayles GD, Kaskassian S. Numerical modeling of oxygen exclusion experiments of anaerobic bioventing. J Contam Hydrol 2002; 58(3-4): 209-20.
[http://dx.doi.org/10.1016/S0169-7722(02)00037-2] [PMID: 12400833]

[26]   Mihopoulos P, Sayles GD, Suidan MT, Shah J, Bishop DF. Vapor phase treatment of PCE in a soil column by lab-scale anaerobic bioventing. Water Res 2000; 34(12): 3231-7.
[http://dx.doi.org/10.1016/S0043-1354(00)00023-3]

[27]   Frutos FJG, Escolano O, García S, Babín M, Fernández MD. Bioventing remediation and ecotoxicity evaluation of phenanthrene-contaminated soil. J Hazard Mater 2010; 183(1-3): 806-13.
[http://dx.doi.org/10.1016/j.jhazmat.2010.07.098] [PMID: 20800967]

[28] Philp JC. Bioremediation of contaminated soils and aquifers. Bioremediation: applied microbial solutions for real-world 30 environmental cleanup. Philp JC, Atlas R. Washington: American Society for Microbiology (ASM) Press 2005; pp. 139-236.

[29] Gidarakos E, Aivalioti M. Large scale and long term application of bioslurping: The case of a Greek petroleum refinery site. J Hazard Mater 2007; 149(3): 574-81.
[http://dx.doi.org/10.1016/j.jhazmat.2007.06.110] [PMID: 17709182]

[30] Kao CM, Chen CY, Chen SC, Chien HY, Chen YL. Application of in situ biosparging to remediate a petroleum-hydrocarbon spill site: Field and microbial evaluation. Chemosphere 2008; 70(8): 1492-9.
[http://dx.doi.org/10.1016/j.chemosphere.2007.08.029] [PMID: 17950413]

[31] Richard B. Phytoremediation of toxic elemental and organic pollutants. Curr Opin Plant Biol 2000; 3: 153-62.

[32] Kuiper I, Lagendijk EL, Bloemberg GV, Lugtenberg BJJ. Rhizoremediation: a beneficial plant-microbe interaction. Mol Plant Microbe Interact 2004; 17(1): 6-15.
[http://dx.doi.org/10.1094/MPMI.2004.17.1.6] [PMID: 14714863]

[33] Lee JH. An overview of phytoremediation as a potentially promising technology for environmental pollution control. Biotechnol Bioprocess Eng; BBE 2013; 18(3): 431-9.
[http://dx.doi.org/10.1007/s12257-013-0193-8]

[34] Yancheshmeh JB. Evaluation of inoculation of plant growth-promoting rhizobacteria on cadmium and lead uptake by canola and barley. Afr J Microbiol Res 2011; 5(14): 1747-54.

[35] de-Bashan LE, Hernandez JP, Bashan Y. The potential contribution of plant growth-promoting bacteria to reduce environmental degradation – A comprehensive evaluation. Appl Soil Ecol 2012; 61: 171-89.
[http://dx.doi.org/10.1016/j.apsoil.2011.09.003]

[36] Dowling DN, Doty SL, Dowling DN, Doty SL. Improving phytoremediation through biotechnology Editorial overview. 2009; pp. 204-6.

[37] Jones DM, Douglas AG, Parkes RJ, Taylor J, Giger W, Schaffner C. The recognition of biodegraded petroleum-derived aromatic hydrocarbons in recent marine sediments. Mar Pollut Bull 1983; 14(3): 103-8.
[http://dx.doi.org/10.1016/0025-326X(83)90310-7]

[38] Bartha R, Bossert I. The treatment and disposal of petroleum refinary wastes.Petroleum Microbiology. New York, NY, USA: Macmillan 1984; pp. 553-78.

[39] Atlas RM. Effects of hydrocarbons on micro-organisms and biodegradation in Arctic ecosystems.Petroleum Effects in the Arctic Environment. London, UK: Elsevier 1985; pp. 63-99.

[40] Cooney JJ. The fate of petroleum pollutants in fresh water ecosystems.Petroleum Microbiology. New York, NY, USA: Macmillan 1984; pp. 399-434.

[41] Floodgate G. The fate of petroleum in marine ecosystems.Petroleum Microbiology. New York, NY, USA: Macmillion 1984; pp. 355-98.

[42] Daugulis AJ, McCracken CM. Microbial degradation of high and low molecular weight polyaromatic hydrocarbons in a two-phase partitioning bioreactor by two strains of *Sphingomonas* sp. Biotechnol Lett 2003; 25(17): 1441-4.
[http://dx.doi.org/10.1023/A:1025007729355] [PMID: 14514047]

[43] Singh H. Mycoremediation: Fungal Bioremediation. New York, NY, USA: Wiley Interscience 2006.
[http://dx.doi.org/10.1002/0470050594]

[44] Bogusławska-Wąs E, Dąbrowski W. The seasonal variability of yeasts and yeast-like organisms in water and bottom sediment of the Szczecin Lagoon. Int J Hyg Environ Health 2001; 203(5-6): 451-8.
[http://dx.doi.org/10.1078/1438-4639-00056] [PMID: 11556149]

[45]   Walker JD, Colwell RR, Petrakis L. Degradation of petroleum by an alga, Prototheca zopfii. Appl Microbiol 1975; 30(1): 79-81.
[http://dx.doi.org/10.1128/am.30.1.79-81.1975] [PMID: 1147621]

[46]   Cerniglia CE, Gibson DT, Van Baalen C, Cerniglia CVBCE, Gibson DT. Oxidation of Naphthalene by Cyanobacteria and Microalgae. Microbiology (Reading) 1980; 116(2): 495-500.
[http://dx.doi.org/10.1099/00221287-116-2-495]

[47]   Cerniglia CE. Biodegradation of polycyclic aromatic hydrocarbons. Biodegradation 1992; 3(2-3): 351-68.
[http://dx.doi.org/10.1007/BF00129093]

[48]   Eaton RW, Chapman PJ. Bacterial metabolism of naphthalene: construction and use of recombinant bacteria to study ring cleavage of 1,2-dihydroxynaphthalene and subsequent reactions. J Bacteriol 1992; 174(23): 7542-54.
[http://dx.doi.org/10.1128/jb.174.23.7542-7554.1992] [PMID: 1447127]

[49]   Gibson D T, Parales R E. Aromatic hydrocarbon dioxygenases in environmental biotechnology. Curr Opin Biotech 2000; 11(3): 236-43.
[http://dx.doi.org/10.1016/S0958-1669(00)00090-2]

[50]   Gkorezis P, Daghio M, Franzetti A, Van Hamme J D, Sillen W, Vangronsveld J. The interaction between plants and bacteria in the remediation of petroleum hydrocarbons: An environmental perspective. Front Microbiol 2016; 7: 1836.
[http://dx.doi.org/10.3389/fmicb.2016.01836]

[51]   Md F. Biosurfactant: Production and Application. J Pet Environ Biotechnol 2012; 3(4): 124.
[http://dx.doi.org/10.4172/2157-7463.1000124]

[52]   Jadhav M, Kalme S, Tamboli D, Govindwar S. Rhamnolipid from *Pseudomonas desmolyticum* NCIM-2112 and its role in the degradation of Brown 3REL. J Basic Microbiol 2011; 51(4): 385-96.
[http://dx.doi.org/10.1002/jobm.201000364] [PMID: 21656804]

[53]   Kim SY, Oh DK, Lee KH, Kim JH. Effect of soybean oil and glucose on sophorose lipid fermentation by Torulopsis bombicola in continuous culture. Appl Microbiol Biotechnol 1997; 48(1): 23-6.
[http://dx.doi.org/10.1007/s002530051009] [PMID: 9274044]

[54]   Kurtzman CP, Price NPJ, Ray KJ, Kuo TM. Production of sophorolipid biosurfactants by multiple species of the Starmerella (Candida) bombicola yeast clade. FEMS Microbiol Lett 2010; 311(2): 140-6.
[http://dx.doi.org/10.1111/j.1574-6968.2010.02082.x] [PMID: 20738402]

[55]   Muthusamy K, Gopalakrishnan S, Ravi TK, Sivachidambaram P. Biosurfactants: Properties, commercial production and application. Curr Sci 2008; 94(6): 736-47.

[56]   Yakimov MM, Timmis KN, Wray V, Fredrickson HL. Characterization of a new lipopeptide surfactant produced by thermotolerant and halotolerant subsurface *Bacillus licheniformis* BAS50. Appl Environ Microbiol 1995; 61(5): 1706-13.
[http://dx.doi.org/10.1128/aem.61.5.1706-1713.1995] [PMID: 7646007]

[57]   Banat IM, Franzetti A, Gandolfi I, *et al.* Microbial biosurfactants production, applications and future potential. Appl Microbiol Biotechnol 2010; 87(2): 427-44.

[58]   Matsuyama T, Tanikawa T, Nakagawa Y. Serrawettins and Other Surfactants Produced by *Serratia*. Berlin, Heidelberg: Springer 2011; pp. 93-120.
[http://dx.doi.org/10.1007/978-3-642-14490-5_4]

[59]   Kamal M, Höög JO, Kaiser R, *et al.* Isolation, characterization and structure of subtilisin from a thermostable *Bacillus subtilis* isolate. FEBS Lett 1995; 374(3): 363-6.
[http://dx.doi.org/10.1016/0014-5793(95)01145-5] [PMID: 7589571]

[60]   Cooper DG, Macdonald CR, Duff SJB, Kosaric N. Enhanced Production of Surfactin from *Bacillus*

subtilis by Continuous Product Removal and Metal Cation Additions. Appl Environ Microbiol 1981; 42(3): 408-12.
[http://dx.doi.org/10.1128/aem.42.3.408-412.1981] [PMID: 16345840]

[61] Stansly PG, Schlosser ME. Studies on Polymyxin: Isolation and Identification of *Bacillus polymyxa* and Differentiation of Polymyxin from Certain Known Antibiotics. J Bacteriol 1947; 54(5): 549-56.
[http://dx.doi.org/10.1128/jb.54.5.549-556.1947] [PMID: 16561391]

[62] Cooper DG, Zajic JE, Gerson DF. Production of surface-active lipids by Corynebacterium lepus. Appl Environ Microbiol 1979; 37(1): 4-10.
[http://dx.doi.org/10.1128/aem.37.1.4-10.1979] [PMID: 760639]

[63] Macdonald CR, Cooper DG, Zajic JE. Surface-active lipids from Nocardia erythropolis grown on hydrocarbons. Appl Environ Microbiol 1981; 41(1): 117-23.
[http://dx.doi.org/10.1128/aem.41.1.117-123.1981] [PMID: 16345679]

[64] Desai JD, Banat IM. Microbial production of surfactants and their commercial potential. Microbiol Mol Biol Rev 1997; 61(1): 47-64.
[PMID: 9106364]

[65] Rubinovitz C, Gutnick DL, Rosenberg E. Emulsan production by *Acinetobacter calcoaceticus* in the presence of chloramphenicol. J Bacteriol 1982; 152(1): 126-32.
[http://dx.doi.org/10.1128/jb.152.1.126-132.1982] [PMID: 6896872]

[66] Rosenberg E, Rubinovitz C, Gottlieb A, Rosenhak S, Ron EZ. Production of Biodispersan by *Acinetobacter calcoaceticus* A2. Appl Environ Microbiol 1988; 54(2): 317-22.
[http://dx.doi.org/10.1128/aem.54.2.317-322.1988] [PMID: 16347544]

[67] Cirigliano MC, Carman GM. Purification and characterization of liposan, a bioemulsifier from Candida lipolytica. Appl Environ Microbiol 1985; 50(4): 846-50.
[http://dx.doi.org/10.1128/aem.50.4.846-850.1985] [PMID: 16346917]

[68] Lotfabad TB, Shourian M, Roostaazad R, Najafabadi AR, Adelzadeh MR, Noghabi KA. An efficient biosurfactant-producing bacterium *Pseudomonas aeruginosa* MR01, isolated from oil excavation areas in south of Iran. Colloids Surf B Biointerfaces 2009; 69(2): 183-93.
[http://dx.doi.org/10.1016/j.colsurfb.2008.11.018] [PMID: 19131218]

[69] Borneleit P, Hermsdorf T, Claus R, Walther P, Kleber HP. Effect of hexadecane-induced vesiculation on the outer membrane of Acinetobacter calcoaceticus. J Gen Microbiol 1988; 134(7): 1983-92.
[PMID: 3246592]

[70] Das N, Chandran P. Microbial degradation of petroleum hydrocarbon contaminants: an overview. Biotechnol Res Int 2011; 2011: 1-13.
[http://dx.doi.org/10.4061/2011/941810] [PMID: 21350672]

[71] Zimmer T, Ohkuma M, Ohta A, Takagi M, Schunck WH. The CYP52 multigene family of Candida maltosa encodes functionally diverse n-alkane-inducible cytochromes P450. Biochem Biophys Res Commun 1996; 224(3): 784-9.
[http://dx.doi.org/10.1006/bbrc.1996.1100] [PMID: 8713123]

[72] Steliga T. Role of Fungi in Biodegradation of Petroleum Hydrocarbons. Pol J Environ Stud 2012; 21(2): 471-9.

[73] Lan H, Yang H, Li P, *et al.* Complete genome sequence of *Enterobacter* sp. strain ODB01, a bacterium that degrades crude oil. Genome Announc 2017; 5(10): e01763-16.
[http://dx.doi.org/10.1128/genomeA.01763-16] [PMID: 28280034]

[74] Al-Dhabaan FA. Morphological, biochemical and molecular identification of petroleum hydrocarbons biodegradation bacteria isolated from oil polluted soil in Dhahran, Saud Arabia. Saudi J Biol Sci 2019; 26(6): 1247-52.
[http://dx.doi.org/10.1016/j.sjbs.2018.05.029] [PMID: 31516354]

[75] Kotani T, Yamamoto T, Yurimoto H, Sakai Y, Kato N. Propane monooxygenase and NAD+-

dependent secondary alcohol dehydrogenase in propane metabolism by Gordonia sp. strain TY-5. J Bacteriol 2003; 185(24): 7120-8.
[http://dx.doi.org/10.1128/JB.185.24.7120-7128.2003] [PMID: 14645271]

[76]   Sluis MK, Sayavedra-Soto LA, Arp DJ. Molecular analysis of the soluble butane monooxygenase from '*Pseudomonas butanovora*' The GenBank accession number for the bmoXYBZDC sequence is AY093933. Microbiology (Reading) 2002; 148(11): 3617-29.
[http://dx.doi.org/10.1099/00221287-148-11-3617] [PMID: 12427952]

[77]   Marín MM, Yuste L, Rojo F. Differential expression of the components of the two alkane hydroxylases from *Pseudomonas aeruginosa*. J Bacteriol 2003; 185(10): 3232-7.
[http://dx.doi.org/10.1128/JB.185.10.3232-3237.2003] [PMID: 12730186]

[78]   van Beilen JB, Marín MM, Smits THM, *et al*. Characterization of two alkane hydroxylase genes from the marine hydrocarbonoclastic bacterium *Alcanivorax borkumensis*. Environ Microbiol 2004; 6(3): 264-73.
[http://dx.doi.org/10.1111/j.1462-2920.2004.00567.x] [PMID: 14871210]

[79]   Sabirova JS, Ferrer M, Regenhardt D, Timmis KN, Golyshin PN. Proteomic insights into metabolic adaptations in *Alcanivorax borkumensis* induced by alkane utilization. J Bacteriol 2006; 188(11): 3763-73.
[http://dx.doi.org/10.1128/JB.00072-06] [PMID: 16707669]

[80]   Schneiker S, dos Santos VAPM, Bartels D, *et al*. Genome sequence of the ubiquitous hydrocarbon-degrading marine bacterium *Alcanivorax borkumensis*. Nat Biotechnol 2006; 24(8): 997-1004.
[http://dx.doi.org/10.1038/nbt1232] [PMID: 16878126]

[81]   Baik MH, Newcomb M, Friesner RA, Lippard SJ. Mechanistic studies on the hydroxylation of methane by methane monooxygenase. Chem Rev 2003; 103(6): 2385-420.
[http://dx.doi.org/10.1021/cr950244f] [PMID: 12797835]

[82]   Lieberman RL, Shrestha DB, Doan PE, Hoffman BM, Stemmler TL, Rosenzweig AC. Purified particulate methane monooxygenase from *Methylococcus capsulatus* (Bath) is a dimer with both mononuclear copper and a copper-containing cluster. Proc Natl Acad Sci USA 2003; 100(7): 3820-5.
[http://dx.doi.org/10.1073/pnas.0536703100] [PMID: 12634423]

[83]   Throne-Holst M, Wentzel A, Ellingsen TE, Kotlar HK, Zotchev SB. Identification of novel genes involved in long-chain n-alkane degradation by *Acinetobacter* sp. strain DSM 17874. Appl Environ Microbiol 2007; 73(10): 3327-32.
[http://dx.doi.org/10.1128/AEM.00064-07] [PMID: 17400787]

[84]   Feng L, Wang W, Cheng J, *et al*. Genome and proteome of long-chain alkane degrading *Geobacillus thermodenitrificans* NG80-2 isolated from a deep-subsurface oil reservoir. Proc Natl Acad Sci USA 2007; 104(13): 5602-7.
[http://dx.doi.org/10.1073/pnas.0609650104] [PMID: 17372208]

[85]   Liang JL, JiangYang JH, Nie Y, Wu XL. J. H. JiangYang, Y. Nie, and X. L. Wu, "Regulation of the alkane hydroxylase CYP153 gene in a Gram-positive alkane-degrading bacterium, Dietzia sp. strain DQ12-45-1b,". Appl Environ Microbiol 2016; 82(2): 608-19.
[http://dx.doi.org/10.1128/AEM.02811-15] [PMID: 26567302]

[86]   Lo Piccolo L, De Pasquale C, Fodale R, Puglia AM, Quatrini P. Involvement of an alkane hydroxylase system of Gordonia sp. strain SoCg in degradation of solid n-alkanes. Appl Environ Microbiol 2011; 77(4): 1204-13.
[http://dx.doi.org/10.1128/AEM.02180-10] [PMID: 21183636]

[87]   Maier T, Förster HH, Asperger O, Hahn U. Molecular Characterization of the 56-kDa CYP153 from *Acinetobacter* sp. EB104. Biochem Biophys Res Commun 2001; 286(3): 652-8.
[http://dx.doi.org/10.1006/bbrc.2001.5449] [PMID: 11511110]

[88]   Liu C, Wang W, Wu Y, Zhou Z, Lai Q, Shao Z. Multiple alkane hydroxylase systems in a marine alkane degrader, Alcanivorax dieselolei B-5. Environ Microbiol 2011; 13(5): 1168-78.

[http://dx.doi.org/10.1111/j.1462-2920.2010.02416.x] [PMID: 21261799]

[89]   Sekine M, Tanikawa S, Omata S, *et al.* Sequence analysis of three plasmids harboured in *Rhodococcus erythropolis* strain PR4. Environ Microbiol 2006; 8(2): 334-46.
[http://dx.doi.org/10.1111/j.1462-2920.2005.00899.x] [PMID: 16423019]

[90]   Iwabuchi T, Harayama S. Biochemical and genetic characterization of 2-carboxybenzaldehyde dehydrogenase, an enzyme involved in phenanthrene degradation by *Nocardioides* sp. strain KP7. J Bacteriol 1997; 179(20): 6488-94.
[http://dx.doi.org/10.1128/jb.179.20.6488-6494.1997] [PMID: 9335300]

[91]   Jurelevicius D, Alvarez VM, Peixoto R, Rosado AS, Seldin L. Bacterial polycyclic aromatic hydrocarbon ring-hydroxylating dioxygenases (PAH-RHD) encoding genes in different soils from King George Bay, Antarctic Peninsula. Appl Soil Ecol 2012; 55: 1-9.
[http://dx.doi.org/10.1016/j.apsoil.2011.12.008]

[92]   Lee HJ, Kim JM, Lee SH, *et al.* Gentisate 1,2-dioxygenase, in the third naphthalene catabolic gene cluster of Polaromonas naphthalenivorans CJ2, has a role in naphthalene degradation. Microbiology (Reading) 2011; 157(10): 2891-903.
[http://dx.doi.org/10.1099/mic.0.049387-0] [PMID: 21737495]

[93]   Ma Y, Herson DS. The catechol 2,3-dioxygenase gene and toluene monooxygenase genes from *Burkholderia* sp. AA1, an isolate capable of degrading aliphatic hydrocarbons and toluene. J Ind Microbiol Biotechnol 2000; 25(3): 127-31.
[http://dx.doi.org/10.1038/sj.jim.7000042]

[94]   Silva AS, Camargo FAO, Andreazza R, Jacques RJS, Baldoni DB, Bento FM. Enzymatic activity of catechol 1,2-dioxygenase and catechol 2,3-dioxygenase produced by Gordonia polyisoprenivorans. Quim Nova 2012; 35(8): 1587-92.
[http://dx.doi.org/10.1590/S0100-40422012000800018]

[95]   Singh SN, Kumari B, Upadhyay SK, Mishra S, Kumar D. Bacterial degradation of pyrene in minimal salt medium mediated by catechol dioxygenases: Enzyme purification and molecular size determination. Bioresour Technol 2013; 133: 293-300.
[http://dx.doi.org/10.1016/j.biortech.2013.01.068] [PMID: 23434805]

[96]   Grund E, Denecke B, Eichenlaub R. Naphthalene degradation *via* salicylate and gentisate by *Rhodococcus* sp. strain B4. Appl Environ Microbiol 1992; 58(6): 1874-7.
[http://dx.doi.org/10.1128/aem.58.6.1874-1877.1992] [PMID: 1622263]

[97]   Okoh A. Biodegradation alternative in the cleanup of petroleum. Biotechnol Mol Biol 2006; 1(June): 38-50.

[98]   Leahy JG, Colwell RR. Microbial degradation of hydrocarbons in the environment. Microbiol Rev 1999; 54(3): 305-15.
[http://dx.doi.org/10.1128/mr.54.3.305-315.1990]

[99]   Chen Q, Janssen DB, Witholt B. Growth on octane alters the membrane lipid fatty acids of *Pseudomonas oleovorans* due to the induction of alkB and synthesis of octanol. J Bacteriol 1995; 177(23): 6894-901.
[http://dx.doi.org/10.1128/jb.177.23.6894-6901.1995] [PMID: 7592483]

[100] Simon MJ, Osslund TD, Saunders R, *et al.* Sequences of genes encoding naphthalene dioxygenase in *Pseudomonas putida* strains G7 and NCIB 9816-4. Gene 1993; 127(1): 31-7.
[http://dx.doi.org/10.1016/0378-1119(93)90613-8] [PMID: 8486285]

[101] Ahn E, Choi KY, Kang BS, Zylstra GJ, Kim D, Kim E. Salicylate degradation by a cold-adapted *Pseudomonas* sp. Ann Microbiol 2017; 67(6): 417-24.
[http://dx.doi.org/10.1007/s13213-017-1273-3]

[102] Otenio MH, Silva MTL, Marques MLO, Roseiro JC, Bidoia ED, Bidoia ED. Benzene, toluene and xylene biodegradation by *Pseudomonas putida* CCMI 852. Braz J Microbiol 2005; 36(3): 258-61.

[http://dx.doi.org/10.1590/S1517-83822005000300010]

[103] Tani A, Ishige T, Sakai Y, Kato N. Gene structures and regulation of the alkane hydroxylase complex in *Acinetobacter* sp. strain M-1. J Bacteriol 2001; 183(5): 1819-23.
[http://dx.doi.org/10.1128/JB.183.5.1819-1823.2001] [PMID: 11160120]

[104] van Beilen JB, Panke S, Lucchini S, Franchini AG, Röthlisberger M, Witholt B. Analysis of *Pseudomonas putida* alkane-degradation gene clusters and flanking insertion sequences: evolution and regulation of the alk genes The EMBL accession numbers for the sequences reported in this paper are AJ245436 [*P. putida* (oleovorans) GPo1 alk gene clusters and flanking DNA], AJ233397 (*P. putida* P1 alk gene clusters and flanking DNA), AJ249793 (*P. putida* P1 nahKJ genes), AJ249825 [*P. putida* (oleovorans) GPo1 16S RNA gene] and AJ271219 (*P. putida* P1 16S RNA gene). Microbiology (Reading) 2001; 147(6): 1621-30.
[http://dx.doi.org/10.1099/00221287-147-6-1621] [PMID: 11390693]

[105] Canosa I, Sánchez-Romero JM, Yuste L, Rojo F. A positive feedback mechanism controls expression of AlkS, the transcriptional regulator of the *Pseudomonas oleovorans* alkane degradation pathway. Mol Microbiol 2000; 35(4): 791-9.
[http://dx.doi.org/10.1046/j.1365-2958.2000.01751.x] [PMID: 10692156]

[106] Wang W, Shao Z. The long-chain alkane metabolism network of Alcanivorax dieselolei. Nat Commun 2014; 5(1): 5755.
[http://dx.doi.org/10.1038/ncomms6755] [PMID: 25502912]

[107] Dinamarca MA, Aranda-Olmedo I, Puyet A, Rojo F. Expression of the *Pseudomonas putida* OCT plasmid alkane degradation pathway is modulated by two different global control signals: evidence from continuous cultures. J Bacteriol 2003; 185(16): 4772-8.
[http://dx.doi.org/10.1128/JB.185.16.4772-4778.2003] [PMID: 12896996]

[108] Dinamarca MA, Ruiz-Manzano A, Rojo F. Inactivation of cytochrome o ubiquinol oxidase relieves catabolic repression of the *Pseudomonas putida* GPo1 alkane degradation pathway. J Bacteriol 2002; 184(14): 3785-93.
[http://dx.doi.org/10.1128/JB.184.14.3785-3793.2002] [PMID: 12081947]

[109] Kim SJ, Kweon O, Jones RC, Freeman JP, Edmondson RD, Cerniglia CE. Complete and integrated pyrene degradation pathway in *Mycobacterium vanbaalenii* PYR-1 based on systems biology. J Bacteriol 2007; 189(2): 464-72.
[http://dx.doi.org/10.1128/JB.01310-06] [PMID: 17085566]

[110] Pagnout C, Frache G, Poupin P, Maunit B, Muller JF, Férard JF. Isolation and characterization of a gene cluster involved in PAH degradation in *Mycobacterium* sp. strain SNP11: Expression in *Mycobacterium smegmatis* mc2155. Res Microbiol 2007; 158(2): 175-86.
[http://dx.doi.org/10.1016/j.resmic.2006.11.002] [PMID: 17258432]

[111] Peng RH, Xiong AS, Xue Y, et al. Microbial biodegradation of polyaromatic hydrocarbons. FEMS Microbiol Rev 2008; 32(6): 927-55.
[http://dx.doi.org/10.1111/j.1574-6976.2008.00127.x] [PMID: 18662317]

[112] Iwai S, Johnson TA, Chai B, Hashsham SA, Tiedje JM. "Comparison of the specificities and efficacies of primers for aromatic dioxygenase gene analysis of environmental samples," *Applied and Environmental Microbiology*, vol. 77, no. 11. American Society for Microbiology 2011; (Jun): 3551-7. [ASM].

[113] Kweon O, Kim SJ, Freeman JP, Song J, Baek S, Cerniglia CE. Substrate specificity and structural characteristics of the novel Rieske nonheme iron aromatic ring-hydroxylating oxygenases NidAB and NidA3B3 from *Mycobacterium vanbaalenii* PYR-1. MBio 2010; 1(2): e00135-10.
[http://dx.doi.org/10.1128/mBio.00135-10] [PMID: 20714442]

[114] Shinoda Y, Sakai Y, Uenishi H, et al. Aerobic and anaerobic toluene degradation by a newly isolated denitrifying bacterium, Thauera sp. strain DNT-1. Appl Environ Microbiol 2004; 70(3): 1385-92.
[http://dx.doi.org/10.1128/AEM.70.3.1385-1392.2004] [PMID: 15006757]

[115] Domínguez-Cuevas P, González-Pastor JE, Marqués S, Ramos JL, de Lorenzo V. Transcriptional tradeoff between metabolic and stress-response programs in *Pseudomonas putida* KT2440 cells exposed to toluene. J Biol Chem 2006; 281(17): 11981-91.
[http://dx.doi.org/10.1074/jbc.M509848200] [PMID: 16495222]

[116] Méndez V, Fuentes S, Hernández M, Morgante V, González M, Seeger M. Isolation of hydrocarbon-degrading heavy-metal-resistant bacteria from crude oil-contaminated soil in central chile. J Biotechnol 2010; 150: 287-7.
[http://dx.doi.org/10.1016/j.jbiotec.2010.09.225]

[117] Fuentes S, Méndez V, Aguila P, Seeger M. Bioremediation of petroleum hydrocarbons: Catabolic genes, microbial communities, and applications. Appl Microbiol Biotech 2014; 98: 4781-94.
[http://dx.doi.org/10.1007/s00253-014-5684-9]

[118] Chain PSG, Denef VJ, Konstantinidis KT, *et al. Burkholderia xenovorans* LB400 harbors a multi-replicon, 9.73-Mbp genome shaped for versatility. Proc Natl Acad Sci USA 2006; 103(42): 15280-7.
[http://dx.doi.org/10.1073/pnas.0606924103] [PMID: 17030797]

[119] Pieper DH, Seeger M. Bacterial metabolism of polychlorinated biphenyls. J Mol Microbiol Biotechnol 2008; 15(2-3): 121-38.
[PMID: 18685266]

[120] Seeger M, Timmis KN, Hofer B. Bacterial pathways for the degradation of polychlorinated biphenyls. Mar Chem 1997; 58(3-4): 327-33.
[http://dx.doi.org/10.1016/S0304-4203(97)00059-5]

[121] Wang W, Shao Z. Genes involved in alkane degradation in the Alcanivorax hongdengensis strain A-11-3. Appl Microbiol Biotechnol 2012; 94(2): 437-48.
[http://dx.doi.org/10.1007/s00253-011-3818-x] [PMID: 22207216]

[122] Tani A, Ishige T, Sakai Y, Kato N. Gene structures and regulation of the alkane hydroxylase complex in *Acinetobacter* sp. strain M-1. J Bacteriol 2001; 183(5): 1819-23.
[http://dx.doi.org/10.1128/JB.183.5.1819-1823.2001] [PMID: 11160120]

[123] Khan AA, Wang RF, Cao WW, Doerge DR, Wennerstrom D, Cerniglia CE. Molecular cloning, nucleotide sequence, and expression of genes encoding a polycyclic aromatic ring dioxygenase from *Mycobacterium* sp. strain PYR-1. Appl Environ Microbiol 2001; 67(8): 3577-85.
[http://dx.doi.org/10.1128/AEM.67.8.3577-3585.2001] [PMID: 11472934]

[124] Larkin MJ, Allen CCR, Kulakov LA, Lipscomb DA. Purification and characterization of a novel naphthalene dioxygenase from *Rhodococcus* sp. strain NCIMB12038. J Bacteriol 1999; 181(19): 6200-4.
[http://dx.doi.org/10.1128/JB.181.19.6200-6204.1999] [PMID: 10498739]

[125] Saito A, Iwabuchi T, Harayama S. A novel phenanthrene dioxygenase from Nocardioides sp. Strain KP7: expression in *Escherichia coli*. J Bacteriol 2000; 182(8): 2134-41.
[http://dx.doi.org/10.1128/JB.182.8.2134-2141.2000] [PMID: 10735855]

[126] Auffret M, Labbé D, Thouand G, Greer CW, Fayolle-Guichard F. Degradation of a mixture of hydrocarbons, gasoline, and diesel oil additives by *Rhodococcus aetherivorans* and *Rhodococcus wratislaviensis*. Appl Environ Microbiol 2009; 75(24): 7774-82.
[http://dx.doi.org/10.1128/AEM.01117-09] [PMID: 19837842]

[127] Bosch R, García-Valdés E, Moore ERB. Complete nucleotide sequence and evolutionary significance of a chromosomally encoded naphthalene-degradation lower pathway from *Pseudomonas stutzeri* AN10. Gene 2000; 245(1): 65-74.
[http://dx.doi.org/10.1016/S0378-1119(00)00038-X] [PMID: 10713446]

[128] Brezna B, Khan AA, Cerniglia CE. Molecular characterization of dioxygenases from polycyclic aromatic hydrocarbon-degrading *Mycobacterium* spp. FEMS Microbiol Lett 2003; 223(2): 177-83.
[http://dx.doi.org/10.1016/S0378-1097(03)00328-8] [PMID: 12829283]

[129] Kulakov LA, Allen CCR, Lipscomb DA, Larkin MJ. Cloning and characterization of a novel *cis* - naphthalene dihydrodiol dehydrogenase gene (*narB*) from *Rhodococcus* sp. NCIMB12038. FEMS Microbiol Lett 2000; 182(2): 327-31.
[http://dx.doi.org/10.1111/j.1574-6968.2000.tb08916.x] [PMID: 10620687]

[130] Schuler L, Jouanneau Y, Ní Chadhain SM, *et al.* Characterization of a ring-hydroxylating dioxygenase from phenanthrene-degrading *Sphingomonas* sp. strain LH128 able to oxidize benz[a]anthracene. Appl Microbiol Biotechnol 2009; 83(3): 465-75.
[http://dx.doi.org/10.1007/s00253-009-1858-2] [PMID: 19172265]

[131] Singleton DR, Guzmán Ramirez L, Aitken MD. Characterization of a polycyclic aromatic hydrocarbon degradation gene cluster in a phenanthrene-degrading Acidovorax strain. Appl Environ Microbiol 2009; 75(9): 2613-20.
[http://dx.doi.org/10.1128/AEM.01955-08] [PMID: 19270134]

[132] Wongwongsee W, Chareanpat P, Pinyakong O. Abilities and genes for PAH biodegradation of bacteria isolated from mangrove sediments from the central of Thailand. Mar Pollut Bull 2013; 74(1): 95-104.
[http://dx.doi.org/10.1016/j.marpolbul.2013.07.025] [PMID: 23928000]

[133] Atlas RM. Microbial degradation of petroleum hydrocarbons: an environmental perspective. Microbiol Rev 1981; 45(1): 180-209.
[http://dx.doi.org/10.1128/mr.45.1.180-209.1981] [PMID: 7012571]

[134] Bisht S, Pandey P, Bhargava B, Sharma S, Kumar V, Sharma KD. Bioremediation of polyaromatic hydrocarbons (PAHs) using rhizosphere technology. Braz J Microbiol 2015; 46(1): 7-21.
[http://dx.doi.org/10.1590/S1517-838246120131354] [PMID: 26221084]

[135] von Wedel RJ, Mosquera JF, Goldsmith CD, *et al.* Bacterial biodegradation of petroleum hydrocarbons in groundwater: In situ augmented bioreclamation with enrichment isolates in California. Water Sci Technol 1988; 20(11-12): 501-3.
[http://dx.doi.org/10.2166/wst.1988.0335]

[136] Grishchenkov VG, Townsend RT, McDonald TJ, Autenrieth RL, Bonner JS, Boronin AM. Degradation of petroleum hydrocarbons by facultative anaerobic bacteria under aerobic and anaerobic conditions. Process Biochem 2000; 35(9): 889-96.
[http://dx.doi.org/10.1016/S0032-9592(99)00145-4]

[137] Haritash AK, Kaushik CP. Biodegradation aspects of polycyclic aromatic hydrocarbons (PAHs): a review. J Hazard Mater 2009; 169(1-3): 1-15.
[http://dx.doi.org/10.1016/j.jhazmat.2009.03.137] [PMID: 19442441]

[138] Meng L, Li H, Bao M, Sun P. Metabolic pathway for a new strain *Pseudomonas synxantha* LSH-7': from chemotaxis to uptake of n-hexadecane. Sci Rep 2017; 7(1): 39068.
[http://dx.doi.org/10.1038/srep39068] [PMID: 28051099]

[139] Al-Hawash AB, Dragh MA, Li S, *et al.* "Principles of microbial degradation of petroleum hydrocarbons in the environment," *Egyptian Journal of Aquatic Research*, vol. 44, no. 2. National Institute of Oceanography and Fisheries 2018; (Jun): 71-6.

[140] Bonomo R, Cennamo G, Purrello R, Santoro AM, Zappalà R. Comparison of three fungal laccases from Rigidoporus lignosus and Pleurotus ostreatus: correlation between conformation changes and catalytic activity. J Inorg Biochem 2001; 83(1): 67-75.
[http://dx.doi.org/10.1016/S0162-0134(00)00130-6] [PMID: 11192701]

[141] Hambrick GA III, DeLaune RD, Patrick WH Jr. Effect of Estuarine Sediment pH and Oxidation-Reduction Potential on Microbial Hydrocarbon Degradation. Appl Environ Microbiol 1980; 40(2): 365-9.
[http://dx.doi.org/10.1128/aem.40.2.365-369.1980] [PMID: 16345614]

[142] Qin X, Tang JC, Li DS, Zhang QM. Effect of salinity on the bioremediation of petroleum

hydrocarbons in a saline-alkaline soil. Lett Appl Microbiol 2012; 55(3): 210-7.
[http://dx.doi.org/10.1111/j.1472-765X.2012.03280.x] [PMID: 22725670]

[143] Kerr RP, Capone DG. The effect of salinity on the microbial mineralization of two polycyclic aromatic hydrocarbons in estuarine sediments. Mar Environ Res 1988; 26(3): 181-98.
[http://dx.doi.org/10.1016/0141-1136(88)90026-8]

[144] Ward DM, Brock TD. Hydrocarbon biodegradation in hypersaline environments. Appl Environ Microbiol 1978; 35(2): 353-9.
[http://dx.doi.org/10.1128/aem.35.2.353-359.1978] [PMID: 16345276]

[145] Ebadi A, Khoshkholgh Sima NA, Olamaee M, Hashemi M, Ghorbani Nasrabadi R. Effective bioremediation of a petroleum-polluted saline soil by a surfactant-producing *Pseudomonas aeruginosa* consortium. J Adv Res 2017; 8(6): 627-33.
[http://dx.doi.org/10.1016/j.jare.2017.06.008] [PMID: 28831308]

[146] Kumar G, Prasad JS, Suman A, Pandey G. Bioremediation of petroleum hydrocarbon-polluted soil using microbial enzymes.Smart Bioremediation Technologies: Microbial Enzymes. Elsevier 2019; pp. 307-17.
[http://dx.doi.org/10.1016/B978-0-12-818307-6.00016-0]

[147] Atlas R, Bragg J. Bioremediation of marine oil spills: When and when not - The *Exxon Valdez* experience Microb Biotechnol 2009; 2(2): 213–21.

[148] Atlas RM, Ed. J. Perry, "Microbial metabolism of cyclin alkane.Petroleum Microbiology. New York, NY, USA: Macmillan Publishing 1984; pp. 61-97.

[149] Barathi S, Vasudevan N. Utilization of petroleum hydrocarbons by *Pseudomonas fluorescens* isolated from a petroleum-contaminated soil. Environ Int 2001; 26(5-6): 413-6.
[http://dx.doi.org/10.1016/S0160-4120(01)00021-6] [PMID: 11392760]

[150] Lal B, Khanna S. Degradation of crude oil by *Acinetobacter calcoaceticus* and *Alcaligenes odorans*. J Appl Bacteriol 1996; 81(4): 355-62.
[http://dx.doi.org/10.1111/j.1365-2672.1996.tb03519.x] [PMID: 8896350]

[151] Atlas RM. Petroleum microbiology.Encyclopedia of Microbiology. Baltimore, Md, USA: Academic Press 1992; pp. 363-9.

Sustainable Materials, 2023, Vol. 1, 299-331

CHAPTER 10

Application and Major Challenges of Microbial Bioremediation of Oil Spill in Various Environments

Rustiana Yuliasni[1], Setyo Budi Kurniawan[2], Abudukeremu Kadier[3,4,*], Siti Rozaimah Sheikh Abdullah[2], Peng-Cheng Ma[3,4], Bekti Marlena[1], Nanik Indah Setianingsih[1], Dongsheng Song[3,5] and Ali Moertopo Simbolon[1]

[1] *Research Centre for Environmental and Clean Technology, National Research and Innovation Agency Republic of Indonesia, Kawasan Puspitek gd 820, Serpong, 15314, Tangerang Selatan, Indonesia*

[2] *Department of Chemical and Process Engineering, Faculty of Engineering and Built Environment, National University of Malaysia (UKM), 43600 UKM Bangi, Selangor, Malaysia*

[3] *Laboratory of Environmental Science and Technology, The Xinjiang Technical Institute of Physics and Chemistry, Key Laboratory of Functional Materials and Devices for Special Environments, Chinese Academy of Sciences, Urumqi, 830011, China*

[4] *Center of Materials Science and Optoelectronics Engineering, University of Chinese Academy of Sciences, Beijing, 100049, China*

[5] *College of Resources and Environment, Xinjiang Agricultural University, Urumqi, 830052, China*

Abstract: Oil spill contamination occurs due to exploration activities in the deep sea and downstream activities such as oil transportation *via* pipelines, oil-tankers (marine and terrestrial), re-fineries, finished product storage, distribution, and retail distribution setup. Physico-chemical technologies are accessible for oil spill clean-up, but oil bioremediation technologies are proven to be more affordable and environmentally friendly. The aim of this book chapter is to give deeper knowledge about the bioremediation technology of oil spills. This chapter discusses the nature and composition of crude oil, bioremediation agents and strategies, bioremediation on different matrices (water, soil sludge), application strategy, and future prospect of bioremediation technology.

Keywords: Bioremediation of oil spill, Biostimulation, Bioaugmentation, *ex-situ*, *in-situ*, Biosurfactant, Total Petroleum Hydrocarbon (TPH).

---

* Corresponding author Abudukeremu Kadier: Laboratory of Environmental Science and Technology, The Xinjiang Technical Institute of Physics and Chemistry, Key Laboratory of Functional Materials and Devices for Special Environments, Chinese Academy of Sciences, Urumqi, 830011, China; Tel: 860991-3677875; E-mail: abudukeremu@ms.xjb.ac.cn

Inamuddin (Ed.)

INTRODUCTION

Fossil fuel (FF) is a significant source of energy. Industries and transportation will collectively share approximately 75% of the total energy consumption from fossil fuels by 2040 [1]. The exploitation of petroleum hydrocarbon (HC) has been complemented by accidental oil spills and severe marine pollution [2]. The release of oil from various sources also causes severe effects on deep ocean organisms. Other than environmental problems, oil spill has also resulted in economic loss. For example, 250 000 seabirds were killed during the Exxon-Valdez oil spill in 1989, and the DWH disaster cost more than US$ 61 billion [2]. Another example was 407 spills of petroleum products that occurred annually in Alaska between 1996 and 1999 [3], caused shrinkage in microbial populations, soil stress, thermal and moisture unbalance as well as soil pH and nutrient unavailability [3]. A more recent case is the 2007 Hebei Spirit oil spill (HSOS) that severely disturbed the entire ecosystem along the western coast of South Korea [4]. The oil spill contamination occurred not only because of exploration activities in the deep sea/ marine, but also during oil transportation *via* pipelines, oil-tankers (marine and terrestrial), re-fineries, finished product storage and distribution, and retail distribution setup [5]. Oil spill pollution in the marine ecosystem could also transfer to the coastline and contaminate soil because of weather and the formation of tar-ball [1].

Physico-chemical technologies are accessible for oil spill clean-up. Despite having many benefits, oils spill remediation techniques also have many limitations, such as low pollutant removal efficiency and by-product disposal challenges [6]. For instance, skimmers are only effective when oil is concentrated in the surface layer, while *in-situ* burning is influenced by weather, oil type, and slick thickness [7]. Burning could also cause localized air pollution. Dispersants break oil into small droplets but could enhance toxicity to aquatic organisms [8]. An alternative technology such as biological remediation then becomes a solution because it is feasible yet effective and environmentally friendly in comparison to several physical and chemical approaches [3]. Bioremediation techniques are based on two different methods. Bio-stimulation is bioremediation method where natural indigenous microorganisms are enhanced, while bio-augmentation is where oil-degrading microorganisms are induced [9, 10]. Bioremediation is affected by several factors such as the presence of suitable microorganisms, concentration of oil, pH, temperature, the presence of electron donors and acceptors and the bioavailability of nutrients [9, 11]. These reasons make biological removal relatively slow [6]. Furthermore, bioremediation is also limited by our lack of knowledge about environmental factors, such as hydro-static pressure, temperature and dispersant toxicity [2].

Present abundantly in nature, many microorganisms have the ability to use hydrocarbons as their sole carbon [12]. Researches related to the utilization of hydrocarbon utilizing, sulphate-, nitrate- and iron- reducing, fermenting, syntrophic and methanogenic bacteria and archaea for hydrocarbon contaminated sites bioremediation have been done [3, 13]. The most used microorganisms species are firmicutes, *Gammaproteo-bacteria*, *Deltaproteo-bacteria,*, *Epsilonproteo-bacteria*, *Betaproteo-bacteria*, *Bacteroidetes*, and *Methanogenic archaebacterial* members (*Methanosarcina*, *Methanosaeta*, etc.) [12]. Phytoremediation, using plants and rhizospheric microorganisms. also quite popular as a bioremediation strategy [14].

As already mentioned earlier, hydrocarbon contamination could not only take place in the deep ocean or marine ecosystem but also could transfer to the coastline and contaminates soil *via* transportation and distribution. The difference in geochemical characteristic between soil, aquatic, or sludge creates a variation of bioremediation techniques [15 - 17]. For instance, only about 33% of moderately contaminated soil has been successfully treated. The limitation of bioremediation application in contaminated soil was because of the toxic inhibitory effect of hydrocarbon metabolites that harm oil-degrading microorganisms in highly contaminated soil [18]. Another bottlenecks were groundwater contamination during the bioremediation process due to various water-soluble oxidized hydrocarbon metabolites being mobile in soil [18].

The aim of this book chapter is to give deeper knowledge about the bioremediation technology of oil spills in water, soil and sludge. In detail, this book chapter discusses the nature and composition of petroleum hydrocarbon, oil bioremediation agent, oil bioremediation methods, matrice to be bioremediated (that include: water, soil and sludge), application strategy and future prospects of bioremediation technology.

NATURE AND COMPOSITION OF PETROLEUM CRUDE OIL

Petroleum crude oil is containing simple to complex mixture of various hydrocarbon compounds in form of aliphatic saturated compounds or paraffins including straight and branched chain alkanes (n-alkanes), cycloalkanes, unsaturated alkenes, alkynes; aromatics including polycyclic aromatic hydrocarbons (PAHs) such as naphthalene, monoaromatic such as benzene, toluene, ethylbenzene, xylene (BTEX); asphaltenes including phenols, fatty acids, ketones, esters, porphyrins; resins including pyridines, quinolines, cardaxoles, sulphonates, amides. While the non-hydrocarbon components of petroleum oil consist of sulphur compounds such as sulphides, thiols, cyclic sulphides, disulphides, dibenzothiophene, and naphthobenzothiophene; oxygen compounds include alcohols, carboxylic acids, esters, ethers, ketones and furans; and nitrogen compounds include pyrrole, pyridine, indole, benzo(a)carbazole, carbazole, benzo(f)quinolone, nitriles, indoline, and quinolone [19].

BIOREMEDIATION AGENTS

In the bioremediation of hydrocarbons, bacteria, fungi and algae played an important role as remediation agents. The reported performance of hydrocarbons by the above-mentioned agent varies from low to complete degradation [20]. It was reported that mixed culture of bacteria performed an intercorrelated complex enzymatic reaction which can create chain-links of hydrocarbon degradation in the environment [21], especially for a complex mixture of hydrocarbon like crude oil. Bacteria are the most well-known agent, at the current time, for bioremediation of hydrocarbon polluted environments [22, 23]. Bacteria can perform well in various contaminated environments, including water (freshwater and seawater) and also soil. Bacteria utilize hydrocarbon as a carbon source, while also performing the degradation of it [24]. Some species of bacteria were proven to have the capability in performing degradation of long chain and even the complex form like polycyclic aromatic hydrocarbon (PAH) [25].

Current findings also mentioned the performance of fungi and algae in remediating hydrocarbon polluted environments [20]. Fungi utilization was mostly shown on soil remediation while algae utilization was on the water. Both fungi and algae were also proven to be able to break-down and accumulate hydrocarbons ranging from a simple chain of alkane to the very complex form of PAH [26, 27]. Similar to bacteria, fungi and algae utilize hydrocarbon as their carbon source by performing cleavage of the carbon chain, resulting in the degradation of hydrocarbon. Current research on the utilization of fungi and algae can be considered as low compared to the bacteria. Mixed culture of fungi, algae, and bacteria to treat hydrocarbon contaminated environment is even very limited [28].

Bacteria as Bioremediation Agents of Hydrocarbon Contaminated Environment

It was known that bacteria can act as probiotics, such as *Lactobacillus* [29]. Bacteria can also function as biofertilizers, such as Azosprillum, Azotobacter, and Rhizobium [30]. There are also bacteria that produce antibiotics, for example, *Streptomyces griseus* which produces streptomycin [31]. In addition, types of food such as cheese, yogurt, soy sauce, and nata de coco were produced with the help of bacteria [32]. In spite of the all mentioned function, bacteria also have an important role in restoring and maintaining the environment [33, 34]. Bacteria help humans in restoring the environment from pollution, which is mostly caused by anthropogenic and industrial activities [35].

Hydrocarbon is one of the biggest contributors to environmental pollution, both in soil and water [36]. Hydrocarbons are compounds that contain the elements hydrogen (H) and carbon (C). Hydrocarbon comes from fossil fuels, which are used in many anthropogenic and industrial activities, and are difficult to dissolve in water, such as gasoline, diesel and crude oil [37]. This compound is also difficult to remove, so it can damage the ecosystem. Hydrocarbon pollution in the environment has negative impacts on other organisms' life, including further prolonged effects that can affect humans [38]. Various efforts have been made to remediate the hydrocarbon polluted environment [39 - 41]. However, the current-utilized methods commonly increase the level of pollution, such as the utilization of chemicals during the processes. Chemical residues after treatment can cause further pollution which often being ignored by the applicant. Hydrocarbon contaminated environment may also be restored by using physical treatment, but the cost to perform this technique is considered to be abundant [42].

Considering the side effects of the chemical treatment and the cost of physical treatment, natural remediation of hydrocarbon polluted environment is currently emerging, in which bacteria played an important role in naturally recovering the polluted environment. Hydrocarbonclastic bacteria are a group of bacteria that are able to utilize hydrocarbon compounds [43]. The process of degradation of hydrocarbons by bacteria begins with the production of a compound called a biosurfactant [44]. These compounds are released to decrease the surface tension of water, so that the hydrocarbons which were difficult to dissolve will be dissolved into water, increasing its bioavailability, then become small granules. This event is known as emulsification [45]. The hydrocarbon compounds that have become the emulsion granules will be utilized by hydrocarbonoclastic bacteria which then release the lipase enzyme. This enzyme has lipolytic abilities, the ability to hydrolyze fat into simpler compounds. Some bacterial species are also known to perform a complete degradation of hydrocarbon into minerals. The end result of the degradation process of these hydrocarbons is water (H_2O) and carbon dioxide (CO_2) which are safe for the surrounding environment [20].

There are various types of bacteria belonging to the hydrocarbonoclastic bacteria group, including *Bacillus subtilis* [46], *Pseudomonas aeruginosa* [47], and *Serratia marcescens* [44]. Each type of bacteria produces a different type of biosurfactant due to the different genes that regulate biosurfactant production. In *B. subtilis*, biosurfactant production is regulated by the sfp gene and the type of biosurfactant produced is a lipopeptide group in the form of surfactant [48]. This type of biosurfactant is composed of amino acid and fatty acid groups.

Meanwhile, in *P. aeruginosa*, the production of biosurfactants is regulated by the rhl gene and a biosurfactant is produced with the type of rhamnolipid which is composed of rhamnose sugars and lipids [49].

The ability of hydrocarbonoclastic bacteria to degrade hydrocarbon is influenced by various factors [23]. Until now, there are still many studies examining hydrocarbonoclastic bacteria in order to be more optimal in the environmental bioremediation process. This is done in order to obtain superior bioremediation agents capable of restoring the polluted environment effectively, efficiently and most importantly environmentally friendly. Several studies in the bacterial remediation of hydrocarbon compounds are summarized in Table **1**.

Table 1. Potential bacteria species in hydrocarbon remediation.

S. No	Species of Bacteria	Finding(s)	References
1	*Achromobacter piechaudii, Bacillus mycoides, Pseudomonas sp., Rhodococcus sp., Rhodococcus qingshengii,* and *Sphingomonas sp.*	All species were tolerance to p-iodonitrotetrazolium and 2,6-dichlorophenolindophenol. *Fusarium oxysporum, Trichoderma tomentosum,* and *Rhodococcus sp.* showed a degradation potency of various components inside PAH up to 90%.	[50]
2	*Acinetobacter baumannii*	The mentioned species was isolated from the diesel-contaminated area. This species showed a capability to remove diesel with a value reaching 99% from the initial concentration of 4%, under pH 7 with the temperature of 35 °C.	[51]
3	*Acinetobacter bouvetii* BP18, *Stenotrophomonas rhizophila* BG32, *Bacillus thuringiensis* BG3, and *Pseudomonas poae* BA1	Isolated bacteria showed a capability in producing biosurfactants under nutrient addition conditions. Removal of diesel ranged between 74 to 96% with the biosurfactant production increased as the increasing of nutrient addition. This research indicated that the production of biosurfactant was highly related to the availability of the nutrient.	[52]
4	*Bacillus anthracis, Bacillus aquimaris,* and *Bacillus cereus*	The utilization of the mixed bacteria culture showed ability in diesel degradation in soil and water. The highest removal of 18% was obtained from contaminated soil while 12% of removal was obtained from water. This research showed the potential of bioaugmentation of potential-diesel degrading bacteria in the treatment of hydrocarbon contaminated sand and water using wetland.	[53]

(Table 1) cont.....

S. No	Species of Bacteria	Finding(s)	References
5	*Bacillus anthracis, Bacillus aquimaris,* and *Bacillus cereus*	These species were isolated from the root area of *Scirpus grossus* plant. These species indicated a performance in cleaving n-alkanes. Total removal of 17.8% was obtained by the addition of the mixed culture of the bacteria. The result showed an evidence that rhizobacterial addition enhances the phytoremediation process of hydrocarbon contaminated environment.	[54]
6	*Bacillus licheniformis, Pseudomonas aeruginosa* HNYM41, *Pseudomonas aeruginosa* 28, and *Serratia marcescens*	Isolated species showed a removal efficiency of gasoline up to 79.7% with *S. marcescens* exhibiting the highest emulsification activity and yield of biosurfactant production of 0.6 g/L.	[55]
7	*Bacillus sp., Pseudomonas sp., Acinetobacter sp.* and *Staphylococcus sp.*	These species were isolated from the sediment of a mangrove ecosystem. Mixed culture of the species showed a capability of performing removal of hexadecane and phenanthrene reaching up to 99%. This research showed that the removal of diesel from the contaminated environment was related to the persistence of the hydrocarbons itself (high molecular weight characteristic).	[56]
8	*Bacillus SQ–2* strain KF453961, *Pseudomonas stutzeri* GQ-4 strain KF453954, and *Pseudomonas SZ-2* strain KF453956	The addition of nutrient enhanced the removal of TPH and alkanes with a value of 48%. This research demonstrated that the degradation of hydrocarbon was highly correlated to the development of the bacterial biomass during the treatment.	[57]
9	*Bacillus subtilis* SPB1, and *Acinetobacter radioresistens* RI7	Both species were proven to have the capability of producing biosurfactants. The produced biosurfactant was varied in different treatment conditions. The highest removal efficiency by mixed culture of both bacteria was 57.77%.	[58]
10	*Brevibacillus sp.* PDM-3	Isolated species showed a capability in degrading phenanthrene up to 93% with the initial concentration of 250 mg/L. This species also exhibit capability in producing biosurfactant with emulsification index of EI_{24}.	[59]
11	*Citrobacter freundii* CCC4DS3, *Stenotrophomonas maltophilia* CCC10S1, *Raoultella ornithinolytica* C5S3, *Serratia marcescens* C7S3A, *Serratia marcescens* C11S1, and *St. pavanii* C5S3FN	The species of *Serratia marcescens* (C7S3A) exhibited high emulsification performance and degradation of diesel components up to 74% and 96%, respectively. Consortium of the bacteria showed the highest removal of 97% with most of the contribution attributed to the *S. marcescens* (C7S3A).	[60]

(Table 1) cont.....

S. No	Species of Bacteria	Finding(s)	References
12	*Citrobacter freundii* HM-2 and *Ochrobactrum anthropi* HM-1	Under the optimum condition of 2% initially used engine oil concentration, the initial culture of 2%, pH 7.5, the temperature of 37 °C and agitation of 150 rpm, isolated species were able to remove up to 80% of the contaminant.	[61]
13	*D8, B7, B12 rhizobacterial strain*	The isolated species were capable of degrading n-alkanes proven by the mass spectroscopy analysis. The highest removal of hydrocarbon was shown by D8 with 47.01% degradation of various lengths of C-chain in hydrocarbon.	[62]
14	*Enterobacter cloacae* (KU923381)	This species was proven to degrade docosane, heptadecane, hexadecane and tridecane from diesel with the value up to 58% under the temperature of 35 °C, Ph 7 and NaCl concentration of 5%.	[63]
15	*Micrococcus sp.* APIO4, *Pseudomonas sp.* APBP1, and *Pseudomonas sp.* APHP9	Mixed culture of the isolated species showed degradation efficiency of benzene and toluene up to 75.4%.	[64]
16	*Pseudomonas aeruginosa* and *Sphingomonas sp.*	Isolated species showed the highest diesel removal efficiency of 35%. This research stated the significance of managing diesel evaporation during the treatment which can be standardized as future protocol. Diesel degradation was shown by analyzing the produced intermediate compounds and the substrate preference during the treatment in this research.	[65]
17	*Pseudomonas sp.*	This species showed the ability in removing diesel from contaminated water up to 49.93% from the initial concentration of 0.5%.	[22]
18	*Serratia marcescens*	This species produced biosurfactant which is highly valuable in the gasoline contaminated water treatment. The addition of 5% glycerol, 4% peptone and 5 g/L $(NH_4)_2SO_4$ showed the optimum condition of biosurfactant production.	[66]
19	*Streptomyces parvus, Bacillus megaterium* B6, *Bacillus megaterium* B7, and *Bacillus pumilus* B1	*Streptomyces parvus* B7 exhibits the capability of removing 82% diesel from contaminated water in a single culture. Mixed culture of all bacteria showed a slightly higher removal (90%) and also showed an indication of high involvement of biosurfactant during the processes.	[67]
20	*Vibrio alginolyticus*	The addition of ionic surfactant and nutrients promoted the degradation of diesel up to 94.22%. This species showed an ability in working under saline conditions of 20‰ with 10 mg/L of surfactant utilization and C, N, P ratio of 100:10:1.	[68]

Fungal Bioremediation of Hydrocarbon Contaminated Environment

Apart from the bacterial group, degradation of hydrocarbons can also be carried out by fungi. Hydrocarbon degrading fungi generally commonly come from the genus of *Phanerochaete* [69], *Cunninghamella* [70], *Penicillium* [71], Candida [72], *Sporobolomyces* [73], and *Cladosporium* [74]. Fungi of these genus were known to degrade PAH. Fungi were able to degrade hydrocarbons through several mechanisms. The involved mechanisms include electron utilization, enzyme degradation, sequential oxidations, and also accumulation [75 - 77]. Hydrocarbon degrading fungi are mostly characterized by their capability in producing lignin peroxidase, which play a very important role in hydrocarbon degradation. Similar to bacteria, several species of fungi are also able to produce biosurfactants [78]. Biosurfactant is known to be an important compound to enhance the solubility of hydrocarbon in the contaminated environment, and make it more bioavailable. Fungi are considered to be more adaptive to the soil contaminated by hydrocarbon due to their physical structure [79]. As compared to the bacteria, research on hydrocarbon degradation by fungi is more likely to focus on soil contamination, which becomes very favorable for sludge treatment or soil remediation. Several potential hydrocarbon degrading fungi species are summarized in Table **2**.

Table 2. Potential hydrocarbon degrading fungi species.

S. No	Species of Fungi	Finding(s)	References
1	*Alternaria alternata, Aspergillus terreus, Cladosporium sphaerospermum, Eupenicillium hirayamae* and *Paecilomyces variotii*	These isolated species from mangroves were able to accumulate hydrocarbons inside their cell. There was no antagonist activity observed during the mixed culture test. A total of 80.7% removal was obtained under mixed culture system.	[80]
2	*Anthracophyllum discolor*	This isolated species originated from contaminated soil. This species was proven to produce lignin degrading enzyme and PAH mineralization. This species was capable of removing various hydrocarbon compounds from soil, including anthracene, fluoranthene, phenanthrene, pyrene and benzo(a)pyrene with the highest removal value reaching 73%.	[81]
3	*Aspergillus niger, Aspergillus ochraceus and Penicillium chrysogenum*	These species were demonstrated to be able to perform n-alkanes degradation. These species exhibited the utilization of C_{15} to C_{18} up to 32%.	[82]
4	*Aspergillus niger, Talaromyces purpurogenus, Trichoderma harzianum, and Aspergillus flavus*	These species were isolated from the rhizosphere of *Megathyrsus maximus*. The existence of rhizosphere fungi boosts diesel degradation by 40% after 90 days of treatment.	[83]

(Table 2) cont.....

S. No	Species of Fungi	Finding(s)	References
5	*Aspergillus polyporicola (MT448790), (A2) Aspergillus spelaeus (MT448791) and (A3) Aspergillus niger (MT459302)*	Isolated species from crude oil polluted soil showed a capability in degrading hydrocarbon up to 58%. *A. niger* showed a capability in performing complete degradation, indicated by the emission of 28.6% CO_2 during the experiment.	[84]
6	*Aspergillus sp., Phanerochaete sp., Trichoderma sp., Cladosporium sp.,* and *Penicillium sp.*	These isolates were able to degrade n-alkanes with carbon number of 9 to 36 with values of removal ranging between 47 to 89%. These species were also able to perform PAH degradation from soil with a removal efficiency of 68%.	[85]
7	*Aspergillus sydowii, Rhizopus sp.,* and *Penicillium funiculosum*	These species were demonstrated to be able in removing TPH from contaminated soil with values of removal ranging from 40% to 47%.	[86]
8	*Candida tropicalis SK21*	The isolated species was able to remove TPH from contaminated soil with the value of removal reaching 83% after 180 days of the incubation period.	[87]
9	*Penicillium decumbens PDX7, Penicillium janthinellum SDX7,* and *Aspergillus terreus PKX4*	These isolated fungi demonstrated the degradation of kerosene and diesel up to 95% after 16 days of treatment. These species are able to perform complete degradation of n-alkanes while aromatic hydrocarbon degradation has a ratio of 0.8 as compared to the aliphatic.	[88]
10	*Penilicilium citrinum*	*P. Citrinum* isolated from marine niches showed a capability of crude oil degradation up to 77%. This species was able to create fractions of long chain hydrocarbon with an average of 95.37%. The produced compound after treatment were lighter in molecular weight and more volatile.	[77]
11	*Phanerochaete chrysosporium*	This species was demonstrated to be PAH degrading fungi by attacking the PAH *via* lignocellulosic enzymatic reaction.	[89]
12	*Phanerochaete chrysosporium*	This species was proven to perform bioaccumulation and biodegradation of hydrocarbons. A total of 99% removal of phenanthrene and pyrene was obtained from the contaminated water by using this species.	[90]
13	*Phellinus sp.,* and *Polyporus sulphureus*	*Phellinus sp.* showed the highest PAHs degradation from soil and also exhibited higher lignin degradation while also performing higher biosurfactant production as compared to the *P. sulphureus*. Coculture with bacteria *(Bacillus pumilus)* showed a positive interaction.	[91]
14	*Pleurotus ostreatus*	This species was able to degrade Benzo[a]pyrene in the presence of heavy metals. This research showed the highest removal of Benzo[a]pyrene up to 83.6% by the addition of vanilin into the system.	[92]

(Table 2) cont.....

S. No	Species of Fungi	Finding(s)	References
15	*Pleurotus ostreatus*	By incubating PAH with isolated species, the researcher was able to remove 32.9% PAH from microcosm. Anthracene, benzo(*a*)anthracene and benzo(*a*)pyrene become the highest degraded compound by this species.	[93]
16	*Pythium ultimum*	This study showed that mycelia played an important role in PAH degradation by transporting into the further metabolisms.	[94]
17	*Trametes versicolor*	This species was proven to be able to degrade petrochemical hydrocarbon from the soil. This species is able to perform lignolytic enzymatic reactions to degrade various types of hydrocarbon compounds.	[95]

Algae as Bioremediation Agent of Hydrocarbon Contaminated Environment

As compared to bacteria and fungi, remediation of hydrocarbon contaminated environment using algae was still limited [96]. This emerging technology comes from the basics of algae's capability which can live in a contaminated environment. Most of the algae applications for hydrocarbon remediation are in the water environment due to their suitability for the living environment [25], thus making research related to remediation bordered only for water pollution. Both macroalgae and microalgae are able to perform hydrocarbon accumulation and degradation [27, 97]. Due to its present state as emerging technology, limited information regarding hydrocarbon bioremediation using algae can be obtained. Several potential species involved in hydrocarbon bioremediation using algae are presented in Table **3**.

Table 3. Potential species of algae in hydrocarbon bioremediation.

S. No	Species of Algae	Finding(s)	References
1	*Chlorella sorokiniana*	The bioreactor test showed the interaction of bacteria and algae to treat phenanthrene. *C. sorokiniana* is able to interact with *Pseudomonas migulae* and enhance the phenanthrene removal from contaminated water.	[98]
2	*Chlorella vulgaris BS1*	This species was demonstrated to remove 98.63% of petroleum hydrocarbon after 14 days of treatment along with the production of 1.76 g/L/d of biomass.	[99]
3	*Chlamydomonas spp., Dunaliella sp., Euglena gracilis,* and *Scenedesmus obliquus*	Isolated species of micro-algae were proven to be able to degrade naphthalene in contaminated environment.	[27]

(Table 3) cont.....

S. No	Species of Algae	Finding(s)	References
4	*Nitzhia sp.*	The isolated micro-algae species of Nitzia sp from a mangrove environment was able to degrade phenanthrene and simultaneously accumulate it inside its cell	[100]
5	*Parachlorella Kessleri*	Benzene, toluene, ethylbenzene and xylenes degradation were achieved after 48 h of incubation with the highest removal of 63% from 100 μ/L of initial concentration.	[101]
6	*Prototheca zopfii*	This species was known as one of the most potential hydrocarbon degrading microalgae species. This species is able to degrade and accumulate various compounds of hydrocarbon.	[102]
7	*Scenedesmus obliquus*	This species demonstrated the capability of performing consortium degradation of crude oil with bacteria (*Sphingomonas GY2B, Burkholderia cepacia GS3C, Pseudomonas GP3A* and *Pandoraea pnomenusa GP3B*). This consortium showed capability in degrading PAH from the contaminated water.	[103]
8	*Scenedesmus obliquus* and *Nitzschia linearis*	These species were isolated from the contaminated river. Both species exhibited capability in degrading n-alkanes and PAH from crude oil.	[104]
9	*Selenastum capicornutum*	*S. Capicornutum* was found to have the capability of accumulating benzene inside the cell. This species was also able to degrade benzo[*a*]pyrene from the contaminated environment.	[105, 106]
10	*Ulva lactuca*	The isolated species was able to produce biosurfactants to enhance the PAH bioavailability. This species is also able to accumulate hydrocarbons inside its cell.	[97]
11	*Ulva prolifera*	This species was able to adsorb 10 μg/L phenanthrene, naphthalene, and benzo[a] pyrene with an absorption capacity of 1.27, 1.97, and 2.49 mg/kg, respectively.	[96]

Commercialized Product of Microbial Agents for Hydrocarbon Remediation

Currently, there are some commercial products utilizing the microbial capability in performing degradation for hydrocarbon pollution. These commercial products vary between ready-to-use microbial agents, enzyme additives and even nutrients to boost the native microorganisms. Some commercial products that are currently available in the market are shown in Table **4**.

Table 4. Commercial products that utilize microbial agents.

S. No	Trademark	Manufacture	Country	Product Type
1	BILGEPRO	International Environmental Products	Pennsylvania	Nutrient additive
2	INIPOL EAP 22	CECA	France	Nutrient additive
3	LAND AND SEA	Land and Sea Restoration	Texas	Nutrient additive
4	OIL SPILL EATER II	Oil Spill Eater International	Texas	Nutrient and enzyme additive
5	OPPENHEIMER FORMULA	Oppenheimer Biotechnology	Texas	Microbial culture
6	PRISTINE SEA II	Marine Systems	Louisiana	Microbial culture
7	RESTORATION MICRO-BLAZE	Verde Environmental	Texas	Microbial culture
8	STEP ONE	B & S Research	Minnesota	Microbial culture
9	SYSTEM E.T. 20.	Quantum Environmental Technologies	California	Microbial culture

APPLICATION STRATEGIES AND PRACTICES

In-situ Bioremediation

In-situ bioremediation is the process of cleaning pollutants without moving the material to another location [107]. This bioremediation process mostly relies on microorganisms present in their original location to remediate the contaminated environment. The *in-situ* stage consists of cleaning the site, promoting the degrading microbe's growth through the bioaugmentation [108] or biostimulation processes [109], and the bioremediation process itself.

In-situ bioremediation has the advantage of being easier and cheaper, especially for areas that are not covered by heavy equipment to dig up contaminated locations [23, 110]. However, there are drawbacks to this technology as well. The remediation process is very dependent on the ability of microorganisms to survive [111]. That way, the degradation and cleaning of pollutants can take longer [39]. In *in-situ* bioremediation, the addition of nutrients and oxygen must continue to be carried out so that microorganisms can survive [23].

There are some techniques for applying *in-situ* bioremediation process, one of them is the bioventing method. Bioventing is an application of in situ bioremediation that is carried out in an unsaturated zone which has good gas permeability [112]. Bioventing is carried out in the treatment of volatile contaminants which are difficult to biodegrade. Bioventing is suitable for contaminants that are degraded through aerobic metabolism and have a vapor pressure of less than 1 atm. In bioventing, air movement that is injected through unsaturated soil or without the addition of nutrients is used to stimulate soil microorganisms to convert organic contaminants such as hydrocarbon vapors greater than 760 mmHg, so evaporation will run faster.

Another method is bioaugmentation. This method is conducted by the addition of certain microorganisms but allows bioremediation to occur independently [53, 87]. Microorganisms that can help to clean up a certain contaminant can be obtained from the contaminated site itself or from the laboratory cultures. This method is the most frequently used method in the remediation process to remove a contaminant. In addition to bioaugmentation, biostimulation is often also conducted. It is a bioremediation process that is carried out by adding nutrients or energy needed by microorganisms, or also by stimulating microorganisms by making a suitable environment for microorganisms so that organisms that are already in the water or soil will develop, and the provision of nutrients will also help their acceleration in carrying out remediation [107, 109].

Ex-situ Bioremediation

If *in-situ* bioremediation is not possible to be conducted, *ex-situ* bioremediation might be an option. It is a bioremediation process carried out outside the original place which is by taking waste at one location and then treating it at another place, after which it is returned to the place of origin. This bioremediation process can be faster and easier to control than *in-situ*, it is also able to remediate a wider variety of contaminants and soil types [113]. *Ex-situ* treatment needs to pay attention to waste management, which includes physically moving contaminated materials to a location for further handling [114]. Examples of this method are the utilization of bioreactors, biopiles, land-farming, composting and several other forms of solid phase treatment [115]. Bioreactor is a treatment process using aqueous reactor on contaminated soil or water. Land-farming is a technique whereby contaminated soil is excavated and moved over a special area of land which is periodically observed until the pollutants are degraded. Composting is a technique that combines contaminated soil with soil containing fertilizers or organic compounds that can increase the population of microorganisms. While biopiles are a combination of land-farming and composting.

FACTOR AFFECTING BIOREMEDIATION

Temperature

The optimum temperature for biodegradation is normally 15–30 °C for aerobic conditions and 25–35 °C for anaerobic ones [3]. Low ground temperatures hinder the evaporation rate of volatile components, and thus delay the activation of oil biodegradation [116], and also the hydrocarbon degradation was faster at 25 °C than at 5 °C [116].

Substances Bioavailability

Bioavailability can be defined as the rate at which a substance is absorbed into microorganism cell or a substance should be formed as a solution in order to be available at the site of the physiological activity of microorganisms before organisms can degrade it. Difficulty in bioavailability delays hydrocarbon transfer into cellulous enzymes, thus limiting the rate of biodegradation. The degree of bioavailability correlates with substance aqueous solubility (expressed as octanol-water partition coefficient (Kow)). The higher the Kow value, the lower bioavailability, hence impeding biodegradation. Specific microorganisms could produce *in-situ* metabolites such as biosurfactants, fatty acids, alcohols and solvents. These metabolites increase bioavailability by decreasing the critical micelle concentration (CMC), interfacial tension and surface tension between substrates and cell membranes [117].

Oxygen or Alternate Electron Acceptors

Oxygen usually acts as the terminal electron acceptor in metabolism. Aerobic processes mostly yield a considerably greater potential energy yield per unit of the substrate and tend to occur considerably more rapidly. Theory suggests that the mass of oxygen necessary to remediate the hydrocarbon load is about 0.3 g of oxygen for each gram of oil oxidized [116]. When oxygen is not present, a number of alternative electron acceptors could be utilized. These electron acceptors are used in an order based on their energy yields per unit of organic carbon oxidation in the following order: O_2, NO_3, Mn^{4+}, Fe^{3+} and SO_2^{-4} [3]. The strongest oxidation is oxygen through aerobic respiration, followed by denitrification, manganese reduction, ferric reduction, sulphate reduction, and finally methanogenesis. When the redox potential is greater than 50 mV, oxygen is preferable as electron acceptor. At low redox potentials (lower than 50 mV), alternative electron acceptors act as electron acceptors.

Nutrients

Theoretically, the desired ratio of C, N, P, and K is 100:15:1:1 [118]. Inorganic such as nitrates, and phosphates are important for bioremediation, as well as low concentrations of some amino acids, vitamins, or other organic molecules [3]. Nutrients or organic compounds are required as a source of carbon or electron donor/acceptor. However, excessively high nitrogen levels with C/N ratios less than 20, may result in inhibited soil microbial activity possibly owing to nitrite toxicity [3].

MATRICES TO BE REMEDIATED

Soil Bioremediation

Soil bioremediation can be done *via* two strategies, *in-situ* and *ex-situ*. In situ bioremediation treats contaminated soil in place through bioventing, injecting oxygen and nutrients directly into the soil. This strategy is less expensive than other conventional remediation because no use of transport to off-site treatment. However, in situ bioremediation has drawbacks, such as: depending on soil type, difficulty to reach complete degradation and difficulty to adjust the environmental condition such as pH and temperature to optimum conditions [119 - 121]. *Ex-situ* bioremediation needs excavation of contaminated soil to be treated elsewhere. *Ex-situ* bioremediation uses bioreactors, land farming and biopiles [121]. In bioreactors, contaminated soil is mixed with water and nutrient, and agitated by a mechanical stirrer. This method is very suitable to treat clay soil. Landfarming is conducted by spreading contaminated soil into collection system with aeration and nutrient supply. Biopiles are done by piling soils and aerating using an injection system [122]. In soil, *actinobacteria*, well known degraders of recalcitrant biomolecules, are generally abundant [10]; they are resistant to drought and good colonizers of organic and inorganic surfaces [11]. A review of soil bioremediation technology is shown in Table **5**.

Table 5. Review of soil bioremediation technology.

S. No.	Type of Soil/ Type of HC	Biodegradation Agents	Method	% Removal	References
1	clean soil, soil artificially contaminated by crude oil to saturation level (17% w/w), crude oil contaminated soil (17% w/w crude oil)	Immobilized *Pseudomonas Putida* (gram negative); *Bacili Subtilis* (gram positive) bacteria on cereal (bran).	*Ex-situ,* biopile (lab scale)	75%	[122]

(Table 5) cont.....

S. No.	Type of Soil/ Type of HC	Biodegradation Agents	Method	% Removal	References
2	Soil contaminated with TPH	petroleum-degrading microorganisms (DM) mix cultures (via bio-stimulation)	In situ (lab scale)	82 – 85%	[18]
3	Clayey acid soil contaminated with diesel oil	Indigenous microorganisms (via biostimulation) with the addition of sawdust as bulk material	*Ex situ* (lab scale)	nA	[123]
4	Mazut (heavy residual fuel oil)-polluted soil	Indigenous microorganisms (via biostimulation) with biosurfactant addition	*Ex-situ*, biopile, (pilot scale)	94%	[124]
5	coal-tar contaminated soil	Four white-rot fungi (*Phanerochaete chrysosporium IMI 232175, Pleurotus ostreatus* from the University of *Alberta Microfungus Collection IMI 341687, Coriolus versicolor IMI 210866 and Wye isolate #7*), bioaugmentation	*Ex-situ* (lab scale)	16.7%	[125]
6	soil from a manufactured gas plant, contaminated with PAH, benzene, toluene, ethyl benzene and xylene (BTEX)	Indigenous microorganisms (via biostimulation)	*Ex situ*, bioreactor (lab scale)	55% PAH and 70% BTEX.	[119]
7	Soil contaminated with diesel and fuel oil	Integration of biostimulation (*Acinetobacter sp.* and *Pseudomonas aeruginosa* predominat) and bioaugmentation (*Gordonia alkanivorans and Rhodococcus erythropolis*)	*Ex situ*, biopile (lab scale)	diesel oil 70% and fuel oil 63	[126]
8	Soil contaminated with high molecular PAHs	microalga, *Chlorella* sp. MM3, and a bacterium, *Rhodococcus wratislaviensis* strain 9	*Ex situ* (lab scale)	95% of 50 mg L−1 phenanthrene in soil under slurry phase within 21 days, while only 88% of 10 mg L−1 pyrene and 75% of 10mgL−1 BaP a	[127]

Water Bioremediation

Aquatic pollution due to an oil spill not only occurs on the sea-shore as an effect of drilling activity, but it can also be transmitted to a beach, sediment, river or become wastewater because of transportation, storage, and distribution. Once the oil is released into the aquatic environment, the oil characteristic (properties and chemical) is changed because of physical, chemical and biological processes. When so called "weathered oil" reaches the shoreline, or has been stranded for many days, its physical form transformation could speed up or slow down its degradation and reduce its toxicity.

In marine environments, PAH was mainly degraded by *hydro-carbonoclastic* bacteria [128] such as Alcanivorax borkumensis SK2 [129] and *Oleibacter marinus*, and two soil, long-chain n-alkane degrading actinobacteria, *Gordonia sp. SoCg* [130] and *Nocardia sp.* To improve the effectiveness of *hydro-carbonoclastic* bacteria, the immobilization of bacteria on carriers was done to preserve their viability and catalytic functions, as well as provide resistance to hostile environmental conditions and high concentrations of pollutants, while displaying a longer half-life [131, 132]. Immobilization techniques can reduce costs by eliminating dispersion and dilution of microbial cells in the environment. The carrier material should be biodegradable, insoluble, non-toxic for the immobilized cells and environment, accessible, affordable, abundant, stable and suitable for regeneration [133]. An example of promising immobilization carriers is natural or synthetic biopolymers. Synthetic biopolymers such as polylactic acid (PLA), and polycaprolactone (PCL) have low density (1.50 mg/cm^3) and have adsorption capacity for different oils (around 145.2–206.8 g/g) [133]. Table **6** provides a review of the current water bioremediation technology.

Table 6. Review of water bioremediation technology.

S. No.	Type of HC	Biodegradation agents	Method	% removal	References
1	Alkanes and polycyclic aromatic hydrocarbons (PAHs), in seawater dan sediment	HC biodegrading bacteria *via* biostimulation	Biostimulation	In sediment, there was no difference compared to the control. In seawater, there was 33% degradation improvement	[134]

(Table 6) cont.....

S. No.	Type of HC	Biodegradation agents	Method	% removal	References
2	Heavy oil, Saturated Hydrocarbon, Aromatic hydrocarbon	immobilized laccase-bacteria consortium	intertidal experimental pools built in the coastal area, *via* bioaugmentation	66.5% after 100 days for heavy oil. Saturated hydrocarbons and aromatic hydrocarbons were 79.2% and 78.7%, which were 64.9% and 65.1% higher than control.	[131]
3	Crude oil polluted seawater bioremediation	chitin and chitosan flakes obtained from shrimp wastes as carrier material for a hydrocarbon-degrading bacterial strain.	Laboratory-scale study	60% of hydrocarbons in the hexanic extract were removed compared with controls	[132]
4	Seawater polluted with Crude oil	*Bacillus subtilis* isolated from polymer dump site *via* bioaugmentation	Lab scale, *ex-situ.*	80% reduction in 10 days	[117]
5	PAH	indigenous microbial consortia UB, bioaugmentation	*In-situ*	80–85% after 60 days	[135]

Sludge Bioremediation

Petroleum industries generate approximately more than one billion tons of waste sludge worldwide per year during various production, extraction transportation, and refining processes [136, 137]. From 500 tons of crude oil processed, one ton of oil sludge is generated. Oil sludge is consisting of the stable water-in-oil emulsion of various petroleum hydrocarbons (PHCs), water, heavy metals, and solid particles [138]. Oil sludge still has a high density of oil in the emulsion. To effectively treat oil sludge, oil recovery should be applied first, after contaminants have been removed, the sludge either could be disposed of to the landfill or treated further with technology such as: incineration, solidification/stabilization, solvent extraction, ultrasonic treatment, pyrolysis, photocatalysis, chemical treatment, and biodegradation [138].

Oil sludge has a pH range between 6.5 and 7.5, while Total petroleum Hydrocarbon (TPH) contents ranging between 15 – 50% [139, 140]. Water and soild contents can range 0-85% and 5-46%, respectively [141]. Usually, oily sludge is composed of 40-52% alkanes, 28-31% aromatics, 8-10% asphaltenes, and 7-22.4% resins by mass [142]. Diverse chemical compositions in oil sludge make density, viscosity, and heat value can vary significantly. Oil sludge treatment management should include three steps, which are: quantity reduction, recovery of valuable fuel [141], and disposed unrecoverable residues [143]. Maximization of hydrocarbon recovery is the primary consideration in handling oil sludge, in order to reduce contamination, and decrease the use of non-renewable energy resources. The type of oil sludge that prefers to be recovered is sludge that has a high concentration (> 50%) of oil and a low concentration of solids (< 30%).

Oil sludge Bioremediation methods, e.q. land treatment, biopile/composting, and bio-slurry treatment, were researched extensively [70, 144, 145]. Land treatment was applied where there is a mixture between oily sludge and soil. Landfarming of oily sludge needs a vast area of land, and is a very time-consuming process (*e.g.*, usually 6 months to 2 years or even longer) due to the lack of environmental control of the contaminated site. It has been proven that land farming methods could remove 70 -90% PHC [35]. Biopile is another oil sludge alternative. Biopiles efficiency can be improved with moisture adjustment, air blowing, and the addition of bulking agent. The balance of nutrients (C:N:P) should be also considered. The TPH removal can be up to 60% and the remediation process can be as long as 1 year [35]. A review of oil sludge bioremediation is shown in Table 7.

Table 7. Review of oil sludge bioremediation.

S. No.	Type of Hydrocarbon	Biodegradation Agents	Methods	% Removal	References
1	Oil refinery sludge containing Total Petroleum Hydrocarbon	Mix culture microorganism, *via* bioaugmentation	Lab scale, *ex-situ*	90	[145]
2	Oily sludge containing TPH	bioaugmentation with biosurfactant plus nutrients addition	Lab scale	57% to 75%	[146]
3	TPH	Anaerobic Microorganism consortium	Bioelectrochemical Treatment (BET)	TPH =41.08%, And 75.54% for aromatic fraction	[140]
4	Hydrocarbon from tank bottom oil sludge	microbial consortia comprising *Shewanalla chilikensis, Bacillus firmus, and Halomonas hamiltonii* w	Lab scale, bioaugmentation	96%	[11]

CONCLUSION AND FUTURE CHALLENGES

Several important topics that will contribute significantly to the present and future knowledge of the biodegradation of hydrocarbon are discussed below:

A. Hydrocarbon solubility is a vital consideration in the application of biodegradation techniques [23]. Developing new methods in the development of biosurfactants and emerging existing technologies to improve biosurfactant development would contribute enormously to the success of treatment. Future researches need to be focused on the development of the biodegradable surfactant [147] as the alternative to the current most-used chemical surfactant.

B. Until now, the anaerobic treatment for hydrocarbon bioremediation has not been explored yet. Since the pollution of hydrocarbon is considered not to decrease for the next decade and its emission is expanding, study on this particular treatment method will be very useful in the future [148]. This particular method can even direct to a new study of energy production from the utilization of the generated methane gases, while also producing valuable material from the remediated sludge which can be utilized in building construction.

C. Many researchers have identified many superior bacterial, fungal and algal species for hydrocarbon biodegradation that are mostly evaluated in laboratory studies [23, 79]. The problem is that some cases note the low adaptability of exogenous microorganisms in polluted sites when the selected species are non-indigenous. Knowing the complex relationship between processes (*i.e.* rivalry amongst species of bacteria, existence of other pathogens, and interrelated mechanisms of degradation) in the microbial population will shed light on this dark position of bioremediation through diesel.

D. Currently, only a few pieces of information on the bioremediation of hydrocarbon using algae are available [27]. This case was obtained due to the limitation of algae application only on the water contamination. Conducting more studies on the utilization of algae may benefit the knowledge of bioremediation, while could also lead to the new resource of biodiesel raw material, in which algae contain many useful components for the production of biodiesel.

E. The presence of numerous different enzymes is essential in hydrocarbon degradation. However, due to the absence of certain enzymes in their metabolism, only a few species can perform complete degradation of hydrocarbons. In addition, research on the functional genes that control the enzymatic process of hydrocarbon degradation needs to be a further priority for presenting new looks in term of molecular understanding of the degradation of hydrocarbon [43]. Knowing the extracellular metabolites in the biodegradation of diesel can further improve the biodegradation of diesel. Using genes edited bacteria/ fungi/ algae to enhance the diesel degradation process can be a future application in the bioremediation of hydrocarbon [149].

F. A comprehensive study of the polluted site is needed for the application of diesel bioremediation. Examination of polluted sites is so necessary because the application of bioremediation has some limitations. This research will include details about how to eliminate contaminants and choose the right agents to eliminate them. Innovative site analysis can contribute immensely not only to the understanding of biodegradation of hydrocarbon but also to other remediation method. Providing and developing a cheaper, simpler, and high-accuracy site characterization can decrease the time needed to decide the most-suitable technological approach to conduct the bioremediation.

CONSENT FOR PUBLICATION

Not applicable.

CONFLICT OF INTEREST

The authors declare no conflict of interest, financial or otherwise.

ACKNOWLEDGEMENT

This book chapter is supported by Research Centre for Environmental and Clean Technology, National Research and Innovation Agency Republic of Indonesia.

REFERENCES

[1]   Imam A, Suman SK, Ghosh D, Kanaujia PK. Analytical approaches used in monitoring the bioremediation of hydrocarbons in petroleum-contaminated soil and sludge. Trends Analyt Chem 2019; 118: 50-64.
        [http://dx.doi.org/10.1016/j.trac.2019.05.023]

[2]   Mapelli F, Scoma A, Michoud G, *et al.* Biotechnologies for Marine Oil Spill Cleanup: Indissoluble Ties with Microorganisms. Trends Biotechnol 2017; 35(9): 860-70.
        [http://dx.doi.org/10.1016/j.tibtech.2017.04.003] [PMID: 28511936]

[3]   Yang SZ. Bioremediation of Oil Spills in Cold Environments: A Review. Pedosphere 2009; 19(3): 371-81.
        [http://dx.doi.org/10.1016/S1002-0160(09)60128-4]

[4]   Yim UH, Hong S, Lee C, et al. Rapid recovery of coastal environment and ecosystem to the Hebei Spirit oil spill's impact. Environ Int 2020; 136: 105438.
        [http://dx.doi.org/10.1016/j.envint.2019.105438]

[5]   Fingas M. Handbook of Oil Spill Science and Technology. Wiley 2014.
        [http://dx.doi.org/10.1002/9781118989982]

[6]   Alabresm A, Chen YP, Decho AW, Lead J. A novel method for the synergistic remediation of oil-water mixtures using nanoparticles and oil-degrading bacteria. Sci Total Environ 2018; 630: 1292-7.
[http://dx.doi.org/10.1016/j.scitotenv.2018.02.277] [PMID: 29554750]

[7]   Mullin JV, Champ MA. Introduction/overview to in situ burning of oil spills. Spill Sci Technol Bull 2003; 8(4): 323-30.
[http://dx.doi.org/10.1016/S1353-2561(03)00076-8]

[8]   Schein A, Scott JA, Mos L, Hodson PV. Oil dispersion increases the apparent bioavailability and toxicity of diesel to rainbow trout (Oncorhynchus mykiss). Environ Toxicol Chem 2009; 28(3): 595-602.
[http://dx.doi.org/10.1897/08-315.1] [PMID: 18939894]

[9]   Zahed MA, Aziz HA, Isa MH, Mohajeri L, Mohajeri S. Optimal conditions for bioremediation of oily seawater. Bioresour Technol 2010; 101(24): 9455-60.
[http://dx.doi.org/10.1016/j.biortech.2010.07.077] [PMID: 20705460]

[10]   Ellis M, Altshuler I, Schreiber L, *et al.* Hydrocarbon biodegradation potential of microbial communities from high Arctic beaches in Canada's Northwest Passage. Mar Pollut Bull 2022; 174: 113288.
[http://dx.doi.org/10.1016/j.marpolbul.2021.113288] [PMID: 35090274]

[11]   Suganthi S H, Murshid S, Sriram S, Ramani K. Enhanced biodegradation of hydrocarbons in petroleum tank bottom oil sludge and characterization of biocatalysts and biosurfactants. J Environ Manag 2018; 220: 87-95.
[http://dx.doi.org/10.1016/j.jenvman.2018.04.120]

[12]   Pal S, Roy A, Kazy SK. Exploring Microbial Diversity and Function in Petroleum Hydrocarbon Associated Environments Through Omics Approaches. Elsevier Inc. 2019.
[http://dx.doi.org/10.1016/B978-0-12-814849-5.00011-3]

[13]   Song B, Tang J, Zhen M, Liu X. Effect of rhamnolipids on enhanced anaerobic degradation of petroleum hydrocarbons in nitrate and sulfate sediments. Sci Total Environ 2019; 678: 438-47.
[http://dx.doi.org/10.1016/j.scitotenv.2019.04.383] [PMID: 31077922]

[14]   Zhang H, Yuan X, Xiong T, Wang H, Jiang L. Bioremediation of co-contaminated soil with heavy metals and pesticides: Influence factors, mechanisms and evaluation methods. Chem Eng J 2020; 398: 125657.
[http://dx.doi.org/10.1016/j.cej.2020.125657]

[15]   Wilkinson S, Nicklin S, Faull JL. Biodegradation of fuel oils and lubricants: Soil and water bioremediation options. Prog Ind Microbiol 2002; 36(C): 69-100.
[http://dx.doi.org/10.1016/S0079-6352(02)80007-8]

[16]   Wei Z, Wang JJ, Gaston LA, *et al.* Remediation of crude oil-contaminated coastal marsh soil: Integrated effect of biochar, rhamnolipid biosurfactant and nitrogen application. J Hazard Mater 2020; 396(March): 122595.
[http://dx.doi.org/10.1016/j.jhazmat.2020.122595] [PMID: 32298868]

[17]   Crisafi F, Genovese M, Smedile F, *et al.* Bioremediation technologies for polluted seawater sampled after an oil-spill in Taranto Gulf (Italy): A comparison of biostimulation, bioaugmentation and use of a washing agent in microcosm studies. Mar Pollut Bull 2016; 106(1-2): 119-26.
[http://dx.doi.org/10.1016/j.marpolbul.2016.03.017] [PMID: 26992747]

[18]   Vasilyeva G, Kondrashina V, Strijakova E, Ortega-Calvo JJ. Adsorptive bioremediation of soil highly contaminated with crude oil. Sci Total Environ 2020; 706: 135739.
[http://dx.doi.org/10.1016/j.scitotenv.2019.135739] [PMID: 31818568]

[19]   Ossai IC, Ahmed A, Hassan A, Hamid FS. Remediation of soil and water contaminated with petroleum hydrocarbon: A review. Environmental Technology & Innovation 2020; 17: 100526.
[http://dx.doi.org/10.1016/j.eti.2019.100526]

[20]   Das N, Chandran P. Microbial degradation of petroleum hydrocarbon contaminants: an overview. Biotechnol Res Int 2011; 2011: 1-13.
[http://dx.doi.org/10.4061/2011/941810] [PMID: 21350672]

[21]   Robinson PK. Enzymes: principles and biotechnological applications. Essays Biochem 2015; 59: 1-41.
[http://dx.doi.org/10.1042/bse0590001] [PMID: 26504249]

[22]   Panda SK, Kar RN, Panda CR. Isolation and identification of petroleum hydrocarbon degrading microorganisms from oil contaminated environment. Int J Enivironmental Sci 2013; 3(1): 1314-21.
[http://dx.doi.org/10.6088/ijes.2013030500001]

[23]   Imron MF, Kurniawan SB, Ismail NI, Abdullah SRS. Future challenges in diesel biodegradation by bacteria isolates: A review. J Clean Prod 2020; 251: 119716.
[http://dx.doi.org/10.1016/j.jclepro.2019.119716]

[24]   Titah HS, Pratikno H, Moesriati A, Imron MF, Putera RI. Isolation and screening of diesel degrading bacteria from ship dismantling facility at Tanjungjati, Madura, Indonesia. Journal of Engineering and Technological Sciences 2018; 50(1): 99-109.
[http://dx.doi.org/10.5614/j.eng.technol.sci.2018.50.1.7]

[25]   Ghosal D, Ghosh S, Dutta TK, Ahn Y. Current state of knowledge in microbial degradation of polycyclic aromatic hydrocarbons (PAHs): A review. Front Microbiol 2016; 7(AUG): 1369.
[http://dx.doi.org/10.3389/fmicb.2016.01369] [PMID: 27630626]

[26]   Agrawal N, Verma P, Shahi SK. Degradation of polycyclic aromatic hydrocarbons (phenanthrene and pyrene) by the ligninolytic fungi Ganoderma lucidum isolated from the hardwood stump. Bioresour Bioprocess 2018; 5(1): 11.
[http://dx.doi.org/10.1186/s40643-018-0197-5]

[27]   Semple KT, Cain RB, Schmidt S. Biodegradation of aromatic compounds by microalgae. FEMS Microbiol Lett 1999; 170(2): 291-300.
[http://dx.doi.org/10.1111/j.1574-6968.1999.tb13386.x]

[28]   Bell TH, El-Din Hassan S, Lauron-Moreau A, *et al.* Linkage between bacterial and fungal rhizosphere communities in hydrocarbon-contaminated soils is related to plant phylogeny. ISME J 2014; 8(2): 331-43.
[http://dx.doi.org/10.1038/ismej.2013.149] [PMID: 23985744]

[29]   Castro-González JM, Castro P, Sandoval H, Castro-Sandoval D. Probiotic lactobacilli precautions. Front Microbiol 2019; 10(MAR): 375.
[http://dx.doi.org/10.3389/fmicb.2019.00375] [PMID: 30915041]

[30]   Naserzadeh Y, Nafchi AM, Mahmoudi N, Nejad DK, Gadzhikurbanov AS. Effect of combined use of fertilizer and plant growth stimulating bacteria Rhizobium, Azospirillum, Azotobacter and *Pseudomonas* on the quality and components of corn forage in Iran. Vestn Ross Univ Druhzby Narodov Ser Agron Zhivotnovod 2019; 14(3): 209-24.
[http://dx.doi.org/10.22363/2312-797X-2019-14-3-209-224]

[31]   de Lima Procópio RE, da Silva IR, Martins MK, de Azevedo JL, de Araújo JM. Antibiotics produced by *Streptomyces*. Braz J Infect Dis 2012; 16(5): 466-71.
[http://dx.doi.org/10.1016/j.bjid.2012.08.014] [PMID: 22975171]

[32]   Doyle MP, Steenson LR, Meng J. Bacteria in food and beverage production. Prokaryotes Appl Bacteriol Biotechnol 2013; 9783642313: 241-56.
[http://dx.doi.org/10.1007/978-3-642-31331-8_27]

[33]   Purwanti IF, Kurniawan SB, Imron MF. Potential of *Pseudomonas aeruginosa* isolated from aluminium-contaminated site in aluminium removal and recovery from wastewater. Environmental Technology & Innovation 2019; 15: 100422.
[http://dx.doi.org/10.1016/j.eti.2019.100422]

[34]   Imron MF, Kurniawan SB, Soegianto A. Characterization of mercury-reducing potential bacteria

isolated from Keputih non-active sanitary landfill leachate, Surabaya, Indonesia under different saline conditions. J Environ Manage 2019; 241: 113-22.
[http://dx.doi.org/10.1016/j.jenvman.2019.04.017] [PMID: 30986663]

[35]   Haritash AK, Kaushik CP. Biodegradation aspects of polycyclic aromatic hydrocarbons (PAHs): a review. J Hazard Mater 2009; 169(1-3): 1-15.
[http://dx.doi.org/10.1016/j.jhazmat.2009.03.137] [PMID: 19442441]

[36]   Imron MF, Kurniawan SB, Titah HS. Potential of bacteria isolated from diesel-contaminated seawater in diesel biodegradation. Environmental Technology & Innovation 2019; 14: 100368.
[http://dx.doi.org/10.1016/j.eti.2019.100368]

[37]   Jemil N, Ben Ayed H, Hmidet N, Nasri M. Characterization and properties of biosurfactants produced by a newly isolated strain *Bacillus methylotrophicus* DCS1 and their applications in enhancing solubility of hydrocarbon. World J Microbiol Biotechnol 2016; 32(11): 175.
[http://dx.doi.org/10.1007/s11274-016-2132-2] [PMID: 27628335]

[38]   Ite AE, Harry TA, Obadimu CO, Asuaiko ER, Inim IJ. Petroleum hydrocarbons contamination of surface water and groundwater in the Niger Delta Region of Nigeria. J Environ Pollut Hum Heal 2018; 6(2): 51-61.
[http://dx.doi.org/10.12691/jephh-6-2-2]

[39]   Abdullah SRS, Al-Baldawi IA, Almansoory AF, Purwanti IF, Al-Sbani NH, Sharuddin SSN. Plant-assisted remediation of hydrocarbons in water and soil: Application, mechanisms, challenges and opportunities. Chemosphere 2020; 247: 125932.
[http://dx.doi.org/10.1016/j.chemosphere.2020.125932] [PMID: 32069719]

[40]   Ghenai C, Bettayeb M, Brdjanin B, Hamid AK. Hybrid solar PV/PEM fuel Cell/Diesel Generator power system for cruise ship: A case study in Stockholm, Sweden. Case Stud Therm Eng 2019; 14: 100497.
[http://dx.doi.org/10.1016/j.csite.2019.100497]

[41]   Ayotamuno JM, Okparanma RN, Amadi F. Enhanced remediation of an oily sludge with saline water. Afr J Environ Sci Technol 2011; 5(4): 262-7.

[42]   Dhar K, Dutta S, Anwar MN. Biodegradation of Petroleum Hydrocarbon by indigenous Fungi isolated from Ship breaking yards of Bangladesh. Int Res J Biol Sci 2014; 3(9): 22-30.

[43]   Mahjoubi M, Jaouani A, Guesmi A, *et al.* Hydrocarbonoclastic bacteria isolated from petroleum contaminated sites in Tunisia: isolation, identification and characterization of the biotechnological potential. N Biotechnol 2013; 30(6): 723-33.
[http://dx.doi.org/10.1016/j.nbt.2013.03.004] [PMID: 23541698]

[44]   Almansoory AF, Hasan HA, Abdullah SRS, Idris M, Anuar N, Al-Adiwish WM. Biosurfactant produced by the hydrocarbon-degrading bacteria: Characterization, activity and applications in removing TPH from contaminated soil. Environmental Technology & Innovation 2019; 14: 100347.
[http://dx.doi.org/10.1016/j.eti.2019.100347]

[45]   Taofeeq Adekunle A, Bolatito Ester B, Olabisi Peter A, Solomon Bankole O, Udeme Joshia Joshua I, Alfa S. Characterization of new glycosophorolipid-surfactant produced by *Aspergillus niger* and *Aspergillus flavus*. Eur J Biotechnol Biosci 2015; 3(4): 34-9.

[46]   Priyanka N, Archana T. Biodegradability of Polythene and Plastic By The Help of Microorganism: A Way for Brighter Future. J Environ Anal Toxicol 2011; 1: 2.
[http://dx.doi.org/10.4172/2161-0525.1000111]

[47]   Cerqueira VS, Hollenbach EB, Maboni F, *et al.* Biodegradation potential of oily sludge by pure and mixed bacterial cultures. Bioresour Technol 2011; 102(23): 11003-10.
[http://dx.doi.org/10.1016/j.biortech.2011.09.074] [PMID: 21993328]

[48]   Quadri LEN, Weinreb PH, Lei M, Nakano MM, Zuber P, Walsh CT. Characterization of Sfp, a *Bacillus subtilis* phosphopantetheinyl transferase for peptidyl carrier protein domains in peptide

synthetases. Biochemistry 1998; 37(6): 1585-95.
[http://dx.doi.org/10.1021/bi9719861] [PMID: 9484229]

[49]   Reis RS, Pereira AG, Neves BC, Freire DMG. Gene regulation of rhamnolipid production in *Pseudomonas aeruginosa* – A review. Bioresour Technol 2011; 102(11): 6377-84.
[http://dx.doi.org/10.1016/j.biortech.2011.03.074] [PMID: 21498076]

[50]   Marchand C, St-Arnaud M, Hogland W, Bell TH, Hijri M. Petroleum biodegradation capacity of bacteria and fungi isolated from petroleum-contaminated soil. Int Biodeterior Biodegradation 2017; 116: 48-57.
[http://dx.doi.org/10.1016/j.ibiod.2016.09.030]

[51]   Palanisamy N, Ramya J, Kumar S, Vasanthi NS, Chandran P, Khan S. Diesel biodegradation capacities of indigenous bacterial species isolated from diesel contaminated soil. J Environ Health Sci Eng 2014; 12(1): 142.
[http://dx.doi.org/10.1186/s40201-014-0142-2] [PMID: 25530870]

[52]   Ali Khan AH, Tanveer S, Alia S, *et al.* Role of nutrients in bacterial biosurfactant production and effect of biosurfactant production on petroleum hydrocarbon biodegradation. Ecol Eng 2017; 104: 158-64.
[http://dx.doi.org/10.1016/j.ecoleng.2017.04.023]

[53]   Al-Baldawi IA, Abdullah SRS, Anuar N, Mushrifah I. Bioaugmentation for the enhancement of hydrocarbon phytoremediation by rhizobacteria consortium in pilot horizontal subsurface flow constructed wetlands. Int J Environ Sci Technol 2017; 14(1): 75-84.
[http://dx.doi.org/10.1007/s13762-016-1120-2]

[54]   Al-Baldawi IA, Abdullah SRS, Idris M, Suja F, Anuar N. Consortia development of hydrocarbon degrading Rhizobacteria isolated from *Scirpus grossus* in diesel exposure. In: Aris, A., Tengku Ismail, T., Harun, R., Abdullah, A., Ishak, M. (eds) From Sources to Solution. Springer, Singapore 2014; pp. 331–35.
[http://dx.doi.org/10.1007/978-981-4560-70-2_60]

[55]   Almansoory AF, Idris M, Abdullah SRS, Anuar N. Screening for potential biosurfactant producing bacteria from hydrocarbon-degrading isolates. Adv Environ Biol 2014; 8(3): 639-47.

[56]   Tiralerdpanich P, Sonthiphand P, Luepromchai E, Pinyakong O, Pokethitiyook P. Potential microbial consortium involved in the biodegradation of diesel, hexadecane and phenanthrene in mangrove sediment explored by metagenomics analysis. Mar Pollut Bull 2018; 133(May): 595-605.
[http://dx.doi.org/10.1016/j.marpolbul.2018.06.015] [PMID: 30041354]

[57]   Wu M, Li W, Dick WA, *et al.* Bioremediation of hydrocarbon degradation in a petroleum-contaminated soil and microbial population and activity determination. Chemosphere 2017; 169: 124-30.
[http://dx.doi.org/10.1016/j.chemosphere.2016.11.059] [PMID: 27870933]

[58]   Mnif I, Sahnoun R, Ellouz-Chaabouni S, Ghribi D. Application of bacterial biosurfactants for enhanced removal and biodegradation of diesel oil in soil using a newly isolated consortium. Process Saf Environ Prot 2017; 109: 72-81.
[http://dx.doi.org/10.1016/j.psep.2017.02.002]

[59]   Reddy MS, Naresh B, Leela T, *et al.* Biodegradation of phenanthrene with biosurfactant production by a new strain of *Brevibacillus* sp. Bioresour Technol 2010; 101(20): 7980-3.
[http://dx.doi.org/10.1016/j.biortech.2010.04.054] [PMID: 20627713]

[60]   Morales-Guzmán G, Ferrera-Cerrato R, Rivera-Cruz MC, *et al.* Diesel degradation by emulsifying bacteria isolated from soils polluted with weathered petroleum hydrocarbons. Appl Soil Ecol 2017; 121(October): 127-34.
[http://dx.doi.org/10.1016/j.apsoil.2017.10.003]

[61]   Ibrahim H M M. Biodegradation of used engine oil by novel strains of Ochrobactrum anthropi HM-1 and *Citrobacter freundii* HM-2 isolated from oil-contaminated soil. 3 Biotech 2016; 6: 226.

[http://dx.doi.org/10.1007/s13205-016-0540-5]

[62]   Al-Baldawi IA, Abdullah SRS, Anuar N, Suja F, Mushrifah I. Phytodegradation of total petroleum hydrocarbon (TPH) in diesel-contaminated water using Scirpus grossus. Ecol Eng 2015; 74: 463-73.
       [http://dx.doi.org/10.1016/j.ecoleng.2014.11.007]

[63]   Ramasamy S, Arumugam A, Chandran P. Optimization of *Enterobacter cloacae* (KU923381) for diesel oil degradation using response surface methodology (RSM). J Microbiol 2017; 55(2): 104-11.
       [http://dx.doi.org/10.1007/s12275-017-6265-2] [PMID: 28120192]

[64]   Prakash A, Bisht S, Singh J, Teotia P, Kela R, Kumar V. Biodegradation potential of petroleum hydrocarbons by bacteria and mixed bacterial consortium isolated from contaminated sites. TURKISH JOURNAL OF ENGINEERING AND ENVIRONMENTAL SCIENCES 2014; 38(1): 41-50.
       [http://dx.doi.org/10.3906/muh-1306-4]

[65]   Martin-Sanchez PM, Becker R, Gorbushina AA, Toepel J. An improved test for the evaluation of hydrocarbon degradation capacities of diesel-contaminating microorganisms. Int Biodeterior Biodegradation 2018; 129: 89-94.
       [http://dx.doi.org/10.1016/j.ibiod.2018.01.009]

[66]   Fadhile Almansoory A, Abu Hasan H, Idris M, Sheikh Abdullah SR, Anuar N, Musa Tibin EM. Biosurfactant production by the hydrocarbon-degrading bacteria (HDB) *Serratia marcescens*: Optimization using central composite design (CCD). J Ind Eng Chem 2017; 47: 272-80.
       [http://dx.doi.org/10.1016/j.jiec.2016.11.043]

[67]   Parthipan P, Elumalai P, Ting YP, Rahman PKSM, Rajasekar A. Characterization of hydrocarbon degrading bacteria isolated from Indian crude oil reservoir and their influence on biocorrosion of carbon steel API 5LX. Int Biodeter Biodegrad 2018; 129: 67-80.
       [http://dx.doi.org/10.1016/j.ibiod.2018.01.006]

[68]   Imron MF, Titah HS. Optimization of diesel biodegradation by *Vibrio alginolyticus using Box-Behnken* design. Environ Eng Res 2018; 23(4): 374-82.
       [http://dx.doi.org/10.4491/eer.2018.015]

[69]   Ding J, Cong J, Zhou J, Gao S. Polycyclic aromatic hydrocarbon biodegradation and extracellular enzyme secretion in agitated and stationary cultures of Phanerochaete chrysosporium. J Environ Sci (China) 2008; 20(1): 88-93.
       [http://dx.doi.org/10.1016/S1001-0742(08)60013-3] [PMID: 18572528]

[70]   Cutright TJ. Polycyclic aromatic hydrocarbon biodegradation and kinetics using *Cunninghamella echinulata* var. elegans. Int Biodeterior Biodegradation 1995; 35(4): 397-408.
       [http://dx.doi.org/10.1016/0964-8305(95)00046-1]

[71]   Govarthanan M, Fuzisawa S, Hosogai T, Chang YC. Biodegradation of aliphatic and aromatic hydrocarbons using the filamentous fungus *Penicillium* sp. CHY-2 and characterization of its manganese peroxidase activity. RSC Advances 2017; 7(34): 20716-23.
       [http://dx.doi.org/10.1039/C6RA28687A]

[72]   Gargouri B, Mhiri N, Karray F, Aloui F, Sayadi S. Isolation and Characterization of Hydrocarbon-Degrading Yeast Strains from Petroleum Contaminated Industrial Wastewater. BioMed Res Int 2015; 2015: 1-11.
       [http://dx.doi.org/10.1155/2015/929424] [PMID: 26339653]

[73]   Harrison JS. Food and Fodder Yeasts. The Yeasts 1993; pp. 399-433.
       [http://dx.doi.org/10.1016/B978-0-08-092543-1.50021-7]

[74]   Andrykovitch G, Neihof RA. Fuel-soluble biocides for control of *Cladosporium resinae* in hydrocarbon fuels. J Ind Microbiol 1987; 2(1): 35-40.
       [http://dx.doi.org/10.1007/BF01569403]

[75]   Prenafeta-Boldú FX, de Hoog GS, Summerbell RC. Fungal Communities in Hydrocarbon Degradation. Microb. Communities Util. Hydrocarb. Lipids Members, Metagenomics Ecophysiol

2018; pp. 1-36.
[http://dx.doi.org/10.1007/978-3-319-60063-5_8-1]

[76]   Al-Hawash AB, Alkooranee JT, Abbood HA, *et al.* Isolation and characterization of two crude oil-degrading fungi strains from Rumaila oil field, Iraq. Biotechnol Rep (Amst) 2018; 17: 104-9.
[http://dx.doi.org/10.1016/j.btre.2017.12.006] [PMID: 29541603]

[77]   Barnes NM, Khodse VB, Lotlikar NP, Meena RM, Damare SR. Bioremediation potential of hydrocarbon-utilizing fungi from select marine niches of India. 3 Biotech 2018; 8(1): 21.
[http://dx.doi.org/10.1007/s13205-017-1043-8]

[78]   Padmapriya B, Suganthi S, Anishya RS. Screening, Optimization and Production of Biosurfactants by Candida Species Isolated from Oil Polluted Soils. J Agric Environ Sci 2013; 13(2): 227-33.
[http://dx.doi.org/10.5829/idosi.aejaes.2013.13.02.2744]

[79]   Bosco F, Mollea C. Mycoremediation in Soil. In: Saldarriaga-Noreña H, Alfonso Murillo-Tovar M, Farooq R, Dongre R, Riaz S (Eds.) Environmental Chemistry and Recent Pollution Control Approaches. IntechOpen 2019.
[http://dx.doi.org/10.5772/intechopen.84777]

[80]   Ameen F, Moslem M, Hadi S, Al-Sabri AE. Biodegradation of diesel fuel hydrocarbons by mangrove fungi from Red Sea Coast of Saudi Arabia. Saudi J Biol Sci 2016; 23(2): 211-8.
[http://dx.doi.org/10.1016/j.sjbs.2015.04.005] [PMID: 26981002]

[81]   Acevedo F, Pizzul L, Castillo MP, Cuevas R, Diez MC. Degradation of polycyclic aromatic hydrocarbons by the Chilean white-rot fungus Anthracophyllum discolor. J Hazard Mater 2011; 185(1): 212-9.
[http://dx.doi.org/10.1016/j.jhazmat.2010.09.020] [PMID: 20934253]

[82]   Elshafie A, AlKindi AY, Al-Busaidi S, Bakheit C, Albahry SN. Biodegradation of crude oil and n-alkanes by fungi isolated from Oman. Mar Pollut Bull 2007; 54(11): 1692-6.
[http://dx.doi.org/10.1016/j.marpolbul.2007.06.006] [PMID: 17904586]

[83]   Asemoloye MD, Ahmad R, Jonathan SG. Synergistic action of rhizospheric fungi with *Megathyrsus maximus* root speeds up hydrocarbon degradation kinetics in oil polluted soil. Chemosphere 2017; 187: 1-10.
[http://dx.doi.org/10.1016/j.chemosphere.2017.07.158] [PMID: 28787637]

[84]   Al-Dhabaan FA. Mycoremediation of crude oil contaminated soil by specific fungi isolated from Dhahran in Saudi Arabia. Saudi J Biol Sci 2020; 28(1): 73-7.
[http://dx.doi.org/10.1016/j.sjbs.2020.08.033] [PMID: 33424285]

[85]   Steliga T, Jakubowicz P, Kapusta P. Changes in toxicity during in situ bioremediation of weathered drill wastes contaminated with petroleum hydrocarbons. Bioresour Technol 2012; 125: 1-10.
[http://dx.doi.org/10.1016/j.biortech.2012.08.092] [PMID: 23018157]

[86]   Mancera-López ME, Esparza-García F, Chávez-Gómez B, Rodríguez-Vázquez R, Saucedo-Castañeda G, Barrera-Cortés J. Bioremediation of an aged hydrocarbon-contaminated soil by a combined system of biostimulation–bioaugmentation with filamentous fungi. Int Biodeterior Biodegradation 2008; 61(2): 151-60.
[http://dx.doi.org/10.1016/j.ibiod.2007.05.012]

[87]   Fan MY, Xie RJ, Qin G. Bioremediation of petroleum-contaminated soil by a combined system of biostimulation–bioaugmentation with yeast. Environ Technol 2014; 35(4): 391-9.
[http://dx.doi.org/10.1080/09593330.2013.829504] [PMID: 24600879]

[88]   Khan SR, Nirmal Kumar JI, Nirmal Kumar R, Patel J. Enzymatic Evaluation During Biodegradation of Kerosene and Diesel by Locally Isolated Fungi from Petroleum-Contaminated Soils of Western India. Soil Sediment Contam 2015; 24(5): 514-25.
[http://dx.doi.org/10.1080/15320383.2015.985783]

[89]   Aydin S, Karaçay HA, Shahi A, Gökçe S, Ince B, Ince O. Aerobic and anaerobic fungal metabolism

and Omics insights for increasing polycyclic aromatic hydrocarbons biodegradation. Fungal Biol Rev 2017; 31(2): 61-72.
[http://dx.doi.org/10.1016/j.fbr.2016.12.001]

[90]   Ding J, Chen B, Zhu L. Biosorption and biodegradation of polycyclic aromatic hydrocarbons by Phanerochaete chrysosporium in aqueous solution. Chin Sci Bull 2013; 58(6): 613-21.
[http://dx.doi.org/10.1007/s11434-012-5411-9]

[91]   Arun A, Eyini M. Comparative studies on lignin and polycyclic aromatic hydrocarbons degradation by basidiomycetes fungi. Bioresour Technol 2011; 102(17): 8063-70.
[http://dx.doi.org/10.1016/j.biortech.2011.05.077] [PMID: 21683591]

[92]   Bhattacharya S, Das A, Prashanthi K, Palaniswamy M, Angayarkanni J. Mycoremediation of Benzo[a]pyrene by *Pleurotus ostreatus* in the presence of heavy metals and mediators. 3 Biotech 2014; 4: 205–11.
[http://dx.doi.org/10.1007/s13205-013-0148-y]

[93]   Li X, Wu Y, Lin X, Zhang J, Zeng J. Dissipation of polycyclic aromatic hydrocarbons (PAHs) in soil microcosms amended with mushroom cultivation substrate. Soil Biol Biochem 2012; 47: 191-7.
[http://dx.doi.org/10.1016/j.soilbio.2012.01.001]

[94]   Schamfuß S, Neu TR, van der Meer JR, Tecon R, Harms H, Wick LY. Impact of mycelia on the accessibility of fluorene to PAH-degrading bacteria. Environ Sci Technol 2013; 47(13): 6908-15.
[http://dx.doi.org/10.1021/es304378d] [PMID: 23452287]

[95]   Aranda E, Marco-Urrea E, Caminal G, Arias ME, García-Romera I, Guillén F. Advanced oxidation of benzene, toluene, ethylbenzene and xylene isomers (BTEX) by Trametes versicolor. J Hazard Mater 2010; 181(1-3): 181-6.
[http://dx.doi.org/10.1016/j.jhazmat.2010.04.114] [PMID: 20627409]

[96]   Zhang C, Lu J, Wu J. Adsorptive removal of polycyclic aromatic hydrocarbons by detritus of green tide algae deposited in coastal sediment. Sci Total Environ 2019; 670: 320-7.
[http://dx.doi.org/10.1016/j.scitotenv.2019.03.296] [PMID: 30904645]

[97]   Net S, Henry F, Rabodonirina S, *et al.* Accumulation of PAHs, Me-PAHs, PCBs and total mercury in sediments and marine species in coastal areas of Dakar, Senegal: Contamination level and impact. Int J Environ Res 2015; 9(2).

[98]   Muñoz R, Guieysse B, Mattiasson B. Phenanthrene biodegradation by an algal-bacterial consortium in two-phase partitioning bioreactors. Appl Microbiol Biotechnol 2003; 61(3): 261-7.
[http://dx.doi.org/10.1007/s00253-003-1231-9] [PMID: 12698286]

[99]   Das B, Deka S. A cost-effective and environmentally sustainable process for phycoremediation of oil field formation water for its safe disposal and reuse. Sci Rep 2019; 9(1): 15232.
[http://dx.doi.org/10.1038/s41598-019-51806-5] [PMID: 31645605]

[100]   Hong YW, Yuan DX, Lin QM, Yang TL. Accumulation and biodegradation of phenanthrene and fluoranthene by the algae enriched from a mangrove aquatic ecosystem. Mar Pollut Bull 2008; 56(8): 1400-5.
[http://dx.doi.org/10.1016/j.marpolbul.2008.05.003] [PMID: 18597790]

[101]   Takáčová A, Smolinská M, Semerád M, Matúš P. Degradation of btex by microalgae *Parachlorella kessleri*. Pet. Coal 2015; 57(2): 101-7.

[102]   de-Bashan LE, Bashan Y. Immobilized microalgae for removing pollutants: Review of practical aspects. Bioresour Technol 2010; 101(6): 1611-27.
[http://dx.doi.org/10.1016/j.biortech.2009.09.043] [PMID: 19931451]

[103]   Tang X, He LY, Tao XQ, *et al.* Construction of an artificial microalgal-bacterial consortium that efficiently degrades crude oil. J Hazard Mater 2010; 181(1-3): 1158-62.
[http://dx.doi.org/10.1016/j.jhazmat.2010.05.033] [PMID: 20638971]

[104]   Gamila HA, Ibrahim MBM. Algal bioassay for evaluating the role of algae in bioremediation of crude

oil: I--Isolated strains. Bull Environ Contam Toxicol 2004; 73(5): 883-9.
[http://dx.doi.org/10.1007/s00128-004-0509-7] [PMID: 15669733]

[105]   Warshawsky D, Keenan TH, Reilman R, Cody TE, Radike MJ. Conjugation of benzo[a]pyrene metabolites by freshwater green alga Selenastrum capricornutum. Chem Biol Interact 1990; 74(1-2): 93-105.
[http://dx.doi.org/10.1016/0009-2797(90)90061-Q] [PMID: 2108810]

[106]   García de Llasera MP, Olmos-Espejel JJ, Díaz-Flores G, Montaño-Montiel A. Biodegradation of benzo(a)pyrene by two freshwater microalgae Selenastrum capricornutum and Scenedesmus acutus: a comparative study useful for bioremediation. Environ Sci Pollut Res Int 2016; 23(4): 3365-75.
[http://dx.doi.org/10.1007/s11356-015-5576-2] [PMID: 26490911]

[107]   Wang Q, Zhang S, Li Y, Klassen W. Potential Approaches to Improving Biodegradation of Hydrocarbons for Bioremediation of Crude Oil Pollution. J Environ Prot (Irvine Calif) 2011; 2(1): 47-55.
[http://dx.doi.org/10.4236/jep.2011.21005]

[108]   Purwanti IF, Kurniawan SB, Ismail NI, Imron MF, Abdullah SRS. Aluminium removal and recovery from wastewater and soil using isolated indigenous bacteria. J Environ Manage 2019; 249: 109412.
[http://dx.doi.org/10.1016/j.jenvman.2019.109412] [PMID: 31445374]

[109]   Abatenh E, Gizaw B, Tsegaye Z, Wassie M. The Role of Microorganisms in Bioremediation- A Review. Open Journal of Environmental Biology 2017; 2(1): 038-46.
[http://dx.doi.org/10.17352/ojeb.000007]

[110]   Kadir AA, Abdullah SRS, Othman BA, *et al.* Dual function of Lemna minor and Azolla pinnata as phytoremediator for Palm Oil Mill Effluent and as feedstock. Chemosphere 2020; 259: 127468.
[http://dx.doi.org/10.1016/j.chemosphere.2020.127468] [PMID: 32603966]

[111]   Lebeau T, Jézéquel K, Braud A. Bioaugmentation-assisted phytoextraction applied to metal-contaminated soils: state of the art and future prospects.Microbes and Microbial Technology. New York, NY: Springer New York 2011; pp. 229-66.
[http://dx.doi.org/10.1007/978-1-4419-7931-5_10]

[112]   Abatenh E, Gizaw B, Tsegaye Z, Wassie M. Review Article Application of microorganisms in bioremediation-review. J Environ Microbiol 2017; 1(1): 2-9.

[113]   Mnif I, Mnif S, Sahnoun R, *et al.* Biodegradation of diesel oil by a novel microbial consortium: comparison between co-inoculation with biosurfactant-producing strain and exogenously added biosurfactants. Environ Sci Pollut Res Int 2015; 22(19): 14852-61.
[http://dx.doi.org/10.1007/s11356-015-4488-5] [PMID: 25994261]

[114]   Dzionek A, Wojcieszyńska D, Guzik U. Natural carriers in bioremediation: A review. Electron J Biotechnol 2016; 23: 28-36.
[http://dx.doi.org/10.1016/j.ejbt.2016.07.003]

[115]   da Silva LJ, Alves FC, de França FP. A review of the technological solutions for the treatment of oily sludges from petroleum refineries. Waste Manag Res 2012; 30(10): 1016-30.
[http://dx.doi.org/10.1177/0734242X12448517] [PMID: 22751947]

[116]   Atlas RM. Microbial degradation of petroleum hydrocarbons: an environmental perspective. Microbiol Rev 1981; 45(1): 180-209.
[http://dx.doi.org/10.1128/mr.45.1.180-209.1981] [PMID: 7012571]

[117]   Sakthipriya N, Doble M, Sangwai JS. Bioremediation of costal and marine pollution due to crude oil using a microorganism *Bacillus subtilis*. Procedia Eng 2015; 116(1): 213-20.
[http://dx.doi.org/10.1016/j.proeng.2015.08.284]

[118]   Leewis MC, Reynolds CM, Leigh MB. Long-term effects of nutrient addition and phytoremediation on diesel and crude oil contaminated soils in subarctic Alaska. Cold Reg Sci Technol 2013; 96: 129-37.

[http://dx.doi.org/10.1016/j.coldregions.2013.08.011] [PMID: 24501438]

[119] Cassidy DP, Srivastava VJ, Dombrowski FJ, Lingle JW. Combining in situ chemical oxidation, stabilization, and anaerobic bioremediation in a single application to reduce contaminant mass and leachability in soil. J Hazard Mater 2015; 297: 347-55.
[http://dx.doi.org/10.1016/j.jhazmat.2015.05.030] [PMID: 26093352]

[120] Perini BLB, Bitencourt RL, Daronch NA, dos Santos Schneider AL, de Oliveira D. Surfactant-enhanced *in-situ* enzymatic oxidation: A bioremediation strategy for oxidation of polycyclic aromatic hydrocarbons in contaminated soils and aquifers. J Environ Chem Eng 2020; 8(4): 104013.
[http://dx.doi.org/10.1016/j.jece.2020.104013]

[121] *In Situ* and *Ex Situ* Biodegradation Technologies for Remediation of Contaminated Sites (Engineering Issue). EPA/625/R-06/015, 2006. https://clu-in.org/download/contaminantfocus/dnapl/Treatment_ Technologies/epa_2006_engin_issue_bio.pdf

[122] Benyahia F, Abdulkarim M, Zekri A, Chaalal O, Hasanain H. Bioremediation of crude oil contaminated soils. A black art or an engineering challenge? Proc Safety Environ Protect 2005; 83(4): 364-70.
[http://dx.doi.org/10.1205/psep.04388]

[123] Alvim GM, Pontes PP. Aeration and sawdust application effects as structural material in the bioremediation of clayey acid soils contaminated with diesel oil. Int Soil Water Conserv Res 2018; 6(3): 253-60.
[http://dx.doi.org/10.1016/j.iswcr.2018.04.002]

[124] Beškoski VP, Gojgić-Cvijović G, Milić J, *et al.* Ex situ bioremediation of a soil contaminated by mazut (heavy residual fuel oil) – A field experiment. Chemosphere 2011; 83(1): 34-40.
[http://dx.doi.org/10.1016/j.chemosphere.2011.01.020] [PMID: 21288552]

[125] Canet R, Birnstingl JG, Malcolm DG, Lopez-Real JM, Beck AJ. Biodegradation of polycyclic aromatic hydrocarbons (PAHs) by native microflora and combinations of white-rot fungi in a coal-tar contaminated soil. Bioresour Technol 2001; 76(2): 113-7.
[http://dx.doi.org/10.1016/S0960-8524(00)00093-6] [PMID: 11131793]

[126] Lin TC, Pan PT, Cheng SS. Ex situ bioremediation of oil-contaminated soil. J Hazard Mater 2010; 176(1-3): 27-34.
[http://dx.doi.org/10.1016/j.jhazmat.2009.10.080] [PMID: 20053499]

[127] Subashchandrabose SR, Venkateswarlu K, Venkidusamy K, Palanisami T, Naidu R, Megharaj M. Bioremediation of soil long-term contaminated with PAHs by algal–bacterial synergy of *Chlorella* sp. MM3 and *Rhodococcus wratislaviensis* strain 9 in slurry phase. Sci Total Environ 2019; 659: 724-31.
[http://dx.doi.org/10.1016/j.scitotenv.2018.12.453] [PMID: 31096402]

[128] Vadillo Gonzalez S, Johnston E, Gribben PE, Dafforn K. The application of bioturbators for aquatic bioremediation: Review and meta-analysis. Environ Pollut 2019; 250: 426-36.
[http://dx.doi.org/10.1016/j.envpol.2019.04.023] [PMID: 31026689]

[129] dos Santos V, Sabirova J, Timmis KN, Yakimov MM, Golyshin PN. Alcanivorax borkumensis.Handbook of Hydrocarbon and Lipid Microbiology. Berlin, Heidelberg: Springer Berlin Heidelberg 2010; pp. 1265-88.
[http://dx.doi.org/10.1007/978-3-540-77587-4_89]

[130] Zampolli J, Collina E, Lasagni M, Di Gennaro P. Biodegradation of variable-chain-length n-alkanes in *Rhodococcus opacus* R7 and the involvement of an alkane hydroxylase system in the metabolism. AMB Express 2014; 4(1): 73.
[http://dx.doi.org/10.1186/s13568-014-0073-4] [PMID: 25401074]

[131] Dai X, Lv J, Yan G, Chen C, Guo S, Fu P. Bioremediation of intertidal zones polluted by heavy oil spilling using immobilized laccase-bacteria consortium. Bioresour Technol 2020; 309(April): 123305.
[http://dx.doi.org/10.1016/j.biotech.2020.123305] [PMID: 32325376]

[132] Gentili AR, Cubitto MA, Ferrero M, Rodriguéz MS. Bioremediation of crude oil polluted seawater by a hydrocarbon-degrading bacterial strain immobilized on chitin and chitosan flakes. Int Biodeterior Biodegradation 2006; 57(4): 222-8.
[http://dx.doi.org/10.1016/j.ibiod.2006.02.009]

[133] Catania V, Lopresti F, Cappello S, Scaffaro R, Quatrini P. Innovative, ecofriendly biosorbent-biodegrading biofilms for bioremediation of oil- contaminated water. N Biotechnol 2020; 58: 25-31.
[http://dx.doi.org/10.1016/j.nbt.2020.04.001] [PMID: 32485241]

[134] Sakaya K, Salam DA, Campo P. Assessment of crude oil bioremediation potential of seawater and sediments from the shore of Lebanon in laboratory microcosms. Sci Tot Environ 2019; 660: 227-35.
[http://dx.doi.org/10.1016/j.scitotenv.2019.01.025]

[135] Fernández-Álvarez P, Vila J, Garrido JM, Grifoll M, Feijoo G, Lema JM. Evaluation of biodiesel as bioremediation agent for the treatment of the shore affected by the heavy oil spill of the Prestige. J Hazard Mater 2007; 147(3): 914-22.
[http://dx.doi.org/10.1016/j.jhazmat.2007.01.135] [PMID: 17360115]

[136] Gholami-Shiri J, Mowla D, Dehghani S, Setoodeh P. Exploitation of novel synthetic bacterial consortia for biodegradation of oily-sludge TPH of Iran gas and oil refineries. J Environ Chem Eng 2017; 5(3): 2964-75.
[http://dx.doi.org/10.1016/j.jece.2017.05.056]

[137] Cameotra SS, Singh P. Bioremediation of oil sludge using crude biosurfactants. Int Biodeterior Biodegradation 2008; 62(3): 274-80.
[http://dx.doi.org/10.1016/j.ibiod.2007.11.009]

[138] Hu G, Li J, Zeng G. Recent development in the treatment of oily sludge from petroleum industry: A review. Elsevier B.V. 2013; Vol. 261.

[139] Biswal BK, Tiwari SN, Mukherji S. Biodegradation of oil in oily sludges from steel mills. Bioresour Technol 2009; 100(4): 1700-3.
[http://dx.doi.org/10.1016/j.biortech.2008.09.037] [PMID: 18986804]

[140] Mohan SV, Chandrasekhar K. Self-induced bio-potential and graphite electron accepting conditions enhances petroleum sludge degradation in bio-electrochemical system with simultaneous power generation. Bioresour Technol 2011; 102(20): 9532-41.
[http://dx.doi.org/10.1016/j.biortech.2011.07.038] [PMID: 21865036]

[141] Ramaswamy B, Kar DD, De S. A study on recovery of oil from sludge containing oil using froth flotation. J Environ Manage 2007; 85(1): 150-4.
[http://dx.doi.org/10.1016/j.jenvman.2006.08.009] [PMID: 17064842]

[142] Van Hamme JD, Odumeru JA, Ward OP. Community dynamics of a mixed-bacterial culture growing on petroleum hydrocarbons in batch culture. Can J Microbiol 2000; 46(5): 441-50.
[http://dx.doi.org/10.1139/w00-013] [PMID: 10872080]

[143] Pinheiro BCA, Holanda JNF. Processing of red ceramics incorporated with encapsulated petroleum waste. J Mater Process Technol 2009; 209(15-16): 5606-10.
[http://dx.doi.org/10.1016/j.jmatprotec.2009.05.018]

[144] Mishra S, Jyot J, Kuhad RC, Lal B. In situ bioremediation potential of an oily sludge-degrading bacterial consortium. Curr Microbiol 2001; 43(5): 328-35.
[http://dx.doi.org/10.1007/s002840010311] [PMID: 11688796]

[145] Srinivasarao Naik B, Mishra IM, Bhattacharya SD. Biodegradation of Total Petroleum Hydrocarbons from Oily Sludge. Bioremediat J 2011; 15(3): 140-7.
[http://dx.doi.org/10.1080/10889868.2011.598484]

[146] Roy A, Dutta A, Pal S, et al. Biostimulation and bioaugmentation of native microbial community accelerated bioremediation of oil refinery sludge. Bioresour Technol 2018; 253: 22-32.
[http://dx.doi.org/10.1016/j.biortech.2018.01.004] [PMID: 29328931]

[147] Sari GL, Trihadiningrum Y, Wulandari DA, Pandebesie ES, Warmadewanthi IDAA. Compost humic acid-like isolates from composting process as bio-based surfactant: Properties and feasibility to solubilize hydrocarbon from crude oil contaminated soil. J Environ Manage 2018; 225: 356-63.
[http://dx.doi.org/10.1016/j.jenvman.2018.08.010] [PMID: 30119010]

[148] Kimes NE, Callaghan AV, Aktas DF, *et al.* Metagenomic analysis and metabolite profiling of deep–sea sediments from the Gulf of Mexico following the Deepwater Horizon oil spill. Front Microbiol 2013; 4(MAR): 50.
[http://dx.doi.org/10.3389/fmicb.2013.00050] [PMID: 23508965]

[149] Gerhardt P, Wood W A, Krieg N R, Murray R. Methods for General and Molecular Bacteriology American Society for Microbiology, Washington, D.C., 1994: p. 791.

CHAPTER 11

Bioremediation of Hydrocarbons

Grace N. Ijoma[1], Weiz Nurmahomed[1], Tonderayi S. Matambo[1], Charles Rashama[1] and Joshua Gorimbo[1,*]

[1] *Institute for the Development of Energy for African Sustainability (IDEAS) Research Unit, University of South Africa (UNISA), Florida Campus, Private Bag X6, Johannesburg 1710, South Africa*

Abstract: Hydrocarbons are a common contaminant in both terrestrial and aquatic ecological systems. This is most likely due to the widespread use of hydrocarbons as everyday energy sources and precursors in the majority of chemical manufacturing applications. Because of their physical and chemical properties, most hydrocarbons in the environment are resistant to degradation. Although several derivatives are classified as xenobiotics, their persistence in the environment has induced microorganisms to devise ingenious strategies for incorporating their degradation into existing biochemical pathways. Understanding these mechanisms is critical for microbial utilization in bioremediation technologies. This chapter focuses on recalcitrant and persistent hydrocarbons, describing the reasons for their resistance to biodegradation as well as the effects on ecological systems. Furthermore, aerobic and anaerobic degradation pathways, as well as ancillary strategies developed by various microorganisms in the degradation of hydrocarbon pollutants, are discussed.

Keywords: Biodegradation, Fortuitous degradation, Anaerobic degradation, Bioattenuation, Biostimulation, Bioaugmentation, Bioavailability.

INTRODUCTION

The most commonly occurring forms of hydrocarbons in nature are crude oil, coal, and natural gas, which are referred to as fossil fuels. Additionally, there are seemingly limitless variations of saturated and unsaturated hydrocarbons derived from these natural hydrocarbons. Other hydrocarbons arise from processing naturally occurring hydrocarbon forms into domestic, agricultural, and industrial end products. Hydrocarbon processing generates hydrocarbon derivatives and modifies chemical structures by introducing elements such as sulfur, chlorine, bromine, and many more, with the intention of improving structural bonds, incre-

* **Corresponding author Joshua Gorimbo:** Institute for the Development of Energy for African Sustainability (IDEAS) research unit, University of South Africa (UNISA), Florida Campus, Private Bag X6, Johannesburg 1710, South Africa; E-mail: joshuagorimbo@gmail.com

Inamuddin (Ed.)

asing compound saturation, and transforming aromatic branched hydrocarbons. Examples of such derivative compounds include asphaltenes (phenols, fatty acids, porphyrin, ketones, and esters) and resins (pyridines, quinolones, sulfoxides, amides, and carbazoles). Other notable examples include benzene, toluene, alkyl ketones, halogenated heterocyclics, nitroaromatics, styrenes, chorophenols, chloroaromatic hydrocarbons, polychlorinated biphenyls (PCBs), polycyclic aromatic hydrocarbons (PAHs), and several pesticides. Although these compounds are applied in several aspects of human endeavors, their persistence in the environment and recalcitrance to degradation [1] prove detrimental to all ecological systems. More concerns have been raised, especially with established data that demonstrate that hydrocarbon derivatives exhibit various carcinogenic, teratogenic, and neurotoxic properties [2, 3].

The impact of environmental pollution from these hydrocarbon-derived compounds (HCs) is dependent on the location, concentration, and composition [4]. In the terrestrial environment, mobile and viscous HCs tend to absorb into the soil, often penetrating several strata and contaminating groundwater [5, 6]. These compounds are also capable of mixing with the soil, affecting permeability and porosity [1], thereby changing the structure and composition of the soil with impacts readily demonstrated in the changed ratios of C/N, C/P. Moreover, there are increased concentrations of toxic elements (lead, nickel, and vanadium), as well as changes in pH, salinity, and conductivity of the soil [7]. These changes damage the soil and environment [8]. Several studies have also shown that HCs impact the microbial population, often causing irreversible changes to the community composition and consequently the enzyme systems of the soil usually, through substrate stimulation, catabolite repression, and inhibitory actions [5]. Crops that are cultivated on these contaminated soils and water are exposed to deleterious conditions, including a reduction in germination rates and fertility. Furthermore, growing plants, through biosorption and bioaccumulation, experience toxemia and sometimes mutagenic effects that can be passed onto their off springs [9]. Interactions along the food chain imply that these effects are later transferred to humans with other health consequences. Moreover, in terrestrial environments, the burning of hydrocarbons as fuels pollute the air, causing inhalation and skin-related health challenges [10, 11]. Combustion of HCs increases the atmospheric concentrations of nitrous oxides and sulfur dioxide which are precursors of acid rain. Acid rain in the aquatic environment reduces the water pH (acidification) which negatively affects organisms present in this ecosystem, such as fish and coral reefs [12].

Hydrocarbon pollution in the aquatic environment, although to a certain extent is mitigated by dilution, is no less consequential in comparison to the terrestrial environments. Pollution in aquatic environments is exacerbated by HCs cleaning and recovery difficulties arising from several factors. A typical factor is tidal wave movement which promotes the mobility of HCs and their rapid spread from the original location of contamination [13]. Often it takes years if ever, to restore the environment to near-pristine conditions.

The microbial population is by far the greatest in number in all environments, and their role in all food chains and the geochemical cycles is considered the most important. When contaminants such as HCs are introduced to an ecosystem, there is an immediate change in concentrations of several elements, generally reaching toxic levels for some. Microorganisms present in the polluted site, respond to these toxic stressors, in a variety of ways. Their most important characteristic that aids in the development of tolerance and adaptation along the food chain, is the rapid generational times common to bacteria, fungi, and algae. Moreover, they perform the important function of biodegrading contaminants and nutrient recycling that ensures an effective and efficient ecosystem [14]. Pollutant degradation is achieved using basically two alternative routes, aerobic and anaerobic pathways. However, these pathways have adaptive variations that have been evolved to cope with substrate changes and availability. Microorganisms have also developed strategies to handle substrates, especially within consortia interactions. This chapter will review the fundamental microbial pathways as well as the strategies including fortuitous degradation, synergistic, multi-phasic kinetics, cometabolism, and genetic exchange, applied by these microorganisms in the degradation of HCs pollutants.

The understanding of microbial mechanisms and strategies in the degradation of pollutants in general has been exploited in the development of bioremediation programs. The overarching goal of bioremediation is to effectively reclaim polluted soil and waste water using sustainable processes that apply organisms and their enzymes. In particular, the reclamation process for wastewater ensures a reduction in pressure on freshwater resources and maintains aquatic biodiversity. This chapter will further review recalcitrant and emerging xenobiotic hydrocarbons, with a focus on how microorganisms have evolved mechanisms to degrade these compounds as well as the exploitation of this knowledge in both *in situ* and *ex situ* bioremediation programs and the future of prospects in the treatment of these pollutants.

Hydrocarbon Pollution Effects on Macrobiota

Hydrocarbons penetrate the plant through the leaves, stomata and roots and adversely affect various metabolic activities including transpiration, respiration and photosynthetic rates. Moreover, the heavy metals often present in crude oil and petroleum derivatives affect the porphyrins in the pyrrolic structure of chlorophyll in the leaves, which plays an important role in electron transport necessary for the critical step of energy conversion during photosynthesis. Arellano *et al.* (2015) studied the effects of hydrocarbon pollution in the Amazon forest using hyperspectral satellite images and they observed several growth problems within the tropical forest flora. These problems included general vegetation stress, biochemical leaf alterations, lower level of chlorophyll content as a symptom of vegetation stress, increased foliar water content of vegetation at polluted sites as well as seepages in the rainforest [15]. Crude oil and petroleum derivatives coat the surfaces of water, soil, plant, and animals thus preventing efficient gaseous exchange. Previous studies have also demonstrated that the fumigation of cherry trees over one growing season with approximately 100ppbv propene caused deleterious changes in soluble protein concentration and glutathione reductase in leaf extracts following analysis [16]

In aquatic environments, hydrocarbons prevent the penetration of light and the exchange of gases in the water. By blocking light and gaseous exchange required for photosynthesis and respiration, oil on water surfaces cause plant deaths [17]. The absorption of these pollutants by plants, from contaminated soil and water, is transferred along the food chain to animals and humans. The most prevalent effect of toxicity due to hydrocarbon and associated heavy metals pollution to wildlife is oxidative stress leading to the development of various diseases in these animals. Isaksson (2010) performed a meta-analysis on oxidative stress and found that the oxidative damage caused a reduction in glutathione (GSH) and other enzymes [18]. Glutathione is an important antioxidant, that assists in the elimination of free radicals often associated with cell damage and weakened immunity. The increased concentration of PAH in aquatic environments has also been shown to fatally affect fish embryo development [19]. Moreover, in a recent review, researchers alluded to a causal link between increased toxicity of PAHs and endocrine disruption and tissue-specific toxicity in aquatic animals [20]. Studies have demonstrated that hydrocarbon pollution of the air has both acute and chronic effects on human health particularly within the upper respiratory tract [21] as well as the possible link to a variety of cancers. In the majority of cases, the extent of effect on humans is dependent on the extent of exposure and absorbed concentrations. Hydrocarbons are known to impact mental health and induce physical and physiological effects, with toxicity leading to genetic, immune and endocrine systems disorders [22].

Hydrocarbon Pollution Effects on Microbiota

Within the marine environment, hydrocarbon pollution threatens microbial biodiversity. Studies have shown the effects of bioaccumulation and the increased concentration of aliphatic hydrocarbons and persistent organic pollutants on the coral species and the symbiotic zooxanthellae algae found in the Persian Gulf region, an area known for the high crude oil exploration activities. The effects of this increase in HCs were demonstrated in the significant reduction in the zooxanthellae algae and chlorophyll levels in the reefs. Such a reduced population of the primary producers within this food chain and ecosystem subsequently affects the other inhabitants of the environment [12]. Studies on the long-term effect of HCs contaminated soil on microalgae, other microorganisms and soil enzymes (dehydrogenase and urease) showed a decline in the algal population but seemingly had no effect on the other microorganisms and enzymes, indicating higher susceptibility to increased toxicity by the microalgae in comparison to other microbes. This allowed the researchers to propose a possibility for the use of microalgae as an indicator of increased HCs pollution in soil [23]. Toxic effects of HCs on algal metabolism can be classified into two categories: metabolic impact due to the coating of the organisms and secondly, consequences due to HCs uptake and its subsequent disruption of cellular metabolism. Both toxic mechanisms have significant effects on CO_2 absorption, and reduction in light penetration cumulatively affecting respiration and photosynthesis and in general, negatively impacting cell metabolism [24]. In hypersaline aquatic environments, hydrocarbonoclastic bacteria require the presence of halophilic photoautotrophs to achieve efficient hydrocarbon biodegradation as their photosynthetic activity compensates for the absence of oxygen caused by this hypersalinity [25].

Both fungi and bacteria, by virtue of their important role as decomposers and recyclers in the food chain, tend to be the least affected by HCs. However, the degradation process for microorganisms in terrestrial environments (soil sediments) is easier compared to aquatic environments. This is because there are increased concentrations of other participating microorganisms in the former [26]. Pertinently, the greatest negative impacts of HCs pollution in any given environment on these microorganisms are reductions in microbial biomass, species richness, evenness and phylogenetic diversity, often resulting in the dominance of some species who rapidly respond to changes in nutrient concentrations and toxicities [27 - 29]. Studies highlight the ability of fungi to colonise HCs polluted soils, for example, the arbuscular mycorrhizal fungi, known for their interspecific interactions with plants roots and other soil microorganisms [30]. Another study investigated the effects of various concentrations of crude oil on fungal populations in soil. Species of fourteen fungal genera were isolated from the soils and included *Alternaria, Aspergillus, Candida, Cephalosporium, Cladosporium, Fusarium, Geotrichum, Mucor, Penicillium, Rhizopus, Rhodotolura, Saccharomyces, Torulopsis* and *Trichoderma*. The authors demonstrated that higher concentrations of crude oil negatively impacted fungal diversity, however, a few members of the fungal population actually thrived under these negative conditions [31]. Recent studies have demonstrated the effectiveness of the filamentous fungus in the removal and biodegradation of different petroleum hydrocarbons [32].

Bacteria have a predominant role in the geochemical nutrient cycling, with the decomposition of complex compounds and the release of these nutrients back into the environment. Some notable genera have demonstrated a propensity for the degradation of various branched-chained, straight-chained, saturated and even polycyclic aromatic hydrocarbons often using these compounds as their sole carbon source. These include, *Alcanivorax*, *Cycloclasticus* and *Thalassolituus* [33]. An *in situ* study investigated the effect of hydrocarbon on soil nitrification, but the same study also included *in vitro* observation of ammonia-oxidizing bacteria, where the findings showed that these bacteria are capable of incomplete oxidation of hydrocarbons. Such transformation reactions are a consequence of a competitive rate-limiting enzyme, ammonia monooxygenase which is the first enzyme in the nitrification process [34]. This is evident in the often associated decline in richness and phylogenetic diversity linked to the disruption of the nitrogen cycle, in these polluted environments, accompanied by a reduction in species and functional genes involved in nitrification [28, 35]. Using 16S rRNA sequencing an increased predominance of *Saccharibacteria* and *Alcanivorax*, in low to moderate concentrations was observed. However, *Saccharibacteria* and *Desulfuromonas* were prevalent in high concentrations of crude oil. A generalized observation made in the same study was that nitrogen cycling bacteria are sensitive to the total nitrogen, total phosphorus, ammonia nitrogen, nitrate nitrogen and pH of the soil. Sulfur cycling bacteria are sensitive to aromatic hydrocarbons, saturated hydrocarbons, and asphaltene in soil.

The Fate of Hydrocarbon Pollution in the Environment

Weathering, Physical and Chemical Interactions with the Terrestrial Environment

The fate of hydrocarbon contamination within a given ecological system is rather complex. It undergoes both abiotic and biotic interaction with varying transformations in the natural attenuation processes. The focus of this section will be to describe the abiotic influences on hydrocarbons once they contaminate the soil. These pollutants experience weathering and biodegradation mostly in tandem. Natural weathering includes physical, chemical and biological processes. Once hydrocarbons contaminate soil, they change the composition, toxicity, availability and distribution (dispersion) within the ecological system. This happens through, volatilisation, adsorption, dissolution, oxidation, hydrolysis, chemical transformation and photodegradation. The extent of physical and chemical weathering in attenuating hydrocarbon contamination determines the extent of recalcitrance compounds that continue to persist in the environment, and as such would likely require further microbial degradation. Often the normal alkanes in the range of C_{10} to C_{26} are viewed as the most readily degraded, but low-molecular-weight aromatics, such as benzene, toluene and xylene, which are among the toxic compounds found in petroleum, are also very readily degraded. The more complex structures are least likely to be susceptible to weathering and biodegradation; meaning that fewer microorganisms can degrade these complex hydrocarbons. This complexity in hydrocarbon structures is a consequence of high numbers of methyl branched substituents or condensed aromatic rings, making for poor degradation rates, for example, asphaltene is considered to be one of the most recalcitrant hydrocarbons [36]. Table 1 reports some of the common recalcitrant hydrocarbons.

Table 1. Examples of recalcitrant hydrocarbons.

Name	Molecular Formula	*Estimated Half-Lives (in Days)
Naphthalene	$C_{10}H_8$	5.66
Acenaphthene	$C_{12}H_{10}$	18.77
Acenaphthylene	$C_{12}H_8$	30.7
Anthracene	$C_{14}H_{10}$	123
Phenanthrene	$C_{14}H_{10}$	14.97
Fluorene	$C_{13}H_{10}$	15.14
Fluoranthene	$C_{16}H_{10}$	191.4
Benzo[a]anthracene	$C_{18}H_{12}$	343.8
Chrysene	$C_{18}H_{12}$	343.8
Pyrene	$C_{16}H_{10}$	283.4
Benzo[a]pyrene	$C_{20}H_{12}$	421.6
Benzo[b]fluoranthene	$C_{20}H_{12}$	284.7
Benzo[k]fluoranthene	$C_{20}H_{12}$	284.7
Dibenzo[a,h]anthracene	$C_{22}H_{14}$	511.4
Benzo[g,h,i]perylene	$C_{22}H_{12}$	517.1
Indole[1,2,3-cd]pyrene	$C_{22}H_{12}$	349.2

*Estimation done using BioHCwin software v1.01 on EPI Suite software develop by [43].

When contamination is on the top layer, some hydrocarbons especially the short chain alcohols and alkanes easily undergo volatilization [37, 38]. For example, Sanscartier *et al.* (2009) observed that in the bioremediation of weathered medium to high molecular weight hydrocarbons in cold regions, although, low moisture and temperature are factors to consider in the degradation process, volatilization was adjudged to be more pertinent to the natural attenuation identified [39]. It is also possible that some hydrocarbons adsorb to surfaces of biomasses present in the soil. This adsorption often limits the hydrocarbon's mobility and reduces further contamination in soil. However, it should be noted that earthworms can hinder the binding of hydrocarbons and in general organic contaminants in soil, and thereby promote dispersion and bioavailability of these organic pollutants to hydrocarbon-degrading microorganisms [40]. The extent of dissolution of hydrocarbons in the soil is premeditated upon the moisture content within the environment and this miscibility in water improves most chemical reactions that could take place including oxidation, hydrolysis and other chemical transformations. On land the photodegradation of hydrocarbons is mediated by sunlight, leading to the transformation of these contaminants in the environment, especially with PAHs [41, 42].

Weathering, Physical and Chemical Interactions within the Terrestrial Environment

In the aquatic environment, oil spillage is the most common accidents that introduce hydrocarbon contamination. One of the most cited examples of such devastating oil spillage occurred on March 23, 1989, when at least 11 million gallons of crude oil was spilled from the tanker, Exxon Valdez, in Prince William Sound, Alaska; even 25 years after the incident the clean-up is still on-going [44]. A significant amount of knowledge and insights gained into the behaviour of oils after spillage was as a result of this incident. Water is a more sensitive medium than soil when contaminated by petroleum spillage, however, spills on the surface of the water are easier to clean up due to the hydrophobic nature of oils. Furthermore, weather conditions and turbulence affect the surface of the water and these surface oils. Weathering including photodegradation, tidal and wave actions in aquatic environments, promote the chemical breakdown of oil molecules. These weathering processes include evaporation (volatilization), dissolution, dispersion, photochemical oxidation, water-in-oil emulsification, adsorption onto suspended particulate materials, sinking and sedimentation.

In the first 48 hours of oil spillage, 15 – 20% of the initial weathering can be achieved due to volatilization of low boiling point compounds. These are mostly low- to medium-weight crude oil components (aliphatic hydrocarbons) with less than 12 carbon atoms; some aromatic hydrocarbons such as benzene, toluene, xylene and methyl-substituted naphthalenes can also be lost in this way [45]. Several factors influence evaporative loss such as the composition of the hydrocarbon, the surface area covered, physical properties, wind velocity, air and sea temperatures, weather conditions, and intensity of solar radiation. After this

48-hour period, the rate of evaporation slows down leaving mostly fractions of the hydrocarbons that are rich in metals (*e.g.* nickel and vanadium), as well as waxes and asphaltenes. The decrease in evaporation also leaves components of the hydrocarbons that have higher specific gravity and viscosity (heavy oils). During this time also, undisturbed as well as slightly dissolved oils present in the top layer of water also undergo photodegradation to the extent of solar penetration [46]. The dissolution of oils is a less important weathering process in the aquatic environment. Most fractions of crude oil are insoluble in water, with few exceptions such as unsaturated and aromatic hydrocarbons which are slightly soluble, for example, the solubility of toluene is ~500mg/L. It is important to note that most of the disappearance of an oil slick that occurs is a consequence of dispersion and the break-up of the heavy oils into smaller particles from the surface to the water column and its sinking as sediments to the water bed [47]. Dispersion is influenced by surface turbulence, and increased turbulence also promotes greater dispersion. Moreover, once the oils settle onto the water bed, adsorption to other organic compounds can occur allowing for the adhesion of microorganisms and the formation of slime that facilitates the process of microbial degradation [46].

Reasons for Hydrocarbon Recalcitrance to Biodegradation

When physical and chemical conditions in any ecological system, do not lead to the degradation of hydrocarbons, they persist in the environment and can only be degraded by microorganisms. Microbial degradation is achieved through metabolic pathways operating within an organism or in the combination with organisms working in a consortium. This understanding is the basis of bioremediation programs. Recalcitrant compounds including hydrocarbons are able to resist biodegradation, sometimes due to the absence of organisms in the contaminated site that can degrade such compounds. Table **2** gives an overview of the most common reasons for recalcitrance in hydrocarbons.

Table 2. Common reasons for recalcitrance to biodegradation in hydrocarbons.

Reasons	Description	Examples of recalcitrant hydrocarbons
Lack of oxygenated functional groups	Associated with kinetic inertness, consequently, there are few enzymes that can overcome the problem. An example is the anaerobic biodegradation of benzene, lacking oxygen containing functional groups. This is generally regarded as extremely difficult (aromatic rings hard to break), although occasionally in some anaerobic consortia its anaerobic degradation has been observed, including under methanogenic conditions.	Linear straight-chain alkanes, branched alkanes, cycloalkanes, benzenes

(Table 2) cont.....

Reasons	Description	Examples of recalcitrant hydrocarbons
Presence/absence of electron withdrawing groups/electron donating groups	Halogen groups interfere with ring cleavage. As a general rule increasing the number of electron withdrawing functional groups on a xenobiotic molecule tends to increase the ease of the initial reductive biotransformation of the molecule.	Bromopropane, methylene chloride, chloroform, tetrachloroethylene, carbon tetrachloride, chlorofluorocarbons, methyl bromide, chloroethane
Highly branched hydrocarbons	Historically branched alkane sulfonates were used in detergents and were slowly replaced with linear alkane sulfonates to facilitate aerobic degradation in wastewater treatment systems. Branching interferes with degradation due to steric hindrance, thus substrates will not readily fit into active sites of common degradative enzymes	Alkane sulfonate, methyl-tertbutyl ether (MTBE)
Bioavailability	Many hydrocarbons can be characterized as highly non-polar compounds with extremely low aqueous solubilities. Compounds may occur predominantly in the solid phase or are tightly adsorbed to soil, sediment, or sludge particles. In such cases, the kinetic limitations of dissolution and or desorption play an important role in dictating the biodegradation rates.	PCBs, PAHs, dioxins, asphaltenes
Recently synthesized hydrocarbons	Microorganisms have not developed enzyme machinery/pathways necessary to degrade these hydrocarbon derivatives	Resins, asphalthenes, trichloroethylene, perchloroethylene, atrazine, carbaryl, aiazinon, glycophosphate, parathion

Ecotoxicology: Consortia Stress Responses, Tolerance and Adaptation

Rate-limiting Nutrients: Changes in Nitrogen Flux

The unicellular nature of microorganisms implies that survival in nature is dependent on consortia interactions. Such interactions promote efficient territorial colonisation, substrate management and the degradation of complex chemical compounds, as well as rapid responses and adaptations to changes in environmental conditions [48]. The spillage and increased concentrations of hydrocarbons is such an example of an environmental change and trigger a cascade of microbial responses that registers with single species and the surrounding complex microbial communities. These microbial communities utilize the increased quantities of oil as carbon source; however, there is a concomitant demand for inorganic nitrogen, a rate-limiting nutrient necessary for metabolism and biomass production [49]. The implication of this demand is that the process of nitrification in the presence of environmental stressors such as hydrocarbon contamination is significantly diminished to the extent that the behaviour and trend of nitrifying microorganisms have been exploited as bio-indicators to assess the health condition of environments and ecosystems [49]. Previously, studies have demonstrated that the restoration of this lost activity of nitrification can take up from 2 to 20 years of recovery depending on the type of ecosystems that experienced the hydrocarbon contamination [49 - 51]. However, denitrifying microorganisms, because of their heterotrophic mode of nutrition, tend to be more resistant to hydrocarbon toxicity [52, 53]. This was observed with studies were done measuring sediments' potential denitrification in the Chandeleur Islands, where a series of barrier islands were impacted by the oil spill in the Gulf of Mexico. The researchers were able to compare their results to nearby coastal Louisiana sediments, where data was available prior to oil spillage contamination. Reports appear to suggest very little change in the denitrification capacities between both sets of data [54]. It was also noted that a consequence of toxicity was an increase in oxygen demand and concomitant increase in hydrogen sulphide concentration, which also has inhibitory effects on various steps in the nitrogen cycle. Generally, in hydrocarbon polluted marine environments, the decrease in the levels of nitrogen and phosphorus is more pronounced as compared to terrestrial soil environments [55]. The deficiency in these nutrients in freshwater wetlands is rather similar to terrestrial environment as there is also a demand from the inhabiting plants which metabolically utilise these nutrients [56].

Changes in Microbial Population Dynamics

Microbial consortia adaptation is critical in the development of resistance to extreme environmental pollution. However, initial high levels of toxicity as is common to oil spills, tend to affect even adapted microorganisms and will cause a decline in the population of both susceptible and some resistant communities of microorganisms [57]. Lim *et al.* (2016) observed that earthworms demonstrated susceptibility to oil contamination in soil of above 3% and at 1% oil contamination and that nearly all bacterial activity was inhibited [58]. Lahel *et al.* (2016) explain that exposure to hydrocarbon contamination provokes an adaptation response of selective genetic enrichment, which is demonstrated by an increase in the exchange of plasmids that encode the hydrocarbon catabolic genes within the population. Therefore, under favourable environmental conditions, including near ideal temperatures, nutrient balance and oxygen saturation, there is usually a period of acclimation to the increased toxicity, often with the associated decline in the population of susceptible communities of bacteria and fungi [59]. The end of the period of acclimation is signalled by the surge in biomass population of the most resistant species that continue the process of degradation of the pollutants. Such increases in the population of oil degrading microorganisms have been observed for *Alcanivorax* sp. and *Cycloclasticus pugetii*. However, in a recent study conducted on the Indian Marine environment after an oil spill, Neethu *et al.* (2019) used high throughput metagenomics analysis and instrumentation that incorporated metabolomics to capture signature variations among the microbial communities in sediments, water and laboratory enrichments [60]. They observed a contrary result to the previously held assertions of the prevalence of *Alcanivorax*, as their study showed that Pseudomonadales (specifically genus *Acinetobacter*) in oiled sediment and Methylococcales in oiled water outnumbered the relative abundance of *Alcanivorax* in response to hydrocarbon contamination. This study is important as it highlights the need to consider the unique dynamics of each polluted environment in determining the suitability of bioremediation approaches. Especially, when considering that often the existing microbial population in the polluted environment is not always capable or sufficient to efficiently degrade the pollutants, and it is usually necessary to introduce or add oil-degrading microorganisms in a strategy referred to as bioaugmentation.

Microbial Consortia Interactions Employed in the Degradation of Hydrocarbons

Degradation of substrates by microorganisms is premised on the availability of enzymes that can recognize the substrate and therefore catalyse its breakdown. Furthermore, most enzymes are highly specific in their reactions and as such have quite a narrow range of substrates that they can degrade. Moreover, microorganisms prefer to degrade substrates that require the least amount of enzymes or energy investments, therefore they selectively degrade the compounds that are the easiest to digest. If there are options of substrate variability in the environment, it is not likely that these organisms will select the one requiring more energy investment in its degradation. As such a selection would require a diauxic shift and a lag phase of growth or biomass production and time to produce the necessary enzymes. Nevertheless, some oxygenases (*e.g.* monoxygenases and dioxygenases) seemingly have a broad substrate specificity which allows them to catalyse compounds other than their primary substrate- needed for energy production and growth [61]. These enzymes may achieve varied results, leading to either complete (mineralisation) or incomplete (transformation) degradation.

Fortuitous Degradation

The introduction of hydrocarbons in the environment may be regarded as a situation where degradation could be fortuitously achieved. Fortuitous degradation of hydrocarbons happens because the compound closely resembles the compound (substrate) used primarily as a food source by the microbial inhabitants of the given environment, as such the hydrocarbon can serve as an electron donor in primary metabolism. The outcome of such a degradation process is complete mineralisation, where the hydrocarbon is incorporated as a carbon source within the microorganisms' biochemical pathway, leading to its utilisation for ATP production. This energy is used for metabolic activities and growth. Generally, such degradation will lead to the removal of the pollutant from the environment with the production of innocuous compounds such as CO_2 and water. One example is the complete mineralization of pyrethroid pesticide, cypermethrin, by *Micrococcus* sp., a bacterium that was isolated from soil, and demonstrated the capability of utilising cypermethrin as a source of carbon. This microorganism was able to degrade this compound by the hydrolysis of the ester linkage, to produce 3-phenoxybenzoate, leading to the loss of its insecticidal activity. The metabolism of phenoxybenzoate by diphenyl ether cleavage yielded protocatechuate and phenol [62]. Microbial monoxygenases have demonstrated ability in the transformation of Methyl *tert*-butyl ether (MTBE) to *tert*-Butyl alcohol (TBA) which is subsequently mineralised [63]. There are also several evidence of the fortuitous degradation of chlorinated aliphatic hydrocarbons [64 - 67] referred to as reductive dehalogenation. Such microbial-catalysed reactions are halogen atom replacement, with a hydrogen atom, within the structure of the hydrocarbon, usually with the net addition of two electrons to the organic compound. These degrading microorganisms utilise oxygenases produced, to metabolise substrates that hitherto served as a primary electron donor. Some aerobic microorganisms have demonstrated capabilities for the degradation of chlorinated ethenes [68 - 70]. The fortuitous oxidation of these chlorinated hydrocarbons tends to produce unstable chlorinated epoxide, that is readily broken down through abiotic processes. However, it requires the addition of oxygen serving as the electron acceptor and a growth substrate which will serve as the electron donor (hydrocarbon). Several hydrocarbons including methane, phenols and toluene, have demonstrated effectiveness as electron donors for the transformation of chlorinated hydrocarbons [64, 71]. Also, some studies have indicated that halogenated aliphatic hydrocarbons can be mineralized through anaerobic methanogenesis, particularly the tetrahalogenated methanes, ethanes and ethenes [72]. However, reductive biotransformation, rarely achieves mineralization of the pollutant and in this anaerobic condition, degradation requires cometabolism.

Cometabolism

Often, in the natural environment, complete metabolic breakdown of complex hydrocarbons is near impossible with a single species of microorganisms; therefore, it is usual that two or more species endeavour to breakdown the pollutant. The first species of microorganisms achieves incomplete degradation, possibly because, the hydrocarbon is recognised by the enzymatic machinery but it does not provide energy as such cannot be integrated into the biochemical pathway as a carbon source for energy production. For example, *Mycobacterium vaccae* oxidizes cyclohexane during growth on propane, other species use cyclohexanol [73]. Nonetheless, the hydrocarbon is transformed through a series of biochemical reactions, sometimes even with several species involved (cometabolic transformation). The process of transformation ceases in the absence of other catalytic enzymes that can catalyse further degradation. It is noteworthy, that intermediate compounds of transformation reactions are highly unstable and sometimes more toxic than the parent compound. This transformed toxic compounds are sometimes not processed further and build up in the environment, for example, chlorocatechols. It is also possible that toxic intermediates build up, even when they can be degraded, because the microbial-mediated reactions are slow, due to inadequate biomass of these microorganisms and consequently fewer enzymes compared to the rate of accumulation of the intermediate needing degradation. This has been observed with the carcinogenic compound, vinyl chloride, which often builds up during the degradation of trichloroethylene (TCE). The conversion of TCE to vinyl chloride is rapid, but the subsequent degradation of the transformed compound is usually very slow. Intermediate compounds may be cometabolised by other organisms that have the necessary enzymatic machinery to complete the mineralisation process. Cometabolism is the transformation of a substrate that is not required for growth by a microorganism. However, it will only occur in the presence of substrates that primarily support growth and/or other transformable compounds. The term cometabolism only applies when mineralisation is achieved for the compound by other microorganisms contributing to the process of degradation.

Synergism

The process of degradation in nature may also be achieved through the synergistic interactions of microorganisms. Sometimes, synergistic effects in a series of reactions is required, rather than a single reaction or the action of one microorganism to achieve rapid degradation [74]. In a study carried out by Dejonghe *et al.* (2003), to identify the roles of bacteria within a community in the degradation of linuron, they observed that *Variovorax* sp. strain WDL1 was able to use linuron as the sole source of carbon, nitrogen and energy [75]. However, in this mono-degradation set up by *Variovorax* sp. the rates were lower when compared to those achieved by the consortium. Pollutant degradation was increased by co-inoculation with four other members of the consortium. It was also ascertained that *Delftia acidovorans* WDL34 and *Comamonas testosteroni* WDL7 were responsible for the degradation of the intermediate 3,4-dichloroaniline, and *Hyphomicrobium sulfonivorans* WDL6 was the only strain able to degrade N, O-dimethylhydroxylamine. Studies with such unique insights are important in determining bioaugmentation strategies. Asemoloye *et al.* (2017) studied synergistic interactions between fungi and the root rhizophere bacteria for the degradation of lindane polluted soil, used to grow *Megathyrsus maximus* Jacq for a period of 3 months [76]. It was observed that the combined rhizospheric action of *M. maximus* Jacq and four fungi species, *Talaromyces atroroseus*, *Talaromyces purpurogenus*, *Yarrowia lipolytica* and *Aspergillus flavus* significantly increased the lindane degradation rate.

Multi-phasic Degradation

The initial presence of high concentrations of hydrocarbon pollutants poses toxicity that only a few members of any given population of microorganisms can tolerate. However, some organisms can and have demonstrated peculiarly such affinities for high concentrations of hydrocarbons. Such organisms thrive only under these harsh conditions, and activities will cease, once concentrations are low. The implication is that degradation is incomplete before acceptable pollutant levels are achieved in the environment. This phenomenon should not be confused with cometabolism, which involves the first participating population to transform the parent compound into a new compound recognised by the subsequent microbial population which then finally completes the breakdown. In multi-phasic degradation, degradation is of the same compound, but microbial population is able to break down this pollutant only at specific concentrations. This may be a consequence of the cells' internal regulatory mechanisms, for both the organisms that respond to high and those that will only respond at low concentrations. Some cells may require a higher concentration of the substance for sustenance while the other organism may only require the same compounds in minute concentrations. This phenomenon where one microbial population is active in the degradation only at high concentrations, and the other only at low concentrations are referred to as multi-phasic degradation. For example, three bacterial isolates were obtained from a highly saline and alkaline Lake site in South-western Oregon, these bacterial strains were similar in their degradation pathways for 2,4-dichlorophenoxyacetic acid but only different in the concentrations they could tolerate and degrade [77]. The biodegradation of dinitrophenols in soil was observed to occur through multiphasic mineralization kinetics, with slow and partial mineralization of 2,4-dinitrophenol, taking place at low concentrations (<20mg/kg) and conversion to methane under anaerobic conditions after a significant adaptation period. It was also observed that although the rate of conversion of 2,4-dintrophenol is slow; the biodegradation will occur as long as the initial concentration of dinitrophenol in soil or sediments does not exceed the level of ≈100 mg/kg. Above this level, dinitrophenol may be toxic to the degrading microorganisms [78]. The understanding of this type of interactive interplay within the population of microorganism is useful in customizing microbial consortia used in bioaugmentation strategy of bioremediation.

Genetic Exchange

The presence of hydrocarbon contamination is known to provoke adaptation responses that include selective genetic enrichment such as an increase in plasmids exchange among resident organisms. These plasmids encode catabolic genes necessary for the degradation of the hydrocarbon pollutants [59]; even under extreme environmental conditions such as those experienced in the Arctic and Antarctica [79 - 81]. These genes encode several metabolizing enzyme systems capable of degrading hydrocarbons and are important in aerobic degradation processes. However, the most predominant genes known are the alkane monooxygenase gene, *AlkB* and naphathalene dioxygenase gene, *Nah* associated with the first-step hydroxylases involved in the metabolism of alkanes and aromatic hydrocarbons. Moreover, naphathalene dioxygenase can add both atoms of molecular oxygen to the aromatic ring as a first step in the aerobic pathway of degradation, with the dihydric alcohol produced being converted to catechol [61, 82]

Of all soil microbes, Pseudomonads are the most highly developed with the ability to exchange genetic material with other species. Examples of such genetic exchange include; TOL plasmid for toluene degradation and several types of transposons. The complex structure of most hydrocarbon implies that several components within its structure can only be broken with different strategies, for example, the lipid characteristics require surfactant production to promote emulsification. Not all of the microorganisms present in the population may have the genes required to achieve emulsification, but it is possible to confer others with these abilities through horizontal gene transfer (HGT). Table **3** shows some previously isolated plasmids found in hydrocarbon bacteria, conferring the genetic ability to degrade hydrocarbon compounds. For example, *Franconibacter pulveris* strain DJ34 was isolated at the Duliajan oil fields in Assam, and its molecular characterisation, demonstrated various genes related to hydrocarbon degradation including, metal transport and resistance, dissimilatory nitrate, nitrite and sulphite reduction, chemotaxis and biosurfactant synthesis. However, a comparison of this strain with other *Franconibacter* spp. showed that these are not common features within the species implying that these genes are a consequence of environmental adaptation and genetic exchanges with other organisms in their unique environment [103]. In another instance, studies showed three hydrocarbon-degrading psychrotrophic bacteria that were isolated from hydrocarbon contaminated Arctic soils. It was demonstrated that two of the strains, identified as *Pseudomonas* spp., degraded mainly toluene and naphthalene (C_5 –C_{12}*n*-

alkanes) at the temperatures 5 and 25°C because both possessed the alkane (*alk*) and naphthalene (*nah*) catabolic pathways. Remarkably, one of the strains contained both a plasmid slightly smaller than the *P. oleovorans* OCT plasmid, which hybridized to an alkB gene probe, and a NAH plasmid similar to NAH7, demonstrating that both catabolic pathways, located on separate plasmids, can naturally coexist in the same bacterium [104]. Similarly, Liu *et al.* (2015) investigated the distribution of both *AlkB* and *Nah* catabolic genes in soils obtained from three zones of the Dagang Oilfield, Tianjin, China. Such environmental monitoring studies are important in expanding our understanding of the abilities of microorganisms and their potential application in bioremediation [105].

Table 3. Examples of hydrocarbon degrading plasmid and their host.

Strain	Plasmid	Substrate	References
Acinetobacter sp. DSM17874	OCT	C10 – C40, *n*-alkanes	[83]
Acinetobacter baumannii S30	pJES	Crude oil	[84]
Beijerinkia sp.	pKG2	Phenanthrene	[85]
Burkholderia sp. JS15	pRO1727	Benzene, Toluene, Ethylbenzene and Xylenes (BTEX)	[86]
Burkholderia strain R007		Naphthalene, phenanthrene	[87]
Cycloclasticus sp. 78-ME	p7ME01	Polyaromatic hydrocarbon	[88]
Pseudomonas fluorescens strain LP6a	pLP6a	Naphthalene anthracene, phenanthrene	[89]
Pseudomonas putida G7	NAH7	Naphthalene	[90]
Pseudomonas putida strains	NAH7	Naphthalene	[91]
Pseudomonas strain C18	pUC18	Dibenzothiophene naphthalene, phenanthrene	[92]
Pseudomonas putida NCIB9816	pDTG1	Naphthalene	[93]
Pseudomonas sp. GPo1	OCT	C5 – C12, *n*-alkanes	[94]
Pseudomonas sp. VI4.1 *Pseudomonas veronii* VI4T1	pAK5	Naphthalene and BTEX	[95]
Pseudomonas sp. Strain 112		Naphthalene	[96]
Rhodococcus sp. Q15	OCT	C12 – C32, *n*-alkanes	[97]
Sphingomonas aromaticivorans F1999	pNL	Phenanthrene	[98]
Sphingomonas strain Ks14	pKS14	Phenanthrene	[99]
Sphingomonas sp. HS362	p4	Phenanthrene	[100]
Staphylococcus sp. PN/Y	pPNY	Phenanthrene	[101]
Terrabacter sp. DBF63	pDBF1	Flourene	[102]

Genes involved in the degradation of hydrocarbons are employed as biomarkers, for the characterization of the extent of pollutants in soils, and in general the microbial ecology [106]. It is possible to identify and sequence these genes with the capacity of hydrocarbon degradation, even in the absence of any obvious strain specific discrimination applying techniques of PCR and hybridization. The characterisation of indigenous hydrocarbon-degrading bacteria is useful in assessing the intrinsic biodegradation potential of a given site. Usually, the PCR amplification products in the plasmid DNA of an organism are transformed using competent cells such as *E. coli*, thereby conferring the competent cells with properties present in the plasmid. They are grown on suitable media with hydrocarbon of interest and the catabolic activity is further assessed [107]. In the past, the common practice was to use *pahAc*, which encodes the α-subunit of PAH ring hydroxylating dioxygenase, as a functional marker gene [108 - 110]. However, there are known variations in the *pahAc* sequences in PAH-degrading bacteria that make it difficult to inclusively target PAH degraders with highly specific primers. This leads to an underestimation of these types of bacteria at contaminated sites [111]. This non-specificity of *pahAc* as a functional marker for PAH-degrading bacteria distorts the prediction of PAH degradation potential as well as accuracy in the assessment of bioremediation potential. In a recent study done by Liang *et al.* (2019), they proposed the use of *pahE* gene which encodes PAH hydratase-aldolase, as an alternative marker gene as opposed to the use of *pahAc* [112]. These researchers allude to *pahE* having a higher specificity and more genotype as well as resolution based on reads derived from Illumina sequencing platforms.

Mechanisms of Microbial Biodegradation of Hydrocarbons

The ability of microorganisms to thrive in hydrocarbon contaminated environments led to the understanding of biodegradation processes employed by these microbes. Copious reports demonstrate the abilities of bacteria [113], fungi [114] and even algae [115], in the degradation of various hydrocarbons in both soil and water [116, 117]. It is possible to classify these processes into three categories: deterioration of the pollutant [118], fragmentation of the polymers and assimilation of monomers during cellular respiration [119].

Deterioration is mostly associated with food grade hydrocarbons, and it is the result of the combination of both abiotic including air, moisture, UV light and temperature, and biotic factors where microbial influences are predominant [120, 121]. Catalytic activities cause the degradation, often in the case of fatty acids, particularly where a lipid-water interface exists and there is UV light as factor that leads to lipid rancidity [118]. The extent of hydrophobicity significantly influences the degradation rates of hydrocarbons and the degree varies among aliphatic, cyclic alkanes and PAHs [122]. Both the fragmentation and assimilation features in the biodegradation process are integral components of microbial pathways [123]. However, for a hydrocarbon to be incorporated into either the aerobic or anaerobic pathways, its functional groups must be activated as demonstrated in Fig. (1).

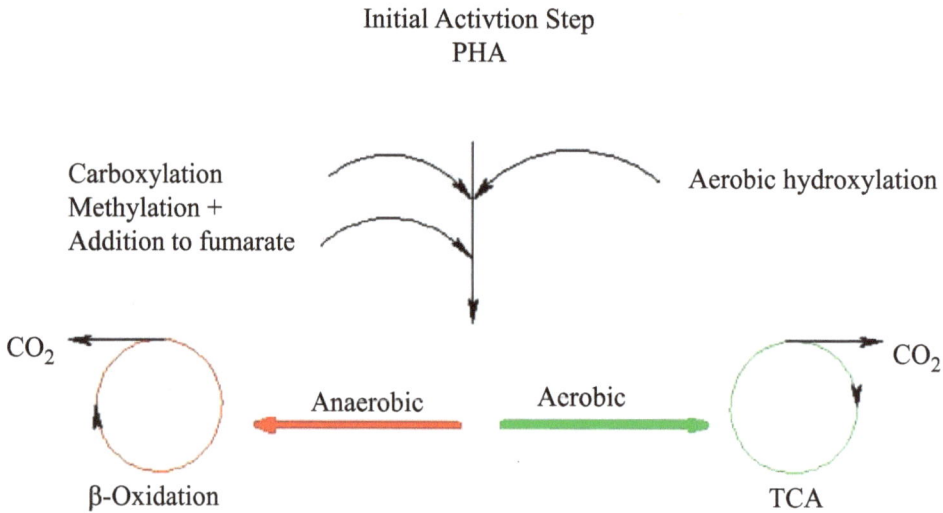

Fig. (1). Hydrocarbon activation in aerobic and anaerobic pathways, adapted from [124].

The assimilation process involves the uptake of the fragmented pollutants and the transformed hydrocarbons are either converted to innocuous compounds such as CO_2 [125] or may at least be reduced to less toxic forms. Hydrocarbons can serve the function of being, electron donor or acceptor in cellular respiration but this varies depending on both the microorganisms and the types of hydrocarbons involved [126]. However, microorganisms have evolved and adapted ancillary pathways of conversion to enable them to incorporate hydrocarbons into the traditional citric acid cycle in aerobic pathways [127] using beta-oxidation, and as for the anaerobic respiration pathways, several routes of integrating hydrocarbons are also possible [128]. Fig. (2) illustrates the three different strategies microorganisms could employ for assimilation. With some hydrocarbons, the chemical structure may allow for an induced fit with microbial enzymes, in this case, it is readily broken down and used in ATP synthesis [129] as is the case with fortuitous biodegradation previously described in this chapter.

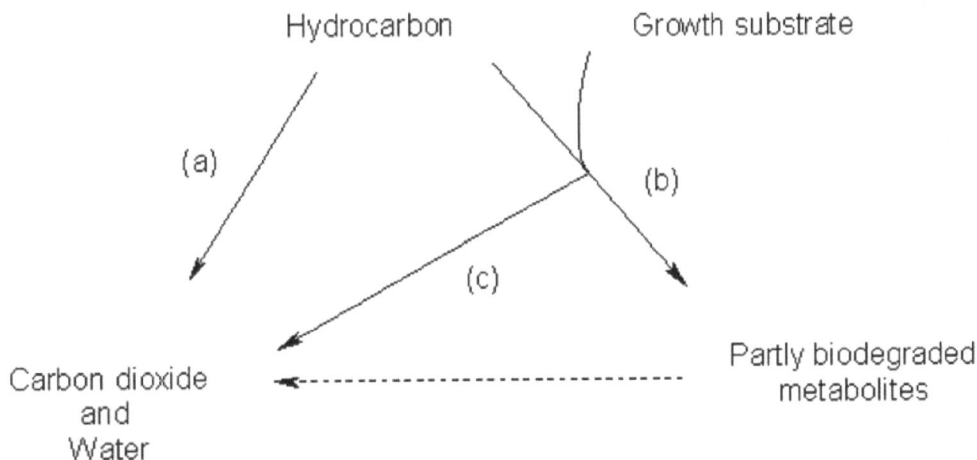

Fig. (2). Approaches to biotransformation adapted from [130] [**(a)** - Gratuitous degradation; **(b)** - Cometabolic Reductive Biotransformation; **(c)** - Cometabolic Biotransformation].

Aerobic Microbial Pathways for the Degradation of Hydrocarbons

Microorganisms utilize carbohydrates converted to simple glucose for cellular respiration, they can equally convert lipids *via* beta-oxidation and directly incorporate them into the TCA cycle for the generation of ATP. Hydrocarbons are lipids and as such can be integrated with modifications into the TCA cycle [131]. The ATP synthesis happens in three stages, glycolysis (conversion of glucose to pyruvate), and the TCA cycle where the activities of oxidoreductases convert pyruvate the byproduct of glycolysis, into acetyl CoA. This stage generates electrons (H^+), that are used in the oxidative phosphorylation of ATP [132]. With the beta-oxidation step, most lipids including hydrocarbons can be converted to acetyl CoA bypassing the glycolytic steps but with the same final outcome of ATP generation from electrons [133]. Often an ancillary pathway of hydrocarbon functional group activation is necessary before the beta-oxidation step. This oxidative biodegradation of lipids is shown in Fig. (3).

Fig. (3) illustrates the capability of bacteria, fungi and algae to activate hydrocarbon using their enzymatic machinery before incorporating these hydrocarbons into traditional biochemical pathways. Although there are different enzymatic strategies the end goal of converting the pollutants to CO_2 an easier gas to be recycled remains the same and achieves the ultimate goal of pollutants removed from the environment [134].

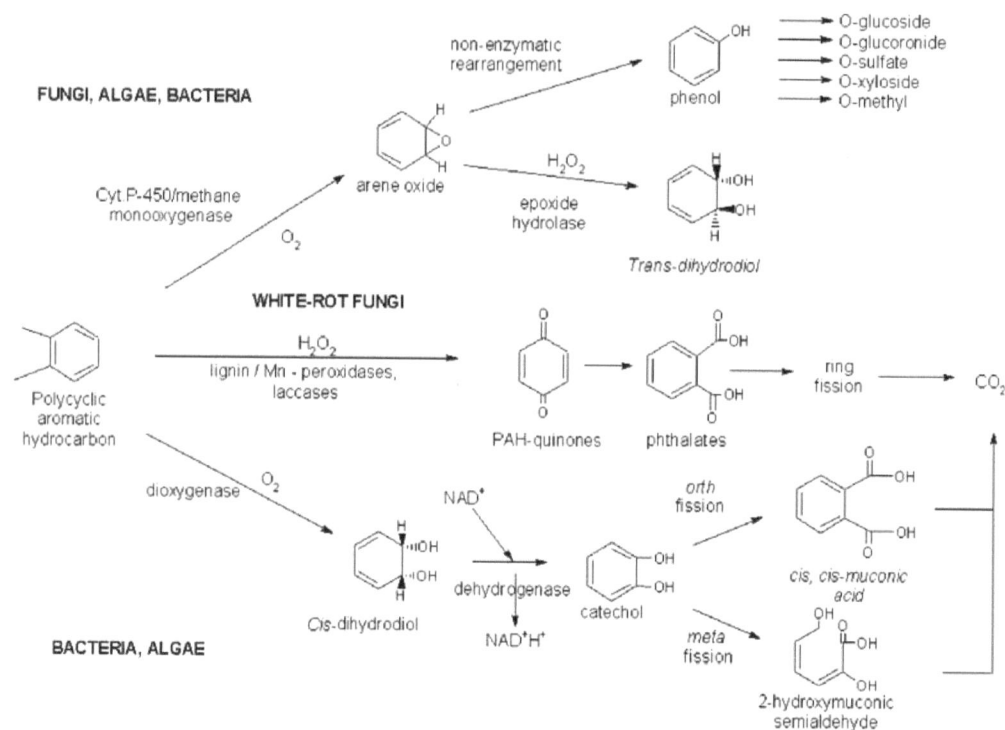

Fig. (3). Aerobic biodegradation of hydrocarbons, adapted from [87].

Biodegradation of hydrocarbons begins with the activities of enzymes such as the mono- and dioxygenases that perform sub-terminal and terminal hydroxylation of aliphatic hydrocarbons and the hydroxylation of benzene-based compounds [135]. In aerobic pathways of hydrocarbon degradation, O_2 acts as a co-substrate providing aid to the hydroxylation of alkanes and aromatic compounds. Hennessee and Li (2016) and Al-Hawash *et al.*, (2018) explain that the O_2 molecules act as an electron acceptor in the reaction, thus behaving as a nucleophile abstracting electrons from the electrophile hydrocarbon that donates electrons [136, 137]. This reaction is quite efficient as it requires lesser energy of activation increases the rate of reaction and provides a higher energy turnover [138].

Aerobic Degradation of Aliphatic Hydrocarbons

Aliphatic hydrocarbons are generally broken down either by terminal or sub-terminal oxidation pathways (Fig. **4**). With terminal oxidation, oxygenase enzymes introduce oxygen atoms to the hydrocarbon, converting it to primary alcohol which is then transformed into an aldehyde followed by further breakdown into a carboxylic acid (fatty acids) [139]. This fatty acid is then conjugated to a coenzyme A (CoA) and processed to form molecules of acetyl CoA by β-oxidation reactions [140]. The acetyl CoA is a precursor in the citric acid cycle [141] where the molecule donates carbon atoms from the acetyl group to the cycle for the production of adenosine triphosphate (ATP) in the mitochondrial cells. The end product of phosphorylation is the ATPs generated, CO_2 and water [142]. However, with sub-terminal oxidation, the alkane is converted to a secondary alcohol, followed by conversion to the corresponding ketone, and then into an ester by the Baeyer-Villinger monooxygenase [143]. This ester is then processed by an esterase forming an alcohol molecule as well as the fatty acid used to form the acetyl CoA.

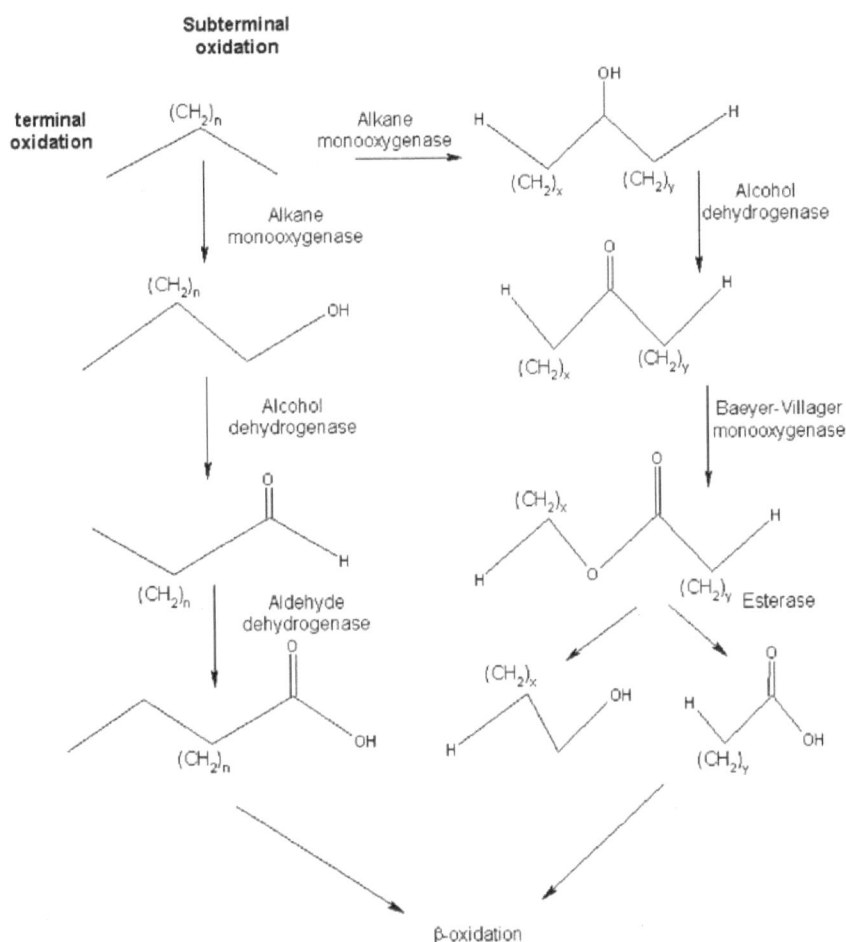

Fig. (4). Aliphatic hydrocarbon sub-terminal and terminal beta oxidation, adapted from [149].

Several classes of hydroxylating enzymes have been found to be able to cleave the methyl group in the alkanes during activation [144], depending on the length of the chain of the hydrocarbons the enzymes may be propane monooxygenase (C_3) butane mononxygenases (C_3-C_9), cytochrome P450 monooxygenases ($C_5 - C_{12}$) [145], AlkB related non-heme iron monooxygenase (C_3-C_{10}; $C_{10} - C_{20}$) [146], Flavin binding monooxygenase (AlmA) ($C_{20} - C_{36}$) Flavin dependent monooxygenase (ladA) ($C_{10} - C_{30}$) and copper Flavin dependent dioxygenase ($C_{10} - C_{30}$) [147, 148]. There is a relative ease to the degradation of alkanes, however, aromatic hydrocarbons pose a greater concern as these molecules are less reactive due to their steric inertness in the environment.

Aerobic Degradation of Aromatic Hydrocarbons

There are several peripheral and central pathways that are known to work for the degradation of aromatic hydrocarbons [150]. Most begin with the activation of the aromatic hydrocarbons through the formation of diol molecules, such as catechol facilitated by *cis*-hydroxylation of the aromatic molecule employing dioxygenase enzymes [151]. Activation is then followed by the cleavage of the rings structure by dehydrogenase enzymes resulting in the formation of carboxylic acid and more than one carboxyl functional group [152]. At this point, a diol degrading enzyme, such as catechol dioxygenase is needed to breakdown the di or tri-carboxylated molecules to yield acetyl CoA, succinate and/or pyruvate, the enzyme catalyzes the insertion of oxygen molecules to the substrate causing the further breakdown of the ring structure [153]. Diol degrading enzymes can be classified into two types: intradiol and extradiol dioxygenases [154]. The intradiol dioxygenases are grouped under the same family of enzymes. However, the extradiol dioxygenases are grouped under different families. An example of one of these families can be the vicinal oxygen chelate family containing enzymes such as 1,2 catechol dioxygenase or 1,3 catechol dioxygenase [155]. The choice of route for entry into the aerobic pathways is premised on the rate of reaction and other environmental conditions. Fig. (**5**) illustrates the reaction pathway for biodegradation of a simple aromatic hydrocarbon (benzene).

Anaerobic Microbial Pathways for the Degradation of Hydrocarbons

Even in anaerobic conditions, recalcitrant hydrocarbons can also be degraded. The enzymatic reactions, in the absence of O_2, use other types of nucleophiles, depending on the environment. To facilitate the degradation of the hydrocarbon, electron acceptors such as ferric ion (Fe^{3+}), nitrates (NH_3), or even sulfates (SO_4), may be used to replace O_2 [157, 158]. Anaerobic biochemical pathways are slow, due to low (2) ATPs generated as compared to the 32 ATP molecules, produced in aerobic pathways. Nonetheless, it is an equally important degradation process in nature, especially in environments characterized by low O_2 content, for example, Mangroves and sludge digesters [159].

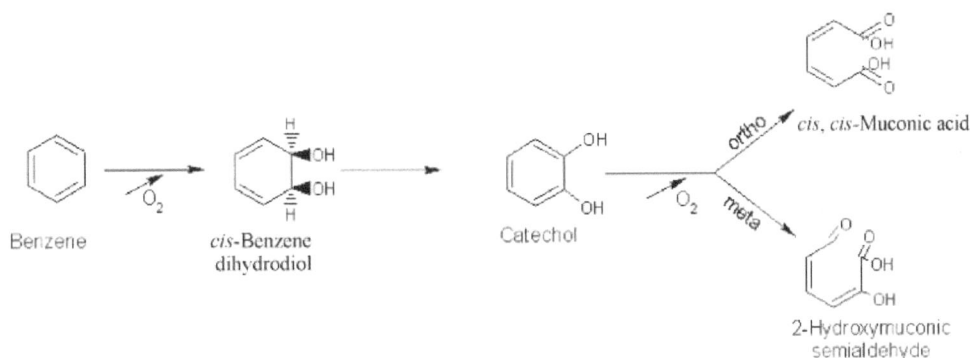

Fig. (5). Bioactivation of aromatic hydrocarbons, adapted from [156].

For bioremediation programs and to improve degradation, it has been suggested that the addition of methyl groups into sterically inert aromatic hydrocarbons may increase polarity [160]. In addition, other approaches that are useful include the combination of water based enzymatic hydroxylation with cofactors (molybdenum) which has proved to aid certain types of oxidoreductase enzymes and the introduction of fumarate supplemented with glycl radical enzymes as well as induction of carboxylation reaction to the addition of methyl groups to the fumarate [159].

Anaerobic Degradation of Aliphatic Hydrocarbons

Widdel and Grundmann (2010) proposed an anaerobic pathway for typical aliphatic hydrocarbons, in which alkylsuccinate synthase (ASS), a glycyl radical enzyme, catalyzes the functionalization of aliphatic hydrocarbons by the addition of fumarate. This reaction results in the production of (1-methylalkyl)- succinates that undergo further reactions to yield 4-methyl branched fatty acid thioesters with the rearrangement of the carbon backbone and decarboxylation of the succinate derivatives, these thioesters are then degraded through β – oxidation and the fumarate is recycled. The ASS enzyme has been isolated from microorganisms of the phylum *Proteobacteria*. Bian *et al.* (2015) support the assertion made by Widdel and Grundmann (2010) but add that it is likely that the addition of fumarate is a much more common route of hydrocarbon activation in anaerobic degradation [161, 162]. This scheme is presented in Fig. (**6**).

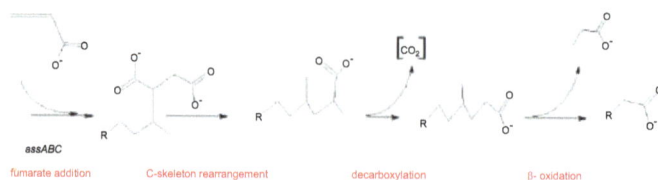

Fig. (6). Proposed pathway for anaerobic degradation of alkanes, adapted from [162].

Anaerobic Degradation of Aromatic Hydrocarbons

The process of activation of the aromatic molecule occurs, through the addition of fumarate to the methyl group. For example, with toluene, its activation is dependent on the presence of benzylsuccinate synthase [163]. Young and Phelps, (2005) and Weelink *et al.*, (2010) explained that this yields benzylsuccinic acid which is converted to benzylsuccinyl-CoA by the action of succinyl-CoA:(R)-benzylsuccinate CoA-transferase and further reduced to phenylitaconyl-CoA by (R)-benzylsuccinyl-CoA dehydrogenase. The phenylitaconyl-CoA hydratase hydrolyzes phenylitaconyl-CoA to produce 2-[hydroxyl(phenyl)-methy-]-succinyl-CoA in the presence of water. The 2-[hydroxy(phenyl)methyl--succinyl-CoA dehydrogenase abstracts the hydrogen from the –OH forming benzoyl-succinyl CoA. Benzoyl CoA is formed by the action of benzoylsuccinyl-CoA thiolase which allows for fumarate recycling. Once benzoil CoA is activated benzoil CoA reductase is responsible for the hydrolytic cleavage of the ring structure resulting in de-aromatization of the molecule, followed by terminal oxidation yielding CO_2 [164]. However, in another example with, ethylbenzene, there is the oxidation of the carbon attached to the methylene to form acetophenone that undergoes carboxylation to be converted to benzoil CoA [165, 166]. This sequence of reactions is depicted in Fig. (7).

Microbial Adaptive Features Developed for the Degradation of Hydrocarbons

Bacteria

The most predominant bacterial genera identified for their involvement in the degradation of hydrocarbons include *Proteobacteria, Actinobacteria, Firmicutes, Bacteroidetes, Chlamydiae* and *Gammaproteobacteria* [3, 5, 167, 168]. Moreover, other bacteria isolates have been identified especially for their ability to break down recalcitrant PAHs such as *Mycobacterium, Pseudomonas, Sphingomonas, Burkholderia, Rhodococcus and Arthrobacter* genera [169, 170]. Examples of characterized strains capable of anaerobic degradation of hydrocarbons include *Desulfococcus alkenivoras,* a sulfate reducing bacteria [171], *Thauera aromatica,*

Clostridiales, Geobacillus metallireducens, Magnetospirillum, Alphaproteo bacteria, Betaproteobacteria [172].

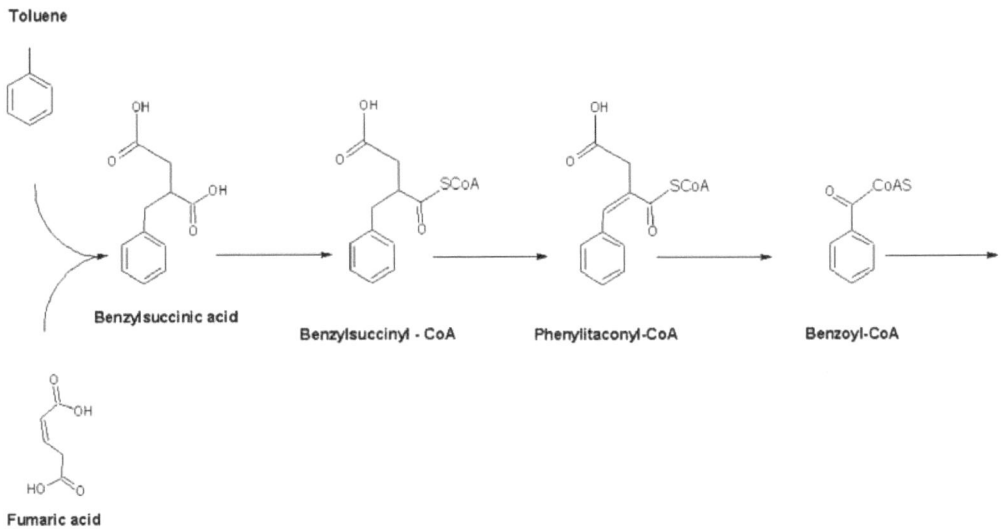

Fig. (7). Anaerobic biodegradation of toluene, adapted from [166].

Generally, it is easier for bacteria to degrade lower molecular weight hydrocarbons, and difficulty is experienced as the molecular weight increases [173]. But there have been a few observed exceptions with naphthalene biodegradation being faster as compared to 10-carbon chain alkanes in water sediments [174]. The bacteria often found in contaminated soils are from genera of *Flavobacterium* and *Pseudomonas spp* [175]. Certain bacteria can express more than one of the hydrocarbon-degrading enzymes which makes it useful in environments with different hydrocarbon pollutants [176]. The extensive body of knowledge in bacterial degradation of hydrocarbons has provided insights into the several ingenious strategies exploited by bacteria to facilitate the degradation of hydrocarbons both as individual microorganisms and within a consortium. Some of these strategies will be discussed in this section.

Biosurfactants Production by Bacteria

Bacteria, produce biosurfactants to aid in the emulsification of lipids including hydrocarbons, thereby allowing for these compounds' assimilation from the environment. Due to the hydrophobic characteristic of petroleum and its derivatives, it is difficult for the bacterium cell to assimilate these substrates [177]. Thus, the first step in the utilization of lipids is to create water-lipid interface (emulsion). These biosurfactants are heterogeneous amphiphilic molecules that act as a chelating agent, for petroleum molecules, by decreasing surface tension, of the environment surrounding the bacteria and the tension between cell wall and petroleum, which enhances the solubility of the substrate facilitating the emulsification of the hydrocarbons [178]. Biosurfactants, therefore, make hydrocarbons more bioavailable for the degradation mechanisms possessed by the bacterium and enhances the process. An example of a bio surfactants is the glycolipid type bio surfactants usually expressed by *Pseudomonas* strains such as *P. aeruginosa, P. chlroraphis* and *P. putida* [179]. These have been observed during the analysis of biodegradation products of contaminated soil. In studies conducted by Patowary *et al.*, (2017) a strain of *Pseudomonas aeruginosa* PG1 was shown to be capable of producing rhamnolipids, a type of biosurfactants that decreased the surface tension in the medium by 22.2 mN/m while using crude oil as the main carbon source for respiration [180]. Marchut-Mikolajczyk *et al.*, (2018) isolated a biosurfactant producing strain of *Bacillus subtilis* 2A that was able to degrade diesel oil (137%) and waste engine oil (120%) which facilitated the growth of *Sinapis alba* in contaminated soil [181].

Bacteria Consortia Formation and Cooperation

The symbiotic physical association among various species of bacteria that are organised into specialised aggregates for the purpose of executing complex metabolic activities is referred to as a consortium [182]. Consortium formation and interaction in response to trophic changes, is more efficient than single strain colonization [183], because these organisms provide a broader spectrum of enzymes, thereby improving the degradation efficiency in handling a variety of hydrocarbon pollutants. Participating organisms bring not only their enzyme machinery but also varied strategic features including biosurfactant production, metal transport and resistance, dissimilatory nitrate, nitrite and sulphite reduction as well as chemotaxis. Patowary *et al.*, (2017) studied bacterial consortia in hydrocarbon contaminated soil in Assam, India. In their study they isolated a consortium composed of *Bacillus pumilus* KS2 and *Bacillus cereus* R2 that was found to be the top degraders of hydrocarbons with a rate of 84.15% after 5 weeks [180]. Potts *et al.*, (2018) conducted biodegradation studies on microbial consortia from deep sea sediments (135 – 1000m), the FSC135 consortia at 135m was seen to be more efficient in the degradation of diesel (+/-40%)and crude oil (+/- 50%) [184]. It can be inferred that there was faster degradation at shallow water due to higher abundance of O_2. A consortium consisting of *Klebsiella* spp., *Enterobacter* sp. and *Pantoea* sp isolated from a 2% hydrocarbon enriched medium was able to completely biodegrade (1-(1,5-Dimethylhexyl)-4- (4-Methylpentyl--Cyclohexane), reduce 63.28% of naphthalene derivatives, degraded 21.87% of aliphatic acids and 75.84% of benzene derivatives [185]. One critical characteristic that allows for consortium cooperation is the formation of biofilms during biodegradation.

Biofilm Formation by Bacteria Consortia

Biofilm constitutes of extracellular polymeric substances (EPS) extruded from individual bacteria cells that forms a support matrix to encase participants in a bacteria consortium [186]. It is very common in aquatic environments but can also be formed in waterlogged terrestrial environments. It is a strategic feature produced to protect the bacteria population from extreme environments by conferring the ability to resist changes in concentrations and toxicity in the external surrounding environment that will otherwise not be possible to free-floating bacteria [187]. It is for this reason biofilms have found extensive application in wastewater treatment plants. Biofilm dwelling bacteria are able to quickly acclimatize to changes due to high organic loads.

Usually in nature, the beginning of biofilm starts with the colonial cell anchoring onto a surface. Free floating bacteria will attach to a surface; aggregate cell will further adhere close to the colonial cell with the beginning of biofilm formation, once this biofilm matures, parts of the biofilm disperses, releasing free-floating bacteria for further colonization and the cycle is repeated. The common organisms found in biofilms are *Listeria, Pseudomonas, Bacillus, Salmonella, Campylobacter* and *Escherichia coli* [188, 189]. Biofilms are characteristically hydrophobic but members produce biosurfactants to create the oil-water interface necessary for the biodegradation of hydrocarbons and for the uptake of nitrogen and phosphorus compounds from the surrounding aqueous environment, while using the oil as the carbon source [190].

The use of biofilms in hydrocarbon biodegradation offer a promising solution when dealing with large polluted sites. *Ex situ* biofilm technology can be applied by using inoculation method, as it will spread in time. The simplistic approach is effective for excavated soils and once spread thinly to increase surface area and improve degradation, bioremediation treatment will be relatively fast and soil can be later returned to the original site of excavation. Fouad and Bhargava, (2005) reviewed, quite comprehensively, the model for the steady-state biofilm-activated sludge reactor [191]. Dasgupta *et al.*, (2013), isolated *Pseudomonas* sp. including *P. otitidis* capable of biofilm formation to degrade crude oil contamination [192]. Studies on a consortia consisting of *Pseudomonas putida, P. fluorescens* and *P. aeruginosa* showed xylene breakdown in hydrocarbons mixture [193]. Using, *Alcanivorax borkumensis* as model organism, Omarova *et al.*, (2019) were able to degrade 90% hexadecane through biofilm dispersion of the oil slicks that was aided by a biosurfactant which decreases by 2 fold the oil-water surface tension [194].

Quorum Sensing by Bacteria Consortia

Biofilms also serve dual functions for cell-to-cell chemical signal communication used in processes such as efficient nutrient utilisation, excretion and distribution of degradation products [188]. This communication is referred to as to as quorum sensing (QS) and, in some cases, quorum quenching (QQ) [195, 196]. QS assists with transcriptional regulation of genes that directly influence resistance, virulence, expression of secondary metabolites and even growth [197]. Usually, bacteria secrete the following signal chemicals, to the extracellular matrix of the biofilm, being acyl-homoserine lactone, autoinducing peptide, and autoinducer-2 [198] and these then bind to receptors that regulates the gene expression [199]. When the chemical signal molecules are released by bacteria, it is employed to gauge population density and control gene expression [196]. Both QS and QQ play significant roles in colonization, biofilm formation, bacterial aggregation and biotransformation of pollutants [195].

The understanding of these survival strategies are pivotal to advance research and large scale application of bioremediation to hydrocarbon contamination. Although the phenomena of biofilm formation and QS are not thoroughly understood, it is known that they play beneficial roles in wastewater treatment by providing the necessary matrix and platform for biocatalysis, biotransformation and biodegradation of pollutants [200]. Moreover, they have found application in granular sludge, trickling filters, moving bed biofilm reactors and constructed wetlands [201 - 203]. However, biofilms can also have a detrimental effect to wastewater treatment facilities and cause biofouling from excessive biofilm formation [201]. When not routinely managed, slime formed, can block small pipes and air vents, by the build-up on membrane and filters, and when this happens poor performance of bioreactors and nanofiltration systems will be triggered [204].

Fungi

Fungi are a diverse group of organisms, and several species have shown capabilities for the biodegradation of recalcitrant hydrocarbons. They also inhabit both aquatic and terrestrial environment [205, 206]. Some hydrocarbon degrading fungi found in both terrestrial and aquatic ecological systems include, *Amorphoteca, Neosartorya, Talaromyces,* and *Graphium Candida, Yarrowia, Pichia, Aspergillus, Cephalosporium* and *Pencillium* [207 - 209]. Yeast species are however prevalent in aquatic environments and examples are *Candida lipolytica, Rhodotorula mucilaginosa, Geotrichum sp* and *Trichosporon mucoides* [210].

Fungi species are capable of aerobic biodegradation of hydrocarbon using both mono and dioxygenase catalysis, similar to those in bacteria. In a study conducted by Al-Naswari, (2012) to evaluate the degradation rates of crude oil by fungi species isolated from the Gulf of Mexico, it was observed that the highest weight reduction of crude oil was achieved by *Penicillium decumbens* Thom strain 4 (on average 8.65% in 21 days) followed by *Aspergillus niger* Tiegh strain 3 (7.85% in 21 days), although, the most abundant strain was *Aspergillus niger* [211]. Olukunle *et al.*, (2016) isolated several fungi species from cow dung and highest degradation of crude oil was observed in *Trichoderma viridae* (66.2%) while *Varicosporium elodeae* exhibited the least degradation (40%) [212]. Babaei and Habibi, (2018) observed an 82.1% degradation of diesel with *Candida catenulata* KP324968 facilitated by the biosynthesis of sophorolipid chelating molecules [213]. Fungi species use basically two main strategies in handling hydrocarbon substrates. These strategies are production of biosurfactants and multi-specific enzymes for the degradation of complex hydrocarbon structures.

Fungal Biosurfactants

Biosurfactants produced by fungi act similarly to those produced by bacteria, they reduce the hydrophobic surface tension and emulsifies the hydrocarbons [214]. The emulsification action tends to reduce the biochemical reactions but it assists in stabilizing the harsh environment created by the presence of the hydrocarbons [215]. These non-toxic biosurfactants are capable of performing their activity efficiently at low concentration and are easily biodegrade due to their low molecular weight [216]. Other features that promotes efficiency and adaptability includes the ease of compatibility with various types of hydrocarbons, ability to perform work at various pH, temperature and salinities in a given environment [217]. Chandran and Das, (2010) isolated a strain of *Trichosporon asahii* from oil contaminated soil from India [3]. A mineral salt media enriched with diesel was used for *T. asahii* and this fungus demonstrated the capability to produce, sphorolipid biosurfactant while degrading 95% of the diesel in 10 days. Experiments conducted by Chaprao *et al.*, (2015) showed that in the presence of biosurfactants biodegradation of crude oil in contaminated sand by *Candida sphaerica* reached 90%, they further add that the biosurfactant increased the degradation by 10 – 20% [218]. Ghunter *et al.*, (2017) suggested that biosurfactant producers can be the members of the orders Saccharomycetales and Ustilaginales [219].

Multispecificity of Ligninolytic enzymes in White Rot Fungi

Ligninolytic enzymes are naturally present in nearly all white rot fungi, but some of these enzymes have also been found in some other fungi species and even bacteria. These enzymes are exploited by white rot fungi in the degradation of recalcitrant lignin, a highly complex structure. These enzymes include some peroxidases and laccases with known capability for multispecific participation in chemical reactions [48]. The most studied enzymes produced by white rot are phenol oxidase, laccase and three heme peroxidases namely lignin peroxidise, manganese dependant peroxidase and versatile peroxidase [220, 221]. This ability to degrade lignin is most likely the reason these group of fungi have also shown affinity for a variety of related hydrocarbons including, aromatics, alkenes and alkynes. Prenafeta-Boldú *et al.*, (2019) explains that in the place of the CYP450 monooxygenases lignolytic fungi detoxifies the hydrocarbons with the catalytic assistance of peroxidases and laccases to produce the PAHs' quinones that behave as the substrate for ring fission [130]. Laccases are a group of isoenzymes capable of oxidizing a wide range of organic or inorganic compounds including phenols, aromatic amines and even ascorbate [222]. Lignin peroxidase is a glycosylated protein containing heme. In the presence of cellular-produced peroxidc, lignin peroxidase catalyzes the oxidation of aromatic non-phenolic lignin structures to give aryl-cation radicals [223]. The oxidative degradation of lignin requires the presence of veratryl alcohol which is also a substrate for lignin peroxidase and is secreted as a fungal metabolite [224]. Manganese dependent peroxidase is a heme-dependent enzyme able to catalyse the hydrogen peroxide-dependent oxidation of Mn^{2+} to highly reactive Mn^{3+} Afterwards, the cation Mn^{3+} subsequently oxidizes phenolic parts of lignin to produce free radicals [225]. The high reactivity of Mn^{3+} is stabilized by chelating molecules (oxalate, malonate, malate) secreted by fungi [226]. Versatile peroxidase has similar catalytic properties of lignin peroxidase and manganese dependent peroxidase. Versatile peroxidase shows high sequence and structural homology with lignin peroxidase but also uniquely, comprises a binding site for Mn^{2+} [227]. Some white rot basidiomycetes may express all of these lignin-modifying enzymes while others may express only one or two of these enzymes [228]. *Phanerochaete chrysosporium* BKM-F-1767, *Trametes versicolor* Paprican 52, and *Bjerkandera adusta* CBS 595.78 were observed to degrade anthracene [229]. Clemente *et al.* (2001) conducted a study on thirteen strains of deuteromycete ligninolytic fungi and the results showed highest degradation of naphthalene (69%) by the Mn-peroxidase, several strains showed varying degrees of degradation for the hydrocarbons, phenanthrene [230]. Similarly, Govarthanan *et al.*, (2017) observed highest degradation of decane (49%) using *Penicillium* sp. CHY-2 [231].

Algae

Algae are photosynthetic organisms and can be classified on the basis of size as micro and macro algae. They convert CO_2 into sugars in their primary metabolism and are capable of lipid utilization, including, hydrocarbons, in their secondary metabolisms. They are predominantly resident in aquatic environments. Microalgae have demonstrated the ability to biodegrade aromatic pollutants under heterotrophic condition, even in the absence of light [232]. Among the strains of algae involved in biodegradation of hydrocarbons *Cyanobacteria sp* [3]. and *Prototheca zopfi* [220] and *Chlorella sp* [233] have been studied quite extensively. Qari and Hassan (2017) observed that the alga, *Padina boryana,* showed a remarkable ability to accumulate pollutants into their cells [234]. However [235], noted that *Selenastrum capricornutum* actually demonstrated the capacity of transforming the pollutants into less toxic compounds. Similar, to bacteria and fungi, an activation step is necessary in the degradation pathways used by microalgae for incorporating pollutants into their metabolism. This is achieved by the hydroxylation of the ring molecules by the addition of an –OH group that replaces the double bond in the benzene rings, *via* the activity of mono and dioxygenase, thus, increasing the reactivity of PAH for further breakdown.

With phenol as depicted in the Fig. (**8**), hydroxylation begins with the activity of phenol monooxygenase to form a catechol followed by oxygenation by catechol 2,3-dioxygenase to form the 2-hydroxymuconic aldehyde. A 2-hydroxymuconic semialdehyde-NAD^+-dependant dehydrogenase catalyzes the conversion of 2-hydroxymuconic aldehyde, in the presence of water, to form 4-oxalocrotonate followed by certain reactions steps that will yield pyruvate and CO_2. Since these pollutants can act as substrates in cellular respiration, producing pyruvate that is a precursor to acetyl CoA), and can even may even be integrated into photosynthesis (CO_2) it could be inferred that these metabolites may be useful for cellular proliferation, along with further reduction in greenhouse gas emissions and production of secondary metabolites. For example, *Prototheca zopfii*, was observed to degrade crude oil by up to 40% and motor oil, by 10% [236]. The poor degradation of motor oil may be as a consequence of a build-up of cytotoxicity and may mean that effective degradation may only be achieved through cometabolic reduction instead of mineralization [237]. Using 18S ribosomal RNA based identification Das and Deka, (2019) identified a strain of *Chlorella vulgaris* BS1 capable of degrading 98.63% of hydrocarbons in 14 days from contaminated water, and they also observed a 75% reduction of COD as well as sulphate removal. While studying biodegradation of hydrocarbon by *Chlorella* sp. Shafi, (2020) observed a total reduction of >180mg/L of oil when incubated at 30°C for 4 days [238]. The influx of contaminants into the natural habitats of microalgae, has motivated the necessity to develop strategies for adaptation to the stress of changing trophic states. The two major strategies employed by microalgae are the formation of consortia and interspecific interaction to assist in the removal of pollutants from their ecological systems.

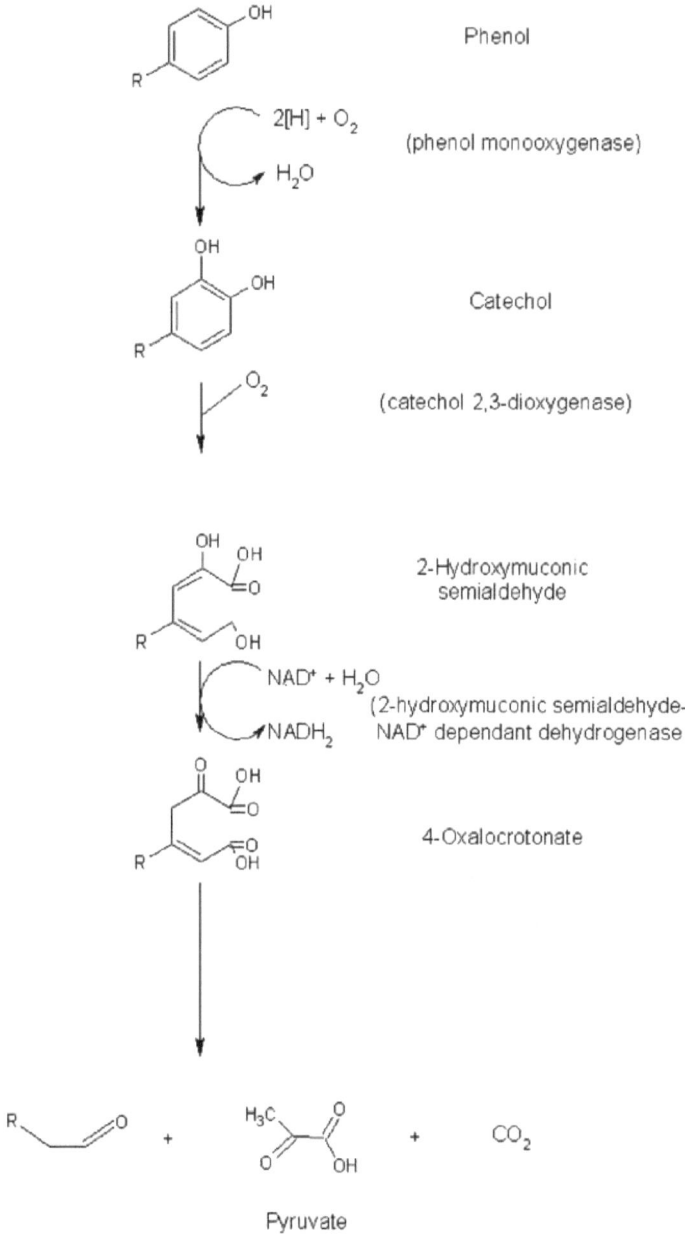

Fig. (8). Microalgae biodegradation of phenol by *Ochromonas danica* adapted from [232].

Microalgae Consortium

Like bacteria, microalgae can also degrade hydrocarbons in a consortium, using similar mechanisms of EPS production. Eregie and Jamal-Ally, (2019) compared the biodegradation of lubricant waste between a microalgae consortium (*Chlamydomonas pitschmannii, Trebouxia australis* and *Pectinodesmus pectinatus)* and *Scenedesmus vacuolatus* [239]. However, the results show that at 5 weeks both experimental set ups had 100% biotransformation of the hydrocarbons. There are very few experiments that focus strictly on consortia with microalgae only, thus this finding is inadequate to make determination as to whether consortia interaction is superior to free floating microalgae in terms of hydrocarbon degradation. It would appear that most biodegradation studies have focused on investigating the cooperative benefits of using consortia of microalgae and bacteria. The reasoning behind this being that it will be highly unlikely that a pure consortium of microalgae will exist in the environment without some form of bacterial contamination [240, 241]

Interspecific Interactions

The aquatic environment is inhabited by a motley of organisms, as such interspecific interaction is a mundane feature of this ecological system [242]. Microalgae form mutually beneficial relationship with bacteria and fungi. In a review by Yao *et al.* (2019), they describe such a symbiotic relationship between microalgae and bacteria [243]. The latter increases the bioavailability of limiting nutrients, N, P, S to the microalgae, while there is an exchange of C-substrates necessary for bacterial cellular respiration and ATP generation. Patel *et al.*, (2015) compared the degradation rate of pyrene by an interspecific consortia consisting of *Synechocystis* sp. and two bacterial species being *Pseudomonas* sp. and *Bacillus* sp [244]. The results of the experiment reveal that after 16 days the microalgae-bacteria consortia had a greater degradation efficiency of 94.1% at 50 mg/L of pollutant, as compared to individual degradation done by *Synechocystis* sp., which only achieved 36% at 1.5 mg/L. The variability in the substrate concentration leaves the findings, questionable and is also not a conclusive support for interspecific interaction superiority in degradation. Omojevwe and Ezekiel, (2016) studied the effect of microalgae-bacteria consortium in the degradation of petroleum effluents [245]. The consortia variations consisted of *Chlorella minutissimma* and *Aphanocapsa* sp.*, Citrobacter* sp. SB9, *Pseudomonas aeruginosa* SA3 and *Bacillus subtilis* SA7; but a higher degradation rate of 92.09% was seen in the consortium composed of *Citrobacter* sp. SB9, *Pseudomonas aeruginosa* SA3, *Bacillus subtilis* SA7 and *Chlorella minutissimma*. This showed that selection of the biodegraders can influence the biodegradation rates perhaps duc to the enzymatic profile contributed by participating organisms. Both species are capable of bio surfactant production [246]. It can be inferred that synergistic activity may have taken place thereby increasing hydrocarbon bioavailability for the microalgae metabolism through chelation. In light of the currently sparse data and findings, it is important that attention be given to this aspect of research to further our understanding of this currently overlooked but important aquatic interspecific interactions for bioremediation programs.

Hydrocarbon Bioremediation Strategies

To understand the process of bioremediation, it was important in this chapter to thoroughly explain the microbial degradation processes because, bioremediation is the conscious application of the natural process of biodegradation. Biodegradation relies on biologically catalysed reactions of organic chemical structures. For hydrocarbons, it is to a great extent oil weathering, where microorganisms break down organic molecules into other substances including fatty acids and CO_2. However, bioremediation involves human intervention and sometimes with relative success in the acceleration of natural degradation processes. Approaches such as stimulation of indigenous microbial populations with the inclusion of nutrients and other chemicals as well as the manipulation of contaminated media with techniques including aeration and temperature control are employed including the addition of exogenous microbial population. Bioremediation would typically involve the actions of a variety of microorganisms working within their individual populations or a sequence of activities to complete the degradation process. There are two alternative approaches to pollutant remediation: *in situ* (in place) and *ex situ* (removal and treatment away, from the site of contamination). *In situ* bioremediation is considered less disruptive to the environment as compared to *ex situ*, with regards to terrestrial pollution. In general, bioremediation approaches are less expensive and more sustainable than other remediation options [247]. Often bioremediation processes involve oxidation-reduction reactions which entail an electron acceptor (usually oxygen), to stimulate oxidation of a reduced pollutant (hydrocarbons); or an electron donor (mostly an organic substrate), to reduce oxidized pollutants (examples, nitrates, perchlorate, chlorinated solvents, oxidized metals, propellants and explosives).

Approaches to Bioremediation

Bioattenuation

The natural biodegradation process without human intervention is regarded as bioattenuation or intrinsic bioremediation. It relies on the combination of naturally occurring biological and physical-chemical processes such as biodegradation, dispersion, adsorption, volatilization and dilution for the reduction of mass or concentration of compounds in groundwater and soil over time or distance from the point of initial pollutant discharge. Bioattenuation stems from our extensive understanding of natural biodegradation processes and how microorganisms in time convert toxic compounds to innocuous forms [248]. Remarkably, most of the early research focused on bioattenuation of polyaromatic hydrocarbons, especially benzene, toluene, ethylbenzene and xylene (BTEX) [249] which are components of petroleum that are commonly found at leaking underground storage tank sites. However, it is debatable whether bioattenuation is an appropriate strategy for managing contaminated sites. Although attenuation is by far the cheapest approach, opponents to this approach argue that its lack of human intervention ('do nothing' approach), prolongs the time for this natural process to restore the environment to its original condition [250].

Biostimulation

The criticism of the poor efficiency and rather slow reaction time associated with the bioattenuation strategy brought about the development of the biostimulation strategy. It aims at stimulating indigenous microorganism through addition of nutrients (nutrient enrichment) such as carbon and nitrogen sources, electron donors and acceptors. Biostimulation may also incorporate a co-substrate with the initial pollutant. Conceptually, it may be likened to the addition of fertilizer to a lawn. The rationale for this approach is that the environment (either aquatic or terrestrial) should possess a significant population of oil-degrading microorganisms that could very well degrade hydrocarbons and are well adapted to resisting local and unique environmental stresses. However, the increased concentrations of pollutants significantly affect the ratios of other growth limiting nutrients. Thus, the addition of nitrogen, phosphorus and other nutrients is intended to overcome these deficiencies and promote the growth of microbial biomass needed to accelerate the hydrocarbon degradation processes [251]. To facilitate and improve the more efficient anaerobic degradation pathways, it is often necessary to stimulate the microorganisms, by increasing oxygen content of the soil, this may be achieved by injecting air into the soil. For example, Moshkovich *et al.* (2018) implemented and monitored *in situ* bioremediation of a gasoline-contaminated deep vadose zone, using enhanced biostimulation that incorporated both nutrient an O_2^- amended water infiltration [252]. They observed improved biostimulation conditions for the degradation of the target compounds. Moreover, biostimulation may also incorporate the addition of chelating agents for metals or surfactants for hydrocarbons [253].

The main challenge associated with biostimulation in oil-contaminated coastal areas is the tidal influence exerted on these environments, which makes it difficult to maintain optimal concentration. Effective bioremediation requires that nutrients are in contact with hydrocarbon, and concentrations should sufficiently support maximum growth rate of the oil-degrading bacteria. However, the intertidal zones in aquatic environments, especially the marine habitat causes rapid washout of dissolved nutrients [254].

One of the solutions currently employed is the development of a range of oleophilic, slow-release formulations containing nutrients when they come in contact with the oil, usually they are designed to dissolve these nutrients into the aqueous phase forming a mixture that can be utilized by hydrocarbon-degrading microorganisms. Research is evolving in the design of carrier and delivery systems that are integrated to bioremediation strategies for these intertidal zones in aquatic environments. For example, Christoph (2017) developed encapsulation technique using a porous biopolymer to encase microorganisms capable of degrading PAHs [255]. This research addressed the challenge of exogeneous microorganisms that are usually opposed in natural environment. The aim was to limit the dispersion of these exogenous organisms within a given ecosystem to reduce the ecological distortions created by their introduction into the environment.

Bioaugmentation

It is possible that bioremediation is approached from the natural attenuation perspective, with little to no external intervention or the processes may be fully engineered employing bioaugmentation. This is made necessary as a consequence of the reduced population of hydrocarbon-degrading microorganisms. Low microbial populations are usually caused by the high initial toxicity of contaminated sites or may be a natural attribute of the ecosystem that lack the organisms with unique abilities to degrade specific pollutants. Bioaugmentation aims at introducing microorganisms to a contaminated site. The process of inoculation is referred to as 'seeding' and it can be achieved either by applying endogenous or exogenous microorganisms, selected for their efficiency in degrading the contaminating hydrocarbons.

Seeding with Naturally Occurring (Endogenous) Microorganisms

The use of 'naturally occurring' microorganisms as a term of description of the bioaugmentation process, may be considered anomalous. Sometimes, the inoculum maybe a blend of both indigenous and non-indigenous microorganisms derived from various polluted sites possessing similar chemical profiles to the current site needing remediation. These inocula are usually mass-cultured in the laboratory or on-site bioreactors with inclusion of nutrients. These custom-designed seed cultures have usually, been studied extensively, with a clear understanding of their nutrient requirements and limitations, making mass-production relatively easy. Scientists recommend the application of seed cultures to *in situ* bioaugmentation, only under extenuating conditions, where native organisms are either extremely slow in growth or are unable to degrade the hydrocarbon present in the site of contamination. This is to avoid potential problems of loss of biodiversity that is ultimately the outcome of bioaugmentation process.

Seeding with Genetically Engineered Microorganisms

Our understanding of genetic exchange of catabolic degradative genes has facilitated the creation of genetically engineered microorganisms (GEMs) [256]. These GEMs are designed to be able to degrade specific fractions of petroleum that are especially recalcitrant. Moreover, these GEMs tend to have superior hydrocarbon-degrading abilities. However, to be effective such organisms must be able to overcome problems of acclimatization and be robust to withstand environmental conditions and compete with native species.

The advocates of GEMs application recommend using specially selected mixtures of microorganism to bio-remediate oil spills that combine all the requisite enzymes and pathways through recombinant DNA technology [256]. The use of GEMs is limited by scientific, economic, regulatory, and public perception obstacles; as such the wide availability of naturally occurring microorganisms capable of degrading components of petroleum will likely deter consideration of GEMs for remediating marine oil spills. However, *in situ* remediation where pre-emptive actions can be taken to avoid their introduction to natural environments, is a consideration in its application. Recently, Garbisu *et al.* (2017) proposed the addition of naked plasmids to contaminated soil, with the understanding that they can be integrated into resident bacteria *via* horizontal gene transfer, a process they referred to as plasmid-mediated bioaugmentation [257]. It is believed that this process will stimulate the dissemination of the catabolic genes amongst indigenous bacteria, thereby, improving the host's fitness and efficiency of degradation. These researchers, however, highlighted challenges of environmental factors, such as soil moisture, temperature, organic matter content, contributing significantly to the efficiency of transfer of these plasmids, as well as the expression and functioning of the genes of interest.

Regardless of whether the microorganisms are native or newly introduced to the site, biostimulation is usually necessary to increase and maintain the population of these bioaugmenting microorganisms to ensure bioremediation is effective. The type of microbial processes that will be involved in the clean-up process determines the nutritional supplements that must be supplied. Successful outcomes of the bioremediation processes are assessed using the by-products transformation and mineralisation as indicators.

It is noteworthy that a study done by Agary *et al.* (2010) demonstrated the efficacy of the three approaches to bioremediation using kerosene contamination of soil, and after five weeks, they observe for bioattenuation, biostimulation and combined biostimulation and bioaugmentation, respectively, percentages of kerosene degradation as 44.1%, 67.8%, 83.1%, and 87.3%. A clear indication using kinetic order of efficiency was: bioattenuation < bioaugmentation < biostimulation < combined biostimulation and bioaugmentation [258]. The relative merits and demerits of these bioremediation approaches are reported in Table **4**.

Table 4. Principal features of alternative bioremediation approaches.

Bioattenuation	*Biostimulation*	*Bioaugmentation (Seeding)*	
		Seeding with endo- and exogenous microorganisms	*Seeding with genetically engineered microorganisms:*
• Excludes any form of intervention, relies on natural attenuation processes • Relatively slow compared to other approaches	• Intended to overcome the chief limitation on the rate of the natural biodegradation of oil • Preferred and necessary method for accelerated turn-around time for remediation time • Possibility that fertilizer use causes algal blooms or other significant adverse impacts • Tests indicated, fertilizer use appeared to increase biodegradation rate by at least a factor of two.	• Intended to take advantage of the properties of the most efficient species of oil degrading microorganisms • Results of field tests of seeding have thus far been inconclusive • May not be necessary at most sites because there are few locales where oil-degrading microbes do not exist • Requirements for successful seeding more demanding than those for nutrient enrichment • In some cases, seeding may help biodegradation get started faster	• Probably not needed in most cases because of wide availability of naturally occurring microbes • Potential use for components of petroleum not degradable by naturally occurring microorganisms • Development and use could face major regulatory hurdles

Hydrocarbon Microbial Bioremediation Technologies

There are generally two alternatives to the bioremediation treatment of hydrocarbon polluted site; *in situ* and *ex situ* techniques. This section will only focus on only microbial enhanced techniques. The choice of technique is premeditated, especially in the terrestrial ecosystems, on the extent of pollution and the mobility of the hydrocarbon. *In situ* techniques are carried out at the original location of contamination, it avoids excavation of topsoil, and therefore, the integrity of the area remains undisturbed. It is often preferred and is also more cost effective as compared to the *ex situ* techniques of remediation where treatment is done outside original location of contamination. The *ex situ* methods involve excavation of the surface soil which compromises the integrity of the environment being bioremediated. Table **5** summarises the attributes of various in-situ and *ex-situ* bioremediation techniques. This information is further depicted in Fig. (**9**).

Table 5. Brief description of various bioremediation technologies.

Bioremediation Technologies		Types	Description and Examples
In situ	Intrinsic Bioremediation	Natural Attenuation	Conversion of pollutants to innocuous compounds by naturally occurring bacteria with no human intervention
	Projected Bioremediation	Permeable Reactive Barriers (PRB)	Used in groundwater remediation, applies PRB that selectively allows the passage of some substances only. Therefore, it captures a plume of contaminants puts them in contact with microorganisms that are embedded for biocatalytic degradation; and releases uncontaminated water. *E.g.* granular zero-valent iron PRBs, clay PRBs
		Bioaugmentation	Introduction of microorganisms to a contaminated site. Seeding with: o exogenous microorganisms o endogenous microorganism o genetically modified microorganisms o naked plasmids
		Biostimulation	Nutrient enrichment to increase biomass and efficiency of degradation o addition of rate-limiting nutrient (C, N, P) o addition of electron donors and acceptors (oxygen, hydrocarbons) Use of encapsulation technologies for: o slow release of nutrients o protection of microorganisms from environment
		Improved Techniques	o Bioventing: used in groundwater treatment and employs microorganisms for the degradation of hydrocarbon contaminant by inducing air and O_2 flow into unsaturated zones, sometime with nutrient enrichment. Effective for volatile organic compounds (VOCs) as vapours move slowly towards the biologically active soil. o Biosparging: used in groundwater treatment, applies indigenous microorganisms to biodegrade hydrocarbons in the saturated zone.

(Table 5) cont.....

Bioremediation Technologies		Types	Description and Examples
			Air (O_2) and sometimes nutrients is injected into the saturated zone to stimulate metabolic activity. o Bioslurping: combination of aspects of bioventing and vacuum-enhanced pumping of lighter hydrocarbons from water interphase (light non-aqueous phase liquid (LNAPL)) to recover free-products from soil or ground water. It uses a slurp tube to suck up contaminant and this can later be separated from air at the surface. He actual bioslurping is the degradation activity of aerobic bacteria when air is introduced into the unsaturated zone of contaminated soil or sludge. o Mycoremediation: application of fungi to treatment of both soil and water. It has been applied to wastewater, hydrocarbons (PAHs), tannery wastes, textile azo dyes e.t.c
Ex situ	Semi Solid Phase Bioremediation	Sludge Bioreactor	o Moving Bed Biofilm Reactors: biological systems containing mobile suspended carriers that serves as anchors for biofilms and promotes the growth, through adequate mixing and aeration. Consist of aerobic and anoxic microorganisms, which aid in the treatment of a wide range of pollutants including hydrocarbons o Sludge Bioreactor: contains microorganisms that can decompose pollutants o Activated sludge: includes flocs of decomposing microorganism that rapidly decomposes organic pollutants rapidly under aerobic conditions
	Solid Phase Bioremediation		o Land farming: remediation technology for soils that reduces concentrations of hydrocarbons through biodegradation. It involves spreading excavated contaminated soils in a thin layer on the ground surface and stimulating aerobic microbial activity within the soils through aeration and/or the addition of minerals, nutrients, and moisture. The enhanced microbial activity results in degradation of adsorbed petroleum product constituents through microbial respiration
			o Biopiles: Biopiles are similar to land farming, both above-ground, engineered systems that use oxygen, generally from air, to stimulate the growth and reproduction of aerobic bacteria. But in this case air is forced into the soil through slotted or perforated piping placed throughout the pile. o Composting: An aerobic microbial driven process that converts solid organic wastes into stable, sanitary, humus-like material that has considerable reduced bulk and can be safely returned to the environment. *E.g.* Industrial composting preferred alternative to landfilling Biofiltration: used to control pollution, a bioreactor that contains a column of microorganisms used to treat wastewater but also adsorbs chemicals or silts from surface runoff, possible biooxidation of contamination of air is involved.

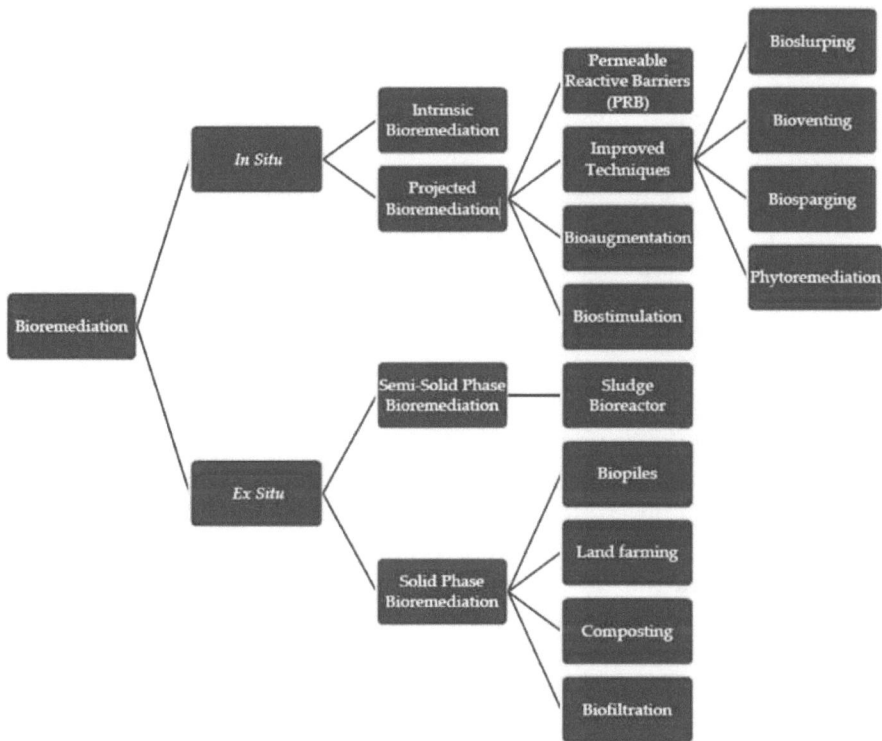

Fig. (9). Different bioremediation technologies, adapted from [259].

Factors Affecting the Application of Bioremediation Technologies

The rate at which hydrocarbons are degraded by microorganisms can be affected by the structure of the hydrocarbons, physical state and concentration of the pollutant [260]. Factors such as temperature, pH of medium, oxygen and nutrient availability, influence the efficiency of microbial degradation [261].

In evaluating factors that affect the industrial application of bioremediation treatments, it is important to consider three separate features: the nature of the hydrocarbon pollutant, the surrounding environment and how these factors, significantly determines the best fit in terms of *in situ* or *ex situ* approaches. These features are important consideration to be reviewed holistically as they determine the success or failure of the designed bioremediation program. Many field tests normally fail, even though laboratory and green house upscaling tests would have demonstrated good potential. This is so because the scientists would not have factored in the real-life natural conditions to which these organisms will be eventually exposed.

Nature of Hydrocarbon Pollutants

Bioavailability

Bioavailability is related to nutrient absorption. Living cells absorb nutrients in their simplest dissociated states. However, these nutrients including C, N, O, P, S, K, Na, Mg are not usually found in their ionic or elemental states but are often locked with complex structures of compounds usually exchanging valence electrons to ensure the stability of the compounds. To maintain the stability of these compounds there is usually a high energy requirement for dissociation. Microorganisms can invest their energy and enzyme machinery towards the breakdown of these compounds in their environment, with their net gain being the ATPs and other nutrients derived from these compounds through this activity. However, hydrocarbons are chemical structures and recalcitrance to degradation is a consequence of the compound stability. When nutrients cannot be reached or absorbed, they are not biologically available for use by the microorganism. In the context of bioremediation, it would therefore, refer to the percentage of the hydrocarbon pollutant that is readily absorbed by the degrading microorganism for their own metabolic activities. Where there are kinetic difficulties in uptake and breakdown, these impact the biodegradation rates and removal of the hydrocarbon pollutant from the environment. PAHs, for example, are very hydrophobic compounds, these are predominantly found in the solid phase or bound to soils, sludge and even sediments [262]. Hydrocarbons differ in their susceptibility to microbial, in general, linear alkanes are the most susceptible to degradation, whilst cyclic alkanes are the least susceptible and therefore are the most persistent in the environment [263, 264]. Some high molecular weight PAHs are very recalcitrant and may not be easily degraded, their physico-chemical characteristics have made their bioavailability to microorganisms impossible.

Dissolvability

Dissolvability refers to the capacity of a hydrocarbon to solubilize within the medium for a facilitated uptake by the degradative microorganisms [265]. Most hydrocarbons are highly hydrophobic and do not easily dissolve in water, if at all. Interaction with amphiphilic cells for the purpose of absorption are often facilitated in living organisms with the production of biosurfactants, as previously discussed. The Dissolvability of a hydrocarbon is also influenced by the presence of reacting functional groups [266]. Functional groups are molecules that facilitate the attachment of microbial enzymes to the hydrocarbon substrates thus, affecting the reactivity of the hydrocarbon [267]. Therefore, the absence of functional groups usually results in kinetic inertness, consequently there are few enzymes that can overcome the problem; for example, anaerobic biodegradation of benzene, lacking oxygen containing functional groups, is generally regarded as extremely difficult (aromatic rings hard to break), although occasionally in some anaerobic consortia its anaerobic degradation has been observed, including under methanogenic conditions [268]. However, Jindrova *et al.*, (2002) provided in their review, a credible explanation of this factor in relation to aerobic biodegradation of benzene, toluene, ethylbenzene and xylene, but they favour natural attenuation as the best option of bioremediation of these compounds [269].

Redox potential

Hydrocarbons can serve as electron donors or electron acceptors [270]. Electron donors act as reducing agents, electron acceptors behave as oxidizing agents during biodegradative reactions as these facilitate the functioning of the redox reactions during hydrocarbon breakdown [271]. Some of the most common electron acceptor groups are chlorine, nitro and azo-groups [272]. As a general rule, increasing the number of electron withdrawing functional groups on recalcitrant hydrocarbons tends to facilitate the initial reductive biotransformation of the molecule.

Stereochemistry

Recalcitrance is often a consequence of structural branching characteristic of high molecular weight hydrocarbons [273]. This large structure and shape means that, it is difficult for the compound to fit as a substrate into the active site of an enzyme. Catalytic efficiency of enzymes often relies on the proximity and orientation of the substrate in relation to the enzyme's binding site [274]. Highly branched hydrocarbons struggle to fit in the binding sites of enzymes which lead to steric interferences (see Table 2), and consequently means these compounds cannot be degraded and remain recalcitrant in the environment [275]. Due to this kinetic limitation, the reduced biodegradative rate can also be a precursor for cytotoxicity in the microbial cells [276].

Low-Medium Toxicity Range

One criteria in selecting a strain of organism to be used in bioremediation programs is the ability to resist changes in the environment either from the introduction of high concentration of compounds (pollutants) or the build-up of metabolites during the degradation process of these pollutants within consortia interactions, especially in semi-closed system as found with *ex situ* remediation where there is often a build-up that could lead to microbial successional behaviour and decline due to changing trophic states and toxicity from metabolic activities. This build-up and toxicity in time affect the degradation efficiency of the microorganism. Concentrations of compounds in any environment affect the behaviour of the habitats of the ecological systems [277]. With *in situ* bioremediation, especially in aquatic environments, the tidal currents and water movement tends to dilute the concentration of pollutants to a certain extent. However, more closed systems, like waste water treatment plants and wetlands are more affected by concentration linked toxicity and the actions of inhibitory compounds that influence microbial growth and degradation rates.

ENVIRONMENTAL FACTORS

Temperature

The temperature of the environment is critical to the degradation rates of hydrocarbons, for example, low temperature causes increased viscosity and decreased volatilization, and consequently lower degradation capacities in microorganisms. Yuniati (2018) explains that at temperatures between 30 – 40 °C, it is possible to observe the highest rates of biodegradation since the hydrocarbons become less viscous and are found in the liquid state increasing the overall kinetics and promoting biochemical reactions [278]. However, research by White (1998), showed the highest degradation rates of diesel oil and alkanes at 0 – 5 °C by *Rhodococcus sp.* strain Q15. Takei *et al.* (2008) isolated a strain of *Rhodococcus* that had highest degradation rates at 20°C [279]. Research conducted over a period of 30 – 64 days by Ribicic *et al.* (2018), in which, they mixed naphthenic oil (troll) and asphaltenic oil (grane) in seawater from Norway showed a higher abundance of the *Colwelliaceae* and *Oceanospirillaceae* families when incubated at 5°C rather than 13°C [280]. Ribicic *et al.*, (2018), demonstrated that biodegradation can be carried out in the arctic waters at temperatures of 0 - 2°C however, they also explained that this was made possible by the *Oleispira* bacterium found in the arctic rather than temperate seawaters. Sun and Kostka, (2019) showed a high abundance of *Colwellia* sp. at 4 and 8°C but at 38°C *Actinobacter* sp. was predominant [281]. Biodegradation rates were shown to be the highest at 20°C when using the yeast *Yarrowia lipolytica* in liquid medium containing 5000mg of diesel per kilo of dry soil [282]. In the same experiment it was observed that over a period of time, the biodegradation rates of the yeast decreased while that of native microorganisms increased. The most plausible reason for the varied temperatures and response by organisms to degradation in each of these environments is the uniqueness of individual microbial adaptation to the specific environmental temperature *versus* the metabolic requirements of the organisms. Importantly, with regards to bioremediation, scientists must ascertain especially with bioaugmentation that organisms selected for treatment, must come from environments that are very similar to those earmarked for bioremediation. This understanding of these temperature requirements especially for *in situ* programs of bioremediation is particularly useful to ensure efficient degradation by these microorganisms [283]. However, the ideal option should be, to employ where possible microorganisms that are capable of degrading pollutants in mesophilic temperature ranges particularly field applications where energy cost consideration is critical in biomass cultivation

pH

The pH of the environment affects microorganisms and their enzyme production or activation hence the degradation rates of the hydrocarbons [284]. Although most microorganisms can grow in a wide range of pH, the different intrinsic and extrinsic pH required for microbial growth, tend to affect the protein structure as well as enzymatic activity [285]. Proteins are usually functional at an optimum range of pH and any slight change may lead to denaturation of the protein [286] and subsequently a likely cessation in enzymatic biochemical reactions necessary for the biodegradation process. The pH affects the binding of substrate (pollutant) to the active site of specific enzymes, because the ionization of substrate and amino acids are critical for catalytic activity. The pH changes may lead to a defective electronic complementarity between the binding site and substrate [287]. A suggested optimum pH range between 6 and 9 for biodegradation of most HC pollutants is between [288]. Previous studies have even shown the necessity of neutralising the pH of soil from 5.2 to a neutral range of 6 to 7. This can be achieved using CaO, for example, with *Sphingomonas paucimobilis,* this change in pH increased the efficiency of degradation of pyrene (Kastner *et al.,* 1998). In another study, a 40% degradation of phenanthrene by *Burkholderia cocovenenans* at pH 5.5 was observed and this was increased to 80% at neutral pH. Moreover, *Pseudomonas* demonstrated capability in degrading hydrocarbon under alkaline conditions [289].

Temperature and pH

Temperature and pH often work in tandem, and their combined effects are significant in the function of microbial cellular activities. Al-Hawash *et al.,* (2018) provide a support to this assertion with experiments using *Aspergillus sp.*RFC-1 strain where they observed that the optimum pH ranged from 6 to 7. However, at pH 7 the highest adsorption rates were obtained at an incubation of 30°C [137]. Although the rates decreased with an increase in pH to a value of 8. This work also highlighted that in acidic environments the enzymatic activity may be competitively inhibited due to the interplay between the H^+ and H_3O^+ ions, perhaps also as a consequence of electrostatic constraints. This can have a negative impact on biodegradation rates, because of the pre-requisite need for microbial cell uptake of substrates [137]. The combined effect of pH and temperature also significantly affects even biosorption, although it is an energy passive mechanism. In general, mesophilic range of temperatures must be paired with near neutral pH ideal for the functioning of living organisms for biodegradation to be effective. It is sometimes necessary to alter the pH of the environment (medium) to facilitate bioremediation processes [290].

Soil Type

In the terrestrial environment lasting impact of hydrocarbon pollution is with the residues of compounds left in the soil. The natural physical and chemical characteristics of the soil can impact the biochemical process kinetics of degradation [291]. Sandy soils and gravel are usually the easiest soils for bioremediation processes while heavy clays and soils with high organic content (eg peaty soils) are harder to bioremediate [292]. Sandy soils and gravel allow higher aeration of the medium for cellular respiration, and have better water and nutrient mobility than clay soils, which are so dense and less permeable [293]. Moreover, clay soils' high organic load implies low biodegradation rate due to preferential binding to less complex C-sources that require less energy for degradation (diauxie). This could lead to different observations during the experimental stage such as diauxic growths, cometabolisms or even non-target substrate affinity [294, 295] but, in a closed system eventually the hydrocarbons would be degraded once all potentially simpler carbons have been consumed. A study conducted by Haghollahi *et al.*, (2016) showed higher degradation rates of 70% in sandy soils with initial concentration of hydrocarbons of 69.62g/kg as opposed to clay soil which had the lowest degradation rate of 23.5% with an initial hydrocarbon concentration of 60.70g/kg [296]. Ali *et al.*, (2020) compared the biodegradation rates by hydrocarbonoclastic bacteria of desert and garden soil that was mixed with 17.3% (v/v) of crude oil and clean water [297]. Their findings pointed to no drastic differences in the biodegradation rate which is intriguing since desert sand may lack certain rate limiting nutrients compared to garden soil. Their study provided little explanation as to the added effect likely from water activity from both soils.

Water Activity

In simple terms, water activity is the minimum content of water in a medium needed for effective microbial metabolism [298]. It varies among bacteria, fungi and algae. Water activity is not the same as moisture content. The moisture content is the total water present in any given medium, whilst water activity is how the water in the given medium will react with microorganisms. Bacteria, yeast and mould, thrive at higher water activity as compared to fungi (mushrooms), however filamentous fungi and algae may also be able to function favourable at high water activity [293]. Water not only allows for the solubility of the nutrients in the medium and chemiotaxis [300] but also permits osmoregulation in the microbial cells [301] and is needed during the stage of chemiosmosis where the water molecule splits and the H^+ ions are pumped across the phospholipid membrane by ATP synthase [142]. Moxley *et al.*, (2019) observed the effect of water content on bacterial breakdown of 2-chlorophenol, their results show that at optimal temperatures biodegradation of the 2-chlorophenol by *Rhodococcus opacus* was highest with 0.9 of water activity. However, while degradation by the thermophilic *Parageobacillus* sp. had high degradation rate when the water activity as at 0.5 showing a potential for bioremediation of halogenated hydrocarbons in arid soils [302]. Ataei, (2018) conducted comparison studies on the impact of water content and activity on hydrocarbon biodegradation rates and noticed that there was a higher degradation rate in the sample which had a more frequent water restoration to the medium (addition of water every 2 days as opposed to every 8 days) [303].

Other Factors affecting Bioremediation Treatments

Nutrient Availability

Organisms need a variety of nutrients to facilitate metabolic activities including macronutrients, micronutrients, vitamins and co-factors. Hydrocarbon-degrading microorganisms have similar requirements and special considerations must be given to these requirements to encourage the proliferation of the necessary biomass needed to ensure the success of pollutant removal. Although the hydrocarbon pollutant may serve as a carbon source for cellular respiration, an electron acceptor or donor in the ATP phosphorylation process, the hydrocarbon is by no means the only nutritional requirement necessary for growth and biomass production as well as other metabolic activities needed to maintain the cells [304]. Other nutrients referred to as growth factors (growth limiting nutrients) include nitrogen (N) and phosphorous (P) (in the forms of nitrates, phosphates, ammonium and phosphate salts), as well as O_2 and other electron acceptors, and their absence or presence in non-optimal concentrations can directly impact microbial kinetics in terms of growth and degradation rates [305]. Lang *et al.*, (2016) studied the effect of limiting nutrients on microbial biodegradation in mangrove ecosystems and determined that they obtained higher degradation rates of 84.7% +/- 4.7% in a medium containing 20% nutrient solution made of N and P [306]. They further added that pH can influence the availability of P and NH_4^+ in the medium. 16S rRNA primer amplification of marine sediments showed high gene abundance of *Alcanivorax sp.* in medium with 0.3% N/0.5% P rather than low nutrient media [307]. It is suggested that the use of bio surfactants assist in the even distribution of N and P however, it can reduce the availability of the hydrocarbons as opposed to the bacterial bloom (biofilm based technology) that has a contrary effect causing limitation of N and P [308]. The appropriate use of these growth-limiting nutrients facilitate the carbon assimilation, meaning that without them even biodegradable hydrocarbons cannot be bioremediated.

Salinity

Salt concentrations in the external environment, critically affects the internal osmotic balance for a living organisms cell and its functioning [309]. Organisms that have to live in high salt concentrated environments develop adaptive features that allow them to function optimally in this environment, sometimes these changes in cell structure and physiology may be a consequence of genetic mutations. Nevertheless, organisms that do not have such features are unable to thrive in high salt concentrations. They suffer osmotic pressure and eventually die. In *ex situ* bioremediation, it is possible to ameliorate this condition by dilution as is done in fed-batch and continuous biodigesters to prevent cytotoxicity of the cells [310, 311]. Qin *et al.*(2012) studied the effect of salinity on hydrocarbon bioremediation in alkaline soil [312]. They observed that at minimal concentrations, salinity can enhance the biodegradation process. Qin and colleagues further added that at the start of the experiment the soil pH was alkaline however, it tends to acidify as the biodegradation of hydrocarbons proceeds, most likely as a consequence of metabolites build-up. However, this contradicts previous studies by Rhykerd *et al.*, (1995) who tested the effect of NaCl (at 40, 120 and 200 dS m^{-1} in sandy clay loam and sandy clay soil [313]. Rhykerd's results showed that there was a decrease in the breakdown of the crude oil by 44% for the 200 dS m^{-1} of NaCl containing soil and even at the lowest concentration of 40 dS m^{-1} there was still a decrease of 10% in mineralization of the oil in both media. They suggested that biodegradation rates can be improved if an adequate salt removal process is put in place prior to microbial digestion of the substrate. This may however not be economically feasible with up scaling.

Oxygen Availability

Aerobic respiration is by far the most efficient form of degradation as it promotes a rapid biomass growth, a consequence of higher ATP generation with O_2 being the main electron acceptor in the aerobic pathway of hydrocarbon degradation [314]. However, an increase in the organic load in a polluted environment means there is a corresponding increase in the biological oxygen demand [315]. The levels of dissolved O_2 may impact on the biodegradation rate as was demonstrated in at least one study of hydrocarbon biodegradation rates and efficiency, these two factors were decreased significantly when the concentration of O_2 was below 10% [316]. However, the rates can be enhanced when O_2 is found at ranges such as

10–40% concentration. This low levels of O_2 results in delayed ATP synthesis [317] thus, reducing the rate of cell division and in bioremediation processes the rate of biodegradation is directly proportional to the quantity of biomass and concomitantly the enzyme produced [318].

Advantages and Disadvantages of Bioremediation Technologies

Several authors have provided their insights on the merits and drawbacks as it relates to bioremediation programs. It is often based on individual experiences on specific aspects and their research objectives and focus. However, the majority of these authors agree that bioremediation offers potential solutions to sustainable environmental management, waste mitigation and remediation. Table **6** articulates a summary of opinions from various authors on the advantages and disadvantages of bioremediation technologies.

Table 6. Advantages and Disadvantages of bioremediation technologies.

Advantages of Bioremediation	Disadvantages of Bioremediation
1. Bioremediation is based on integration of natural processes and is considered an Eco- friendly technology. Microbes proliferate proportionally to the hydrocarbon substrate; The ideal products are usually harmless products and include carbon dioxide, water, and cell biomass [319].	1. Bioremediation is limited to those compounds that are biodegradable. However, there is still a wide variety of xenobiotics that need a solution. Microorganisms require long adaptation times [320]
2. Theoretically, bioremediation treatments are capable of complete destruction of a big range of pollutants. This may assist in the elimination of future liabilities associated with treatment and disposal of recalcitrant pollutants [258]	2. At times the products generated by bioremediation treatments may become more persistent or toxic in the environment [321]
3. Pollutants are dealt with in their phase medium avoiding recurrent pollution of water by soil pollutants and vice-versa [247].	3. Bioprocesses are dynamic and at times may not provide constant outputs since these are reliant on abiotic and biotic factors to first ascertain microbial proliferation, enzymatic expression and resistant to the environment (*In situ* bioremediation) [322].
4. *In situ* bioremediation can easily be performed depending on the degree of contamination and possible challenges of the environment and in a passive manner. This could also be a step forward towards avoiding direct human contact with the pollutant thus, preventing diseases that rise from these interactions [323].	4. It may be difficult to extrapolate from bench and pilot scale research to full scale field operations. Bioaugmentation may interfere with the native microorganisms [324].

(Table 6) cont.....

Advantages of Bioremediation	Disadvantages of Bioremediation
5. In the long run, bioremediation treatments may be more cost effective than other technologies that are used for clean-up of hazardous waste that may need ancillary processes to handle to waste of the remediation process [325].	5 The development of remediation techniques that can address xenobiotic mixtures of pollutants not found homogenously in the environment is due perhaps to soil removal or incineration [326]
	6. Bioremediation treatments may be shunted due to antagonistic interactions and the additives in biostimulation may affect the native microorganisms [327]. There is also high start-up costs and desired outcomes may take time appear [328]

CONCLUSION

There have notably been paradigm shifts in global consciousness towards the consideration for our environment and the startling consequences of the previous lack of concern, which we have experienced in the last decade, as seen with catastrophic incidences of climate change, global warming and the rapid deterioration of water quality, which further reiterates the importance of the ecological studies and application of bioremediation. Bioremediation technologies offer environmentally friendly and adaptable cheap approaches towards solving the inevitable challenge of waste management synonymous with industrial advancements. Moreover, there exists within the framework of these bioremediation technologies, several opportunities for new markets and job creations that are socio-economically sustainable. Both *in situ* and *ex situ* bioremediation approaches come with their respective advantages and disadvantages. But the choice of bioremediation technology employed for any given hydrocarbon polluted site must be based on informed research relevant uniquely to the site. Although cost is an important consideration, it should not be an overarching motivation for the choice of technology. It is also important that we use insights and knowledge from various researchers but it must by no means be the absolute basis for developing bioremediation programs. It is important to research and understand the site and its uniqueness, to make a determination of a best-fit approach. It is also important to consider that most remediation programs succeed only with a combination of multiple complementing techniques. Currently, there are several field tests that are ill-designed, lack credible controls and are not managed in a manner that provides replicable results. A wholesome acceptance and implementation of such designs without adequate research and planning is often costly in the long-run. It is advised that adequate consulting before bioremediation technologies application on a large scale is achieved. Although this technology and bioprospecting have been ongoing for at least two decades, there are a wide variety of microorganisms and metabolites that are potentially useful for detoxification of hydrocarbons that are still largely unexplored. A better understanding of their natural role in the environment can further improve their usefulness in bioremediation programs. Thus, microbial diversity studies should routinely be done even on previously investigated sites as adaptation is evolving with substrate composition and concentration changes. Such studies are needed and contribute to the body of knowledge needed for long-term sustainability of preserving our environment.

CONSENT FOR PUBLICATION

Not applicable.

CONFLICT OF INTEREST

The authors declare no conflict of interest, financial or otherwise.

ACKNOWLEDGEMENTS

Authors are grateful to the South African Government's Department of Science and Innovation (DSI) as well as the Technology Innovation Agency (TIA) for the financial support provided through grant numbers DST/CON 0197/2017 and 2018/FUN/0166 respectively. The various South African National Research Foundation (NRF) funding schemes related to this study are also appreciated.

REFERENCES

[1]   Wang S, Xu Y, Lin Z, Zhang J, Norbu N, Liu W. The harm of petroleum-polluted soil and its remediation research. AIP Conf Proc 2017; 1864: 020222.
[http://dx.doi.org/10.1063/1.4993039]

[2]   Hunt LJ, Duca D, Dan T, Knopper LD. Petroleum hydrocarbon (PHC) uptake in plants: A literature review. Environ Pollut 2019; 245: 472-84.
[http://dx.doi.org/10.1016/j.envpol.2018.11.012] [PMID: 30458377]

[3]   Das N, Chandran P. Microbial degradation of petroleum hydrocarbon contaminants: an overview. Biotechnol Res Int 2011; 2011: 1-13.
[http://dx.doi.org/10.4061/2011/941810] [PMID: 21350672]

[4]   Eze VC, Onwuakor CE, Orok FE. Microbiological and Physicochemical Characteristics of Soil Contaminated With Used Petroleum Products in Umuahia, Abia State, Nigeria. J Appl Environ Microbiol 2014; 2: 281-6.
[http://dx.doi.org/10.12691/jaem-2-6-3]

[5]   Liang Y, Zhao H, Deng Y, Zhou J, Li G, Sun B. Long-term oil contamination alters the molecular ecological networks of soil microbial functional genes. Front Microbiol 2016; 7: 60.
[http://dx.doi.org/10.3389/fmicb.2016.00060] [PMID: 26870020]

[6]   Pichtel J. Oil and gas production wastewater: Soil contamination and pollution prevention. Appl Environ Soil Sci 2016; 2016: 1-24.
[http://dx.doi.org/10.1155/2016/2707989]

[7]   Xu R, Zhang K, Liu P, *et al.* A critical review on the interaction of substrate nutrient balance and microbial community structure and function in anaerobic co-digestion. Bioresour Technol 2018; 247:

1119-27.
[http://dx.doi.org/10.1016/j.biortech.2017.09.095] [PMID: 28958888]

[8]   Bas HS, Dindar E. International Biodeterioration & Biodegradation Variations of soil enzyme activities in petroleum-hydrocarbon contaminated soil. 2015; 105: 268-75.

[9]   Ali MHH, Al-Qahtani KM. Assessment of some heavy metals in vegetables, cereals and fruits in Saudi Arabian markets. Egypt J Aquat Res 2012; 38(1): 31-7.
[http://dx.doi.org/10.1016/j.ejar.2012.08.002]

[10]   Lindén O, Pålsson J. Oil contamination in Ogoniland, Niger Delta. Ambio 2013; 42(6): 685-701.
[http://dx.doi.org/10.1007/s13280-013-0412-8] [PMID: 23749556]

[11]   Environmental Assessment of Ogoniland. United Nations Environment Programme 2011; 39: 205.

[12]   Ranjbar Jafarabadi A, Riyahi Bakhtiari A, Aliabadian M, Laetitia H, Shadmehri Toosi A, Yap CK. First report of bioaccumulation and bioconcentration of aliphatic hydrocarbons (AHs) and persistent organic pollutants (PAHs, PCBs and PCNs) and their effects on alcyonacea and scleractinian corals and their endosymbiotic algae from the Persian Gulf, Iran: Inter and intra-species differences. Sci Total Environ 2018; 627: 141-57.
[http://dx.doi.org/10.1016/j.scitotenv.2018.01.185] [PMID: 29426136]

[13]   Nicodem D, Guedes CLB, Correa RJ, Fernandes MCZ. Photochemical processes and the environmental impact of petroleum spills. Biogeochemistry 1997; 39(2): 121-38.
[http://dx.doi.org/10.1023/A:1005802027380]

[14]   Asquith EA, Geary PM, Evans C, *et al.* Comparative bioremediation of petroleum hydrocarbon-contaminated soil by biostimulation, bioaugmentation and surfactant addition. Ind J Environ Health 2012; 1(5): 637-50.

[15]   Arellano P, Tansey K, Balzter H, Boyd DS. Detecting the effects of hydrocarbon pollution in the Amazon forest using hyperspectral satellite images. Environ Pollut 2015; 205: 225-39.
[http://dx.doi.org/10.1016/j.envpol.2015.05.041] [PMID: 26074164]

[16]   Terry GM, Stokes NJ, Lucas PW, Hewitt CN. Effects of reactive hydrocarbons and hydrogen peroxide on antioxidant activity in cherry leaves. Environ Pollut 1995; 88(1): 19-26.
[http://dx.doi.org/10.1016/0269-7491(95)91044-L] [PMID: 15091565]

[17]   Gibson DT, Parales RE. Aromatic hydrocarbon dioxygenases in environmental biotechnology. Curr Opin Biotechnol 2000; 11(3): 236-43.
[http://dx.doi.org/10.1016/S0958-1669(00)00090-2] [PMID: 10851146]

[18]   Isaksson C. Pollution and its impact on wild animals: a meta-analysis on oxidative stress. EcoHealth 2010; 7(3): 342-50.
[http://dx.doi.org/10.1007/s10393-010-0345-7] [PMID: 20865439]

[19]   Cherr GN, Fairbairn E, Whitehead A. Impacts of Petroleum-Derived Pollutants on Fish Development. Annu Rev Anim Biosci 2017; 5(1): 185-203.
[http://dx.doi.org/10.1146/annurev-animal-022516-022928] [PMID: 27959669]

[20]   Honda M, Suzuki N. Toxicities of polycyclic aromatic hydrocarbons for aquatic animals. Int J Environ Res Public Health 2020; 17(4): 1363.
[http://dx.doi.org/10.3390/ijerph17041363] [PMID: 32093224]

[21]   Kampa M, Castanas E. Human health effects of air pollution. Environ Pollut 2008; 151(2): 362-7.
[http://dx.doi.org/10.1016/j.envpol.2007.06.012] [PMID: 17646040]

[22]   Kuppusamy S, Naga R. Impact of Total Petroleum Hydrocarbons on Human Health. Total Petroleum Hydrocarbons. Springer Cham 2020; pp. 139-65.

[23]   Megharaj M, Singleton I, McClure NC, Naidu R. Influence of petroleum hydrocarbon contamination on microalgae and microbial activities in a long-term contaminated soil. Arch Environ Contam Toxicol 2000; 38(4): 439-45.

[http://dx.doi.org/10.1007/s002449910058] [PMID: 10787094]

[24] Lobban CS, Harrison PJ. Light and photosynthesis Cambridge. In: Press CU, Ed. Seaweed Ecology and Physiology. Cambridge 1994; pp. 123-61.
[http://dx.doi.org/10.1017/CBO9780511626210.005]

[25] Martins LF, Peixoto RS. Biodegradation of petroleum hydrocarbons in hypersaline environments. Braz J Microbiol 2012; 43(3): 865-72.
[http://dx.doi.org/10.1590/S1517-83822012000300003] [PMID: 24031900]

[26] Hasan IF, Al-Jawhari . Role of filamentous fungi to remove petroleum hydrocarbons from the environment. Microb Action Hydrocarb 2019; 567-80.
[http://dx.doi.org/10.1007/978-981-13-1840-5_23]

[27] Xu X, Liu W, Tian S, *et al.* Petroleum Hydrocarbon-Degrading Bacteria for the Remediation of Oil Pollution Under Aerobic Conditions: A Perspective Analysis. Front Microbiol 2018; 9: 2885.
[http://dx.doi.org/10.3389/fmicb.2018.02885] [PMID: 30559725]

[28] van Dorst J, Siciliano SD, Winsley T, Snape I, Ferrari BC. Bacterial targets as potential indicators of diesel fuel toxicity in subantarctic soils. Appl Environ Microbiol 2014; 80(13): 4021-33.
[http://dx.doi.org/10.1128/AEM.03939-13] [PMID: 24771028]

[29] Labud V, Garcia C, Hernandez T. Effect of hydrocarbon pollution on the microbial properties of a sandy and a clay soil. Chemosphere 2007; 66(10): 1863-71.
[http://dx.doi.org/10.1016/j.chemosphere.2006.08.021] [PMID: 17083964]

[30] Cabello MN. Hydrocarbon pollution: its effect on native arbuscular mycorrhizal fungi (AMF). FEMS Microbiol Ecol 1997; 22(3): 233-6.
[http://dx.doi.org/10.1111/j.1574-6941.1997.tb00375.x]

[31] Obire O, Anyanwu EC. Impact of various concentrations of crude oil on fungal populations of soil. Int J Environ Sci Technol 2009; 6(2): 211-8.
[http://dx.doi.org/10.1007/BF03327624]

[32] Al-Hawash AB, Zhang X, Ma F. Removal and biodegradation of different petroleum hydrocarbons using the filamentous fungus *Aspergillus* sp. RFC-1. MicrobiologyOpen 2019; 8(1): e00619.
[http://dx.doi.org/10.1002/mbo3.619] [PMID: 29577679]

[33] Head IM, Jones DM, Röling WFM. Marine microorganisms make a meal of oil. Nat Rev Microbiol 2006; 4(3): 173-82.
[http://dx.doi.org/10.1038/nrmicro1348] [PMID: 16489346]

[34] Deni J, Penninckx MJ. Nitrification and autotrophic nitrifying bacteria in a hydrocarbon-polluted soil. Appl Environ Microbiol 1999; 65(9): 4008-13.
[http://dx.doi.org/10.1128/AEM.65.9.4008-4013.1999] [PMID: 10473409]

[35] Yu Y, Zhang Y, Zhao N, *et al.* Remediation of crude oil-polluted soil by the bacterial rhizosphere community of suaeda salsa revealed by 16S rRNA genes. Int J Environ Res Public Health 2020; 17(5): 1471.
[http://dx.doi.org/10.3390/ijerph17051471] [PMID: 32106510]

[36] Whittaker M, Pollard SJT, Fallick TE. Characterisation of refractory wastes at heavy oil-contaminated sites: A review of conventional and novel analytical methods. Environ Technol 1995; 16(11): 1009-33.
[http://dx.doi.org/10.1080/09593331608616339]

[37] Souza TS, Hencklein FA, Angelis DF, Gonçalves RA, Fontanetti CS. The Allium cepa bioassay to evaluate landfarming soil, before and after the addition of rice hulls to accelerate organic pollutants biodegradation. Ecotoxicol Environ Saf 2009; 72(5): 1363-8.
[http://dx.doi.org/10.1016/j.ecoenv.2009.01.009] [PMID: 19285726]

[38] Sanscartier D, Reimer K, Zeeb B, George K. Management of hydrocarbon-contaminated soil through bioremediation and landfill disposal at a remote location in Northern CanadaA paper submitted to the Journal of Environmental Engineering and Science. Can J Civ Eng 2010; 37(1): 147-55.

[http://dx.doi.org/10.1139/L09-130]

[39]   Sanscartier D, Laing T, Reimer K, Zeeb B. Bioremediation of weathered petroleum hydrocarbon soil contamination in the Canadian High Arctic: Laboratory and field studies. Chemosphere 2009; 77(8): 1121-6.
[http://dx.doi.org/10.1016/j.chemosphere.2009.09.006] [PMID: 19781739]

[40]   Hickman ZA, Reid BJ. Earthworm assisted bioremediation of organic contaminants. Environ Int 2008; 34(7): 1072-81.
[http://dx.doi.org/10.1016/j.envint.2008.02.013] [PMID: 18433870]

[41]   Truskewycz A, Gundry TD, Khudur LS, *et al.* Petroleum hydrocarbon contamination in terrestrial ecosystems—fate and microbial responses. Molecules 2019; 24(18): 3400.
[http://dx.doi.org/10.3390/molecules24183400] [PMID: 31546774]

[42]   Lynch JM, Moffat AJ. Bioremediation - prospects for the future application of innovative applied biological research. Ann Appl Biol 2005; 146(2): 217-21.
[http://dx.doi.org/10.1111/j.1744-7348.2005.040115.x]

[43]   Howard P, Meylan W, Aronson D, *et al.* A new biodegradation prediction model specific to petroleum hydrocarbons. Environ Toxicol Chem 2005; 24(8): 1847-60.
[http://dx.doi.org/10.1897/04-453R.1] [PMID: 16152953]

[44]   Lindeberg MR, Maselko J, Heintz RA, Fugate CJ, Holland L. Conditions of persistent oil on beaches in Prince William Sound 26 years after the Exxon Valdez spill. Deep Sea Res Part II Top Stud Oceanogr 2018; 147: 9-19.
[http://dx.doi.org/10.1016/j.dsr2.2017.07.011]

[45]   Andersen RG, Booth EC, Marr LC, Widdowson MA, Novak JT. Volatilization and biodegradation of naphthalene in the vadose zone impacted by phytoremediation. Environ Sci Technol 2008; 42(7): 2575-81.
[http://dx.doi.org/10.1021/es0714336] [PMID: 18504999]

[46]   Goel M, Chovelon JM, Ferronato C, Bayard R, Sreekrishnan TR. The remediation of wastewater containing 4-chlorophenol using integrated photocatalytic and biological treatment. J Photochem Photobiol B 2010; 98(1): 1-6.
[http://dx.doi.org/10.1016/j.jphotobiol.2009.09.006] [PMID: 19914843]

[47]   Hutchinson JH, Simonsen BL. Cleanup operations after the 1976 SS sansinena explosionan industrial perspective. 2005 Int Oil Spill Conf IOSC 2005 1097. 2005.

[48]   Ijoma GN, Tekere M. Potential microbial applications of co-cultures involving ligninolytic fungi in the bioremediation of recalcitrant xenobiotic compounds. Int J Environ Sci Technol 2017; 14(8): 1787-806.
[http://dx.doi.org/10.1007/s13762-017-1269-3]

[49]   Urakawa H, Rajan S, Feeney ME, Sobecky PA, Mortazavi B. Ecological response of nitrification to oil spills and its impact on the nitrogen cycle. Environ Microbiol 2019; 21(1): 18-33.
[http://dx.doi.org/10.1111/1462-2920.14391] [PMID: 30136386]

[50]   Peterson CH, Rice SD, Short JW, *et al.* Long-term ecosystem response to the exxon valdez oil spill. Science 2003; 302(5653): 2082-6.
[http://dx.doi.org/10.1126/science.1084282]

[51]   Burns KA, Garrity SD, Levings SC. How many years until mangrove ecosystems recover from catastrophic oil spills? Mar Pollut Bull 1993; 26(5): 239-48.
[http://dx.doi.org/10.1016/0025-326X(93)90062-O]

[52]   Horel A, Bernard RJ, Mortazavi B. Impact of crude oil exposure on nitrogen cycling in a previously impacted Juncus roemerianus salt marsh in the northern Gulf of Mexico. Environ Sci Pollut Res Int 2014; 21(11): 6982-93.
[http://dx.doi.org/10.1007/s11356-014-2599-z] [PMID: 24510533]

[53] Shi R, Yu K. Impact of exposure of crude oil and dispersant (COREXIT® EC 9500A) on denitrification and organic matter mineralization in a Louisiana salt marsh sediment. Chemosphere 2014; 108: 300-5.
[http://dx.doi.org/10.1016/j.chemosphere.2014.01.055] [PMID: 24582034]

[54] Hinshaw SE, Tatariw C, Flournoy N, *et al.* Vegetation Loss Decreases Salt Marsh Denitrification Capacity: Implications for Marsh Erosion. Environ Sci Technol 2017; 51(15): 8245-53.
[http://dx.doi.org/10.1021/acs.est.7b00618] [PMID: 28616973]

[55] Kator H, Oppenheimer CH, Miget RJ. Microbial degradation of oil pollutants. Biol Conserv 1971; 4(1): 12.
[http://dx.doi.org/10.1016/0006-3207(71)90046-2]

[56] Mitsch WJ, Gosselink JG. Wetlands. 4th ed., J. Wiley & Sons, Inc. 2007.

[57] Bovio E, Gnavi G, Prigione V, *et al.* The culturable mycobiota of a Mediterranean marine site after an oil spill: isolation, identification and potential application in bioremediation. Sci Total Environ 2017; 576: 310-8.
[http://dx.doi.org/10.1016/j.scitotenv.2016.10.064] [PMID: 27788446]

[58] Lim MW, Lau EV, Poh PE. A comprehensive guide of remediation technologies for oil contaminated soil — Present works and future directions. Mar Pollut Bull 2016; 109(1): 14-45.
[http://dx.doi.org/10.1016/j.marpolbul.2016.04.023] [PMID: 27267117]

[59] Lahel A, Fanta AB, Sergienko N, *et al.* Effect of process parameters on the bioremediation of diesel contaminated soil by mixed microbial consortia. Int Biodeterior Biodegradation 2016; 113: 375-85.
[http://dx.doi.org/10.1016/j.ibiod.2016.05.005]

[60] Neethu CS, Saravanakumar C, Purvaja R, Robin RS, Ramesh R. Oil-Spill Triggered Shift in Indigenous Microbial Structure and Functional Dynamics in Different Marine Environmental Matrices. Sci Rep 2019; 9(1): 1354.
[http://dx.doi.org/10.1038/s41598-018-37903-x] [PMID: 30718727]

[61] Ijoma GN, Tonderayi M. The role of Oxidoreductases in the degradation, elimination and monitoring of environmental pollutants. In: Toft AC, Ed. Frontiers in Bacteriology Research. Nova Science, New York 2020; pp. 1-92.

[62] Tallur PN, Megadi VB, Ninnekar HZ. Biodegradation of Cypermethrin by Micrococcus sp. strain CPN 1. Biodegradation 2008; 19(1): 77-82.
[http://dx.doi.org/10.1007/s10532-007-9116-8] [PMID: 17431802]

[63] François A, Mathis H, Godefroy D, Piveteau P, Fayolle F, Monot F. Biodegradation of methyl tert-butyl ether and other fuel oxygenates by a new strain, *Mycobacterium austroafricanum* IFP 2012. Appl Environ Microbiol 2002; 68(6): 2754-62.
[http://dx.doi.org/10.1128/AEM.68.6.2754-2762.2002] [PMID: 12039730]

[64] Atashgahi S, Maphosa F, Doğan E, Smidt H, Springael D, Dejonghe W. Small-scale oxygen distribution determines the vinyl chloride biodegradation pathway in surficial sediments of riverbed hyporheic zones. FEMS Microbiol Ecol 2013; 84(1): 133-42.
[http://dx.doi.org/10.1111/1574-6941.12044] [PMID: 23167955]

[65] Field JA, Sierra-Alvarez R. Biodegradability of chlorinated solvents and related chlorinated aliphatic compounds. Rev Environ Sci Biotechnol 2004; 3(3): 185-254.
[http://dx.doi.org/10.1007/s11157-004-4733-8]

[66] Adamson DT, Parkin GF. Impact of mixtures of chlorinated aliphatic hydrocarbons on a high-rate, tetrachloroethene-dechlorinating enrichment culture. Environ Sci Technol 2000; 34(10): 1959-65.
[http://dx.doi.org/10.1021/es990809f]

[67] Wild A, Hermann R, Leisinger T. Isolation of an anaerobic bacterium which reductively dechlorinates tetrachloroethene and trichloroethene. Biodegradation 1997-1997; 7(6): 507-11.
[http://dx.doi.org/10.1007/BF00115297] [PMID: 9188197]

[68] Mattes TE, Alexander AK, Coleman NV. Aerobic biodegradation of the chloroethenes: pathways, enzymes, ecology, and evolution. FEMS Microbiol Rev 2010; 34(4): 445-75.
[http://dx.doi.org/10.1111/j.1574-6976.2010.00210.x] [PMID: 20146755]

[69] Rui L, Kwon YM, Reardon KF, Wood TK. Metabolic pathway engineering to enhance aerobic degradation of chlorinated ethenes and to reduce their toxicity by cloning a novel glutathione S-transferase, an evolved toluene o-monooxygenase, and γ-glutamylcysteine synthetase. Environ Microbiol 2004; 6(5): 491-500.
[http://dx.doi.org/10.1111/j.1462-2920.2004.00586.x] [PMID: 15049922]

[70] Gerritse J, Renard V, Gottschal JC, Visser J. Complete degradation of tetrachloroethene by combining anaerobic dechlorinating and aerobic methanotrophic enrichment cultures. Appl Microbiol Biotechnol 1995; 43(5): 920-8.
[http://dx.doi.org/10.1007/BF02431929] [PMID: 7576559]

[71] Moe WM, Reynolds SJ, Griffin MA, McReynolds JB. Bioremediation Strategies Aimed at Stimulating Chlorinated Solvent Dehalogenation Can Lead to Microbially-Mediated Toluene Biogenesis. Environ Sci Technol 2018; 52(16): 9311-9.
[http://dx.doi.org/10.1021/acs.est.8b02081] [PMID: 30044084]

[72] Murray WD, Richardson M. Progress toward the biological treatment of C$_1$ and C$_2$ halogenated hydrocarbons. Crit Rev Environ Sci Technol 1993; 23(3): 195-217.
[http://dx.doi.org/10.1080/10643389309388451]

[73] Beam HW, Perry JJ. Microbial degradation of cycloparaffinic hydrocarbons *via* co metabolism and commensalism. J Gen Microbiol 1974; 82(1): 163-9.
[http://dx.doi.org/10.1099/00221287-82-1-163]

[74] Liu C, Wang J, Ji X, Qian H, Huang L, Lu X. The biomethane producing potential in China: A theoretical and practical estimation. Chin J Chem Eng 2016; 24(7): 920-8.
[http://dx.doi.org/10.1016/j.cjche.2015.12.025]

[75] Dejonghe W, Berteloot E, Goris J, *et al.* Synergistic degradation of linuron by a bacterial consortium and isolation of a single linuron-degrading variovorax strain. Appl Environ Microbiol 2003; 69(3): 1532-41.
[http://dx.doi.org/10.1128/AEM.69.3.1532-1541.2003] [PMID: 12620840]

[76] Asemoloye MD, Ahmad R, Jonathan SG. Synergistic rhizosphere degradation of γ-hexachlorocyclohexane (lindane) through the combinatorial plant-fungal action. PLoS One 2017; 12(8): e0183373.
[http://dx.doi.org/10.1371/journal.pone.0183373] [PMID: 28859100]

[77] Maltseva O, McGowan C, Fulthorpe R, Oriel P. Degradation of 2,4-dichlorophenoxyacetic acid by haloalkaliphilic bacteria. Microbiology (Reading) 1996; 142(5): 1115-22.
[http://dx.doi.org/10.1099/13500872-142-5-1115] [PMID: 8704953]

[78] Schmidt SK, Gier MJ. Coexisting bacterial populations responsible for multiphasic mineralization kinetics in soil. Appl Environ Microbiol 1990; 56(9): 2692-7.
[http://dx.doi.org/10.1128/aem.56.9.2692-2697.1990] [PMID: 16348277]

[79] Ciok A, Budzik K, Zdanowski MK, *et al.* Plasmids of Psychrotolerant *Polaromonas* spp. Isolated From Arctic and Antarctic Glaciers – Diversity and Role in Adaptation to Polar Environments. Front Microbiol 2018; 9: 1285.
[http://dx.doi.org/10.3389/fmicb.2018.01285] [PMID: 29967598]

[80] Panicker G, Mojib N, Aislabie J, Bej AK. Detection, expression and quantitation of the biodegradative genes in Antarctic microorganisms using PCR. Antonie van Leeuwenhoek 2010; 97(3): 275-87.
[http://dx.doi.org/10.1007/s10482-009-9408-6] [PMID: 20043207]

[81] Powell SM, Ferguson SH, Bowman JP, Snape I. Using real-time PCR to assess changes in the hydrocarbon-degrading microbial community in Antarctic soil during bioremediation. Microb Ecol

2006; 52(3): 523-32.
[http://dx.doi.org/10.1007/s00248-006-9131-z] [PMID: 16944337]

[82]   Witzig R, Junca H, Hecht HJ, Pieper DH. Assessment of toluene/biphenyl dioxygenase gene diversity in benzene-polluted soils: links between benzene biodegradation and genes similar to those encoding isopropylbenzene dioxygenases. Appl Environ Microbiol 2006; 72(5): 3504-14.
[http://dx.doi.org/10.1128/AEM.72.5.3504-3514.2006] [PMID: 16672497]

[83]   Throne-Holst M, Markussen S, Winnberg A, Ellingsen TE, Kotlar HK, Zotchev SB. Utilization of n-alkanes by a newly isolated strain of *Acinetobacter venetianus*: the role of two AlkB-type alkane hydroxylases. Appl Microbiol Biotechnol 2006; 72(2): 353-60.
[http://dx.doi.org/10.1007/s00253-005-0262-9] [PMID: 16520925]

[84]   Mishra S, Sarma PM, Lal B. Crude oil degradation efficiency of a recombinant *Acinetobacter baumannii* strain and its survival in crude oil-contaminated soil microcosm. FEMS Microbiol Lett 2004; 235(2): 323-31.
[http://dx.doi.org/10.1111/j.1574-6968.2004.tb09606.x] [PMID: 15183881]

[85]   Kiyohara H, Sugiyama M, Mondello FJ, Gibson DT, Yano K. Plasmid involvement in the degradation of polycyclic aromatic hydrocarbons by a Beijerinckia species. Biochem Biophys Res Commun 1983; 111(3): 939-45.
[http://dx.doi.org/10.1016/0006-291X(83)91390-6] [PMID: 6838595]

[86]   Kahng HY, Malinverni JC, Majko MM, Kukor JJ. Genetic and functional analysis of the tbc operons for catabolism of alkyl- and chloroaromatic compounds in *Burkholderia* sp. strain JS150. Appl Environ Microbiol 2001; 67(10): 4805-16.
[http://dx.doi.org/10.1128/AEM.67.10.4805-4816.2001] [PMID: 11571188]

[87]   Obayori OS, Salam LB. Degradation of polycyclic aromatic hydrocarbons: Role of plasmids. Sci Res Essays 2010; 5: 4096-109.

[88]   Yakimov MM, Crisafi F, Messina E, *et al.* Analysis of defence systems and a conjugative IncP-1 plasmid in the marine polyaromatic hydrocarbons-degrading bacterium *Cycloclasticus* sp. 78-ME. Environ Microbiol Rep 2016; 8(4): 508-19.
[http://dx.doi.org/10.1111/1758-2229.12424] [PMID: 27345842]

[89]   Foght JM, Westlake DWS. Transposon and spontaneous deletion mutants of plasmid-borne genes encoding polycyclic aromatic hydrocarbon degradation by a strain of *Pseudomonas fluorescens*. Biodegradation 1996; 7(4): 353-66.
[http://dx.doi.org/10.1007/BF00115749] [PMID: 8987893]

[90]   Mavrodi DV, Kovalenko NP, Sokolov SL, Parfenyuk VG, Kosheleva IA, Boronin AM. [Identification of the key genes of naphthalene catabolism in soil DNA]. Microbiology 2003; 72(5): 597-604.
[http://dx.doi.org/10.1023/A:1026055503274] [PMID: 14679907]

[91]   Dunn NW, Gunsalus IC. Transmissible plasmid coding early enzymes of naphthalene oxidation in *Pseudomonas putida*. J Bacteriol 1973; 114(3): 974-9.
[http://dx.doi.org/10.1128/jb.114.3.974-979.1973] [PMID: 4712575]

[92]   Denome SA, Stanley DC, Olson ES, Young KD. Metabolism of dibenzothiophene and naphthalene in *Pseudomonas* strains: complete DNA sequence of an upper naphthalene catabolic pathway. J Bacteriol 1993; 175(21): 6890-901.
[http://dx.doi.org/10.1128/jb.175.21.6890-6901.1993] [PMID: 8226631]

[93]   Kurkela S, Lehväslaiho H, Palva ET, Teeri TH. Cloning, nucleotide sequence and characterization of genes encoding naphthalene dioxygenase of *Pseudomonas putida* strain NCIB9816. Gene 1988; 73(2): 355-62.
[http://dx.doi.org/10.1016/0378-1119(88)90500-8] [PMID: 3243438]

[94]   Van Beilen JB, Li Z, Duetz WA, Smits THM, Witholt B. Diversity of Alkane Hydroxylase Systems in the Environment. Oil Gas Sci Technol 2003; 58(4): 427-40.
[http://dx.doi.org/10.2516/ogst:2003026]

[95] Imperato V, Portillo-Estrada M, McAmmond BM, *et al.* Genomic diversity of two hydrocarbon-degrading and plant growth-promoting *Pseudomonas* species isolated from the oil field of bóbrka (Poland). Genes (Basel) 2019; 10(6): 443.
[http://dx.doi.org/10.3390/genes10060443] [PMID: 31212674]

[96] Fuenmayor SL, Wild M, Boyes AL, Williams PA. A gene cluster encoding steps in conversion of naphthalene to gentisate in *Pseudomonas* sp. strain U2. J Bacteriol 1998; 180(9): 2522-30.
[http://dx.doi.org/10.1128/JB.180.9.2522-2530.1998] [PMID: 9573207]

[97] Whyte LG, Smits THM, Labbé D, Witholt B, Greer CW, van Beilen JB. Gene cloning and characterization of multiple alkane hydroxylase systems in *Rhodococcus* strains Q15 and NRRL B-16531. Appl Environ Microbiol 2002; 68(12): 5933-42.
[http://dx.doi.org/10.1128/AEM.68.12.5933-5942.2002] [PMID: 12450813]

[98] Romine MF, Stillwell LC, Wong KK, *et al.* Complete sequence of a 184-kilobase catabolic plasmid from *Sphingomonas aromaticivorans* F199. J Bacteriol 1999; 181(5): 1585-602.
[http://dx.doi.org/10.1128/JB.181.5.1585-1602.1999] [PMID: 10049392]

[99] Cho CH, Oh KY, Kim SK, Yeo JG, Sharma P. Pervaporative seawater desalination using NaA zeolite membrane: Mechanisms of high water flux and high salt rejection. J Membr Sci 2011; 371(1-2): 226-38.
[http://dx.doi.org/10.1016/j.memsci.2011.01.049]

[100] Hwa KS, Seung-Bok H, Jeong KH, *et al.* Sphingomonas sp. HS3620 phenanthrene. Korean J Microbiol 2005; 41: 201-7.

[101] Mallick S, Dutta TK. Kinetics of phenanthrene degradation by *Staphylococcus* sp. strain PN/Y involving 2-hydroxy-1-naphthoic acid in a novel metabolic pathway. Process Biochem 2008; 43(9): 1004-8.
[http://dx.doi.org/10.1016/j.procbio.2008.04.022]

[102] Habe H, Chung JS, Ishida A, *et al.* The fluorene catabolic linear plasmid in *Terrabacter* sp. strain DBF63 carries the β-ketoadipate pathway genes, pcaRHGBDCFIJ, also found in proteobacteria. Microbiology (Reading) 2005; 151(11): 3713-22.
[http://dx.doi.org/10.1099/mic.0.28215-0] [PMID: 16272392]

[103] Pal S, Kundu A, Banerjee TD, *et al.* Genome analysis of crude oil degrading *Franconibacter pulveris* strain DJ34 revealed its genetic basis for hydrocarbon degradation and survival in oil contaminated environment. Genomics 2017; 109(5-6): 374-82.
[http://dx.doi.org/10.1016/j.ygeno.2017.06.002] [PMID: 28625866]

[104] Whyte LG, Bourbonniére L, Greer CW. Biodegradation of petroleum hydrocarbons by psychrotrophic *Pseudomonas* strains possessing both alkane (alk) and naphthalene (nah) catabolic pathways. Appl Environ Microbiol 1997; 63(9): 3719-23.
[http://dx.doi.org/10.1128/aem.63.9.3719-3723.1997] [PMID: 9293024]

[105] Liu Q, Tang J, Bai Z, Hecker M, Giesy JP. Distribution of petroleum degrading genes and factor analysis of petroleum contaminated soil from the Dagang Oilfield, China. Sci Rep 2015; 5(1): 11068.
[http://dx.doi.org/10.1038/srep11068] [PMID: 26086670]

[106] Paisse S, Duran R, Coulon F, Goñi-Urriza M. Are alkane hydroxylase genes (alkB) relevant to assess petroleum bioremediation processes in chronically polluted coastal sediments? Appl Microbiol Biotechnol 2011; 92(4): 835-44.
[http://dx.doi.org/10.1007/s00253-011-3381-5] [PMID: 21660544]

[107] Shiju M, Yahya HH. A new method for the detection of oil degrading genes in *Pseudomonas aeruginosa* based on transformation and PCR hybridization. Int J Biotechnol Mol Biol Res 2015; 6(1): 1-6.
[http://dx.doi.org/10.5897/IJBMBR2014.0218]

[108] DeBruyn JM, Chewning CS, Sayler GS. Comparative quantitative prevalence of Mycobacteria and

functionally abundant nidA, nahAc, and nagAc dioxygenase genes in coal tar contaminated sediments. Environ Sci Technol 2007; 41(15): 5426-32.
[http://dx.doi.org/10.1021/es070406c] [PMID: 17822112]

[109]   Kamenos NA, Burdett HL, Aloisio E, *et al*. Coralline algal structure is more sensitive to rate , rather than the magnitude , of ocean acidification. 2013; 3621-8.
[http://dx.doi.org/10.1111/gcb.12351]

[110]   Flocco CG, Gomes NCM, Mac Cormack W, Smalla K. Occurrence and diversity of naphthalene dioxygenase genes in soil microbial communities from the Maritime Antarctic. Environ Microbiol 2009; 11(3): 700-14.
[http://dx.doi.org/10.1111/j.1462-2920.2008.01858.x] [PMID: 19278452]

[111]   Bordenave S, Goñi-urriza M, Vilette C, Blanchard S, Caumette P, Duran R. Diversity of ring-hydroxylating dioxygenases in pristine and oil contaminated microbial mats at genomic and transcriptomic levels. Environ Microbiol 2008; 10(12): 3201-11.
[http://dx.doi.org/10.1111/j.1462-2920.2008.01707.x] [PMID: 18662307]

[112]   Liang C, Huang Y, Wang H. *pahE*, a Functional Marker Gene for Polycyclic Aromatic Hydrocarbon-Degrading Bacteria. Appl Environ Microbiol 2019; 85(3): e02399-18.
[http://dx.doi.org/10.1128/AEM.02399-18] [PMID: 30478232]

[113]   Khanafer M, Al-Awadhi H, Radwan S. Coliform Bacteria for Bioremediation of Waste Hydrocarbons. BioMed Res Int 2017; 2017: 1-8.
[http://dx.doi.org/10.1155/2017/1838072] [PMID: 29082238]

[114]   Rhodes CJ. Mycoremediation (bioremediation with fungi) – growing mushrooms to clean the earth. Chem Spec Bioavail 2014; 26(3): 196-8.
[http://dx.doi.org/10.3184/095422914X14047407349335]

[115]   Lawton RJ, Mata L, de Nys R, Paul NA. Algal bioremediation of waste waters from land-based aquaculture using ulva: selecting target species and strains. PLoS One 2013; 8(10): e77344.
[http://dx.doi.org/10.1371/journal.pone.0077344] [PMID: 24143221]

[116]   Obire O, Nwaubeta O. BioD egradation of Refined Petroleum Hydrocarbons in Soil ABSTRACT. J Appl Environ Manag 2001; 5(1): 43-6.

[117]   Lima SD, Oliveira AF, Golin R, *et al*. Isolation and characterization of hydrocarbon-degrading bacteria from gas station leaking-contaminated groundwater in the Southern Amazon, Brazil. Braz J Biol 2020; 80(2): 354-61.
[http://dx.doi.org/10.1590/1519-6984.208611] [PMID: 31389483]

[118]   Smith MR. The biodegradation of aromatic hydrocarbons by bacteria. In: Ratledge, C. (eds) Physiology of Biodegradative Microorganisms. 1991; Springer, Dordrecht.
[http://dx.doi.org/10.1007/978-94-011-3452-1_9]

[119]   Sivadon P, Grimaud R. Assimilation of Hydrocarbons and Lipids by Means of Biofilm Formation. Cell Ecophysiol Microbe Hydrocarb Lipid Interact. 2018: 47-58.
[http://dx.doi.org/10.1007/978-3-319-50542-8_41]

[120]   St Angelo AJ, Angelo S. Lipid oxidation on foods. Crit Rev Food Sci Nutr 1996; 36(3): 175-224.
[http://dx.doi.org/10.1080/10408399609527723] [PMID: 8744604]

[121]   van der Heul R. Environmental Degradation of petroleum hydrocarbons2009.

[122]   Xia ZY, Zhang L, Zhao Y, *et al*. Biodegradation of the Herbicide 2,4-Dichlorophenoxyacetic Acid by a New Isolated Strain of Achromobacter sp. LZ35. Curr Microbiol 2017; 74(2): 193-202.
[http://dx.doi.org/10.1007/s00284-016-1173-y] [PMID: 27933337]

[123]   Johnston B, Radecka I, Chiellini E, *et al*. Mass spectrometry reveals molecular structure of polyhydroxyalkanoates attained by bioconversion of oxidized polypropylene waste fragments. Polymers (Basel) 2019; 11(10): 1580.
[http://dx.doi.org/10.3390/polym11101580] [PMID: 31569718]

[124] Duran R, Cravo-Laureau C. Role of environmental factors and microorganisms in determining the fate of polycyclic aromatic hydrocarbons in the marine environment. FEMS Microbiol Rev 2016; 40(6): 814-30.
[http://dx.doi.org/10.1093/femsre/fuw031] [PMID: 28201512]

[125] Ajao A, Yakubu S, Umoh V, Ameh H. Enzymatic Studies and Mineralization Potential of *Burkholderia cepacia* and *Corynebacterium kutscheri* Isolated from Refinery Sludge. Interciencia 2012; 4(2): 29-42.
[http://dx.doi.org/10.5923/j.microbiology.20140402.01]

[126] Kronen M, Lee M, Jones ZL, Manefield MJ. Reductive metabolism of the important atmospheric gas isoprene by homoacetogens. ISME J 2019; 13(5): 1168-82.
[http://dx.doi.org/10.1038/s41396-018-0338-z] [PMID: 30643199]

[127] Deeba F, Pruthi V, Negi YS. Aromatic hydrocarbon biodegradation activates neutral lipid biosynthesis in oleaginous yeast. Bioresour Technol 2018; 255: 273-80.
[http://dx.doi.org/10.1016/j.biortech.2018.01.096] [PMID: 29428782]

[128] Heider J, Spormann AM, Beller HR, Widdel F. Anaerobic bacterial metabolism of hydrocarbons. FEMS Microbiol Rev 1998; 22(5): 459-73.
[http://dx.doi.org/10.1111/j.1574-6976.1998.tb00381.x]

[129] Krivoruchko A, Zhang Y, Siewers V, Chen Y, Nielsen J. Microbial acetyl-CoA metabolism and metabolic engineering. Metab Eng 2015; 28: 28-42.
[http://dx.doi.org/10.1016/j.ymben.2014.11.009] [PMID: 25485951]

[130] Prenafeta-Boldú FX, de Hoog GS, Summerbell RC. Fungal communities in hydrocarbon degradation. In: McGenity T (Ed.) Microbial Communities Utilizing Hydrocarbons and Lipids: Members, Metagenomics and Ecophysiology . Handbook of Hydrocarbon and Lipid Microbiology . Springer, Cham 2019; pp. 307-42.

[131] Yetkin-Arik B, Vogels IMC, Nowak-Sliwinska P, *et al.* The role of glycolysis and mitochondrial respiration in the formation and functioning of endothelial tip cells during angiogenesis. Sci Rep 2019; 9(1): 12608.
[http://dx.doi.org/10.1038/s41598-019-48676-2] [PMID: 31471554]

[132] Oexle H, Gnaiger E, Weiss G. Iron-dependent changes in cellular energy metabolism: influence on citric acid cycle and oxidative phosphorylation. Biochim Biophys Acta Bioenerg 1999; 1413(3): 99-107.
[http://dx.doi.org/10.1016/S0005-2728(99)00088-2] [PMID: 10556622]

[133] Schulenburg C, Miller BG. Enzyme recruitment and its role in metabolic expansion. Biochemistry 2014; 53(5): 836-45.
[http://dx.doi.org/10.1021/bi401667f] [PMID: 24483367]

[134] Thakur M, Medintz IL, Walper SA. Enzymatic Bioremediation of Organophosphate Compounds—Progress and Remaining Challenges. Front Bioeng Biotechnol 2019; 7: 289.
[http://dx.doi.org/10.3389/fbioe.2019.00289] [PMID: 31781549]

[135] Bødtker G, Hvidsten IV, Barth T, Torsvik T. Hydrocarbon degradation by Dietzia sp. A14101 isolated from an oil reservoir model column. Antonie van Leeuwenhoek 2009; 96(4): 459-69.
[http://dx.doi.org/10.1007/s10482-009-9359-y] [PMID: 19565350]

[136] Hennessee CT, Li QX. Effects of polycyclic aromatic hydrocarbon mixtures on degradation, gene expression, and metabolite production in four *Mycobacterium* species. Appl Environ Microbiol 2016; 82(11): 3357-69.
[http://dx.doi.org/10.1128/AEM.00100-16] [PMID: 27037123]

[137] Al-Hawash AB, Dragh MA, Li S, *et al.* Principles of microbial degradation of petroleum hydrocarbons in the environment. Egypt J Aquat Res 2018; 44(2): 71-6.
[http://dx.doi.org/10.1016/j.ejar.2018.06.001]

[138] Ulrich AC, Guigard SE, Foght JM, *et al.* Effect of salt on aerobic biodegradation of petroleum hydrocarbons in contaminated groundwater. Biodegradation 2009; 20(1): 27-38.
[http://dx.doi.org/10.1007/s10532-008-9196-0] [PMID: 18437506]

[139] Elumalai P, Parthipan P, Karthikeyan OP, Rajasekar A. Enzyme-mediated biodegradation of long-chain n-alkanes (C32 and C40) by thermophilic bacteria. 3 Biotech 7: 116.
[http://dx.doi.org/10.1007/s13205-017-0773-y]

[140] Hidayat A, Tachibana S. Biodegradation of aliphatic hydrocarbon in three types of crude oil by Fusarium sp. F092 under stress with artificial sea water. J Environ Sci Technol 2011; 5(1): 64-73.
[http://dx.doi.org/10.3923/jest.2012.64.73]

[141] Spydevold O, Davis EJ, Bremer J. Replenishment and depletion of citric acid cycle intermediates in skeletal muscle. Indication of pyruvate carboxylation. Eur J Biochem 1976; 71(1): 155-65.
[http://dx.doi.org/10.1111/j.1432-1033.1976.tb11101.x] [PMID: 1009946]

[142] Manoj KM. Aerobic Respiration: Criticism of the Proton-centric Explanation Involving Rotary Adenosine Triphosphate Synthesis, Chemiosmosis Principle, Proton Pumps and Electron Transport Chain. Biochem Insights 2018; 11
[http://dx.doi.org/10.1177/1178626418818442] [PMID: 30643418]

[143] Rodriguez ACL. Genetic Structure and Crude Oil Associated Transcriptomic Profile of Environmental and Clinical Strains of *Pseudomonas aeruginosa.* Andes (Salta) 2018.

[144] Van Bogaert INA, Groeneboer S, Saerens K, Soetaert W. The role of cytochrome P450 monooxygenases in microbial fatty acid metabolism. FEBS J 2011; 278(2): 206-21.
[http://dx.doi.org/10.1111/j.1742-4658.2010.07949.x] [PMID: 21156025]

[145] Abbasian F, Palanisami T, Megharaj M, Naidu R, Lockington R, Ramadass K. Microbial diversity and hydrocarbon degrading gene capacity of a crude oil field soil as determined by metagenomics analysis. Biotechnol Prog 2016; 32(3): 638-48.
[http://dx.doi.org/10.1002/btpr.2249] [PMID: 26914145]

[146] Aliakbari E, Tebyanian H, Hassanshahian M, Kariminik A. Degradation of Alkanes in Contaminated Sites. Int J Adv Biol Biomed Res 2014; 2: 1620-37.

[147] Bugg TDH, Ramaswamy S. Non-heme iron-dependent dioxygenases: unravelling catalytic mechanisms for complex enzymatic oxidations. Curr Opin Chem Biol 2008; 12(2): 134-40.
[http://dx.doi.org/10.1016/j.cbpa.2007.12.007] [PMID: 18249197]

[148] Bell SG, Spence JTJ, Liu S, George JH, Wong LL. Selective aliphatic carbon–hydrogen bond activation of protected alcohol substrates by cytochrome P450 enzymes. Org Biomol Chem 2014; 12(15): 2479-88.
[http://dx.doi.org/10.1039/C3OB42417K] [PMID: 24599100]

[149] Kristin MS. Characterization of the molecular foundations and biochemistry of alkane and ether oxidation in a filamentous fungus, a Graphium species. Oregon State University 2007.

[150] Gibson DT, Koch JR, Kallio RE. Oxidative degradation of aromatic hydrocarbons by microorganisms. I. Enzymic formation of catechol from benzene. Biochemistry 1968; 7(7): 2653-62.
[http://dx.doi.org/10.1021/bi00847a031] [PMID: 4298226]

[151] Guzik U, Hupert-Kocurek K, Wojcieszysk D. Intradiol Dioxygenases — The Key Enzymes in Xenobiotics Degradation. In: Chamy R (Ed.) Biodegradation of Hazardous and Special Products. IntechOpen 2013.
[http://dx.doi.org/10.5772/56205]

[152] Walsh CT. Biologically generated carbon dioxide: nature's versatile chemical strategies for carboxy lyases. Nat Prod Rep 2020; 37(1): 100-35.
[http://dx.doi.org/10.1039/C9NP00015A] [PMID: 31074473]

[153] Wang Y, Li J, Liu A. Oxygen activation by mononuclear nonheme iron dioxygenases involved in the

degradation of aromatics. J Biol Inorg Chem 2017; 22(2-3): 395-405.
[http://dx.doi.org/10.1007/s00775-017-1436-5] [PMID: 28084551]

[154] Sugimoto K, Senda T, Aoshima H, Masai E, Fukuda M, Mitsui Y. Crystal structure of an aromatic ring opening dioxygenase LigAB, a protocatechuate 4,5-dioxygenase, under aerobic conditions. Structure 1999; 7(8): 953-65.
[http://dx.doi.org/10.1016/S0969-2126(99)80122-1] [PMID: 10467151]

[155] Terrón-González L, Martín-Cabello G, Ferrer M, Santero E. Functional metagenomics of a biostimulated petroleum-contaminated soil reveals an extraordinary diversity of extradiol dioxygenases. Appl Environ Microbiol 2016; 82(8): 2467-78.
[http://dx.doi.org/10.1128/AEM.03811-15] [PMID: 26896130]

[156] Leja K, Lewandowicz G. Polymer biodegradation and biodegradable polymers - A review. Pol J Environ Stud 2010; 19: 255-66.

[157] Jaekel U, Zedelius J, Wilkes H, Musat F. Anaerobic degradation of cyclohexane by sulfate-reducing bacteria from hydrocarbon-contaminated marine sediments. Front Microbiol 2015; 6: 116.
[http://dx.doi.org/10.3389/fmicb.2015.00116] [PMID: 25806023]

[158] Flanagan PV, Kelleher BP, Allen CCR. Assessment of Anaerobic Biodegradation of Aromatic Hydrocarbons: The Impact of Molecular Biology Approaches. Geomicrobiol J 2014; 31(4): 276-84.
[http://dx.doi.org/10.1080/01490451.2013.820237]

[159] Foght J. Anaerobic biodegradation of aromatic hydrocarbons: pathways and prospects. Microbial Physiology 2008; 15(2-3): 93-120.
[http://dx.doi.org/10.1159/000121324] [PMID: 18685265]

[160] Boll M, Fuchs G, Heider J. Anaerobic oxidation of aromatic compounds and hydrocarbons. Curr Opin Chem Biol 2002; 6(5): 604-11.
[http://dx.doi.org/10.1016/S1367-5931(02)00375-7] [PMID: 12413544]

[161] Grundmann O, Widdel F. Biochemistry of the Anaerobic Degradation of Non-Methane Alkanes. In: Timmis KN (Ed.) Handbook of Hydrocarbon and Lipid Microbiology. Springer, Berlin, Heidelberg 2010; pp. 909-24.

[162] Bian XY, Maurice Mbadinga S, Liu YF, et al. Insights into the anaerobic biodegradation pathway of n-Alkanes in oil reservoirs by detection of signature metabolites. Sci Rep 2015; 5(1): 9801.
[http://dx.doi.org/10.1038/srep09801] [PMID: 25966798]

[163] Porter AW, Young LY. Benzoyl-CoA, a universal biomarker for anaerobic degradation of aromatic compounds. Adv Appl Microbiol 2014; 88: 167-203.
[http://dx.doi.org/10.1016/B978-0-12-800260-5.00005-X]

[164] Carmona M, Zamarro MT, Blázquez B, et al. Anaerobic catabolism of aromatic compounds: a genetic and genomic view. Microbiol Mol Biol Rev 2009; 73(1): 71-133.
[http://dx.doi.org/10.1128/MMBR.00021-08] [PMID: 19258534]

[165] Kniemeyer O, Fischer T, Wilkes H, Glöckner FO, Widdel F. Anaerobic degradation of ethylbenzene by a new type of marine sulfate-reducing bacterium. Appl Environ Microbiol 2003; 69(2): 760-8.
[http://dx.doi.org/10.1128/AEM.69.2.760-768.2003] [PMID: 12570993]

[166] Young LY, Phelps CD. Metabolic biomarkers for monitoring in situ anaerobic hydrocarbon degradation. Environ Health Perspect 2005; 113(1): 62-7.
[http://dx.doi.org/10.1289/ehp.6940] [PMID: 15626649]

[167] Militon C, Boucher D, Vachelard C, et al. Bacterial community changes during bioremediation of aliphatic hydrocarbon-contaminated soil. FEMS Microbiol Ecol 2010; 74(3): 669-81.
[http://dx.doi.org/10.1111/j.1574-6941.2010.00982.x] [PMID: 21044099]

[168] Sutton NB, Maphosa F, Morillo JA, et al. Impact of long-term diesel contamination on soil microbial community structure. Appl Environ Microbiol 2013; 79(2): 619-30.
[http://dx.doi.org/10.1128/AEM.02747-12] [PMID: 23144139]

[169] Kanaly RA, Harayama S. Biodegradation of high-molecular-weight polycyclic aromatic hydrocarbons by bacteria. J Bacteriol 2000; 182(8): 2059-67.
[http://dx.doi.org/10.1128/JB.182.8.2059-2067.2000] [PMID: 10735846]

[170] Bacosa HP, Inoue C. Polycyclic aromatic hydrocarbons (PAHs) biodegradation potential and diversity of microbial consortia enriched from tsunami sediments in Miyagi, Japan. J Hazard Mater 2015; 283: 689-97.
[http://dx.doi.org/10.1016/j.jhazmat.2014.09.068] [PMID: 25464311]

[171] Stagars MH, Ruff SE, Amann R, Knittel K. High diversity of anaerobic alkane-degrading microbial communities in marine seep sediments based on (1-methylalkyl)succinate synthase genes. Front Microbiol 2016; 6: 1511.
[http://dx.doi.org/10.3389/fmicb.2015.01511] [PMID: 26779166]

[172] Acosta-González A, Rosselló-Móra R, Marqués S. Diversity of benzylsuccinate synthase-like (bssA) genes in hydrocarbon-polluted marine sediments suggests substrate-dependent clustering. Appl Environ Microbiol 2013; 79(12): 3667-76.
[http://dx.doi.org/10.1128/AEM.03934-12] [PMID: 23563947]

[173] Yan S, Wang Q, Qu L, Li C. Characterization of oil-degrading bacteria from oilcontaminated soil and activity of their enzymes. Biotechnol Biotechnol Equip 2013; 27(4): 3932-8.
[http://dx.doi.org/10.5504/BBEQ.2013.0050]

[174] Cooney JJ, Silver SA, Beck EA. Factors influencing hydrocarbon degradation in three freshwater lakes. Microb Ecol 1985; 11(2): 127-37.
[http://dx.doi.org/10.1007/BF02010485] [PMID: 24221301]

[175] Ebakota OD, Osarueme JO, Gift ON, Odoligie I, Osazee JO. Isolation and Characterization of Hydrocarbon-Degrading Bacteria in Top and Subsoil of selected Mechanic Workshops in Benin City Metropolis, Nigeria. J Appl Sci Environ Manag 2017; 21(4): 641.
[http://dx.doi.org/10.4314/jasem.v21i4.3]

[176] Rao MA, Scelza R, Scotti R, Gianfreda L. Role of enzymes in the remediation of polluted environments. J Soil Sci Plant Nutr 2010; 10(3): 333-53.
[http://dx.doi.org/10.4067/S0718-95162010000100008]

[177] Santos D, Rufino R, Luna J, Santos V, Sarubbo L. Biosurfactants: Multifunctional biomolecules of the 21st century. Int J Mol Sci 2016; 17(3): 401.
[http://dx.doi.org/10.3390/ijms17030401] [PMID: 26999123]

[178] Krasowska A, Sigler K. How microorganisms use hydrophobicity and what does this mean for human needs? Front Cell Infect Microbiol 2014; 4: 112.
[http://dx.doi.org/10.3389/fcimb.2014.00112] [PMID: 25191645]

[179] Thavasi R, Subramanyam Nambaru VRM, Jayalakshmi S, Balasubramanian T, Banat IM. Biosurfactant Production by *Pseudomonas aeruginosa* from Renewable Resources. Indian J Microbiol 2011; 51(1): 30-6.
[http://dx.doi.org/10.1007/s12088-011-0076-7] [PMID: 22282625]

[180] Patoway K, Patoway R, Kalita MC, Deka S. Characterization of biosurfactant produced during degradation of hydrocarbons using crude oil as sole source of carbon. Front Microbiol 2017; 8: 279.
[http://dx.doi.org/10.3389/fmicb.2017.00279] [PMID: 28275373]

[181] Marchut-Mikolajczyk O, Drożdżyński P, Pietrzyk D, Antczak T. Biosurfactant production and hydrocarbon degradation activity of endophytic bacteria isolated from *Chelidonium majus* L. Microb Cell Fact 2018; 17(1): 171.
[http://dx.doi.org/10.1186/s12934-018-1017-5] [PMID: 30390702]

[182] Zhang S, Merino N, Okamoto A, Gedalanga P. Interkingdom microbial consortia mechanisms to guide biotechnological applications. Microb Biotechnol 2018; 11(5): 833-47.
[http://dx.doi.org/10.1111/1751-7915.13300] [PMID: 30014573]

[183] Wu X, Gu Y, Wu X, *et al.* Construction of a tetracycline degrading bacterial consortium and its application evaluation in laboratory-scale soil remediation. Microorganisms 2020; 8(2): 292.
[http://dx.doi.org/10.3390/microorganisms8020292] [PMID: 32093355]

[184] Potts LD, Perez Calderon LJ, Gontikaki E, *et al.* Effect of spatial origin and hydrocarbon composition on bacterial consortia community structure and hydrocarbon biodegradation rates. FEMS Microbiol Ecol 2018; 94(9): 1-12.
[http://dx.doi.org/10.1093/femsec/fiy127] [PMID: 29982504]

[185] Poddar K, Sarkar D, Sarkar A. Construction of potential bacterial consortia for efficient hydrocarbon degradation. Int Biodeterior Biodegradation 2019; 144: 104770.
[http://dx.doi.org/10.1016/j.ibiod.2019.104770]

[186] Abu Bakar A, Mohd Ali MKF, Md Noor N, Yahaya N, Ismail M, Abdullah A. Bio-corrosion of carbon steel by sulfate reducing bacteria consortium in oil and gas pipelines. J Mech Eng Sci 2017; 11(2): 2592-600.
[http://dx.doi.org/10.15282/jmes.11.2.2017.3.0237]

[187] Jefferson KK. What drives bacteria to produce a biofilm? FEMS Microbiol Lett 2004; 236(2): 163-73.
[http://dx.doi.org/10.1111/j.1574-6968.2004.tb09643.x] [PMID: 15251193]

[188] Jamal M, Ahmad W, Andleeb S, *et al.* Bacterial biofilm and associated infections. J Chin Med Assoc 2018; 81(1): 7-11.
[http://dx.doi.org/10.1016/j.jcma.2017.07.012] [PMID: 29042186]

[189] Garnett JA, Matthews S. Interactions in bacterial biofilm development: a structural perspective. Curr Protein Pept Sci 2012; 13(8): 739-55.
[http://dx.doi.org/10.2174/138920312804871166] [PMID: 23305361]

[190] Fenibo EO, Ijoma GN, Selvarajan R, Chikere CB. Microbial surfactants: The next generation multifunctional biomolecules for applications in the petroleum industry and its associated environmental remediation. Microorganisms 2019; 7(11): 581.
[http://dx.doi.org/10.3390/microorganisms7110581] [PMID: 31752381]

[191] Fouad M, Bhargava R. A simplified model for the steady-state biofilm-activated sludge reactor. J Environ Manage 2005; 74(3): 245-53.
[http://dx.doi.org/10.1016/j.jenvman.2004.09.005] [PMID: 15644264]

[192] Dasgupta D, Ghosh R, Sengupta TK. Biofilm-mediated enhanced crude oil degradation by newly isolated pseudomonas species. ISRN Biotechnol 2013; 2013: 1-13.
[http://dx.doi.org/10.5402/2013/250749] [PMID: 25937972]

[193] Meliani A. Enhancement of Hydrocarbons Degradation by Use of *Pseudomonas* Biosurfactants and Biofilms. J Pet Environ Biotechnol 2014; 5(1): 1-7.
[http://dx.doi.org/10.4172/2157-7463.1000168]

[194] Omarova M, Swientoniewski LT, Mkam Tsengam IK, *et al.* Biofilm Formation by Hydrocarbon-Degrading Marine Bacteria and Its Effects on Oil Dispersion. ACS Sustain Chem& Eng 2019; 7(17): 14490-9.
[http://dx.doi.org/10.1021/acssuschemeng.9b01923]

[195] Maddela NR, Sheng B, Yuan S, Zhou Z, Villamar-Torres R, Meng F. Roles of quorum sensing in biological wastewater treatment: A critical review. Chemosphere 2019; 221: 616-29.
[http://dx.doi.org/10.1016/j.chemosphere.2019.01.064] [PMID: 30665091]

[196] Shrout JD, Nerenberg R. Monitoring bacterial twitter: does quorum sensing determine the behavior of water and wastewater treatment biofilms? Environ Sci Technol 2012; 46(4): 1995-2005.
[http://dx.doi.org/10.1021/es203933h] [PMID: 22296043]

[197] Barnard AML, Bowden SD, Burr T, Coulthurst SJ, Monson RE, Salmond GPC. Quorum sensing, virulence and secondary metabolite production in plant soft-rotting bacteria. Philos Trans R Soc Lond B Biol Sci 2007; 362(1483): 1165-83.

[http://dx.doi.org/10.1098/rstb.2007.2042] [PMID: 17360277]

[198]   Jiang Q, Chen J, Yang C, Yin Y, Yao K. Quorum Sensing: A Prospective Therapeutic Target for Bacterial Diseases. BioMed Res Int 2019; 2019: 1-15.
[http://dx.doi.org/10.1155/2019/2015978] [PMID: 31080810]

[199]   Castillo-Juárez I, Maeda T, Mandujano-Tinoco EA, *et al.* Role of quorum sensing in bacterial infections. World J Clin Cases 2015; 3(7): 575-98.
[http://dx.doi.org/10.12998/wjcc.v3.i7.575] [PMID: 26244150]

[200]   Wu Y, Liu J, Rene ER. Periphytic biofilms: A promising nutrient utilization regulator in wetlands. Bioresour Technol 2018; 248(Pt B): 44-8.
[http://dx.doi.org/10.1016/j.biortech.2017.07.081] [PMID: 28756125]

[201]   Salgot M, Folch M. Wastewater treatment and water reuse. Curr Opin Environ Sci Health 2018; 2: 64-74.
[http://dx.doi.org/10.1016/j.coesh.2018.03.005]

[202]   Beccari M, Majone M, Dionisi D, *et al.* CrgA Protein Represses AlkB2 Monooxygenase and Regulates the Degradation of Medium-to-Long-Chain n-Alkanes in *Pseudomonas aeruginosa* SJTD-1. Front Microbiol 2007; 10: 1-15.

[203]   Magnabosco C, Tekere M, Lau MCY, *et al.* Comparisons of the composition and biogeographic distribution of the bacterial communities occupying South African thermal springs with those inhabiting deep subsurface fracture water. Front Microbiol 2014; 5: 679.
[http://dx.doi.org/10.3389/fmicb.2014.00679] [PMID: 25566203]

[204]   Ji N, Wang X, Yin C, Peng W, Liang R. CrgA Protein Represses AlkB2 Monooxygenase and Regulates the Degradation of Medium-to-Long-Chain *n*-Alkanes in *Pseudomonas aeruginosa* SJTD-1. Front Microbiol 2019; 10: 400.
[http://dx.doi.org/10.3389/fmicb.2019.00400] [PMID: 30915046]

[205]   Gargouri B, Mhiri N, Karray F, Aloui F, Sayadi S. Isolation and Characterization of Hydrocarbon-Degrading Yeast Strains from Petroleum Contaminated Industrial Wastewater. BioMed Res Int 2015; 2015: 1-11.
[http://dx.doi.org/10.1155/2015/929424] [PMID: 26339653]

[206]   Cutright TJ. Polycyclic aromatic hydrocarbon biodegradation and kinetics using *Cunninghamella echinulata* var. elegans. Int Biodeterior Biodegradation 1995; 35(4): 397-408.
[http://dx.doi.org/10.1016/0964-8305(95)00046-1]

[207]   Adeleye AO, Nkereuwem ME, Omokhudu GI, Amoo AO, Shiaka GP, Yerima MB. Effect of microorganisms in the bioremediation of spent engine oil and petroleum related environmental pollution. J Appl Sci Environ Manag 2018; 22(2): 157.
[http://dx.doi.org/10.4314/jasem.v22i2.1]

[208]   Ahamed F, Hasibullah M, Ferdouse J, Anwar MN. Microbial Degradation of Petroleum Hydrocarbon. Bangladesh J Microbiol 1970; 27(1): 10-3.
[http://dx.doi.org/10.3329/bjm.v27i1.9161]

[209]   Chikere CB, Azubuike CC. Characterization of hydrocarbon utilizing fungi from hydrocarbon polluted sediments and water. Niger J Biotechnol 2014; 27: 49-54.

[210]   Abraham J, Kumari M. Biodegradation of diesel oil using yeast rhodosporidium toruloides. Research Journal of Environmental Toxicology 2011; 5(6): 369-77.
[http://dx.doi.org/10.3923/rjet.2011.369.377]

[211]   Nasrawi HA. Biodegradation of Crude Oil by Fungi Isolated from Gulf of Mexico. J Bioremediat Biodegrad 2012; 3(4).
[http://dx.doi.org/10.4172/2155-6199.1000147]

[212]   Olukunle OF, Oyegoke TS. Biodegradation of Crude-oil by Fungi Isolated from Cow Dungcontaminated soils. Niger J Biotechnol 2016; 31(1): 46.

[http://dx.doi.org/10.4314/njb.v31i1.7]

[213] Babaei F, Habibi A. Fast biodegradation of diesel hydrocarbons at high concentration by the sophorolipid-producing yeast Candida catenulata KP324968. Microbial Physiology 2018; 28(5): 240-54.
[http://dx.doi.org/10.1159/000496797] [PMID: 30852573]

[214] Da Silva C, Astals S, Peces M, Campos JL, Guerrero L. Biochemical methane potential (BMP) tests: Reducing test time by early parameter estimation. Waste Manag 2018; 71: 19-24.
[http://dx.doi.org/10.1016/j.wasman.2017.10.009] [PMID: 29033134]

[215] Perfumo A, Smyth TJ, Banat IM. Production and roles of Biosurfactants and bioemulsifiers in accessing hydrophobic substances. In: Timmis KN (Ed.) Handbook of Hydrocarbon and Lipid Microbiology. Springer, Berlin, Heidelberg 2010; pp. 1501-12.

[216] Bustamante M, Durán N, Diez MC. Biosurfactants are useful tools for the bioremediation of contaminated soil: a review. J Soil Sci Plant Nutr 2012; 12 (4): 667- 87.
[http://dx.doi.org/10.4067/S0718-95162012005000024]

[217] Uzoigwe C, Burgess JG, Ennis CJ, Rahman PKSM. Bioemulsifiers are not biosurfactants and require different screening approaches. Front Microbiol 2015; 6: 245.
[http://dx.doi.org/10.3389/fmicb.2015.00245] [PMID: 25904897]

[218] Chaprão MJ, Ferreira INS, Correa PF, *et al.* Application of bacterial and yeast biosurfactants for enhanced removal and biodegradation of motor oil from contaminated sand. Electron J Biotechnol 2015; 18(6): 471-9.
[http://dx.doi.org/10.1016/j.ejbt.2015.09.005]

[219] Günther M, Zibek S, Rupp S. Fungal Glycolipids as Biosurfactants. Curr Biotechnol 2017; 6(3): 205-18.
[http://dx.doi.org/10.2174/2211550105666160822170256]

[220] Janusz G, Pawlik A, Sulej J, *et al.* Lignin degradation : microorganisms , enzymes involved , genomes analysis and evolution. 2017; 941-62.
[http://dx.doi.org/10.1093/femsre/fux049]

[221] Kumar A, Chandra R. Ligninolytic enzymes and its mechanisms for degradation of lignocellulosic waste in environment. Heliyon 2020; 6(2): e03170.
[http://dx.doi.org/10.1016/j.heliyon.2020.e03170] [PMID: 32095645]

[222] Upadhyay P, Shrivastava R, Agrawal PK. Bioprospecting and biotechnological applications of fungal laccase. 3 Biotech 6: 1-12.2016;
[http://dx.doi.org/10.1007/s13205-015-0316-3]

[223] Bansal N, Kanwar SS. Peroxidase(s) in environment protection. ScientificWorldJournal 2013; 2013: 1-9.
[http://dx.doi.org/10.1155/2013/714639] [PMID: 24453894]

[224] Houtman CJ, Maligaspe E, Hunt CG, Fernández-Fueyo E, Martínez AT, Hammel KE. Fungal lignin peroxidase does not produce the veratryl alcohol cation radical as a diffusible ligninolytic oxidant. J Biol Chem 2018; 293(13): 4702-12.
[http://dx.doi.org/10.1074/jbc.RA117.001153] [PMID: 29462790]

[225] Datta R, Kelkar A, Baraniya D, *et al.* Enzymatic degradation of lignin in soil: A review. Sustainability (Basel) 2017; 9(7): 1163.
[http://dx.doi.org/10.3390/su9071163]

[226] Hatakka A, Hammel KE. Fungal Biodegradation of Lignocelluloses. Ind Appl 319-40.2011;
[http://dx.doi.org/10.1007/978-3-642-11458-8_15]

[227] Camarero S, Sarkar S, Ruiz-Dueñas FJ, Martínez MJ, Martínez ÁT. Description of a versatile peroxidase involved in the natural degradation of lignin that has both manganese peroxidase and lignin peroxidase substrate interaction sites. J Biol Chem 1999; 274(15): 10324-30.

[http://dx.doi.org/10.1074/jbc.274.15.10324] [PMID: 10187820]

[228] D'Souza TM, Merritt CS, Reddy CA. Lignin-modifying enzymes of the white rot basidiomycete Ganoderma lucidum. Appl Environ Microbiol 1999; 65(12): 5307-13.
[http://dx.doi.org/10.1128/AEM.65.12.5307-5313.1999] [PMID: 10583981]

[229] Field JA, de Jong E, Feijoo Costa G, de Bont JA. Biodegradation of polycyclic aromatic hydrocarbons by new isolates of white rot fungi. Appl Environ Microbiol 1992; 58(7): 2219-26.
[http://dx.doi.org/10.1128/aem.58.7.2219-2226.1992] [PMID: 1637159]

[230] Clemente AR, Anazawa TA, Durrant LR. Biodegradation of polycyclic aromatic hydrocarbons by soil fungi. Braz J Microbiol 2001; 32(4): 255-61.
[http://dx.doi.org/10.1590/S1517-83822001000400001]

[231] Govarthanan M, Fuzisawa S. RSC Advances Biodegradation of aliphatic and aromatic hydrocarbons using the fi lamentous fungus *Penicillium sp.* CHY-2 and characterization of its 2017; 20716-23.
[http://dx.doi.org/10.1039/C6RA28687A]

[232] Semple KT, Cain RB, Schmidt S. Biodegradation of aromatic compounds by microalgae. FEMS Microbiol Lett 1999; 170(2): 291-300.
[http://dx.doi.org/10.1111/j.1574-6968.1999.tb13386.x]

[233] Aldaby ESE, Mawad AMM. Pyrene biodegradation capability of two different microalgal strains. Glob NEST J 2018; 21: 290-5.
[http://dx.doi.org/10.30955/gnj.002767]

[234] Qari H, Hassan I. Bioaccumulation of PAHs in padina boryana alga collected from a contaminated site on the red sea, Saudi Arabia. Pol J Environ Stud 2017; 26(1): 435-9.
[http://dx.doi.org/10.15244/pjoes/63937]

[235] Warshawsky D, Cody T, Radike M, *et al.* Biotransformation of benzo[a]pyrene and other polycyclic aromatic hydrocarbons and heterocyclic analogs by several green algae and other algal species under gold and white light. Chem Biol Interact 1995; 97(2): 131-48.
[http://dx.doi.org/10.1016/0009-2797(95)03610-X] [PMID: 7606812]

[236] Walker JD, Colwell RR, Petrakis L. Degradation of petroleum by an alga, Prototheca zopfii. Appl Microbiol 1975; 30(1): 79-81.
[http://dx.doi.org/10.1128/am.30.1.79-81.1975] [PMID: 1147621]

[237] Syrmanova KK, Kovaleva AY, Kaldybekova ZB, *et al.* Chemistry and recycling technology of used motor oil. Orient J Chem 2017; 33: 3195-9.
[http://dx.doi.org/10.13005/ojc/330665]

[238] Shafi M. Remediation of crude petroleum oil-water emulsions using microalgae. United Arab Emirates University College of Engineering 2020.

[239] Eregie SB, Jamal-Ally SF. Comparison of biodegradation of lubricant wastes by *Scenedesmus vacuolatus* vs a microalgal consortium. Bioremediat J 2019; 23(4): 277-301.
[http://dx.doi.org/10.1080/10889868.2019.1671792]

[240] Ichor T, Okerentugb PO, Okpokwasil GC. Biodegradation of total petroleum hydrocarbon by a consortium of cyanobacteria isolated from crude oil polluted brackish waters of bodo creeks in Ogoniland, Rivers State. Research Journal of Environmental Toxicology 2016; 10(1): 16-27.
[http://dx.doi.org/10.3923/rjet.2016.16.27]

[241] Gontikaki E, Potts LD, Anderson JA, Witte U. Hydrocarbon-degrading bacteria in deep-water subarctic sediments (Faroe-Shetland Channel). J Appl Microbiol 2018; 125(4): 1040-53.
[http://dx.doi.org/10.1111/jam.14030] [PMID: 29928773]

[242] Mora-Salguero D, Vives Florez MJ, Husserl Orjuela J, Fernández-Niño M, González Barrios AF. Evaluation of the phenol degradation capacity of microalgae-bacteria consortia from the bay of Cartagena, Colombia. TecnoLógicas 2019; 22(44): 149-58.
[http://dx.doi.org/10.22430/22565337.1179]

[243] Yao S, Lyu S, An Y, Lu J, Gjermansen C, Schramm A. Microalgae-bacteria symbiosis in microalgal growth and biofuel production: a review. J Appl Microbiol 2019; 126(2): 359-68.
[http://dx.doi.org/10.1111/jam.14095] [PMID: 30168644]

[244] Patel JG, Nirmal Kumar JI, Kumar RN, Khan SR. Enhancement of pyrene degradation efficacy of *Synechocystis* sp., by construction of an artificial microalgal-bacterial consortium. Cogent Chem 2015; 1(1): 1064193.
[http://dx.doi.org/10.1080/23312009.2015.1064193]

[245] Godsgift Omojevwe E, Obasola Ezekiel F. Microalgal-bacterial consortium in polyaromatic hydrocarbon degradation of petroleum based effluent. J Bioremediat Biodegrad 2016; 7(4).
[http://dx.doi.org/10.4172/2155-6199.1000359]

[246] Ndlovu T, Rautenbach M, Vosloo JA, Khan S, Khan W. Characterisation and antimicrobial activity of biosurfactant extracts produced by *Bacillus amyloliquefaciens* and *Pseudomonas aeruginosa* isolated from a wastewater treatment plant. AMB Express 2017; 7(1): 108.
[http://dx.doi.org/10.1186/s13568-017-0363-8] [PMID: 28571306]

[247] Azubuike CC, Chikere CB, Okpokwasili GC. Bioremediation techniques–classification based on site of application: principles, advantages, limitations and prospects. World J Microbiol Biotechnol 2016; 32(11): 180.
[http://dx.doi.org/10.1007/s11274-016-2137-x] [PMID: 27638318]

[248] Chapelle FH. Bioremediation of Petroleum Hydrocarbon-Contaminated Ground Water: The Perspectives of History and Hydrology. Ground Water 1999; 37(1): 122-32.
[http://dx.doi.org/10.1111/j.1745-6584.1999.tb00965.x]

[249] Atteia O, Franceschi M. Kinetics of natural attenuation of BTEX: review of the critical chemical conditions and measurements at bore scale. ScientificWorldJournal 2002; 2: 1338-46.
[http://dx.doi.org/10.1100/tsw.2002.299] [PMID: 12805917]

[250] Lach D, Sanford S. Eliciting Public Attitudes Regarding Bioremediation Cleanup Technologies: Lessons Learned from a Consensus Workshop in Idaho. ostigov 2003.

[251] Suttinun O, Luepromchai E, Mu R. *In Situ* bioremediation of a gasoline- contaminated vadose zone : implications from direct observations. 2013; 99-114.
[http://dx.doi.org/10.1007/s11157-012-9291-x]

[252] Moshkovich E, Ronen Z, Gelman F, Dahan O. *In Situ* Bioremediation of a Gasoline- Contaminated Vadose Zone : Implications from Direct Observations. Vadose Zone Journal 2018; 17(1).
[http://dx.doi.org/10.2136/vzj2017.08.0153]

[253] Mulligan CN, Yong RN, Gibbs BF. Surfactant-enhanced remediation of contaminated soil : a review. 2001; 60: 371-80.

[254] Nikolopoulou M, Kalogerakis N. Biostimulation strategies for fresh and chronically polluted marine environments with petroleum hydrocarbons. J Chem Technol Biotechnol 2009; 84(6): 802-7.
[http://dx.doi.org/10.1002/jctb.2182]

[255] Christoph S. Elaboration of cellularized hybrid macroporous materials by freeze-casting for soil bioremediation. Pierre et Marie Curie - Paris IV 2017.

[256] Timmis KN, Pieper DH. Bacteria designed for bioremediation. 1999; 17: 201-4.

[257] Garbisu C, Garaiyurrebaso O, Epelde L, Grohmann E, Alkorta I. Plasmid-mediated bioaugmentation for the bioremediation of contaminated soils. Front Microbiol 2017; 8: 1966.
[http://dx.doi.org/10.3389/fmicb.2017.01966] [PMID: 29062312]

[258] Agarry SE, Owabor CN, Yusuf RO. Studies on Biodegradation of Kerosene in Soil under Different Bioremediation Strategies. Bioremediat J 2010; 14(3): 135-41.
[http://dx.doi.org/10.1080/10889868.2010.495364]

[259] Sales da Silva IG, Gomes de Almeida FC, Padilha da Rocha e Silva NM, Casazza AA, Converti A,

Asfora Sarubbo L. Soil Bioremediation: Overview of Technologies and Trends. Energies 2020; 13(18): 4664.
[http://dx.doi.org/10.3390/en13184664]

[260] Snezana M, Bozo D, Srdan R. Petroleum Hydrocarbon Biodegradability in Soil – Implications for Bioremediation. In: Vladimir K, Kolesnikov A (Eds.) Hydrocarbon. IntechOpen. 2013.

[261] Varjani SJ. Microbial degradation of petroleum hydrocarbons. Bioresour Technol 2017; 223: 277-86.
[http://dx.doi.org/10.1016/j.biortech.2016.10.037] [PMID: 27789112]

[262] Lawal AT. Polycyclic aromatic hydrocarbons. A review. Cogent Environ Sci 2017; 3(1): 1339841.
[http://dx.doi.org/10.1080/23311843.2017.1339841]

[263] Vidali M. Bioremediation. An overview. Pure Appl Chem 2001; 73(7): 1163-72.
[http://dx.doi.org/10.1351/pac200173071163]

[264] Owen CR. Water budget and flow patterns in an urban wetland. J Hydrol (Amst) 1995; 169(1-4): 171-87.
[http://dx.doi.org/10.1016/0022-1694(94)02638-R]

[265] Mokraoui S, Coquelet C, Valtz A, Hegel PE, Richon D. New solubility data of hydrocarbons in water and modeling concerning vapor-liquid-liquid binary systems. Ind Eng Chem Res 2007; 46(26): 9257-62.
[http://dx.doi.org/10.1021/ie070858y]

[266] Hutacharoen P, Dufal S, Papaioannou V, *et al.* Predicting the Solvation of Organic Compounds in Aqueous Environments: From Alkanes and Alcohols to Pharmaceuticals. Ind Eng Chem Res 2017; 56(38): 10856-76.
[http://dx.doi.org/10.1021/acs.iecr.7b00899]

[267] López-Cortés N, Beloqui A, Ghazi A, Ferrer M. Enzymatic Functionalization of Hydrocarbon-like Molecules. In: Timmis KN, Ed. Handbook of Hydrocarbon and Lipid Microbiology. Berlin, Heidelberg: Springer 2010.
[http://dx.doi.org/10.1007/978-3-540-77587-4_211]

[268] Da Silva MLB, Alvarez PJJ. Enhanced anaerobic biodegradation of benzene-toluene-ethylbenze-e-xylene-ethanol mixtures in bioaugmented aquifer columns. Appl Environ Microbiol 2004; 70(8): 4720-6.
[http://dx.doi.org/10.1128/AEM.70.8.4720-4726.2004] [PMID: 15294807]

[269] Jindrová E, Chocová M, Demnerová K, Brenner V. Bacterial aerobic degradation of benzene, toluene, ethylbenzene and xylene. Folia Microbiol (Praha) 2002; 47(2): 83-93.
[http://dx.doi.org/10.1007/BF02817664] [PMID: 12058403]

[270] Li J, Peng K, Zhang D, *et al.* Autochthonous bioaugmentation with non-direct degraders: A new strategy to enhance wastewater bioremediation performance. Environ Int 2020; 136: 105473.
[http://dx.doi.org/10.1016/j.envint.2020.105473] [PMID: 31999970]

[271] Kerstin ES. Extracellular Electron Transfer *in situ* Petroleum Hydrocarbon Bioremediation. IntechOpen 2012.

[272] Parales RE, Bruce NC, Schmid A, Wackett LP. Biodegradation, biotransformation, and biocatalysis (b3). Appl Environ Microbiol 2002; 68(10): 4699-709.
[http://dx.doi.org/10.1128/AEM.68.10.4699-4709.2002] [PMID: 12324310]

[273] Fan CY, Krishnamurthy S. Enzymes for enhancing bioremediation of petroleum-contaminated soils: a brief review. J Air Waste Manag Assoc 1995; 45(6): 453-60.
[http://dx.doi.org/10.1080/10473289.1995.10467375] [PMID: 7788508]

[274] Harris DA. A unified approach to enzyme catalysis. Biochem Educ 1986; 14(1): 2-6.
[http://dx.doi.org/10.1016/0307-4412(86)90003-8]

[275] Van Hamme JD, Singh A, Ward OP. Recent advances in petroleum microbiology. Microbiol Mol Biol

Rev 2003; 67(4): 503-49.
[http://dx.doi.org/10.1128/MMBR.67.4.503-549.2003] [PMID: 14665675]

[276] Wali Alwan S. Cytotoxicity study and bioremediation of petroleum contaminated soil. J Phys Conf Ser 2019; 1294(9): 092049.
[http://dx.doi.org/10.1088/1742-6596/1294/9/092049]

[277] Adelaja O, Keshavarz T, Kyazze G. The effect of salinity, redox mediators and temperature on anaerobic biodegradation of petroleum hydrocarbons in microbial fuel cells. J Hazard Mater 2015; 283: 211-7.
[http://dx.doi.org/10.1016/j.jhazmat.2014.08.066] [PMID: 25279757]

[278] Yuniati MD. Bioremediation of petroleum-contaminated soil: A Review. IOP Conf Ser Earth Environ Sci 2018; 118: 012063.
[http://dx.doi.org/10.1088/1755-1315/118/1/012063]

[279] Takei D, Washio K, Morikawa M. Identification of alkane hydroxylase genes in *Rhodococcus* sp. strain TMP2 that degrades a branched alkane. Biotechnol Lett 2008; 30(8): 1447-52.
[http://dx.doi.org/10.1007/s10529-008-9710-9] [PMID: 18414802]

[280] Ribicic D, McFarlin KM, Netzer R, *et al.* Oil type and temperature dependent biodegradation dynamics - Combining chemical and microbial community data through multivariate analysis. BMC Microbiol 2018; 18(1): 83.
[http://dx.doi.org/10.1186/s12866-018-1221-9] [PMID: 30086723]

[281] Sun X, Kostka JE. Hydrocarbon-degrading microbial communities are site specific, and their activity is limited by synergies in temperature and nutrient availability in surface ocean waters. Appl Environ Microbiol 2019; 85(15): e00443-19.
[http://dx.doi.org/10.1128/AEM.00443-19] [PMID: 31126938]

[282] Margesin R, Schinner F. Effect of temperature on oil degradation by a psychrotrophic yeast in liquid culture and in soil. FEMS Microbiol Ecol 1997; 24(3): 243-9.
[http://dx.doi.org/10.1111/j.1574-6941.1997.tb00441.x]

[283] Harish S, Bewoor A. Capacitive sensor for engine oil deterioration measurement. AIP Conference Proceedings 2018; 1943(1): 020099.

[284] Kadri T, Rouissi T, Brar SK, *et al.* ScienceDirect Biodegradation of polycyclic aromatic hydrocarbons (PAHs) by fungal enzymes. RE:view 2016; 1.

[285] Torretta V, Katsoyiannis I, Viotti P, Rada E. Critical Review of the Effects of Glyphosate Exposure to the Environment and Humans through the Food Supply Chain. Sustainability (Basel) 2018; 10(4): 950.
[http://dx.doi.org/10.3390/su10040950]

[286] Kishore D, Kundu S, Kayastha AM. Thermal, chemical and pH induced denaturation of a multimeric β-galactosidase reveals multiple unfolding pathways. PLoS One 2012; 7(11): e50380.
[http://dx.doi.org/10.1371/journal.pone.0050380]

[287] Flores R, Serra P, Minoia S, *et al. Viroids* : from genotype to phenotype just relying on RNA sequence and structural motifs. 2012; 3: 1-13.
[http://dx.doi.org/10.3389/fmicb.2012.00217]

[288] Chen Q, Li J, Liu M, *et al.* Study on the biodegradation of crude oil by free and immobilized bacterial consortium in marine environment. 2017; 1-23.

[289] Bamforth SM, Singleton I. Bioremediation of polycyclic aromatic hydrocarbons : current knowledge and future directions. 2005; 736: 723-36.
[http://dx.doi.org/10.1002/jctb.1276]

[290] Ratzke C, Gore J. Modifying and reacting to the environmental pH can drive bacterial interactions. PLoS Biol 2018; 16(3): e2004248.
[http://dx.doi.org/10.1371/journal.pbio.2004248] [PMID: 29538378]

[291] Riveroll-Larios J, Escalante-Espinosa E, Fócil-Monterrubio RL, Díaz-Ramírez IJ. Biological Activity Assessment in Mexican Tropical Soils with Different Hydrocarbon Contamination Histories. Water Air Soil Pollut 2015; 226(10): 353.
[http://dx.doi.org/10.1007/s11270-015-2621-1] [PMID: 26478633]

[292] Rizzo ACL, Cunha CD, Santos RLC, *et al.* Preliminary identification of the bioremediation limiting factors of a clay bearing soil contaminated with crude oil. J Braz Chem Soc 2008; 19(1): 169-74.
[http://dx.doi.org/10.1590/S0103-50532008000100024]

[293] Ivica K, Zeljka Z, Aleksandra P. Soil treatment engineering. Physical Sciences Reviews 2017; 2(11): 1-26.
[http://dx.doi.org/10.1515/psr-2016-0124]

[294] Sayara T, Sánchez A. Bioremediation of PAH-contaminated soils: Process enhancement through composting/compost. Appl Sci (Basel) 2020; 10(11): 3684.
[http://dx.doi.org/10.3390/app10113684]

[295] Lee K, Park JW, Ahn IS. Effect of additional carbon source on naphthalene biodegradation by *Pseudomonas putida* G7. J Hazard Mater 2003; 105(1-3): 157-67.
[http://dx.doi.org/10.1016/j.jhazmat.2003.08.005] [PMID: 14623425]

[296] Haghollahi A, Fazaelipoor MH, Schaffie M. The effect of soil type on the bioremediation of petroleum contaminated soils. J Environ Manage 2016; 180: 197-201.
[http://dx.doi.org/10.1016/j.jenvman.2016.05.038] [PMID: 27233045]

[297] Ali N, Dashti N, Khanafer M, Al-Awadhi H, Radwan S. Bioremediation of soils saturated with spilled crude oil. Sci Rep 2020; 10(1): 1116.
[http://dx.doi.org/10.1038/s41598-019-57224-x] [PMID: 31980664]

[298] Fernandez-Salguero J, Gómez R, Carmona MA. Water Activity in Selected High-Moisture Foods. Journal of Food Composition and Analysis 1993; 6. 364-369.
[http://dx.doi.org/10.1006/jfca.1993.1040]

[299] Schultz C. Water activity as related to microorganisms in the manufacturing environment. General Internal Medicine and Clinical Innovations 2016; 1(6): 1-2.
[http://dx.doi.org/10.15761/GIMCI.1000133]

[300] Ojuederie O, Babalola O. Microbial and plant-assisted bioremediation of heavy metal polluted environments: A Review. Int J Environ Res Public Health 2017; 14(12): 1504.
[http://dx.doi.org/10.3390/ijerph14121504] [PMID: 29207531]

[301] Wood JM. Bacterial responses to osmotic challenges. J Gen Physiol 2015; 145(5): 381-8.
[http://dx.doi.org/10.1085/jgp.201411296] [PMID: 25870209]

[302] Moxley E, Puerta-Fernández E, Gómez EJ, Gonzalez JM. Influence of Abiotic Factors Temperature and Water Content on Bacterial 2-Chlorophenol Biodegradation in Soils. Front Environ Sci 2019; 7: 41.
[http://dx.doi.org/10.3389/fenvs.2019.00041]

[303] Bahmani F, Ahmad Ataei S, Ali Mikaili M. The Effect of Moisture Content Variation on the Bioremediation of Hydrocarbon Contaminated Soils: Modeling and Experimental Investigation. Journal of Environmental Analytical Chemistry 2018; 5(2).
[http://dx.doi.org/10.4172/2380-2391.1000236]

[304] Ucar D, Zhang Y, Angelidaki I. An overview of electron acceptors in microbial fuel cells. Front Microbiol 2017; 8: 643.
[http://dx.doi.org/10.3389/fmicb.2017.00643] [PMID: 28469607]

[305] Ferreira AS, Santos MAD, Corrêa GF. Soil microbial response to glucose and phosphorus addition under agricultural systems in the Brazilian Cerrado. An Acad Bras Cienc 2013; 85(1): 395-403.
[http://dx.doi.org/10.1590/S0001-37652013005000021] [PMID: 23460427]

[306] Semboung Lang F, Tarayre C, Destain J, *et al.* The effect of nutrients on the degradation of hydrocarbons in mangrove ecosystems by microorganisms. Int J Environ Res 2016; 10: 583-92.
[http://dx.doi.org/10.22059/ijer.2016.59903]

[307] Singh AK, Sherry A, Gray ND, Jones DM, Bowler BFJ, Head IM. Kinetic parameters for nutrient enhanced crude oil biodegradation in intertidal marine sediments. Front Microbiol 2014; 5: 160.
[http://dx.doi.org/10.3389/fmicb.2014.00160] [PMID: 24782848]

[308] Ławniczak Ł, Woźniak-Karczewska M, Loibner AP, Heipieper HJ, Chrzanowski Ł. Microbial degradation of hydrocarbons—basic principles for bioremediation: A review. Molecules 2020; 25(4): 856.
[http://dx.doi.org/10.3390/molecules25040856] [PMID: 32075198]

[309] Bremer E. Coping with osmotic challenges: osmoregulation through accumulation and release of compatible solutes in *B. subtilis*. Comp Biochem Physiol A Mol Integr Physiol 2000; 126: 17.
[http://dx.doi.org/10.1016/S1095-6433(00)80031-8]

[310] Reichert CC, Reinehr CO, Costa JAV. Semicontinuous cultivation of the cyanobacterium *Spirulina platensis* in a closed photobioreactor. Braz J Chem Eng 2006; 23(1): 23-8.
[http://dx.doi.org/10.1590/S0104-66322006000100003]

[311] Devarapalli M, Lewis R, Atiyeh H. Continuous ethanol production from synthesis gas by *Clostridium ragsdalei* in a trickle-bed reactor. Fermentation (Basel) 2017; 3(2): 23.
[http://dx.doi.org/10.3390/fermentation3020023]

[312] Qin X, Tang JC, Li DS, Zhang QM. Effect of salinity on the bioremediation of petroleum hydrocarbons in a saline-alkaline soil. Lett Appl Microbiol 2012; 55(3): 210-7.
[http://dx.doi.org/10.1111/j.1472-765X.2012.03280.x] [PMID: 22725670]

[313] Rhykerd RL, Weaver RW, McInnes KJ. Influence of salinity on bioremediation of oil in soil. Environ Pollut 1995; 90(1): 127-30.
[http://dx.doi.org/10.1016/0269-7491(94)00087-T] [PMID: 15091510]

[314] Babcock GT. How oxygen is activated and reduced in respiration. Proc Natl Acad Sci USA 1999; 96(23): 12971-3.
[http://dx.doi.org/10.1073/pnas.96.23.12971] [PMID: 10557256]

[315] Baharvand S, Mansouri Daneshvar MR. Impact assessment of treating wastewater on the physiochemical variables of environment: a case of Kermanshah wastewater treatment plant in Iran. Environ Syst Res 2019; 8(1): 18.
[http://dx.doi.org/10.1186/s40068-019-0146-0]

[316] Abubakar Clarkson M, Isa Abubakar S. Bioremediation and Biodegradation of Hydrocarbon Contaminated Soils: A Review. IOSR J Environ Sci Ver I 2015; 9: 2319-99.
[http://dx.doi.org/10.9790/2402-091113845]

[317] Weinberg F, Hamanaka R, Wheaton WW, *et al.* Mitochondrial metabolism and ROS generation are essential for Kras-mediated tumorigenicity. Proc Natl Acad Sci USA 2010; 107(19): 8788-93.
[http://dx.doi.org/10.1073/pnas.1003428107] [PMID: 20421486]

[318] Sayel H, Bahafid W, Tahri Joutey N, *et al.* Cr(VI) reduction by Enterococcus gallinarum isolated from tannery waste-contaminated soil. Ann Microbiol 2012; 62(3): 1269-77.
[http://dx.doi.org/10.1007/s13213-011-0372-9]

[319] Sutar H, Das CK. A Review on : Bioremediation International Journal of Research in Chemistry and Environment A Review on : Bioremediation. Int J Reasearch Chem Environ 2012; 2: 13-21.

[320] Tekere M. Microbial bioremediation and different bioreactors designs applied. Intech 2019; i: 38.
[http://dx.doi.org/10.5772/intechopen.83661]

[321] Tingle CCD, Rother JA, Dewhurst CF, Lauer S, King WJ. Fipronil: environmental fate, ecotoxicology, and human health concerns. Rev Environ Contam Toxicol 2003; 176: 1-66.

[http://dx.doi.org/10.1007/978-1-4899-7283-5_1] [PMID: 12442503]

[322] Fesenko S, Howard BJ, Brenda J. International Atomic Energy Agency. Guidelines for remediation strategies to reduce the radiological consequences of environmental contamination 2012; 167.

[323] Niti C, Sunita S, Kamlesh K, Rakesh K. Bioremediation: An emerging technology for remediation of pesticides. Res J Chem Environ 2013; 17: 88-105.

[324] Herrero M, Stuckey DC. Bioaugmentation and its application in wastewater treatment: A review. Chemosphere 2015; 140: 119-28.
[http://dx.doi.org/10.1016/j.chemosphere.2014.10.033] [PMID: 25454204]

[325] Vidonish JE, Zygourakis K, Masiello CA, Sabadell G, Alvarez PJJ. Thermal Treatment of Hydrocarbon-Impacted Soils: A Review of Technology Innovation for Sustainable Remediation. Engineering (Beijing) 2016; 2(4): 426-37.
[http://dx.doi.org/10.1016/J.ENG.2016.04.005]

[326] Rayu S, Karpouzas DG, Singh BK. Emerging technologies in bioremediation: constraints and opportunities. Biodegradation 2012; 23(6): 917-26.
[http://dx.doi.org/10.1007/s10532-012-9576-3] [PMID: 22836784]

[327] Frey-Klett P, Burlinson P, Deveau A, Barret M, Tarkka M, Sarniguet A. Bacterial-fungal interactions: hyphens between agricultural, clinical, environmental, and food microbiologists. Microbiol Mol Biol Rev 2011; 75(4): 583-609.
[http://dx.doi.org/10.1128/MMBR.00020-11] [PMID: 22126995]

[328] Sasikumar CS, Papinazath T. 2003; Environmental management:-bioremediation of polluted environment. Proc Third Int Conf Environ Heal 465 469.

CHAPTER 12

Microbial Bioremediation of Microplastics

Manish Kumar Singh¹, Younus Raza Beg², Gokul Ram Nishad² and **Priyanka Singh²,***

¹ Krishi Vigyan Kendra, Rajnandgaon-491441, Chhattisgarh, India

² Department of Chemistry, Govt. Digvijay PG Autonomous College, Rajnandgaon-491441, Chhattisgarh, India

Abstract: Plastic is being used over the entire globe in the form of capsules, microbeads, fibers or microplastics. The waste thus generated has gained concern due to the loss of aesthetic value, the presence of various toxic chemicals such as plasticizers, antioxidants, *etc.*, and the release of greenhouse gases. The small size and slow degradability of microplastics are responsible for their accumulation in the environment and organisms. Plastic degradability can be improved by altering its chemical and physical structure or using better degrading agents. Different types of microorganisms and enzymes are being designed and employed for degrading plastic waste. This chapter gives an overview of the degradation mechanism along with different microbial, plant and animal species responsible for this process.

Keywords: Microplastics, Microbial degradation, Animal-mediated degradation, Plant-mediated degradation, Environmental pollutants.

INTRODUCTION

Plastic has become essential evil due to their indispensable necessity. Heaps of plastic waste are continually being added and accumulated in spite of its ill effects. Plastic gets accumulated in the form of macroplastics, mesoplastics, microplastics and nanoplastics in the environment. Generally, several additives such as plasticizers, antioxidants, UV stabilizers, curing agents, *etc.* are added to plastics during their processing which further increases environmental pollution. Moreover, dumping plastics in landfills leads to aesthetic degradation and the release of toxic chemicals and greenhouse gases [1].

* **Corresponding author Priyanka Singh:** Department of Chemistry, Govt. Digvijay PG Autonomous College, Rajnandgaon-491441, Chhattisgarh, India; E-mail: priyankasingh121@yahoo.com

Inamuddin (Ed.)

Microplastics are extremely small and harmful plastic pieces or fragments produced from industrial and consumer products. They can be commonly observed all around the globe and can be in the form of microbeads, capsules, fibers or pellets or may be produced from larger plastic items through their breakdown. This term was coined by a marine biologist, Richard Thompson. Polymers like phthalates, polybrominated diphenyl ethers, tetrabromobisphenol A, polyvinyl chloride, polystyrene, polyvinyl alcohol, *etc.* are common examples of microplastics [2, 3].

They have a size less than 5 mm and pose danger to the environment, flora and fauna. Degradation of plastics is a very slow process that may exceed a period of over hundreds or thousands of years. During this process, their ingestion and incorporation into the environment and living organisms are highly probable. They possess a large surface-to-volume ratio and therefore, interact with organic pollutants like polycyclic aromatic hydrocarbons, polychlorinated biphenyls, *etc.* depending on the properties of the latter [4, 5]. Ingestion of microplastics leads to oxidative stress, reduced growth and reproductive ability, false satiation, *etc* [6, 7]. Toussaint *et al.* (2019) have reviewed the contamination of microplastic and nanoplastic in the food chain. They have studied about 200 animal species and food products forming an element of the human food chain [8]. Zhang *et al.* (2020) have reviewed the direct human exposure to microplastics through the air, table salt and drinking water consumption [9]. Moreover, microplastics act as carrier for toxic chemicals in the environment [10].

New technologies and strategies need to be developed in order to combat the issue of ever-increasing plastic waste. Various methods have been utilized for the physical and chemical degradation of plastics. Different types of microorganisms and enzymes are being designed and employed for degrading plastic waste. Experiments are carried out all over the world in order to increase the catalytic efficiency of these enzymes. Canada, the United States of America, the United Kingdom, *etc.* have banned products containing microbeads [11]. Jenkins *et al.* (2020) and Masia *et al.* (2020) have reported the microbial degradation of plastics and bioremediation of microplastics, respectively [12, 13]. Functional groups capable of affecting the hydrophobicity, molecular weight, morphology, chemical and physical structure, structural complexity, molecular composition and density of polymers affect their degradability [14]. This review deals with microbial, plant and animal or their combinations mediated degradation of microplastics. The mechanism of biodegradation of some polymers has also been discussed briefly.

Types

On the basis of source, microplastics are mainly classified as primary and secondary microplastics.

Primary microplastics: They are the plastic pieces or fragments having size less than 5.0 mm. Microbeads, microfibers and plastic pellets are some of its examples. They are used in face washes, toothpaste, cosmetics [15], scrubbers [16], air blasting technology, biomedical research, *etc.* (Fig. **1**). Polyethylene, polyethylene terephthalate, polypropylene and nylon are generally used in these formulations or technologies.

Secondary microplastics: They are produced through the breakdown of larger macroplastic materials such as water bottles, plastic bags, plastic furniture, tyres, nets, *etc.* (Fig. **1**). UV radiations, wind and wave action, *etc.* may be held responsible for such breakdown.

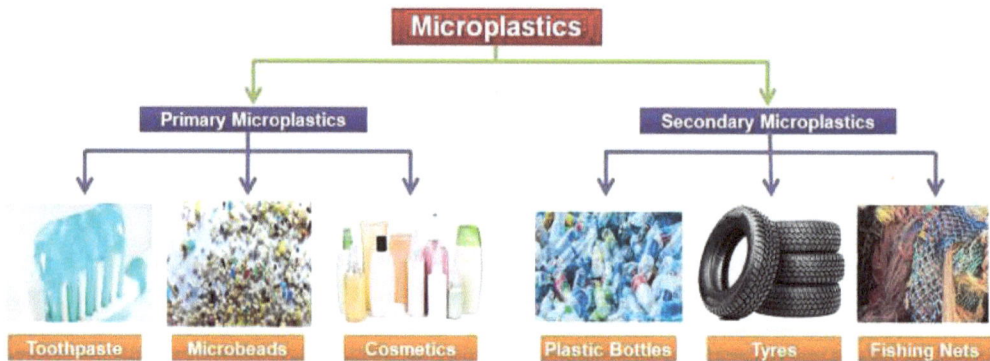

Fig. (1). Classification of plastics.

Sources: There are several sources of microplastics, some of which are being discussed here.

- Microplastics have been identified in primary as well as secondary water treatment stages which are subsequently added to the oceans and surface water bodies.
- Sewage sludge containing microplastics also contaminates the water bodies.
- Automobile tyres undergo wear and tear, thus adding microplastics into the surroundings.
- Natural exfoliating materials have been continually replaced with microplastics in cosmetic industries.
- Microfibers are released during handling and washing of clothing made up of

synthetic fibers like polyester, nylon, spandex, *etc.*

- Plastic manufacturing uses small granules and pellets as raw materials. Accidental spillage during transport adds these micropastics to the environment.
- Marine industries, shipping and commercial fishing also act as a source of microplastic to the marine environment.
- Microplastics are also added from plastic water bottles and their packaging.

Effects

Microplastics are omnipresent in soil, water and air. Effect of microplastics still needs to be studied in an elaborative manner in order to mother earth. Microplastics enter the living systems *via* respiratory, dermal and oral routes. The presence of microplastics in lugworms (*Arenicola marina)* in their gastrointestinal tracts has been detected. It was found to be present in *Carcinus maenas*'s (shore crab) respiratory as well as digestive system [17]. Microplastics can affect soil structure, nutrient immobilization, microbial composition, water evaporation rate, *etc.* leading to alteration of biophysical characteristics of the soil. The effect of Microplastics on plant performance depends on plant species as well as plastic types. Machado *et al.* (2019) have discussed the plant biomass as well as elemental composition of spring onion *(Allium fistulosum) in* reference to the effect of microplastics [18].

Small animals are in big trouble due to microplastics as it takes 14 days to pass through the digestive system as compared to 2 days for normal food; which leads to starvation [19]. *Tetrabromobisphenol A* (TBBPA) is used as a flame retardant in plastics; it disturbs the hormonal balance of the thyroid in Wistar rats [20]. Lantern fish are reported to be laden with microplastics. Parrotfish, eaten by humans, eat seagrass which is embedded with microplastics . Accumulation of microplastics in human organs is also possible through respiration, and drinking of water which may cause cytotoxicity [21]. The microplastics ingested by fish and crustaceans can be subsequently consumed by humans at the end of the food chain [22, 23].

Polyvinyl chloride (PVC) was the most commonly used plastics as pipes for plumbing as well as to wrap meat. Toxic chemicals are produced when PVC is exposed to water. Bisphenol A (BPA) is a very common additive used to harden plastics. It is responsible for various disorders in human beings [24]. The average person consumes 50000 particles of microplastics every year through both eating and respiration [25, 26]. Polyethylene and polystyrene induce oxidative stress and endoplasmic reticulum stress, respectively. This might lead to cytotoxicity [27]. Workers exposed to airborne microplastics have been found to be victims of dyspnea, interstitial inflammation, occupational asthma as well as allergic alveolitis [28]. Microplastics are a snake in the grass. Long exposure due to bioaccumulation of these particles may cause various disorders in human beings. It is poisoning the planet earth.

BIODEGRADATION OF PLASTICS/MICROPLASTICS

Biodegradation of plastic can be categorized into two types based on the mechanisms involved. The first is direct action which involves the utilization of plastics by organisms for their growth leading to their deterioration. The second being indirect action which involves the degradation of plastic due to the action of metabolic products of organisms. The organisms involved in biodegradation of plastics or microplastics may be bacteria, fungi, algae, animals or plants. Their biodegradation action will be discussed in this section.

Microbial Degradation

Microbial degradation is an ecofriendly technique to get rid of the vast heaps of plastic produced everyday and that too without adverse effects. Microbes use enzymes, either extracellular or intracellular, categorized as lipases, proteinases and dehydrogenases, for this degradation process [29].

Bacteria-mediated Degradation

Bacterial species are commonly used for biodegradation of plastics. They are found prominantly in soil, water, and atmosphere. Several bacterial species such as *Bacillus subtilis* [30], *Pseudomonas sp* [31], *Gordonia sp.* strain QH-11 [32], *Pseudomonas citronellolis* and *Bacillus flexus* [33], *Gordonia terrae* RL-JC02 [34], *Microbulbifer hydrolyticus* IRE-31 [35], *etc.* have been used for degrading plastics/microplastics (Table **1**). Park *et al.* (2019) used bacterial culture isolates acquired from municipal landfill residue for the degradation of polyethylene microplastics [36]. Syranidou *et al.* (2019) used tailored marine consortia for the treatment of plastic film mixture [37]. Polycaprolactone (PCL) can be degraded by *Alcaligenes faecalis* and *Clostridium botulinum* [38, 39]. Polylactic acid (PLA) and Polyvinyl chloride (PVC) can be degraded by *Bacillus brevis* and *Pseudomonas putida*, respectively [40, 41] Polyhydroxyalkanoates (PHA) can be biodegraded due to the PHA depolymerases produced by *Alcaligenes faecalis* and *Streptomyces sp* [42, 43].

Narciso-Ortiz *et al.* (2020) studied the biodegradation of polyethylene terephthalate (PET) through airlift bioreactor using bacterial species for 3 days [44]. Li *et al.* (2020) used *Microbulbifer hydrolyticus* IRE-31 for the biodegradation of low-density polyethylene with copper sulfate for 30 days; SEM studies revealed surface alteration such as cavities, cracks, erosion, *etc.* on polymer surfaces as a result of microbial action in both the cases. FTIR spectroscopy revealed transmittance percentage reduction in the carboxylic group due to microbial enzymatic action as a result of additional carbonyl group formation. The formation of new hydroxyl and carbonyl functional groups occurs at 3392 cm^{-1} and 1644 cm^{-1}, respectively due to bacterial treatment. This indicates bio-oxidation of a linear low-density polyethylene surface. The peak at 3195 cm^{-1} is due to C=C bond which is weaker than the C–C bond. Ether groups and multiple peaks were not observed in 1100 cm^{-1} and 1600-1800 cm^{-1} regions, respectively. Bacterium-treated linear low-density polyethylene samples did not show ester and aromatic groups as well in the FTIR spectrum [35].

Table 1. List of different polymers and the bacterial species involved in their biodegradation.

S.No.	Substrate	Bacterial Species	Biodegradation Efficiency (% Weight Loss)	Reference
1	Polyethylene	*Bacillus subtilis*	9.26% in 30 days	Vimala *et al.* (2016) [30]
2	Dibutyl phthalate (DBP)	*Gordonia phthalatica* sp.	-	Kong *et al.* (2019) [32]
3	Polyvinyl chloride (PVC)	*Pseudomonas citronellolis*	10%	Giacomucci *et al.* (2019) [33]
4	Polyethylene (PE) and polyphenylene sulfide (PPS)	*Pseudomonas* sp.	9.71%	Li *et al.* (20220) [35]
5	Di-(2-ethylhexyl) phthalate (DEHP)	*Gordonia terrae* RL-JC02	14.7% after 60 days	Park *et al.* (2019) [36]
6	Polyethylene (PE) and polystyrene (PS)	*,Rhodobacterales Oceanospirillales* and *Burkholderiales* communities and *Bacillus* and *Pseudonocardia* genera	33% and 27% in indigenous (INDG) and bioaugmented (BIOG) treatment, respectively	Syranidou *et al.* (2019) [37]
7	Polyethylene terephthalate (PET)	*Vibrio sp.*	35%	Sarkhel *et al.* (2020) [56]

Fungi-Mediated Degradation

Fungi are the eukaryotic organisms that can be classified as aerobic or anaerobic based on oxygen requirement and yeasts, filamentous or dimorphic fungi based on their morphological structure. They are heterotrophs and possess extraordinary adaptability. They are capable of releasing digestive enzymes outside their hyphae through exocytosis. Epoxidases and transferases are the Phase I and II enzymes catalyzing the oxidation and conjugation reactions, respectively [45]. Therefore, they can absorb nutrients from the exterior of their cells through the degradation of the organic compounds into CO_2 and H_2O or CH_4 [46]. These enzymes can decrease polyethylene chain length. The ascomycetes, basidiomycetes and zigomycetes classes of fungi can deteriorate petroleum based plastics in decreasing order. Polyethylene and polyurethane are degraded by *Penicillium simplicissimum and* Fusarium, respectively [47, 48]. *Fusarium solani* and *Aureobasidium pullulans* sp. are capable of degrading polyurethane [49, 50]. *Fusarium moniliforme* and *Penicillium roqueforti* can deteriorate polylactic acid (PLA) [39]. Fungi produce hydrophobins which aid in the attachment of hyphae to the hydrophobic substrates. Fungal hyphae are capable of penetrating three dimensional substrates [51]. Table **2** enlists the polymers and fungal species capable of degrading them.

Table 2. List of different polymers and the fungal species involved in their biodegradation.

S.No.	Substrate	Fungal Species	Biodegradation Efficiency (% Weight Loss)	Reference
1	Plastic mulch film	yeast Pseudozyma antarctica	-	Sameshima-Yamashita *et al.* (2018) [53]
2	Gamma irradiated low density polyethylene and polypropylene	*Aspergillus sp., Paecilomyces lilacinus* isolated from *H. brunonis* and *Lasiodiplodia theobromae* isolated from *Psychotria flavida* indicate	-	Sheik *et al.* (2015) [54]
3	Poly-(butylene succinate-co-butylene adipate) (PBSA) and poly-(butylene succinate) (PBS) films	Phylloplane fungi (B47-9 a strain)	23.7 wt%, and 14.6 wt% for PBSA and PBS, respectively	Koitabashi *et al.* (2012) [55]
4	Polyethylene terephthalate (PET)	*Aspergillus sp*	22%	Sarkhel *et al.* (2020) [56]
5	Phthalate acid esters	Diatom *Cylindrotheca closterium*	-	Li *et al.* (2015) [61]
6	Polyethylene (PE)	*Aspergillus flavus* from the guts of wax moth *Galleria mellonella*	-	Zhang *et al.* (2019) [72]

Paco *et al.* (2017) studied the role of fungus *Zalerion maritimum* on the degradation of polyethylene pellets and their use as substrate by *Z. maritimum*. The FTIR, NMR and mass variations studies suggest that the fungus uses polyethylene microplastics as substrate. FTIR study revealed that peak intensity at 3700-3000cm^{-1}, 1700-1500 cm^{-1} and 1200-950 cm^{-1} was found to be increased regularly when microplastics were exposed *to Zalerion maritimum* for various intervals (for day 7, 14, 21 and 28). Increased intensities of these peaks were associated to hydroperoxide / hydroxyl group, carbonyl groups and double bonds respectively, which leads toward the conclusion of various degradation mechanisms of polyethylene. The fungi devoid of microplastics showed enhanced carbohydrate and decreased protein and lipid concentration to a larger extent [52]. An enzyme produced by yeast *Pseudozyma antarctica* upon cultivation with xylose is capable of degrading commercial biodegradable plastic mulch films within 5 weeks [53]. Endophytic fungi *Aspergillus sp.*, *Paecilomyces lilacinus* obtained from *Humboldtia brunonis Lasiodiplodia theobromae* obtained from *Psychotria flavida* produce laccase enzymes and were studied for degradation of gamma irradiated low density polyethylene and polypropylene [54]. Several fungal strains obtained from phylloplane (plant surface) were used against biodegradable plastic mulch films. B47-9 strain obtained from barley leaf surface proved to be most capable of degrading poly-(butylene succinate-co-butylene adipate) (PBSA) and poly-(butylene succinate) (PBS) films [55].

Sarkhel *et al.* (2020) studied the biodegradability of plastic bottle synthetic polymer, polyethylene terephthalate (PET) using bacterium (*Vibrio sp*) and fungus (*Aspergillus sp*) obtained from marine environment near the Bay of Bengal. FTIR spectroscopy, SEM, XRD and weight loss studies show 35% and 22% degradation of plastic bottle polymers using bacterial and fungal strains, respectively within 6 weeks. Variation in temperature, pH and inoculum concentration also shows better performance by bacteria. C-H bond shows breakage whereas, no change in the peak due to -OCH$_3$ (250 cm^{-1}) group shows that it remained intact during the degradation process. The peaks at 769 and 806 cm^{-1} show ester stre*tc*hing of PET before and after the degradation process. The peak due to C=O stre*tc*hing and C-H bond showed gradual disappearance and intensification, respectively after degradation. SEM images show microbial growth and an increase in roughness of the surface after degradation. X-ray diffraction study reveals a decrease in crystallinity of the polymer films and an increase in the intensity of the sample [56].

Algae-mediated Degradation

Algae are the autotrophic organisms and find application in the field of food, chemicals, biofuel, biofertilizers, pharmaceuticals, *etc.* The EPS mucilage facilitates algal colonization on the polymeric surfaces (Boney, 1981). Therefore, they are capable of degrading different organic pollutants and plastics through the secretion of mucilaginous extracellular polymeric substance (EPS). *Phormidium tenue, Monoraphidium contortum, Oscillatoria tenuis, Microcystis aeruginosa, Chlorella vulgaris, Closterium constatum, etc.* were reported to form colonies on polythene materials in aquatic systems [57]. *Scenedesmus dimorphus, Anabaena spiroides*, and *Navicula pupula* have been reported as successful degraders of low-density polyethylene. SEM studies show the adherence of *Scenedesmus dimorphus* to the polymer surface without significant changes. *Anabaena spiroides* showed minute hole formation on the surface and were most efficient in degrading the polymer (8.18%). Whereas, partial erosion of the polymer surface was shown by *Navicula pupula* [58]. *Phormidium lucidum* and *Oscillatoria subbrevis* named cyanobacterial species can also degrade the low-density polyethylene [59].

Freshwater *Cyanothece sp.* produces an EPS with high bioflocculant activity on being exposed to nanoplastics and microplastics. The bioflocculant capacity of the biopolymer produced was evaluated. The microalgae forms aggregates consisting of nanoplastics and microplastics. Fluorescence microscopic studies reveal the formation of heteroaggregates containing microalgae, EPS and polystyrene nanoplastics and microplastics. Microalgal-based biopolymers can be effectively used in place of toxic synthetic flocculants for wastewater treatment [60]. A marine benthic diatom, *Cylindrotheca closterium*, was found capable of degrading diethyl phthalate (DEP) and dibutyl phthalate (DBP) from the surface of sediment effectively as compared to its bottom layer [61].

Biofilm-mediated Degradation

In the aquatic environment, microplastics get into touch with organic and inorganic materials along with microorganisms such as bacteria, fungi, protists, algae, *etc* [62]. These microorganisms get attached to the microplastic surfaces and result in biofilm formation [63]. Microplastics serve as an adhesion media, carbon and energy sources for the microorganisms forming the biofilm. The stages of biofilm based degradation of microplastics have been depicted in Fig. (**2**). Hossain *et al.* (2019) examined the biodegradation of polypropylene and the biofilm mainly consisted of *Acinetobacter calcoaceticus, Burkholderia cepacia*, and *Escherichia coli* [64].

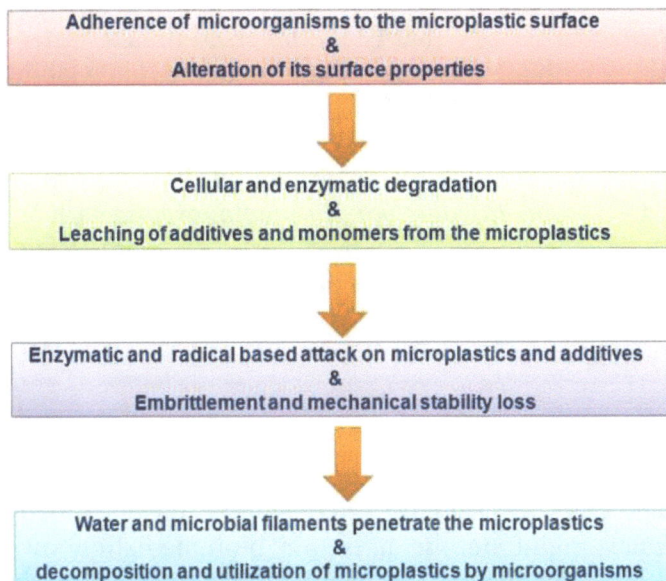

Fig. (2). Stages of biofilm based degradation of microplastics.

Bacterial strain *Pseudomonas aeruginosa* ISJ14 shows LDPE degradation through biofilm formation. Studies regarding hydrophobicity, viability and total protein content were performed. Surface modification, cracks on the LDPE surface and an increase in carbonyl index intensity confirm biodegradation of the polymer [65]. Laccase, a bacterial copper-binding enzyme, is capable of oxidizing and degrading polyethylene. RT-PCR studies show that laccase mRNA quantity increases by about thirteen times as compared to control. Copper increased polyethylene degradation of PE by 75. FTIR analysis shows carbonyl peak increase [66]. The effect of the addition of carbon sources glucose, peptone and their combination to periphytic biofilm for the biodegradation of polypropylene (PP), polyethylene (PE) and polyethylene terephthalate (PET) microplastics were also studied. Glucose addition increased the microplastic biodegradation whereas, peptone and the combination of glucose and peptone had inhibitory effects [67]. Biodegradation of polyurethane (PU) coating depends on biotic as well as abiotic factors because microorganisms cannot stay alive on PU-based coatings. But, a strain of *Papiliotrema laurentii* is capable of deteriorating polyester-based PU coating through biofilm formation [68].

Biodegradability of polylactide (PLA) was studied in lake water, compost and soil. OxiTop was used as a control. Bacterial species were isolated from the biofilm and identified as *Acidovorax sp.*, *Chryseobacterium sp.*, *Aeromonas veronii*, *Arthrobacteraurescens*, *Arthrobacter sp.*, *A. aurescens*, *Elizabethkingia meningoseptica, A. aurescens, A. aurescens* on the basis of 16S rRNA gene sequencing. Experiments showed that PLA immersed in lake water had the most profuse biofilm along with a high population of diatoms. The viability of bacteria was high and more than 80% of cells were found to be living through LIVE/DEAD and SEM analysis. PLA incubated in lake water soil and compost showed the highest profused biofilm formation by *Acidivorax sp.* LW9, *Arthrobacter sp.* LG12 and *Elizabethkingia meningoseptica* LK3, respectively. Whereas, hydrolytic activity was highest for *Chryseobacterium sp.* LW2, *Arthrobacter sp.* LG12 and *Elizabethkingia meningoseptica* LK3, respectively. Therefore, it was concluded that biofilm abundance does not depend on their hydrolytic activity of specific bacterial strains [69].

Animal-mediated Degradation

Several organisms such as zooplankton and predators consume microplastics [70 - 72]. Several species might not be used for bioremediation owing to lower retention. However, some species like sandworm *Arenicola marina* do have a higher retention ability of 240-700 microplastics throughout its life without notable effects on metabolism [73]. Several worms and their larvae are capable of converting macroplastics to microplastics (Table **3**). But, raising them is a very slow and costly process. The exact mechanism of degradation is still a matter of research. They generally feed on food wastes through composting. Yellow mealworms larvae (*Tenebrio molitor*) have been found to be effective against the biodegradation of polystyrene and low density polyethylene over 60 days as evidenced by a decrease in Mw and Mn. Limited degradation of the former was seen. Antibiotic gentamicin shows inhibitory action on the degradation of polymers. Depolymerization and formation of new functional groups containing oxygen were confirmed by gel permeation chromatography, Fourier transform infrared spectroscopy and thermogravimetric studies [74]. The microorganisms *Bacillus sp.* YP1 and *Enterobacter asburiae* YT1 residing in the gut of Indian meal moth (*Plodia interpunctella*) larvae showed fast degradation of poly(ethylene terephthalate) (PET) [75].

Table 3. List of different polymers and the animal species involved in their biodegradation.

S.No.	Substrate	Animal Species	Consumption Rate	References
1	Polystyrene (PS) and low density polyethylene (LDPE)	Yellow mealworms (*Tenebrio molitor*)	4.27±0.09, 3.33±0.02, and 3.45±0.04 mg 100 larvae^{-1}d^{-1} for PS, PE-1 and PE-2, respectively	Yang *et al.* (2020) [74]
2	Poly(ethylene terephthalate) (PET)	Indian meal moth (*Plodia interpunctella*)	-	Yoshida *et al.* (2016) [75]

(Table 3) cont.....

S.No.	Substrate	Animal Species	Consumption Rate	References
3	Polyethylene	Wax moth (*Galleria mellonella*)	-	Bombelli *et al.* (2017) [76]
4	Polystyrene (PS) and low density polyethylene (LDPE)	Yellow mealworms (*Tenebrio molitor*) and Wax moth (*Galleria mellonella*)	0.38 mg worm^{-1}day^{-1}	Billen *et al.* (2020) [77]
5	Polyethylene	Earthworms (*Lumbricus terrestris*)	-	Zhang *et al.* (2018) [78]
6	Low-density polyethylene and polystyrene foams	Superworm (*Zophobas atratus*)	58.7 ± 1.8 mg and 61.5 ± 1.6 mg 100 larvae^{-1}d^{-1} for LDPE and PS foams, respectively	Peng *et al.* (2020) [79]
7	Polystyrene foams	Superworm (*Zophobas atratus*)	0.58 mg larvae^{-1}d^{-1}	Yang *et al.* (2020) [82]
8	Polystyrene	Snails (*Achatina fulica*)	18.5 ± 2.9 mg per snail	Song *et al.* (2020) [83]

Rapid deterioration of polyethylene has been reported by the caterpillar larva of wax moth *Galleria mellonella* belonging to the Pyralidae family. Hole formation was observed in the polymer just after 40 minutes. FTIR spectra of untreated samples demonstrated through black line show peaks specific for polyethylene at 2,921 and 2,852 cm^{-1}. Whereas, worm homogenate treated sample demonstrated red line shows an extra peak at about 3,350 cm^{-1} and 1,700 cm^{-1} specific for ethylene glycol and carbonyl bond, respectively. Peak owing to ethylene glycol showed around the holes of polyethylene but disappeared at distant points. Atomic Force Microscopy shows a change in morphology of polymer surface. The wax worms lay their eggs in the honeycomb where they grow to the pupa stage and eat beeswax. Beeswax is a mixture of lipids, hydrocarbons, fatty acids and esters, *etc.* Ethylene is the commonest hydrocarbon present in the beeswax. The hydrocarbon-digesting ability of the wax worms might be due to its own enzyme system or its intestinal flora dominated by *Plodia interpunctella* [76].

Another study involved the use of mealworms (*Tenebrio molitor*) and the larvae of wax moth (*Galleria mellonella*) for biodegrading polyethylene through chewing. Live specimens as well as homogenated paste were used. The former was found to be more effective as compared to the later. The plasticizing effect was shown by liquid excreted by the live specimen. Streamlined life cycle assessment as well as techno-economic analysis (LCA/TEA) techniques were also used [77]. Earthworms (*Lumbricus terrestris*) are also capable of combating macroscopic polyethylene in Petri dishes and mesocosms. Four types of biodegradable plastic mulches including PLA/PHA [polylactic acid/polyhydroxy alkanoate], BioAgri, Organix and Nature cycle were also used. Biodegradable paper mulch, Weed Guard Plus and poplar litter were used as control. Earthworms were not capable of ingesting polyethylene or field-weathered biodegradable plastic mulches. But, biodegradable plastics were ingested by the earthworm following soil burial along with composting [78].

Zophobas atratus larvae degraded low-density polyethylene and foams of expanded polystyrene [79]. Yellow mealworms, *Tenebrio molitor* larvae, superworm, *Zophobas atratus* larvae and soil invertebrate snails, *Achatina fulica* also degrade polystyrene [80 - 83]. The consumption rate of styrofoam polystyrene by *Zophobas atratus* was found to be about 4 times more (0.58 mg/d) than *Tenebrio molitor*. Polystyrene degrading ability of *Zophobas atratus* depends on the gut microbiota as proven by the antibiotic inhibition [82]. *Achatina fulica* degraded 18.5 ± 2.9 mg polystyrene to 1.343 ± 0.625 mm microplastic in faeces within 4 weeks. Mass loss of about 30.7% was observed. Gel permeation chromatography shows limited depolymerization, as increased Mw and Mn of polystyrene were observed in the residue of faeces. Oxidation of polymers was confirmed by FTIR and proton NMR. Oxytetracycline inhibited gut microbes but depolymerization was not affected. Ingestion of polystyrene resulted in an increase in the *Sphingobacteriaceae*, *Enterobacteriaceae* and *Aeromonadaceae* families [84].

Plant-mediated Degradation

Microplastics do not affect plants much [84]. Plants can absorb and accumulate nanoplastics and microplastics without affecting the ecosystem. Herbivores feeding on these plants introduce microplastics in the food chain which are then transferred to higher trophic levels. But, if these plants are grown under controlled conditions, they can be used for removing microplastics from the environment. *Fucus vesiculosus*, a seaweed can hold and thus act as a sink for microplastcs on their surface [85, 86]. Seagrasses can retain microplastics through encrustation to the epibionts of macrophytes. Adherence can also take place in the polysaccharide mucousal layer [87]. Microplastics are encrusted on blades of caribbean angiosperm, *Thalassia testudinum* by epibionts [88]. Leaves of different plants can retain atmospheric microplastics and thus act as their sink as well as source. 28% of the natural materials retained on plant leaves were found to be plastics. Microplastics deposited plant leaves ranged from 0.07-0.19 n/cm^2 (pieces per area of leaves) [89]. Electrostatic forces, the morphology of leaves and periphyton affect the adsorption and retention of microplastics. Such type of plant and microplastic interactions can prove effective in the removal of microplastics from the environment through phytostabilization [84]. Microplastics can be attached to the cellulose part of plant cells depending on surface roughness owing to the electrostatic forces [90].

MECHANISM OF BODEGRADATION OF PLASTICS/ MICROPLASTICS

Biodegradation of Polyethylene

Albertsson *et al.* (1987) studied synergistic effect between photo-oxidation and biodegradation. LDPE was exposed to biotic and abiotic environmental conditions and ultraviolet light was irradiated on it for different periods of time. For abiotic photooxidation; Norish Type I and Norish Type II degradation was compared with biotic paraffin degradation. UV light and oxidizing agents play an important role. During degradation, carbonyl group was attacked by microorganisms to produce a small segment of the polymeric chain. Furthermore, as final products, water and carbon dioxide were obtained. Non sterile soil was used as a medium for biotic condition and with the help of silver nitrate abiotic condition was prepared. Due to photo-oxidation, cleavage of polyethylene molecule takes place. As an additive, photosensitizer increases the production of CO_2. UV irradiation and photooxidation lead to radical formation as shown in Scheme (**1**) [91].

Extra exposure to UV light leads to Norish Type I and Norish Type II degradation as shown in Scheme (**2**). Carbonyl radical is formed due to Norish type I cleavage, which reacts with alkoxy free radical (shown in Scheme (**1**)) to form ester (IR = 1740 cm^{-1}) (Scheme **3**). The terminal double bond peak can be easily seen in the IR spectrum of polymer if cleavage leads to Norish Type II degradation. Paraffin chain (carbon 10-20) can be oxidized to carboxylic acid group by microorganisms like Microbacterium, Candida, Carynebacterium and Nocardia. Coenzyme A from microorganisms removes two carbon. As a result, carbondioxide and water molecules are formed (Scheme **4**). The peak for double bond was not observed for abiotic conditions. Hence, they concluded that Norish type II degradation was not possible [91].

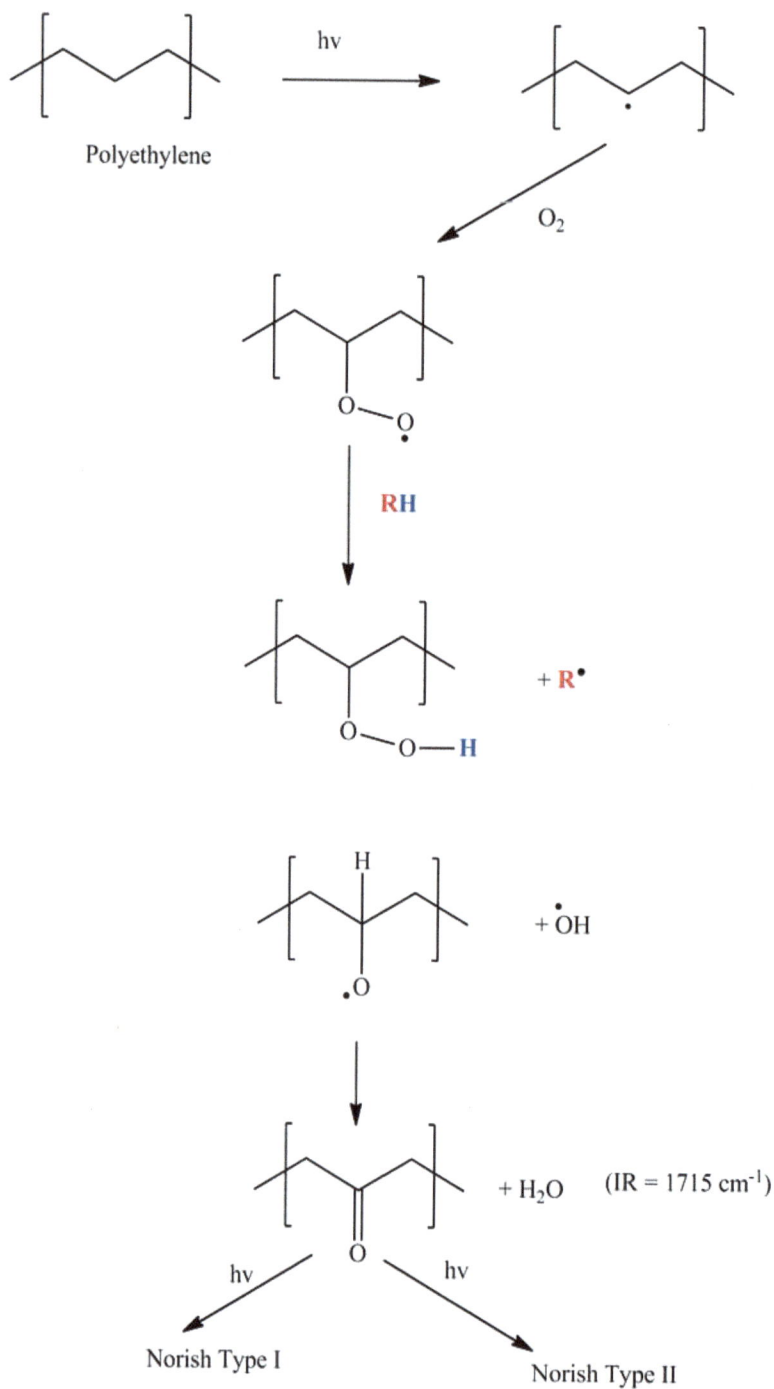

Scheme 1. Photooxidation of polyethylene [91].

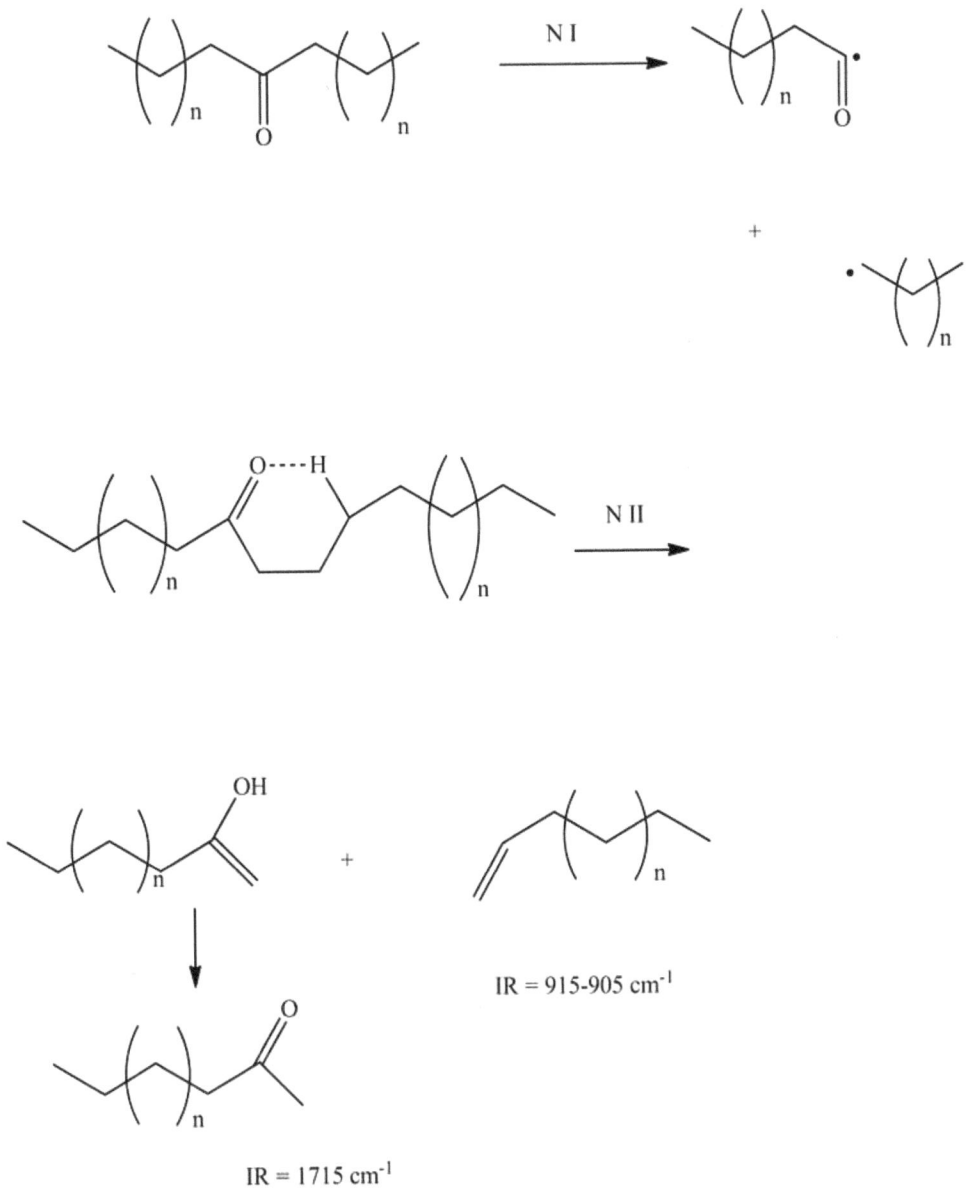

Scheme 2. Norish type I and II reaction [91].

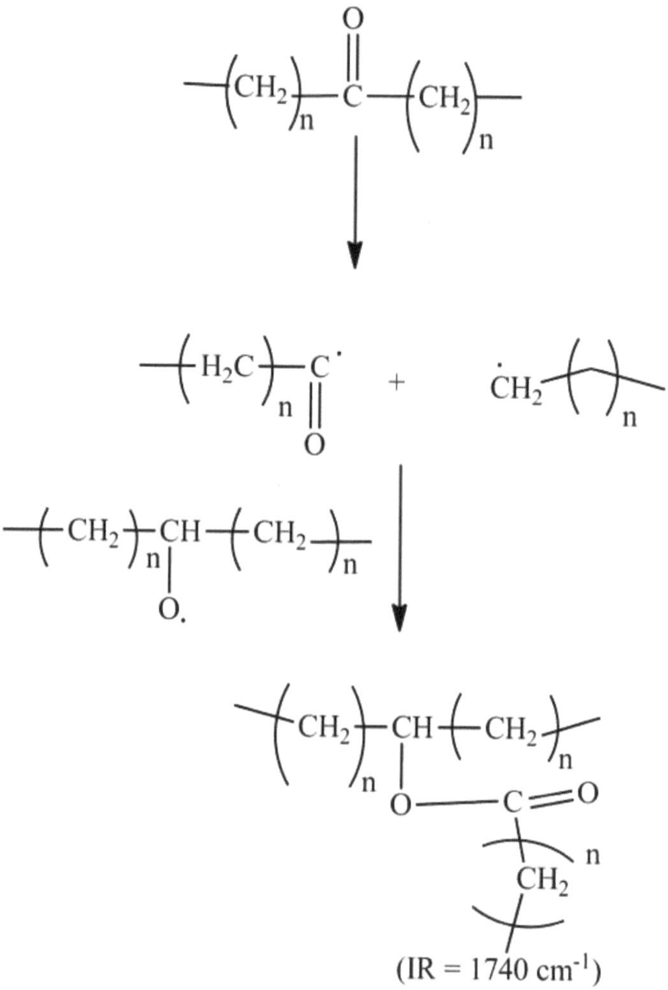

Scheme 3. Abiotic ester formation [91].

Biodegradation of Nylon

Nylons, the synthetic poly(amino acids) are degraded by nylon oligomer hydrolyzing enzymes (hydrolalse), manganese peroxidase obtained from *Flavobacterium sp.* KI72 and white rot fungus strain, respectively [92, 93]. Scheme (5) shows the degradation mechanism of nylon by fungus peroxidase. Enzyme attacks the methylene group next to the nitrogen atom of the nylon chain leading to its auto-oxidation [94].

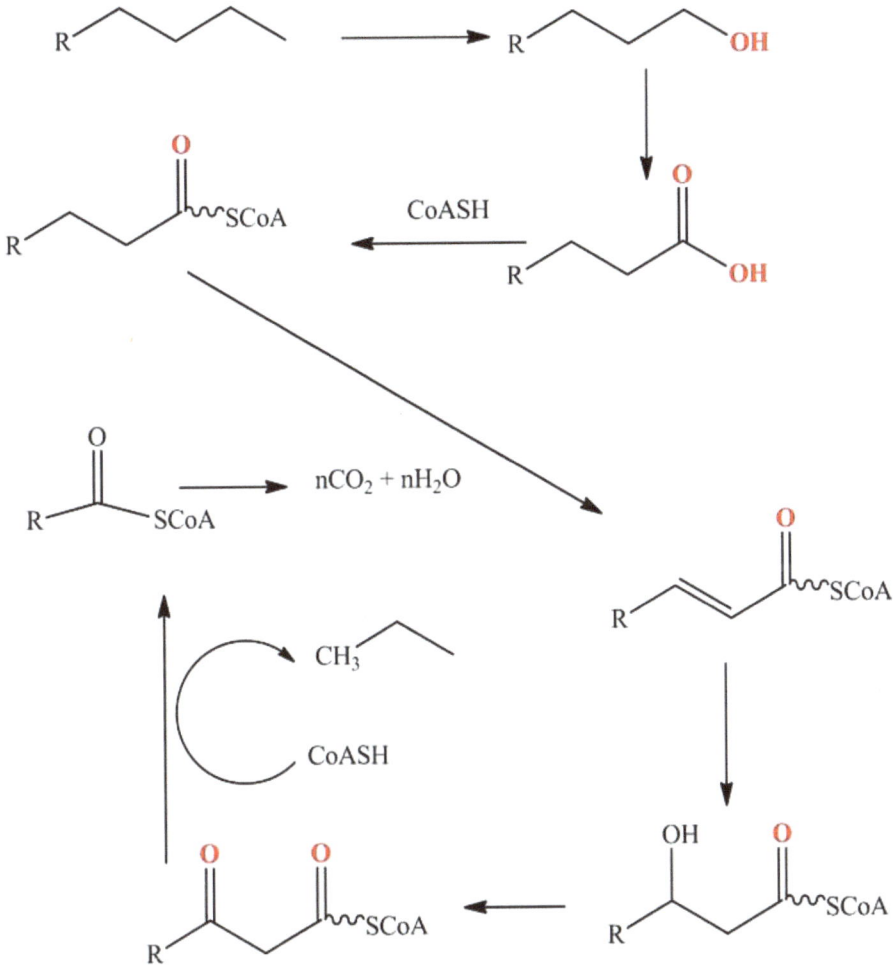

Scheme 4. Biotic paraffin degradation [91].

Biodegradation of Polyester, Poly(ε-caprolactone) (PCL)

The degradation of polyester by lipase enzyme was studied by Tokiwa and Suzuki, (1977a) [95]. It breaks the ester bond randomly along the main chain [96]. The polymer chain gets hydrolysed to 6-hydroxyhexanoic acid which is an ω-oxidation intermediate. This is followed by β-oxidation to yield acetyl-CoA. Acetyl-CoA enters the TCA cycle for further degradation [97].

$$R\text{-}CO\text{-}NH\text{-}CH_2\text{-}CH_2\text{-}R'$$

$$\downarrow \quad \rightarrow H$$

$$R\text{-}CO\text{-}NH\text{-}CH\text{-}CH_2\text{-}R'$$

$$O_2 \downarrow$$

$$R\text{-}CO\text{-}NH\text{-}CH\text{-}CH_2\text{-}R'$$
$$\overset{|}{O\text{-}\overset{\bullet}{O}}$$

$$O_2 \downarrow \quad \rightarrow OH^{\bullet}$$

$$R\text{-}CO\text{-}NH\text{-}CH\text{-}CH_2\text{-}R'$$
$$\overset{|}{\overset{\bullet}{O}}$$

$$H \downarrow$$

$$R\text{-}CO\text{-}NH_2 + R'\text{-}CH_2\text{-}CHC$$

$$R\text{-}CO\text{-}NH\text{-}CHO + R'\text{-}CH_3$$

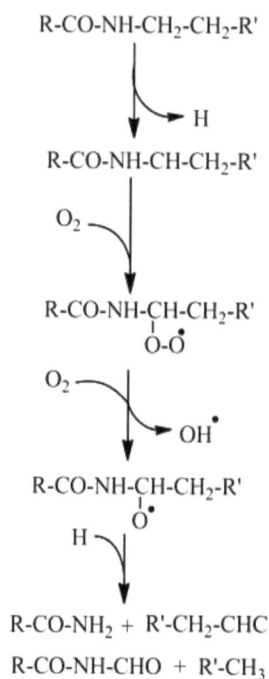

Scheme 5. Mechanism of nylon degradation by peroxidase [94].

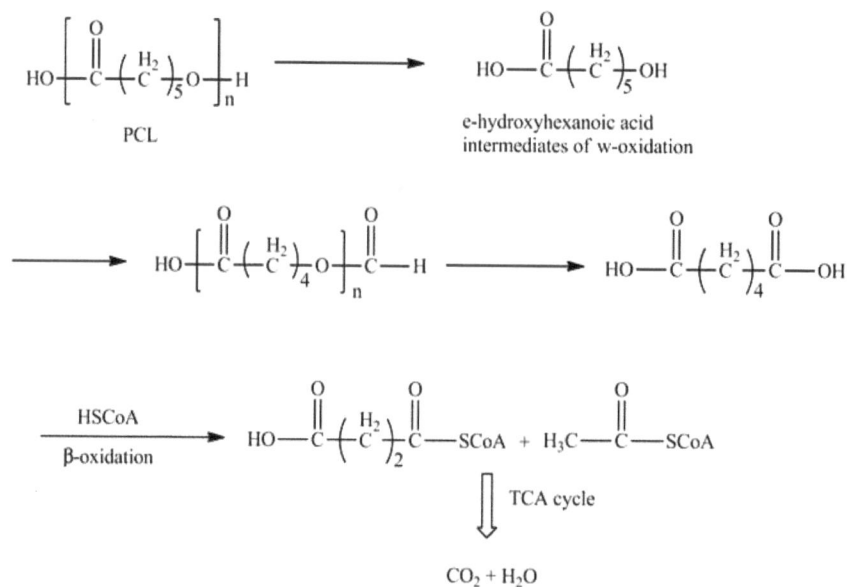

PCL

e-hydroxyhexanoic acid
intermediates of w-oxidation

HSCoA
β-oxidation

TCA cycle

$$CO_2 + H_2O$$

Scheme 6. Degradation mechanism of Poly(ε-caprolactone) PCL [97].

CONCLUSION

Plastics or microplastics deposited in landmasses or water bodies pose harmful effects on the environment and living organisms. The movement and cycling of microplastics in the environment along with their impact are under study. Their chemical constituents leach into the environment and get accumulated in the living bodies. Studies have revealed the presence of microplastics in the stomachs of organisms. Public awareness might result in reducing, reusing and recycling waste materials. The development of newer strategies and technologies is a must for the prevention of discharge of microplastics in the environment. It would also prove to be helpful for tertiary treatment of water and sludges. The addition of photosensitizers and avoiding the use of antioxidant stabilizers and biocides during plastic processing can also aid in biodegradation. Interdisciplinary studies on the removal of microplastics and their impact on organisms employed for their remediation along with ecofriendly manufacturing of plastics are required which will definitely improve the situation.

CONSENT FOR PUBLICATION

Not applicable.

CONFLICT OF INTEREST

The authors declare no conflict of interest, financial or otherwise.

ACKNOWLEDGEMENT

Declared none.

REFERENCE

[1]   Souza Machado AA, Kloas W, Zarfl C, Hempel S, Rillig MC. Microplastics as an emerging threat to terrestrial ecosystems. Glob Change Biol 2018; 24(4): 1405-16.
[http://dx.doi.org/10.1111/gcb.14020] [PMID: 29245177]

[2]   Carr SA, Liu J, Tesoro AG. Transport and fate of microplastic particles in wastewater treatment plants. Water Res 2016; 91: 174-82.
[http://dx.doi.org/10.1016/j.watres.2016.01.002] [PMID: 26795302]

[3]   Avio CG, Gorbi S, Milan M, *et al.* Pollutants bioavailability and toxicological risk from microplastics to marine mussels. Environ Pollut 2015; 198: 211-22.
[http://dx.doi.org/10.1016/j.envpol.2014.12.021] [PMID: 25637744]

[4]   Engler RE. The complex interaction between marine debris and toxic chemicals in the ocean. Environ Sci Technol 2012; 46(22): 12302-15.
[http://dx.doi.org/10.1021/es3027105] [PMID: 23088563]

[5]   Mato Y, Isobe T, Takada H, Kanehiro H, Ohtake C, Kaminuma T. Plastic resin pellets as a transport medium for toxic chemicals in the marine environment. Environ Sci Technol 2001; 35(2): 318-24.
[http://dx.doi.org/10.1021/es0010498] [PMID: 11347604]

[6]   Fossi MC, Marsili L, Baini M, *et al.* Fin whales and microplastics: The Mediterranean Sea and the Sea of Cortez scenarios. Environ Pollut 2016; 209: 68-78.
[http://dx.doi.org/10.1016/j.envpol.2015.11.022] [PMID: 26637933]

[7]   Sutton R, Mason SA, Stanek SK, Willis-Norton E, Wren IF, Box C. Microplastic contamination in the San Francisco Bay, California, USA. Mar Pollut Bull 2016; 109(1): 230-5.
[http://dx.doi.org/10.1016/j.marpolbul.2016.05.077] [PMID: 27289280]

[8]   Toussaint B, Raffael B, Angers-Loustau A, *et al.* Review of micro- and nanoplastic contamination in the food chain. Food Addit Contam Part A Chem Anal Control Expo Risk Assess 2019; 36(5): 639-73.
[http://dx.doi.org/10.1080/19440049.2019.1583381] [PMID: 30985273]

[9]   Zhang Q, Xu EG, Li J, *et al.* A Review of Microplastics in Table Salt, Drinking Water, and Air: Direct Human Exposure. Environ Sci Technol 2020; 54(7): 3740-51.
[http://dx.doi.org/10.1021/acs.est.9b04535] [PMID: 32119774]

[10]   Verla AW. C.E. Enyoh, E.N. Verla and K.O. Nwarnorh, "Microplastic-toxic chemical interaction: a review study on quantified levels, mechanism and implication," *SN appl.* Sci 2019; 1: 1400.

[11]   Prata JC. Microplastics in wastewater: State of the knowledge on sources, fate and solutions. Mar Pollut Bull 2018; 129(1): 262-5. a
[http://dx.doi.org/10.1016/j.marpolbul.2018.02.046] [PMID: 29680547]

[12]   Jenkins S, Quer AM, Fonseca C, Varrone C. Microbial Degradation of Plastics: New Plastic Degraders, Mixed Cultures and Engineering Strategies.Soil Microenvironment for Bioremediation and Polymer Production. Scrivener Publishing LLC 2019; pp. 213-38.
[http://dx.doi.org/10.1002/9781119592129.ch12]

[13]   Masiá P, Sol D, Ardura A, *et al.* Bioremediation as a promising strategy for microplastics removal in wastewater treatment plants. Mar Pollut Bull 2020; 156: 111252.
[http://dx.doi.org/10.1016/j.marpolbul.2020.111252] [PMID: 32510394]

[14]   Alshehrei F. Biodegradation of Synthetic and Natural Plastic by Microorganisms. J Appl Environ Microbiol 2017; 5: 8-19.

[15]   Fendall LS, Sewell MA. Contributing to marine pollution by washing your face: Microplastics in facial cleansers. Mar Pollut Bull 2009; 58(8): 1225-8.
[http://dx.doi.org/10.1016/j.marpolbul.2009.04.025] [PMID: 19481226]

[16]   Rochman CM, Kross SM, Armstrong JB, *et al.* Scientific evidence supports a ban on microbeads. Environ Sci Technol 2015; 49(18): 10759-61.
[http://dx.doi.org/10.1021/acs.est.5b03909] [PMID: 26334581]

[17]   Watts AJR, Lewis C, Goodhead RM, *et al.* Uptake and retention of microplastics by the shore crab *Carcinus maenas.* Environ Sci Technol 2014; 48(15): 8823-30.
[http://dx.doi.org/10.1021/es501090e] [PMID: 24972075]

[18]   de Souza Machado AA, Lau CW, Kloas W, *et al.* Microplastics Can Change Soil Properties and Affect Plant Performance. Environ Sci Technol 2019; 53(10): 6044-52.
[http://dx.doi.org/10.1021/acs.est.9b01339] [PMID: 31021077]

[19]   McAlpine KJ. Have your plastic and eat it too. Bostonia 2019; 36-7.

[20]   Van der Ven LTM, Van de Kuil T, Verhoef A, *et al.* Endocrine effects of tetrabromobisphenol-A (TBBPA) in Wistar rats as tested in a one-generation reproduction study and a subacute toxicity study. Toxicology 2008; 245(1-2): 76-89.
[http://dx.doi.org/10.1016/j.tox.2007.12.009] [PMID: 18255212]

[21]   The State of World Fisheries and Aquaculture 2010.

[22]   De-la-Torre GE, Gabriel E. Microplastics: an emerging threat to food security and human health. J Food Sci Technol 2020; 57(5): 1601-8.
[http://dx.doi.org/10.1007/s13197-019-04138-1] [PMID: 32327770]

[23]　Cózar A, Echevarría F, González-Gordillo JI, *et al.* Plastic debris in the open ocean. Proc Natl Acad Sci USA 2014; 111(28): 10239-44.
[http://dx.doi.org/10.1073/pnas.1314705111] [PMID: 24982135]

[24]　Thompson RC, Moore CJ, vom Saal FS, Swan SH. Plastics, the environment and human health: current consensus and future trends. Philos Trans R Soc Lond B Biol Sci 2009; 364(1526): 2153-66.
[http://dx.doi.org/10.1098/rstb.2009.0053] [PMID: 19528062]

[25]　Cox KD, Covernton GA, Davies HL, Dower JF, Juanes F, Dudas SE. Human Consumption of Microplastics. Environ Sci Technol 2019; 53(12): 7068-74.
[http://dx.doi.org/10.1021/acs.est.9b01517] [PMID: 31184127]

[26]　Health effects of ingestion of microplastics *via* food, water and breathing still unknown. 2019.

[27]　Prata JC, da Costa JP, Lopes I, Duarte AC, Rocha-Santos T. Environmental exposure to microplastics: An overview on possible human health effects. Sci Total Environ 2020; 702: 134455.
[http://dx.doi.org/10.1016/j.scitotenv.2019.134455] [PMID: 31733547]

[28]　Prata JC. Airborne microplastics: Consequences to human health? Environ Pollut 2018; 234: 115-26.
[http://dx.doi.org/10.1016/j.envpol.2017.11.043] [PMID: 29172041]

[29]　Caruso G. Plastic degrading microorganisms as a tool for bioremediation of plastic contamination in aquatic environments. J Pollut Eff Cont 2015; 3: 3.
[http://dx.doi.org/10.4172/2375-4397.1000e112]

[30]　Vimala PP, Mathew L. Biodegradation of Polyethylene using *Bacillus subtilis.* Procedia Technol 2016; 24: 232-9.
[http://dx.doi.org/10.1016/j.protcy.2016.05.031]

[31]　Li J, Kim HR, Lee HM, *et al.* Rapid biodegradation of polyphenylene sulfide plastic beads by *Pseudomonas sp.* Sci Total Environ 2020; 720: 137616.
[http://dx.doi.org/10.1016/j.scitotenv.2020.137616] [PMID: 32146401]

[32]　Kong X, Jin D, Tai X, *et al.* Bioremediation of dibutyl phthalate in a simulated agricultural ecosystem by *Gordonia* sp. strain QH-11 and the microbial ecological effects in soil. Sci Total Environ 2019; 667: 691-700.
[http://dx.doi.org/10.1016/j.scitotenv.2019.02.385] [PMID: 30849609]

[33]　Giacomucci L, Raddadi N, Soccio M, Lotti N, Fava F. Polyvinyl chloride biodegradation by *Pseudomonas citronellolis* and *Bacillus flexus.* N Biotechnol 2019; 52: 35-41.
[http://dx.doi.org/10.1016/j.nbt.2019.04.005] [PMID: 31026607]

[34]　Zhang H, Lin Z, Liu B, *et al.* Bioremediation of di-(2-ethylhexyl) phthalate contaminated red soil by *Gordonia terrae* RL-JC02: Characterization, metabolic pathway and kinetics. Sci Total Environ 2020; 733: 139138.
[http://dx.doi.org/10.1016/j.scitotenv.2020.139138] [PMID: 32446058]

[35]　Li Z, Wei R, Gao M, *et al.* Biodegradation of low-density polyethylene by *Microbulbifer hydrolyticus* IRE-31. J Environ Manage 2020; 263: 110402.
[http://dx.doi.org/10.1016/j.jenvman.2020.110402] [PMID: 32174537]

[36]　Park SY, Kim CG. Biodegradation of micro-polyethylene particles by bacterial colonization of a mixed microbial consortium isolated from a landfill site. Chemosphere 2019; 222: 527-33.
[http://dx.doi.org/10.1016/j.chemosphere.2019.01.159] [PMID: 30721811]

[37]　Syranidou E, Karkanorachaki K, Amorotti F, *et al.* Biodegradation of mixture of plastic films by tailored marine consortia. J Hazard Mater 2019; 375: 33-42.
[http://dx.doi.org/10.1016/j.jhazmat.2019.04.078] [PMID: 31039462]

[38]　Oda Y, Oida N, Urakami T, Tonomura K. Polycaprolactone depolymerase produced by the bacterium *Alcaligenes faecalis.* FEMS Microbiol Lett 1997; 152(2): 339-43.
[http://dx.doi.org/10.1111/j.1574-6968.1997.tb10449.x] [PMID: 9273313]

[39]   Ghosh SK, Pal S, Ray S. Study of microbes having potentiality for biodegradation of plastics. Environ Sci Pollut Res Int 2013; 20(7): 4339-55.
[http://dx.doi.org/10.1007/s11356-013-1706-x] [PMID: 23613206]

[40]   Tomita K, Kuroki Y, Nagai K. Isolation of thermophiles degrading poly(l-lactic acid). J Biosci Bioeng 1999; 87(6): 752-5.
[http://dx.doi.org/10.1016/S1389-1723(99)80148-0] [PMID: 16232549]

[41]   Danko AS, Luo M, Bagwell CE, Brigmon RL, Freedman DL. Involvement of linear plasmids in aerobic biodegradation of vinyl chloride. Appl Environ Microbiol 2004; 70(10): 6092-7.
[http://dx.doi.org/10.1128/AEM.70.10.6092-6097.2004] [PMID: 15466555]

[42]   Mabrouk MM, Sabry SA. Degradation of poly (3-hydroxybutyrate) and its copolymer poly (3-hydroxybutyrate-co-3-hydroxyvalerate) by a marine *Streptomyces* sp. SNG9. Microbiol Res 2001; 156(4): 323-35.
[http://dx.doi.org/10.1078/0944-5013-00115] [PMID: 11770850]

[43]   Kita K, Mashiba S, Nagita M, *et al.* Cloning of poly(3-hydroxybutyrate) depolymerase from a marine bacterium, *Alcaligenes faecalis* AE122, and characterization of its gene product. Biochim Biophys Acta Gene Struct Expr 1997; 1352(1): 113-22.
[http://dx.doi.org/10.1016/S0167-4781(97)00011-0] [PMID: 9177489]

[44]   Narciso-Ortiz L, Coreño-Alonso A, Mendoza-Olivares D, Lucho-Constantino CA, Lizardi-Jiménez MA. Baseline for plastic and hydrocarbon pollution of rivers, reefs, and sediment on beaches in Veracruz State, México, and a proposal for bioremediation. Environ Sci Pollut Res Int 2020; 27(18): 23035-47.
[http://dx.doi.org/10.1007/s11356-020-08831-z] [PMID: 32333346]

[45]   Schwartz M, Perrot T, Deroy A, *et al. Trametes versicolor* glutathione transferase Xi 3, a dual Cys□GST with catalytic specificities of both Xi and Omega classes. FEBS Lett 2018; 592(18): 3163-72.
[http://dx.doi.org/10.1002/1873-3468.13224] [PMID: 30112765]

[46]   Pathak VM, Navneet . Review on the current status of polymer degradation: a microbial approach. Bioresour Bioprocess 2017; 4(1): 15.
[http://dx.doi.org/10.1186/s40643-017-0145-9]

[47]   Yamada-Onodera K, Mukumoto H, Katsuyaya Y, Saiganji A, Tani Y. Degradation of polyethylene by a fungus, *Penicillium simplicissimum* YK. Polym Degrad Stabil 2001; 72(2): 323-7.
[http://dx.doi.org/10.1016/S0141-3910(01)00027-1]

[48]   Murphy CA, Cameron JA, Huang SJ, Vinopal RT. Fusarium polycaprolactone depolymerase is cutinase. Appl Environ Microbiol 1996; 62(2): 456-60.
[http://dx.doi.org/10.1128/aem.62.2.456-460.1996] [PMID: 8593048]

[49]   Shimao M. Biodegradation of plastics. Curr Opin Biotechnol 2001; 12(3): 242-7.
[http://dx.doi.org/10.1016/S0958-1669(00)00206-8] [PMID: 11404101]

[50]   Nakajima-Kambe T, Shigeno-Akutsu Y, Nomura N, Onuma F, Nakahara T. Microbial degradation of polyurethane, polyester polyurethanes and polyether polyurethanes. Appl Microbiol Biotechnol 1999; 51(2): 134-40.
[http://dx.doi.org/10.1007/s002530051373] [PMID: 10091317]

[51]   Sánchez C. Fungal potential for the degradation of petroleum-based polymers: An overview of macro- and microplastics biodegradation. Biotechnol Adv 2020; 40: 107501.
[http://dx.doi.org/10.1016/j.biotechadv.2019.107501] [PMID: 31870825]

[52]   Paço A, Duarte K, da Costa JP, *et al.* Biodegradation of polyethylene microplastics by the marine fungus *Zalerion maritimum.* Sci Total Environ 2017; 586: 10-5.
[http://dx.doi.org/10.1016/j.scitotenv.2017.02.017] [PMID: 28199874]

[53]   Sameshima-Yamashita Y, Ueda H, Koitabashi M, Kitamoto H. Pretreatment with an esterase from the

yeast *Pseudozyma antarctica* accelerates biodegradation of plastic mulch film in soil under laboratory conditions. J Biosci Bioeng 2019; 127(1): 93-8.
[http://dx.doi.org/10.1016/j.jbiosc.2018.06.011] [PMID: 30054060]

[54] Sheik S, Chandrashekar KR, Swaroop K, Somashekarappa HM. Biodegradation of gamma irradiated low density polyethylene and polypropylene by endophytic fungi. Int Biodeterior Biodegradation 2015; 105: 21-9.
[http://dx.doi.org/10.1016/j.ibiod.2015.08.006]

[55] Koitabashi M, Noguchi MT, Sameshima-Yamashita Y, *et al.* Degradation of biodegradable plastic mulch films in soil environment by phylloplane fungi isolated from gramineous plants. AMB Express 2012; 2(1): 40.
[http://dx.doi.org/10.1186/2191-0855-2-40] [PMID: 22856640]

[56] Sarkhel R, Sengupta S, Das P, Bhowal A. Comparative biodegradation study of polymer from plastic bottle waste using novel isolated bacteria and fungi from marine source. J Polym Res 2020; 27(1): 16.
[http://dx.doi.org/10.1007/s10965-019-1973-4]

[57] Sarmah P, Rout J. Role of algae and cyanobacteria in bioremediation: prospects in polyethylene biodegradation. Advances in Cyanobacterial Biology 2020; pp. 333-49.

[58] Kumar RV, Kanna GR, Elumalai S. Biodegradation of polyethylene by green photosynthetic microalgae. J Bioremediat Biodegrad 2017; 8: 381-8.

[59] Sarmah P, Rout J. Efficient biodegradation of low-density polyethylene by cyanobacteria isolated from submerged polyethylene surface in domestic sewage water. Environ Sci Pollut Res Int 2018; 25(33): 33508-20. a
[http://dx.doi.org/10.1007/s11356-018-3079-7] [PMID: 30267347]

[60] Cunha C, Silva L, Paulo J, Faria M, Nogueira N, Cordeiro N. Microalgal-based biopolymer for nano- and microplastic removal: a possible biosolution for wastewater treatment. Environ Pollut 2020; 263(Pt B): 114385.
[http://dx.doi.org/10.1016/j.envpol.2020.114385] [PMID: 32203858]

[61] Li Y, Gao J, Meng F, Chi J. Enhanced biodegradation of phthalate acid esters in marine sediments by benthic diatom *Cylindrotheca closterium*. Sci Total Environ 2015; 508: 251-7.
[http://dx.doi.org/10.1016/j.scitotenv.2014.12.002] [PMID: 25481253]

[62] Parrish K, Fahrenfeld NL. Microplastic biofilm in fresh- and wastewater as a function of microparticle type and size class. Environ Sci Water Res Technol 2019; 5(3): 495-505.
[http://dx.doi.org/10.1039/C8EW00712H]

[63] Oberbeckmann S, Löder MGJ, Labrenz M. Marine microplastic-associated biofilms – a review. Environ Chem 2015; 12(5): 551-62.
[http://dx.doi.org/10.1071/EN15069]

[64] Hossain MR, Jiang M, Wei Q, Leff LG. Microplastic surface properties affect bacterial colonization in freshwater. J Basic Microbiol 2019; 59(1): 54-61.
[http://dx.doi.org/10.1002/jobm.201800174] [PMID: 30370668]

[65] Gupta KK, Devi D. Characteristics investigation on biofilm formation and biodegradation activities of *Pseudomonas aeruginosa* strain ISJ14 colonizing low density polyethylene (LDPE) surface. Heliyon 2020; 6(7): e04398.
[http://dx.doi.org/10.1016/j.heliyon.2020.e04398] [PMID: 32671274]

[66] Santo M, Weitsman R, Sivan A. The role of the copper-binding enzyme – laccase – in the biodegradation of polyethylene by the actinomycete *Rhodococcus ruber*. Int Biodeterior Biodegradation 2013; 84: 204-10.
[http://dx.doi.org/10.1016/j.ibiod.2012.03.001]

[67] Shabbir S, Faheem M, Ali N, *et al.* Periphytic biofilm: An innovative approach for biodegradation of microplastics. Sci Total Environ 2020; 717: 137064.

[http://dx.doi.org/10.1016/j.scitotenv.2020.137064] [PMID: 32070890]

[68]   Hung CS, Barlow DE, Varaljay VA, *et al.* The biodegradation of polyester and polyester polyurethane coatings using *Papiliotrema laurentii.* Int Biodeterior Biodegradation 2019; 139: 34-43.
[http://dx.doi.org/10.1016/j.ibiod.2019.02.002]

[69]   Walczak M, Swiontek Brzezinska M, Sionkowska A, Michalska M, Jankiewicz U, Deja-Sikora E. Biofilm formation on the surface of polylactide during its biodegradation in different environments. Colloids Surf B Biointerfaces 2015; 136: 340-5.
[http://dx.doi.org/10.1016/j.colsurfb.2015.09.036] [PMID: 26433346]

[70]   Frydkjær CK, Iversen N, Roslev P. Ingestion and egestion of microplastics by the cladoceran *Daphnia magna*: effects of regular and irregular shaped plastic and sorbed phenanthrene. Bull Environ Contam Toxicol 2017; 99(6): 655-61.
[http://dx.doi.org/10.1007/s00128-017-2186-3] [PMID: 29027571]

[71]   Masiá P, Ardura A, Garcia-Vazquez E. Microplastics in special protected areas for migratory birds in the Bay of Biscay. Mar Pollut Bull 2019; 146: 993-1001.
[http://dx.doi.org/10.1016/j.marpolbul.2019.07.065] [PMID: 31426247]

[72]   Zhu J, Yu X, Zhang Q, *et al.* Cetaceans and microplastics: First report of microplastic ingestion by a coastal delphinid, *Sousa chinensis.* Sci Total Environ 2019; 659: 649-54.
[http://dx.doi.org/10.1016/j.scitotenv.2018.12.389] [PMID: 31096394]

[73]   Van Cauwenberghe L, Claessens M, Vandegehuchte MB, Janssen CR. Microplastics are taken up by mussels (*Mytilus edulis*) and lugworms (*Arenicola marina*) living in natural habitats. Environ Pollut 2015; 199: 10-7.
[http://dx.doi.org/10.1016/j.envpol.2015.01.008] [PMID: 25617854]

[74]   Yang L, Gao J, Liu Y, *et al.* Biodegradation of expanded polystyrene and low-density polyethylene foams in larvae of *Tenebrio molitor* Linnaeus (Coleoptera: Tenebrionidae): Broad *versus* limited extent depolymerization and microbe-dependence *versus* independence. Chemosphere 2021; 262: 127818.
[http://dx.doi.org/10.1016/j.chemosphere.2020.127818] [PMID: 32771707]

[75]   Yoshida S, Hiraga K, Takehana T, *et al.* A bacterium that degrades and assimilates poly(ethylene terephthalate). Science 2016; 351(6278): 1196-9.
[http://dx.doi.org/10.1126/science.aad6359] [PMID: 26965627]

[76]   Bombelli P, Howe CJ, Bertocchini F. Polyethylene bio-degradation by caterpillars of the wax moth Galleria mellonella. Curr Biol 2017; 27(8): R292-3.
[http://dx.doi.org/10.1016/j.cub.2017.02.060]

[77]   Billen P, Khalifa L, Van Gerven F, Tavernier S, Spatari S. Technological application potential of polyethylene and polystyrene biodegradation by macro-organisms such as mealworms and wax moth larvae. Sci Total Environ 2020; 735: 139521.
[http://dx.doi.org/10.1016/j.scitotenv.2020.139521] [PMID: 32470676]

[78]   Zhang L, Sintim HY, Bary AI, *et al.* Interaction of *Lumbricus terrestris* with macroscopic polyethylene and biodegradable plastic mulch. Sci Total Environ 2018; 635: 1600-8.
[http://dx.doi.org/10.1016/j.scitotenv.2018.04.054] [PMID: 29678255]

[79]   Peng BY, Li Y, Fan R, *et al.* Biodegradation of low-density polyethylene and polystyrene in superworms, larvae of *Zophobas atratus* (Coleoptera: Tenebrionidae): Broad and limited extent depolymerization. Environ Pollut 2020; 266(Pt 1): 115206.
[http://dx.doi.org/10.1016/j.envpol.2020.115206] [PMID: 32682160]

[80]   Yang SS, Brandon AM, Andrew Flanagan JC, *et al.* Biodegradation of polystyrene wastes in yellow mealworms (larvae of *Tenebrio molitor Linnaeus*): Factors affecting biodegradation rates and the ability of polystyrene-fed larvae to complete their life cycle. Chemosphere 2018; 191: 979-89.
[http://dx.doi.org/10.1016/j.chemosphere.2017.10.117] [PMID: 29145143]

[81]   Yang SS, Wu WM, Brandon AM, *et al.* Ubiquity of polystyrene digestion and biodegradation within yellow mealworms, larvae of Tenebrio molitor Linnaeus (Coleoptera: Tenebrionidae). Chemosphere 2018; 212: 262-71.
       [http://dx.doi.org/10.1016/j.chemosphere.2018.08.078] [PMID: 30145418]

[82]   Yang Y, Wang J, Xia M. Biodegradation and mineralization of polystyrene by plastic-eating superworms *Zophobas atratus.* Sci Total Environ 2020; 708: 135233.
       [http://dx.doi.org/10.1016/j.scitotenv.2019.135233] [PMID: 31787276]

[83]   Song Y, Qiu R, Hu J, *et al.* Biodegradation and disintegration of expanded polystyrene by land snails Achatina fulica. Sci Total Environ 2020; 746: 141289.
       [http://dx.doi.org/10.1016/j.scitotenv.2020.141289] [PMID: 32745868]

[84]   Kalčíková G. Aquatic vascular plants – A forgotten piece of nature in microplastic research. Environ Pollut 2020; 262: 114354.
       [http://dx.doi.org/10.1016/j.envpol.2020.114354] [PMID: 32193083]

[85]   Gutow L, Eckerlebe A, Giménez L, Saborowski R. Experimental evaluation of seaweeds as a vector for microplastics into marine food webs. Environ Sci Technol 2016; 50(2): 915-23.
       [http://dx.doi.org/10.1021/acs.est.5b02431] [PMID: 26654910]

[86]   Huang Y, Xiao X, Xu C, Perianen YD, Hu J, Holmer M. Seagrass beds acting as a trap of microplastics - Emerging hotspot in the coastal region? Environ Pollut 2020; 257: 113450.
       [http://dx.doi.org/10.1016/j.envpol.2019.113450] [PMID: 31679874]

[87]   Seng N, Lai S, Fong J, *et al.* Early evidence of microplastics on seagrass and macroalgae. Mar Freshw Res 2020; 71(8): 922-8.
       [http://dx.doi.org/10.1071/MF19177]

[88]   Goss H, Jaskiel J, Rotjan R. *Thalassia testudinum* as a potential vector for incorporating microplastics into benthic marine food webs. Mar Pollut Bull 2018; 135: 1085-9.
       [http://dx.doi.org/10.1016/j.marpolbul.2018.08.024] [PMID: 30301005]

[89]   Liu K, Wang X, Song Z, Wei N, Li D. Terrestrial plants as a potential temporary sink of atmospheric microplastics during transport. Sci Total Environ 2020; 742: 140523.
       [http://dx.doi.org/10.1016/j.scitotenv.2020.140523] [PMID: 32721722]

[90]   Bhattacharya P, Lin S, Turner JP, Ke PC. Physical adsorption of charged plastic nanoparticles affects algal photosynthesis. J Phys Chem C 2010; 114(39): 16556-61.
       [http://dx.doi.org/10.1021/jp1054759]

[91]   Albertsson AC, Andersson SO, Karlsson S. The mechanism of biodegradation of polyethylene. Polym Degrad Stabil 1987; 18(1): 73-87.
       [http://dx.doi.org/10.1016/0141-3910(87)90084-X]

[92]   Kinoshita S, Terada T, Taniguchi T, *et al.* Purification and characterization of 6-aminohexanoic-acd-oligomer hydrolase of Flavobacterium sp. Ki72. Eur J Biochem 1981; 116(3): 547-51.
       [http://dx.doi.org/10.1111/j.1432-1033.1981.tb05371.x] [PMID: 7262074]

[93]   Deguchi T, Kitaoka Y, Kakezawa M, Nishida T. Purification and characterization of a nylon-degrading enzyme. Appl Environ Microbiol 1998; 64(4): 1366-71.
       [http://dx.doi.org/10.1128/AEM.64.4.1366-1371.1998] [PMID: 9546174]

[94]   Nomura N, Deguchi T, Shigeno-Akutsu Y, Nakajima-Kambe T, Nakahara T. Gene structures and catalytic mechanisms of microbial enzymes able to biodegrade the synthetic solid polymers nylon and polyester polyurethane. Biotechnol Genet Eng Rev 2001; 18(1): 125-47.
       [http://dx.doi.org/10.1080/02648725.2001.10648011] [PMID: 11530686]

[95]   Tokiwa Y, Suzuki T. Hydrolysis of polyesters by lipases. Nature 1977; 270(5632): 76-8. a
       [http://dx.doi.org/10.1038/270076a0] [PMID: 927523]

[96]   Tokiwa Y, Suzuki T. Microbial degradation of polyesters. Part III. Purification and some properties of

polyethylene adipate-degrading enzyme produced by *Penicillium* sp. strain 14-3. Agric Biol Chem 1977; 41: 265-74. b

[97]   Pitt CG, Chasalow FI, Hibionada YM, Klimas DM, Schindler A. Aliphatic polyesters. I. The degradation of poly(ϵ-caprolactone) *in vivo*. J Appl Polym Sci 1981; 26(11): 3779-87. [http://dx.doi.org/10.1002/app.1981.070261124] [PMID: 7326315]

Sustainable Materials, 2023, *Vol. 1*, 433-450

CHAPTER 13

Microbial Degradation of Plastics

Geetanjali[1], **Vikram Singh**[2] and **Ram Singh**[3,*]

[1] *Department of Chemistry, Kirori Mal College, University of Delhi, Delhi – 110 007, India*

[2] *Department of Chemistry, NREC College, Khurja, Uttar Pradesh – 203131, India*

[3] *Department of Applied Chemistry, Delhi Technological University, Delhi – 110 042, India*

Abstract: The essentiality of plastics in our daily life is inseparable. Almost all industrial sectors utilize plastics either directly or indirectly. But the downside of plastics also increased simultaneously. These materials increased water and soil pollution due to unmanaged discharge. Hence, plastic waste treatment becomes essential for a sustainable and efficient environment. Plastic recycling and degradation are two processes to deal with plastic waste. Out of the three degradation processes, physical, chemical, and biological, biological degradation is near to a sustainable environment. Recent studies revolve around the use of micro-organisms for the degradation of plastics. The present chapter reports the microbial degradation of plastic waste using bacteria and fungi. The discussion also includes the impact of plastic properties and environmental factors on biodegradation.

Keywords: Plastics, Plastic waste, Degradation, Microbes, Bacteria, Fungus, Waste management.

INTRODUCTION

Plastics are derived from petrochemicals or synthetic monomers and possess high molecular weight. These polymeric materials have valuable industrial applications for all walks of life due to user-friendly properties like low cost, high durability, lightweight, and relatively better strength [1 - 5]. Their production has increased manifolds in the last few decades [6, 7]. According to Statista, the global production of plastic has increased from 245 million metric tons in 2008 to 359 million metric tons in 2018 [8]. This huge production, low circular use, long period of degradation or no degradation, and poor recycling lead to the accumulation of plastics in the environment, especially marine and terrestrial causing adverse effects in all ecosystems [3, 9]. The plastics are present in the environment in three main fractions: mismanaged plastic waste, post-consumer

---

* **Corresponding author Ram Singh:** Department of Applied Chemistry, Delhi Technological University, Delhi – 110 042, India; E-mail: ramsingh@dtu.ac.in

Inamuddin (Ed.)

managed plastic waste and plastic in use [10, 11]. The first one, mismanaged plastic waste, is mainly responsible for environmental pollution. The packaging-related and urban litter plastics are mostly mismanaged category which also include open dump plastics. The last two categories are mostly accountable.

The land and soil environment are directly affected by the pollution of plastics [6, 12]. According to a study by the UN Environment Programme, in the current scenario, we are generating about 300 million tonnes of plastic waste every year [13]. This has been estimated that till now, about 8.3 billion tonnes of plastic have been generated, and approximately 60% of them are mismanaged plastic waste. As per the data on plastic waste management, only 9% has been recycled, and about 12% are incinerated and the rest 79% have been put to the natural environment as landfills or dumps [13]. The ocean receives about 8 million tonnes of plastics every year, either through rivers or human activities at beaches, *etc* [13]. The countries like Philippines, Vietnam, Indonesia, Thailand, and China are some of the major plastic contributors to the ocean [14]. The plastic pieces are entering the marine food chain and further to humans causing several diseases [15].

There have been reviews and reports published regarding the degradation and recycling of plastic materials with their limitations and advantages [16 - 20]. The recycling efforts started after 1980 with only non-fiber plastics whose recycling and incineration in 2014 reached 18 and 24%, respectively of total non-fiber plastics waste generated worldwide [3]. Plastic degradation takes place using different paths (Fig. 1) and leads to chemical or physical changes in plastic causing deterioration of functionality, discolouration, reduction in strength, changes in optical and electrical properties, and so on [21]. The changes mainly take place due to the formation of new functional groups, changes in bond properties, and chemical transformations [22]. The degradation processes such as UV-treatment, heating, hydrolysis, trans-esterification, ammonolysis, *etc.* require severe conditions and also produce toxic by-products [23 - 25].

The biological method of degradation is better for the sustainable environment. The present chapter discusses the microbial degradation of plastic waste using bacteria and fungus. This chapter also includes the impact of plastic properties and environmental factors on biodegradation.

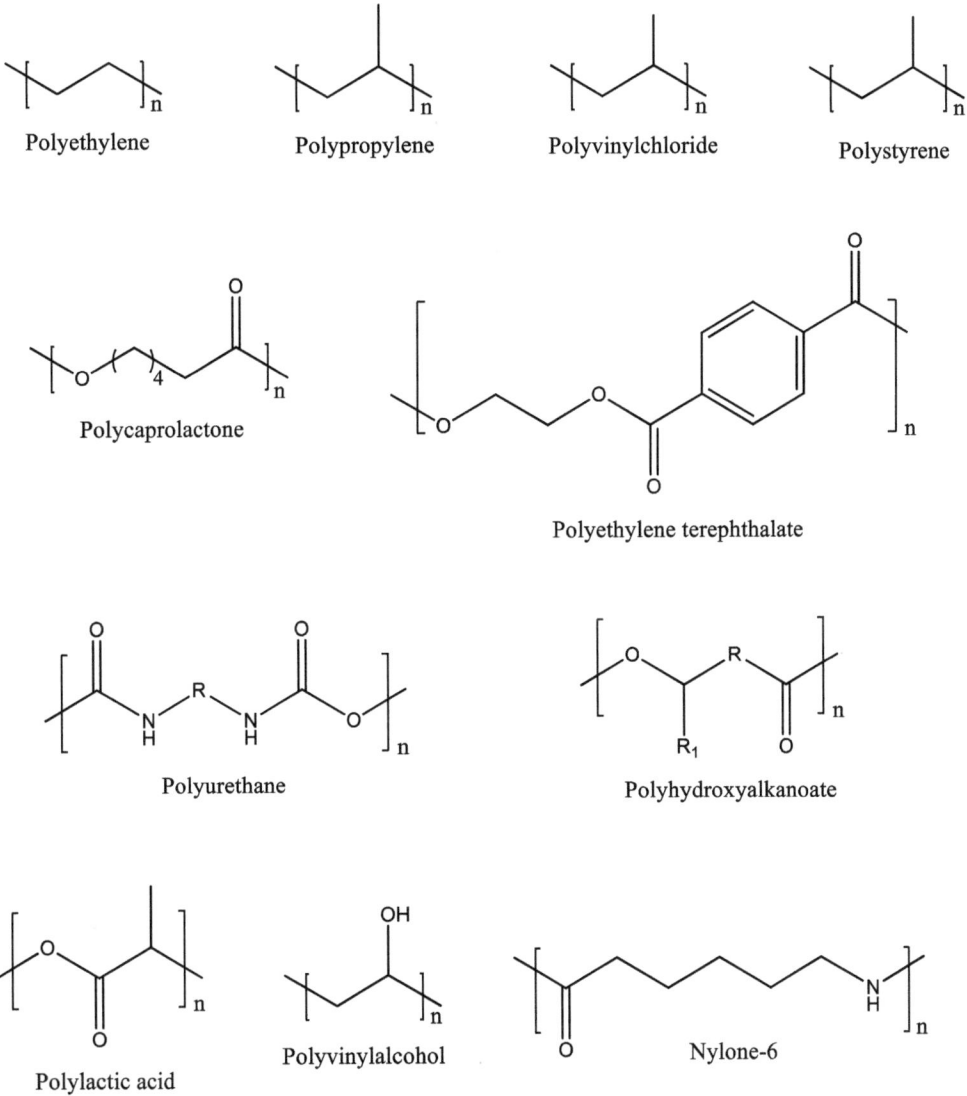

Fig. (1). Methods of plastic degradation.

MICROORGANISM FOR PLASTIC DEGRADATION

The degradation of plastics with the help of microbes is known as microbial degradation. The microbes mainly include bacteria and fungi which transform the structure of plastics and are released into the environment [26 - 28]. This is an environmental-friendly procedure that destroys the gathering of harmful metabolic by-products of plastics [29]. The plastics that are degraded by biological agents are known as biodegradable plastics [30, 31]. The chemical and physical properties of plastics play an important role in any type of degradation including microbial degradation (Table **1**) [32 - 39]. This has been observed that the high molecular weight resists biodegradation [34, 40]. The high molecular weight (Mn > 4,000) PCL was degraded slowly by *Rhizopus delemar* lipase in comparison to low molecular weight PCL [40]. The plastic degradation has also been affected by environmental factors such as pH, temperature and humidity (Table **2**) [41 - 43].

Table 1. Plastic properties affecting its biodegradation.

S. No.	Plastic Properties	Effect on Biodegradation	References
1	Molecular weight	The higher molecular weight of plastics reduces the degradation potential. The low-molecular weight plastics are better soluble and hence enhances the biodegradation.	[34]
2	Crystallinity	The crystalline portion in plastics produce resistance to biodegradation. The amorphous part is more susceptible to microbial enzymatic attack due to its loosely pack nature.	[35]
3	Hydrophobic nature	Moisture is required for biodegradation and hence anything that makes the plastic water-repellent or hydrophobic nature restricts the microbial action. However, they can be overcome by biofilm formation.	[36]
4	Additives	The additives used in plastics like dyes or filler affect the degradation process.	[37]
5	Biosurfactants	The Biosurfactants enhances biodegradation of plastic materials. Their action is due to the functional group possessed by them, which also allow degradation under severe environmental conditions like salinity, extreme pH, high temperature *etc.*	[38, 39]

Table 2. Environmental factors affecting plastic biodegradation.

Environmental Factors	Effect on Biodegradation	References
Moisture	Moisture is required for growth and multiplication of microbes. Hence, humid environment enhances the microbial action on plastics.	[37, 41]
pH	Every microbe has its own optimum pH condition to do optimum action. The pH affects the rate of hydrolysis reactions due to microbial action in acidic and basic condition.	[42]
Temperature	Temperature also affects the microbial degradation. The plastics possessing higher melting point has less possibility of biodegradation. However, enzymatic degradation works under optimum temperature and tends to decreases with the increase in temperature.	[43]

During the plastic degradation by microorganisms, the plastics acts as a nutrient in different ways [44 - 47]. When plastic as whole do not get attacked or come in contact with microbes, they do not get degraded. But due to the action of oxygen, and UV-radiation, the carbon-carbon backbone gets cleaved and become smaller fragments which are susceptible of getting attacked by the microbes. In this case, the abiotic degradation is followed by biodegradation [24]. In biodegradation, extracellular and intracellular depolymerases enzymes are involved where former enzyme yields oligomers or monomers that enter through the cells to further get metabolised [46]. The plastic surface gets attacked by microbes either in a direct or indirect way [48]. In the indirect way, the plastics get degraded by the action of microbial metabolic products. This is a consecutive method that goes step-wise as biodeterioration, followed by fragmentation, assimilation and mineralization [49]. The direct degradation method involves microbe attack for its nutrition and growth [44].

On the basis of the degradation environment, the indirect method is grouped into aerobic and anaerobic degradation. They differ in their metabolic product formation. The anaerobic mode gives carbon dioxide (CO_2), methane (CH_4), and water (H_2O). The aerobic degradation produces only CO_2 and H_2O [49]. The main central metabolic pathway is the tricarboxylic acid (TCA) cycle for energy generation from mostplastics [50].

Bacterial Degradation

The bacterial genera *Pseudomonas* and *Rhodococcus* have been involved in plastic biodegradation and studied for various plastics. The polyethylene degradation has been studied with the *Pseudomonas* sp and it has been observed that the efficiency of degradation differs with strain [51, 52]. The studies showed that the enhanced interaction between the PE film and bacteria enhances the degradation process. The early microbial degradation of PE was studied with ^{14}C-labeled PE and soil microbe by Albertsson in 1978 [53]. The degradation process was very slow and in only 2 years 0.36–0.39% degradation was observed in terms of [14]CO_2 release [53]. Further studies revealed that low molecular weight PE fraction degraded like straight-chain n-alkanes and [14]CO_2 release was mainly from low molecular weight PE fraction [54]. The high molecular weight PE fraction is the major bottleneck in its microbial degradation. Hence, it has been recommended to pre-treat this to convert it into low-molecular weight to facilitate microbial degradation [55, 56]. A strain of *Serratia marcescens* degraded PE (un-pretreated) in 70 days with 36% weight loss [57]. The polymer, PE (Fig. **2**), has also been used as a carbon source for *Rhodococcus ruber* (C208) where 30-days of incubation leads to 8% decrease in the weight of the polymer [58]. The PE degradation by the *Staphylococcus* sp has been found to be 52% [59].

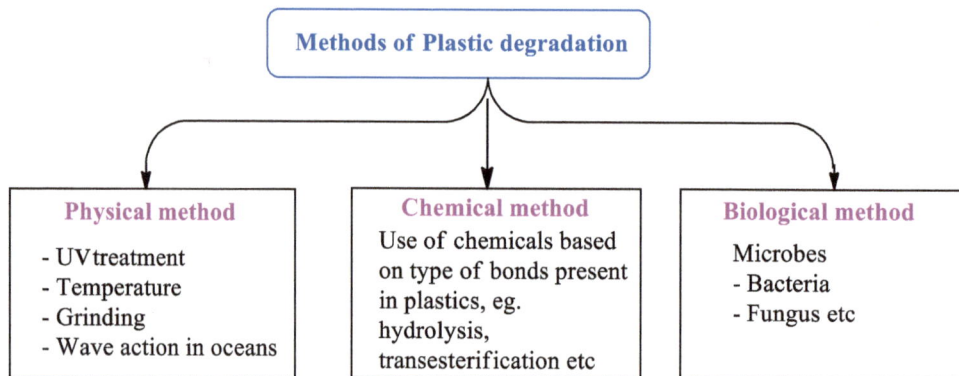

Fig. (2). Chemical structures of polymers.

The UV and nitric acid (HNO$_3$) treated polypropylene (PP) have been degraded by the combination of *Pseudomonas* sp and *Actinomycetes* sp [60]. These studies have further shown that two species of *Pseudomonas*, *P. azotoformans* and *P. stutzeri* secrete biosurfactants that helped in the degradation of PP [60]. The PP has also been degraded with *Rhodococcus* sp 36 isolated from soil sediments up to 6.4%. The incubation time was taken as 40 days. During the same incubation time, *Bacillus* sp degraded PP up to 4.0% [61].

Poly(ethylene adipate) (PEA) is an aliphatic polyester and degraded by *Bacillus* sp. YP1 and *Enterobacter asburiae* YT1 [52]. Another polyester, poly(ethylene terephthalate) (PET) has been degraded with the help of Saprospiraceae, Cryomorphaceae, and Flavobacteriaceae [62].

Poly(ethylene terephthalate) (PET) (Fig. **2**) degradation to oligomers or monomers has been achieved by very few bacteria and fungi [63, 64]. Most of the bacteria for PET degradation are the Gram-positive *phylum Actinobacteria* [65] and the genera *Thermobifida* and *Thermomonospora* [66]. The enzyme associated with the PET degradation are typical serine hydrolases possessing catalytic triad of a serine, a histidine, and an aspartate residue and have low turnover rates [67]. Several sulphide bonds are also present to promote specific binding to PET and thermal stability as found in PETase from *Ideonella sakaiensis* 201-F6 [68]. They are capable of degrading the PET to mono(2-hydroxyethyl)terephthalic acid which is used for bacterial metabolism to yield ethylene glycol and terephthalic acid [64].

Polyamide (PA) is an aliphatic, semi-aromatic, or aromatic polymer with repeating units of molecules linked *via* amide bonds. Various bacteria use oligomers from nylon industries for their growth [69]. The bacterial strain of *Flavobacterium* sp. KI72 is one of the early examples that grew on oligomers of polyamide [69].

Polyurethanes (PUR) (Fig. **2**) have also been degraded by either bacteria or fungi. A Gram-negative *Betaproteobacteria* from the genus *Pseudomonas* have been studied to degrade PUR with the PueB lipase from *Pseudomonas chlororaphis* [70]. Another *Pseudomonas* sp, *P. putida* and other microorganisms also biodegraded PUR at relatively high rates [71]. The PUR-active enzyme was also isolated from *Comamonas acidovorans* TB-35 [72], which produced diethylene glycol and adipic acid at an optimum pH of 6.5 and 45°C temperature. This worked on surface-binding effect. Schmidt *et al.* employed polyester hydrolases LC-cutinase and observed substantial weight loss of the tested PUR foils [73]. The PUR has also been degraded by *B. subtilis* and *Alicycliphilus* sp. isolates [74]. A list of plastic and their microbial degrading agent has been summarized in Table **3**.

Table 3. Plastic degrading microorganisms.

S. No.	Plastic Degraded	Microorganisms (References)
1	Polyethylene (PE)	*Zalerion maritimum* from marine environment [91]; *Pseudomonas* sp [92]; *Rhodococcus ruber* (C208) [58, 93]; *Brevibacillus* sp [94].
2	Low-density polyethylene (LDPE)	*Aspergillus versicolor, Aspergillus* sp. from Kovalam coast [95]; from Soil of disposal site, *Rhodococcus ruber* C208 [58]; *Pseudomonas* sp. AKS2 from waste dumping soil [96]; *Bacillus subtilis* H1584 from marine water [97]; from gut of waxworm *Enterobacter asburiae* YT1 and *Bacillus* sp [52, 98]; *Alcanivorax borkumensis* from mediterranean Sea LDPE film [99]; *A. niger* and *A. japonicas* [77].
3	High-Density Polyethylene (HDPE)	*Klebsiella pneumonia CH001, Bacillus* sp. BCBT21, *Arthrobacter sp.*, *Pseudomonas putida* S3A, *Comamonas acidovorans, Rhodococcus, Aspergillus flavus, Penicillium oxalicum* NS4 [100, 101].
4	Polypropylene (PP)	*Brevibacillus, Aneurinibacillus, Bacillus, Rhodococcus, Rhizopus oryzae, Phanerochaete chrysosporium* [102, 103]; *Pseudomonas* sp and *Actinomycetes* sp [104]; *Stenotrophomonas panacihumi* from soil of waste storage yard [105]; soil consortia [106]; *Bacillus* sp. strain 27; *Rhodococcus* sp. strain 36 from mangrove environments [61, 107].

(Table 3) cont.....

S. No.	Plastic Degraded	Microorganisms (References)
5	Polyvinyl Chloride (PVC)	*Staphylococcus, Pseudomonas putida* AJ, *lebsiella, Micrococcus, Chaetomium* [108, 109]; *Aureobasidium pullulans* from atmosphere [110]; *Phanerochaete chrysosporium; Lentinus tigrinus; Aspergillus niger; Aspergillus sydowii PVC film* from *soil* [111]; *Pseudomonas otitidis, Acanthopleurobacter pedis, Bacillus aerius, B. cereus,* from plastic disposal sites [112]; *Pseudomonas citronellolis* from soil [113].
6	Polystyrene (PS)	*Sphingobacterium* sp., *Xanthomonas* sp., and *Bacillus* sp. STR-YO from field soil [114]; *Rhodococcus ruber* C208 from soil of disposal site [115]; *Exiguobacterium* sp. YT2 from mealworm's gut [116].
7	Poly-caprolactone (PCL)	*Shewanella, Moritella* sp., *Psychrobacter* sp., *Pseudomonas* sp. from deep sea [27, 88, 89]
8	Polyvinyl alcohol-low linear density polyethylene	*Vibrio alginolyticus*, Vibrio, Parahemolyticus from benthic zones of marine environments [90]
9	Polyethylene terephthalate (PET)	*Ideonella sakaiensis* 201-F6 [64, 117, 118]
10	Polyurethanes (PUR)	PueB lipase from *Pseudomonas chlororaphis* [70] *Pseudomonas putida* [119]
11	Polyhydroxy alkanoates	*Pseudomonas stutzeri* [37, 120]
12	Polylactic acid (PLA)	*Bacillus brevis, Amycolatopsis* sp., *Penicillium Roquefort* [121]
13	Polyvinyl alcohol	*Pseudomonas O-3* [122]
14	Nylon-6	*Flavobacterium sp., Pseudomonas sp* [43]. *Anoxybacillus rupiensis* Ir3 [123]

Fungal Degradation

Studies have shown that plastic biodegradation has been carried out by both bacteria and fungi either alone or in combination [75]. The fungal species such as *Penicillium, Aspergillus*, and *Fusarium* have been successfully studied for plastic degradation and eroding the surfaces after their attachment [76]. The LDPE has been mainly degraded by *Aspergillus* sp whereas *Penicillium* sp. could degrade both LDPE and HDPE. A polyethylene polluted site revealed the presence of *Aspergillus niger* and *A. japonicus* fungal species. Raaman *et al.* studied their degradation behaviour with LDPE and observed that *A. niger* degraded to a maximum of 5.8%, and *A. japonicas* up to 11.11% in one-month time [77]. The fungi have the capability to release hydrophobic proteins that helps in degrading low-density polyethylene (LDPE), which is considered a highly resistant plastic [78]. Ojha *et al.* reported that *Penicillium chrysogenum* and *P. oxalicum* degraded the plastics like HDPE (55.60%) and LDPE (34.35%) after 90 days of incubation period [79]. Another fungus, *Mucor* sp. along with the support of other microbes has been useful in PE biodegradation. The LDPE was also degraded with the help of organisms like *Mucor circinilloides* and *Aspergillus flavus* obtained from municipal landfill areas [80]. These plastics showed weight loss of 18.1 and 6%, respectively when mixed with cow dung and poultry dropping after a time period of nine months [80].

In some cases, plastic degradation by the genus *Penicillium* uses prior conversion to oligomers with the help of their degrading enzymes, followed by complete degradation [79]. For example, the strains of *P. simplicissimum* produces laccase and manganese peroxidase [79]. Polyhydroxybutyrate (PHB) is initially degraded by PHB depolymerase secreted by both fungi and bacteria [79].

The PUR was degraded with a 21-kDa metallo-hydrolase from *Pestalotiopsis microspore* [81]. Other species were also tried for PUR degradation including *Fusarium solani, Candida ethanolica* [82], *Candida rugosa* [83], *Aspergillus fumigatus, Penicillium chrysogenum* [84], and *Aspergillus flavus* [85].

Degradation in Marine Habitats

The marine species and ecosystems have been affected by floating plastic wastes. Despite many reports, we lack the accurate information about the sources, accumulation pathways, and quantity of plastics in marine habitats [27, 86]. However, different stakeholders such as producers, consumers, *etc.* have been made aware of the plastic accumulations in the marine environment [87]. Many microorganisms are available in the marine environment that has the potential for biodegradation of plastic materials available in marine habitats. Polycaprolactone (PCL) (Fig. **2**) has been found to degrade at a very low temperature of 4°C from a *Pseudomonas* genus PCL-degrading bacterium [88]. These bacteria were isolated from the 320 m deep seawater in Toyama Bay. From the depth of 5000–7000 m, other bacteria belonging to the genera, *Pseudomonas, Moritella, Shewanella*, and *Psychrobacter* were obtained which showed potential for PCL degradation [88, 89]. The other plastic such as polylactic acid (PLA), poly(butylene succinate-co-butylene adipate) (PBSA), poly(butylene succinate) PBS, and polyhydroxybutyrate (PHB) were studied for their degradation using the obtained genera but these polymers were found to be resistant towards them, however, further study showed some degradation [88, 89].

A blend film having the polymer, polyvinyl alcohol-low linear density polyethylene, developed visible cracks and grooves on the surface after bacterial incubation for 15 weeks. The *Vibrio* genus bacteria isolated from a depth of 8 m,

Vibrio alginolyticus and *Vibrio parahaemolyticus* were responsible for the degradation [90].

CONCLUSION

Plastic materials have proved themselves an inseparable part of human life. But unfortunately, they are equally a threat to the environment, mainly due to mismanagement being the primary cause. The demand and use are an ever-increasing process for plastic materials. Their production has increased manifolds and reached more than 350 million metric tons in 2018. This colossal production forced the scientists across the Globe to find scientific solutions to manage the waste plastic materials and also develop biodegradable plastics for sustainable environmental safety. The plastic management revolved around recycling, physical degradation, chemical degradation, and biological degradation with advantages and limitations. Those plastic wastes that do not get included in the above processes are generally found in landfills and dumped in the marine environment.

In recent years, more emphasis has been given to the biological degradation of plastic waste. Microorganisms present in the environment played a key role in this through enzymatic action. They have successfully converted complex plastic waste into simpler products like oligomers, monomers, methane, carbon dioxide, and water through aerobic and anaerobic mechanisms. Visualizing the huge growth, there is a requirement to increase the efficiency of biological degradation of plastic waste. Also, the design and synthesis of biodegradable plastics match with the properties of synthetic non-biodegradable plastics.

CONSENT FOR PUBLICATION

Not applicable.

CONFLICT OF INTEREST

The authors declare no conflict of interest, financial or otherwise.

ACKNOWLEDGEMENT

The authors are thankful to their respective organizations for providing the facilities to complete this chapter.

REFERENCES

[1]   Andrady AL, Neal MA. Applications and societal benefits of plastics. Philos Trans R Soc Lond B Biol Sci 2009; 364(1526): 1977-84.

[http://dx.doi.org/10.1098/rstb.2008.0304] [PMID: 19528050]

[2]   Maddah HA. Polypropylene as a promising plastic: a review. Am J Pol Sci 2016; 6: 1-11.

[3]   Halden RU. Plastics and health risks. Annu Rev Public Health 2010; 31(1): 179-94.
        [http://dx.doi.org/10.1146/annurev.publhealth.012809.103714] [PMID: 20070188]

[4]   Suman S, Singh R. Iodide-selective PVC membrane electrode based on copper complex of 2-acetylthiophene semicarbazone as carrier. Anal Chem Lett 2020; 10(3): 357-65.
        [http://dx.doi.org/10.1080/22297928.2020.1788989]

[5]   Suman S, Singh R. Anion selective electrodes: A brief compilation. Microchem J 2019; 149: 104045.
        [http://dx.doi.org/10.1016/j.microc.2019.104045]

[6]   Jaiswal S, Sharma B, Shukla P. Integrated approaches in microbial degradation of plastics. Environmental Technology & Innovation 2020; 17: 100567.
        [http://dx.doi.org/10.1016/j.eti.2019.100567]

[7]   Devasahayam S, Raman RKS, Chennakesavulu K, Bhattacharya S. Materials. 2019; 12.

[8]   https://www.statista.com/statistics/282732/global-production-of-plastics-since-1950/

[9]   Jambeck JR, Geyer R, Wilcox C, *et al.* Plastic waste inputs from land into the ocean. Science 2015; 347(6223): 768-71.
        [http://dx.doi.org/10.1126/science.1260352] [PMID: 25678662]

[10]   Lebreton L, Andrady A. Future scenarios of global plastic waste generation and disposal. Palgrave Commun 2019; 5(1): 6.
        [http://dx.doi.org/10.1057/s41599-018-0212-7]

[11]   Geyer R, Jambeck JR, Law KL. Production, use, and fate of all plastics ever made. Sci Adv 2017; 3(7): e1700782.
        [http://dx.doi.org/10.1126/sciadv.1700782] [PMID: 28776036]

[12]   Andrady AL. Microplastics in the marine environment. Mar Pollut Bull 2011; 62(8): 1596-605.
        [http://dx.doi.org/10.1016/j.marpolbul.2011.05.030] [PMID: 21742351]

[13]   https://www.unenvironment.org/interactive/beat-plastic-pollution/

[14]   Mrowiec B. Plastic pollutants in water environment. Environmental protection and Natural resources. Journal of Institute of Environmental Protection-National Research Institute 2017; 28: 51-5.

[15]   Horton AA, Walton A, Spurgeon DJ, Lahive E, Svendsen C. Microplastics in freshwater and terrestrial environments: Evaluating the current understanding to identify the knowledge gaps and future research priorities. Sci Total Environ 2017; 586: 127-41.
        [http://dx.doi.org/10.1016/j.scitotenv.2017.01.190] [PMID: 28169032]

[16]   Chamas A, Moon H, Zheng J, *et al.* Degradation rates of plastics in the environment. ACS Sustain Chem& Eng 2020; 8(9): 3494-511.
        [http://dx.doi.org/10.1021/acssuschemeng.9b06635]

[17]   Singh R, Shahi S, Geetanjali . Geetanjali. Chemical degradation of poly(bisphenol A carbonate) waste materials: a review. ChemistrySelect 2018; 3(42): 11957-62.
        [http://dx.doi.org/10.1002/slct.201802577]

[18]   Shahi S, Geetanjali, Singh R. Chemical degradation of post-consumer poly(bisphenol A carbonate) waste materials: a review. ChemistrySelect 2018; 3(42): 11957-62.

[19]   Yu J, Sun L, Ma C, Qiao Y, Yao H. Thermal degradation of PVC: A review. Waste Manag 2016; 48: 300-14.
        [http://dx.doi.org/10.1016/j.wasman.2015.11.041] [PMID: 26687228]

[20]   Al-Salem SM, Antelava A, Constantinou A, Manos G, Dutta A. A review on thermal and catalytic pyrolysis of plastic solid waste (PSW). J Environ Manage 2017; 197: 177-98.
        [http://dx.doi.org/10.1016/j.jenvman.2017.03.084] [PMID: 28384612]

[21]   Shah AA, Hasan F, Hameed A, Ahmed S. Biological degradation of plastics: A comprehensive review. Biotechnol Adv 2008; 26(3): 246-65.
[http://dx.doi.org/10.1016/j.biotechadv.2007.12.005] [PMID: 18337047]

[22]   Pospíšil J, Nešpůrek S. Highlights in chemistry and physics of polymer stabilization. Macromol Symp 1997; 115(1): 143-63.
[http://dx.doi.org/10.1002/masy.19971150110]

[23]   Kamini NR, Iefuji H. Lipase catalyzed methanolysis of vegetable oils in aqueous medium by *Cryptococcus* spp. S-2. Process Biochem 2001; 37(4): 405-10.
[http://dx.doi.org/10.1016/S0032-9592(01)00220-5]

[24]   Gewert B, Plassmann MM, MacLeod M. Pathways for degradation of plastic polymers floating in the marine environment. Environ Sci Process Impacts 2015; 17(9): 1513-21.
[http://dx.doi.org/10.1039/C5EM00207A] [PMID: 26216708]

[25]   Hauenstein O, Agarwal S, Greiner A. Bio-based polycarbonate as synthetic toolbox. Nat Commun 2016; 7(1): 11862.
[http://dx.doi.org/10.1038/ncomms11862] [PMID: 27302694]

[26]   Muthu SS. Roadmap to sustainable textiles and clothing: Environmental and social aspects of textiles and clothing supply chain. Springer: Singapore 2014; p. 297.

[27]   Urbanek AK, Rymowicz W, Mirończuk AM. Degradation of plastics and plastic-degrading bacteria in cold marine habitats. Appl Microbiol Biotechnol 2018; 102(18): 7669-78.
[http://dx.doi.org/10.1007/s00253-018-9195-y] [PMID: 29992436]

[28]   Trivedi P, Hasan A, Akhtar S, Siddiqui MH, Sayeed U, Khan MKA. Role of microbes in degradation of synthetic plastics and manufacture of bioplastics. J Chem Pharm Res 2016; 8: 211-6.

[29]   Restrepo-Flórez JM, Wood JA, Rehmann L, Thompson M, Bassi A. Effect of biodiesel on biofilm biodeterioration of linear low-density polyethylene in a simulated fuel storage tank. J Energy Res Tech. 2015; 137: p. 032211.

[30]   Singh R. Biodegradable Polymers: An overview in fundamentals of plastic waste management. In: Singh , Haritash , Eds. Delhi, India: DBH Publishers and Distributors 2019; Chapter 14: pp. 151-9.

[31]   Singh R. Poly (lactic acid) as degradable resorbable polymer matrices in Biomedical Engineering: An overview. International Journal of Engineering Research and Advanced Technology 2018; 4(5): 56-62.
[http://dx.doi.org/10.31695/IJERAT.2018.3267]

[32]   Das MP, Kumar SA. Influence of cell surface hydrophobicity in colonization and biofilm formation on LDPE biodegradation. Int J Pharm Pharm Sci 2013; 5: 690-4.

[33]   Arutchelvi J, Sudhakar M, Arkatkar A, Doble M, Bhaduri S, Uppara PV. Biodegradation of polyethylene and polypropylene. Int J Biotechnol 2008; 7: 9-22.

[34]   Siracusa V, Rocculi P, Romani S, Rosa MD. Biodegradable polymers for food packaging: a review. Trends Food Sci Technol 2008; 19(12): 634-43.
[http://dx.doi.org/10.1016/j.tifs.2008.07.003]

[35]   Slor G, Papo N, Hananel U, Amir RJ. Tuning the molecular weight of polymeric amphiphiles as a tool to access micelles with a wide range of enzymatic degradation rates. Chem Commun (Camb) 2018; 54(50): 6875-8.
[http://dx.doi.org/10.1039/C8CC02415D] [PMID: 29774332]

[36]   Syranidou E, Karkanorachaki K, Amorotti F, *et al.* Biodegradation of weathered polystyrene films in seawater microcosms. Sci Rep 2017; 7(1): 17991.
[http://dx.doi.org/10.1038/s41598-017-18366-y] [PMID: 29269847]

[37]   Ahmed T, Shahid M, Azeem F, *et al.* Biodegradation of plastics: current scenario and future prospects for environmental safety. Environ Sci Pollut Res Int 2018; 25(8): 7287-98.
[http://dx.doi.org/10.1007/s11356-018-1234-9] [PMID: 29332271]

[38] Wierckx N, Prieto MA, Pomposiello P, Lorenzo V, O'Connor K, Blank LM. Plastic waste as a novel substrate for industrial biotechnology. Microb Biotechnol 2015; 8(6): 900-3.
[http://dx.doi.org/10.1111/1751-7915.12312] [PMID: 26482561]

[39] Kawai F, Watanabe M, Shibata M, Yokoyama S, Sudate Y, Hayashi S. Comparative study on biodegradability of polyethylene wax by bacteria and fungi. Polym Degrad Stabil 2004; 86(1): 105-14.
[http://dx.doi.org/10.1016/j.polymdegradstab.2004.03.015]

[40] Tokiwa Y, Suzuki T. Hydrolysis of polyesters by *Rhizopus delemar* lipase. Agric Biol Chem 1978; 42(5): 1071-2.
[http://dx.doi.org/10.1271/bbb1961.42.1071]

[41] Ho KLG, Pometto AL III, Hinz PN. Effects of temperature and relative humidity on polylactic acid plastic degradation. J Polym Environ 1999; 7(2): 83-92.
[http://dx.doi.org/10.1023/A:1021808317416]

[42] Henton DE, Gruber P, Lunt J, Randall J. Polylactic acid technology. Natural Fibers. Biopolymers, and Biocomposites 2005; pp. 527-77.

[43] Tokiwa Y, Calabia B, Ugwu C, Aiba S. Biodegradability of Plastics. Int J Mol Sci 2009; 10(9): 3722-42.
[http://dx.doi.org/10.3390/ijms10093722] [PMID: 19865515]

[44] Ghosh S, Qureshi A, Purohit HJ. Microbial degradation of plastics: Biofilms and degradation pathways. Contaminants in Agriculture and Environment: Health Risks and Remediation 2019; 1: 184-99.
[http://dx.doi.org/10.26832/AESA-2019-CAE-0153-014]

[45] Ru J, Huo Y, Yang Y. Microbial degradation and valorization of plastic wastes. Front Microbiol 2020; 11: 442.
[http://dx.doi.org/10.3389/fmicb.2020.00442] [PMID: 32373075]

[46] Uttiya Dey UD, Mondal NK, Das K, Dutta S. An approach to polymer degradation through microbes. IOSR J Pharm 2012; 2(3): 385-8.
[http://dx.doi.org/10.9790/3013-0230385388]

[47] Mohanan N, Montazer Z, Sharma PK, Levin DB. Microbial and enzymatic degradation of synthetic plastics. Front Microbiol 2020; 11: 580709.
[http://dx.doi.org/10.3389/fmicb.2020.580709] [PMID: 33324366]

[48] Manjunathan S, Chinnagounder S. Biodegradation of low-density polythene materials using microbial consortium – An overview. Int J Pharm Chem Sci 2015; 4: 507-14.

[49] Singh B, Sharma N. Mechanistic implications of plastic degradation. Polym Degrad Stabil 2008; 93(3): 561-84.
[http://dx.doi.org/10.1016/j.polymdegradstab.2007.11.008]

[50] Ghosh S, Qureshi A, Purohit HJ. Enhanced expression of catechol 1,2 dioxygenase gene in biofilm forming *Pseudomonas mendocina* EGD-AQ5 under increasing benzoate stress. Int Biodeterior Biodegradation 2017; 118: 57-65.
[http://dx.doi.org/10.1016/j.ibiod.2017.01.019]

[51] Kyaw BM, Champakalakshmi R, Sakharkar MK, Lim CS, Sakharkar KR. Biodegradation of Low Density Polythene (LDPE) by *Pseudomonas* Species. Indian J Microbiol 2012; 52(3): 411-9.
[http://dx.doi.org/10.1007/s12088-012-0250-6] [PMID: 23997333]

[52] Yang J, Yang Y, Wu WM, Zhao J, Jiang L. Evidence of polyethylene biodegradation by bacterial strains from the guts of plastic-eating waxworms. Environ Sci Technol 2014; 48(23): 13776-84.
[http://dx.doi.org/10.1021/es504038a] [PMID: 25384056]

[53] Albertsson AC. Biodegradation of synthetic polymers. II. A limited microbial conversion of ^{14}C in polyethylene to $^{14}CO_2$ by some soil fungi. J Appl Polym Sci 1978; 22(12): 3419-33.

[http://dx.doi.org/10.1002/app.1978.070221207]

[54] Albertsson AC, Bánhidi ZG. Microbial and oxidative effects in degradation of polyethene. J Appl Polym Sci 1980; 25(8): 1655-71.
[http://dx.doi.org/10.1002/app.1980.070250813]

[55] Albertsson AC, Barenstedt C, Karlsson S, Lindberg T. Degradation product pattern and morphology changes as means to differentiate abiotically and biotically aged degradable polyethylene. Polymer (Guildf) 1995; 36(16): 3075-83.
[http://dx.doi.org/10.1016/0032-3861(95)97868-G]

[56] Hakkarainen M, Albertsson AC. Environmental degradation of polyethylene. Long term properties of polyolefins. Berlin, Heidelberg: Springer 2004; pp. 177-200.
[http://dx.doi.org/10.1007/b13523]

[57] Azeko ST, Etuk-Udo GA, Odusanya OS, Malatesta K, Anuku N, Soboyejo WO. Biodegradation of linear low-density polyethylene by *Serratia marcescens* subsp. marcescens and its cell free extracts. Waste Biomass Valoriz 2015; 6(6): 1047-57.
[http://dx.doi.org/10.1007/s12649-015-9421-0]

[58] Orr IG, Hadar Y, Sivan A. Colonization, biofilm formation and biodegradation of polyethylene by a strain of *Rhodococcus ruber*. Appl Microbiol Biotechnol 2004; 65(1): 97-104.
[PMID: 15221232]

[59] Vatsaldutt P, Anbuselvi S. Isolation and characterization of polythene degrading bacteria from polythene dumped garbage. Int J Pharm Sci Rev Res 2014; 25: 205-6.

[60] Sepperumal U, Markandan M. Growth of *Actinomycetes* and *Pseudomonas* sp., biofilms on abiotically pre-treated polypropylene surface. Eur J Zoolog Res 2014; 3: 6-17.

[61] Auta HS, Emenike CU, Jayanthi B, Fauziah SH. Growth kinetics and biodeterioration of polypropylene microplastics by *Bacillus sp.* and *Rhodococcus sp.* isolated from mangrove sediment. Mar Pollut Bull 2018; 127: 15-21.
[http://dx.doi.org/10.1016/j.marpolbul.2017.11.036] [PMID: 29475646]

[62] Oberbeckmann S, Osborn AM, Duhaime MB. Microbes on a bottle: substrate, season and geography influence community composition of microbes colonizing marine plastic debris. PLoS One 2016; 11(8): e0159289.
[http://dx.doi.org/10.1371/journal.pone.0159289] [PMID: 27487037]

[63] Wei R, Zimmermann W. Microbial enzymes for the recycling of recalcitrant petroleum□based plastics: how far are we? Microb Biotechnol 2017; 10(6): 1308-22.
[http://dx.doi.org/10.1111/1751-7915.12710] [PMID: 28371373]

[64] Danso D, Chow J, Streit WR. Plastics: environmental and biotechnological perspectives on microbial degradation. Appl Environ Microbiol 2019; 85(19): e01095-19.
[http://dx.doi.org/10.1128/AEM.01095-19] [PMID: 31324632]

[65] Herrero Acero E, Ribitsch D, Steinkellner G, *et al.* Enzymatic surface hydrolysis of PET: effect of structural diversity on kinetic properties of cutinases from *Thermobifida*. Macromolecules 2011; 44(12): 4632-40.
[http://dx.doi.org/10.1021/ma200949p]

[66] Hu X, Thumarat U, Zhang X, Tang M, Kawai F. Diversity of polyester-degrading bacteria in compost and molecular analysis of a thermoactive esterase from *Thermobifida alba* AHK119. Appl Microbiol Biotechnol 2010; 87(2): 771-9.
[http://dx.doi.org/10.1007/s00253-010-2555-x] [PMID: 20393707]

[67] Wei R, Oeser T, Then J, *et al.* Functional characterization and structural modeling of synthetic polyester-degrading hydrolases from Thermomonospora curvata. AMB Express 2014; 4(1): 44.
[http://dx.doi.org/10.1186/s13568-014-0044-9] [PMID: 25405080]

[68] Yoshida S, Hiraga K, Takehana T, *et al.* A bacterium that degrades and assimilates poly(ethylene

terephthalate). Science 2016; 351(6278): 1196-9.
[http://dx.doi.org/10.1126/science.aad6359] [PMID: 26965627]

[69] Tosa T, Chibata I. Utilization of cyclic amides and formation of omega-amino acids by microorganisms. J Bacteriol 1965; 89(3): 919-20.
[http://dx.doi.org/10.1128/jb.89.3.919-920.1965] [PMID: 14273687]

[70] Howard GT, Blake RC. Growth of *Pseudomonas fluorescens* on a polyester–polyurethane and the purification and characterization of a polyurethanase–protease enzyme. Int Biodeterior Biodegradation 1998; 42(4): 213-20.
[http://dx.doi.org/10.1016/S0964-8305(98)00051-1]

[71] Cosgrove L, McGeechan PL, Robson GD, Handley PS. Fungal communities associated with degradation of polyester polyurethane in soil. Appl Environ Microbiol 2007; 73(18): 5817-24.
[http://dx.doi.org/10.1128/AEM.01083-07] [PMID: 17660302]

[72] Shigeno-Akutsu Y, Nakajima-Kambe T, Nomura N, Nakahara T. Purification and properties of culture-broth-secreted esterase from the polyurethane degrader *Comamonas acidovorans* TB-35. J Biosci Bioeng 1999; 88(5): 484-7.
[http://dx.doi.org/10.1016/S1389-1723(00)87663-X] [PMID: 16232649]

[73] Schmidt J, Wei R, Oeser T, *et al.* Degradation of polyester polyurethane by bacterial polyester hydrolases. Polymers (Basel) 2017; 9(12): 65.
[http://dx.doi.org/10.3390/polym9020065] [PMID: 30970745]

[74] Shah Z, Krumholz L, Aktas DF, Hasan F, Khattak M, Shah AA. Degradation of polyester polyurethane by a newly isolated soil bacterium, *Bacillus subtilis* strain MZA-75. Biodegradation 2013; 24(6): 865-77.
[http://dx.doi.org/10.1007/s10532-013-9634-5] [PMID: 23536219]

[75] Yamada-Onodera K, Mukumoto H, Katsuyaya Y, Saiganji A, Tani Y. Degradation of polyethylene by a fungus, *Penicillium simplicissimum* YK. Polym Degrad Stabil 2001; 72(2): 323-7.
[http://dx.doi.org/10.1016/S0141-3910(01)00027-1]

[76] Restrepo-Flórez JM, Bassi A, Thompson MR. Microbial degradation and deterioration of polyethylene – A review. Int Biodeterior Biodegradation 2014; 88: 83-90.
[http://dx.doi.org/10.1016/j.ibiod.2013.12.014]

[77] Raaman N, Rajitha N, Jayshree A, Jegadeesh A. Biodegradation of plastic by *Aspergillus* spp. isolated from polythene polluted sites around Chennai. J Acad Ind Res 2012; 1: 313-6.

[78] Mohan SK, Suresh B. Studies on biodegradation of plastics by *Aspergillus sp.* isolated from dye effluent enriched soil. Indo Am J Pharm Sci 2015; 2: 1636-9.

[79] Ojha N, Pradhan N, Singh S, *et al.* Evaluation of HDPE and LDPE degradation by fungus, implemented by statistical optimization. Sci Rep 2017; 7(1): 39515.
[http://dx.doi.org/10.1038/srep39515] [PMID: 28051105]

[80] Pramila R, Ramesh KV. Biodegradation of low-density polyethylene (LDPE) by fungi isolated from marine water a SEM analysis. Afr J Microbiol Res 2011; 5: 5013-8.
[http://dx.doi.org/10.5897/AJMR11.670]

[81] Russell JR, Huang J, Anand P, *et al.* Biodegradation of polyester polyurethane by endophytic fungi. Appl Environ Microbiol 2011; 77(17): 6076-84.
[http://dx.doi.org/10.1128/AEM.00521-11] [PMID: 21764951]

[82] Zafar U, Houlden A, Robson GD. Fungal communities associated with the biodegradation of polyester polyurethane buried under compost at different temperatures. Appl Environ Microbiol 2013; 79(23): 7313-24.
[http://dx.doi.org/10.1128/AEM.02536-13] [PMID: 24056469]

[83] Gautam R, Bassi AS, Yanful EK. Candida rugosa lipase-catalyzed polyurethane degradation in aqueous medium. Biotechnol Lett 2007; 29(7): 1081-6.

[http://dx.doi.org/10.1007/s10529-007-9354-1] [PMID: 17450322]

[84]   Álvarez-Barragán J, Domínguez-Malfavón L, Vargas-Suárez M, González-Hernández R, Aguilar-Osorio G, Loza-Tavera H. Biodegradative activities of selected environmental fungi on a polyester polyurethane varnish and polyether polyurethane foams. Appl Environ Microbiol 2016; 82(17): 5225-35.
[http://dx.doi.org/10.1128/AEM.01344-16] [PMID: 27316963]

[85]   Mathur G, Prasad R. Degradation of polyurethane by *Aspergillus flavus* (ITCC 6051) isolated from soil. Appl Biochem Biotechnol 2012; 167(6): 1595-602.
[http://dx.doi.org/10.1007/s12010-012-9572-4] [PMID: 22367637]

[86]   Renzi M, Blašković A, Broccoli A, Bernardi G, Grazioli E, Russo G. Chemical composition of microplastic in sediments and protected detritivores from different marine habitats (Salina Island). Mar Pollut Bull 2020; 152: 110918.
[http://dx.doi.org/10.1016/j.marpolbul.2020.110918] [PMID: 32479291]

[87]   Löhr A, Savelli H, Beunen R, Kalz M, Ragas A, Van Belleghem F. Solutions for global marine litter pollution. Curr Opin Environ Sustain 2017; 28: 90-9.
[http://dx.doi.org/10.1016/j.cosust.2017.08.009]

[88]   Sekiguchi T, Ebisui A, Nomura K, Watanabe T, Enoki M, Kanehiro H. Biodegradation of several fibers submerged in deep sea water and isolation of biodegradable plastic degrading bacteria from deep ocean water. Nippon Suisan Gakkaishi 2009; 75(6): 1011-8.
[http://dx.doi.org/10.2331/suisan.75.1011]

[89]   Sekiguchi T, Sato T, Enoki M, Kanehiro H, Uematsu K, Kato C. Isolation and characterization of biodegradable plastic degrading bacteria from deep-sea environments. JAMSTEC Rep Res Dev 2011; 11: 33-41.
[http://dx.doi.org/10.5918/jamstecr.11.33]

[90]   Raghul SS, Bhat SG, Chandrasekaran M, Francis V, Thachil ET. Biodegradation of polyvinyl alcohol-low linear density polyethylene-blended plastic film by consortium of marine benthic vibrios. Int J Environ Sci Technol 2014; 11(7): 1827-34.
[http://dx.doi.org/10.1007/s13762-013-0335-8]

[91]   Paço A, Duarte K, da Costa JP, *et al.* Biodegradation of polyethylene microplastics by the marine fungus *Zalerion maritimum.* Sci Total Environ 2017; 586: 10-5.
[http://dx.doi.org/10.1016/j.scitotenv.2017.02.017] [PMID: 28199874]

[92]   Nanda S, Sahu S, Abraham J. Studies on the biodegradation of natural and synthetic polyethylene by *Pseudomonas* spp. J Appl Sci Environ Manag 2010; 14(2): 57-60.
[http://dx.doi.org/10.4314/jasem.v14i2.57839]

[93]   Gilan I, Sivan A. Effect of proteases on biofilm formation of the plastic-degrading actinomycete *Rhodococcus ruber* C208. FEMS Microbiol Lett 2013; 342(1): 18-23.
[http://dx.doi.org/10.1111/1574-6968.12114] [PMID: 23448092]

[94]   Nanda S, Sahu SS. Biodegradability of polyethylene by *Brevibacillus, Pseudomonas* and *Rhodococcus* spp. N Y Sci J 2010; 3: 95-8.

[95]   Verma N, Gupta S. Assessment of LDPE degrading potential *Aspergillus* species isolated from municipal landfill sites of Agra. SN Applied Sciences 2019; 1(7): 701.
[http://dx.doi.org/10.1007/s42452-019-0746-3]

[96]   Tribedi P, Sil AK. Low-density polyethylene degradation by *Pseudomonas* sp. AKS2 biofilm. Environ Sci Pollut Res Int 2013; 20(6): 4146-53.
[http://dx.doi.org/10.1007/s11356-012-1378-y] [PMID: 23242625]

[97]   Harshvardhan K, Jha B. Biodegradation of low-density polyethylene by marine bacteria from pelagic waters, Arabian Sea, India. Mar Pollut Bull 2013; 77(1-2): 100-6.
[http://dx.doi.org/10.1016/j.marpolbul.2013.10.025] [PMID: 24210946]

[98]   Das MP, Kumar S. An approach to low-density polyethylene biodegradation by *Bacillus amyloliquefaciens*. 3 Biotech 2015; 5(1): 81-6.

[99]   Delacuvellerie A, Cyriaque V, Gobert S, Benali S, Wattiez R. The plastisphere in marine ecosystem hosts potential specific microbial degraders including *Alcanivorax borkumensis* as a key player for the low-density polyethylene degradation. J Hazard Mater 2019; 380: 120899.
[http://dx.doi.org/10.1016/j.jhazmat.2019.120899] [PMID: 31326835]

[100] Awasthi S, Srivastava P, Singh P, Tiwary D, Mishra PK. Biodegradation of thermally treated high-density polyethylene (HDPE) by Klebsiella pneumoniae CH001. 3 Biotech 2017; 7: 332.

[101] Balasubramanian V, Natarajan K, Hemambika B, *et al.* High-density polyethylene (HDPE)-degrading potential bacteria from marine ecosystem of Gulf of Mannar, India. Lett Appl Microbiol 2010; 51(2): no.
[http://dx.doi.org/10.1111/j.1472-765X.2010.02883.x] [PMID: 20586938]

[102] Shimpi N, Borane M, Mishra S, Kadam M, Sonawane SS. Biodegradation of isotactic polypropylene (iPP)/poly (lactic acid) (PLA) and iPP/PLA/nano calcium carbonates using *phanerochaete chrysosporium*. Adv Polym Technol 2018; 37(2): 522-30.
[http://dx.doi.org/10.1002/adv.21691]

[103] Skariyachan S, Patil AA, Shankar A, Manjunath M, Bachappanavar N, Kiran S. Enhanced polymer degradation of polyethylene and polypropylene by novel thermophilic consortia of *Brevibacillus sps.* and *Aneurinibacillus sp.* screened from waste management landfills and sewage treatment plants. Polym Degrad Stabil 2018; 149: 52-68.
[http://dx.doi.org/10.1016/j.polymdegradstab.2018.01.018]

[104] Raziyafathima M, Praseetha PK, Rimal Isaac RS. Microbial Degradation of Plastic Waste: A Review. J Pharmaceut Chem Biol Sci 2016; 4(2): 231-42.

[105] Jeon HJ, Kim MN. Isolation of mesophilic bacterium for biodegradation of polypropylene. Int Biodeterior Biodegradation 2016; 115: 244-9.
[http://dx.doi.org/10.1016/j.ibiod.2016.08.025]

[106] Arkatkar A, Arutchelvi J, Bhaduri S, Uppara PV, Doble M. Degradation of unpretreated and thermally pretreated polypropylene by soil consortia. Int Biodeterior Biodegradation 2009; 63(1): 106-11.
[http://dx.doi.org/10.1016/j.ibiod.2008.06.005]

[107] Auta HS, Emenike CU, Fauziah SH. Screening of *Bacillus* strains isolated from mangrove ecosystems in Peninsular Malaysia for microplastic degradation. Environ Pollut 2017; 231(Pt 2): 1552-9.
[http://dx.doi.org/10.1016/j.envpol.2017.09.043] [PMID: 28964604]

[108] Vivi VK, Martins-Franchetti SM, Attili-Angelis D. Biodegradation of PCL and PVC: *Chaetomium globosum* (ATCC 16021) activity. Folia Microbiol (Praha) 2019; 64(1): 1-7.
[http://dx.doi.org/10.1007/s12223-018-0621-4] [PMID: 29882027]

[109] Wierckx N, Narancic T, Eberlein C, *et al.* Plastic biodegradation: Challenges and opportunities. Consequences of Microbial Interactions with Hydrocarbons, Oils, and Lipids: Biodegradation and Bioremediation. Cham: Springer 2018; pp. 1-29.
[http://dx.doi.org/10.1007/978-3-319-44535-9_23-1]

[110] Webb JS, Nixon M, Eastwood IM, Greenhalgh M, Robson GD, Handley PS. Fungal colonization and biodeterioration of plasticized polyvinyl chloride. Appl Environ Microbiol 2000; 66(8): 3194-200.
[http://dx.doi.org/10.1128/AEM.66.8.3194-3200.2000] [PMID: 10919769]

[111] Ali MI, Ahmed S, Robson G, *et al.* Isolation and molecular characterization of polyvinyl chloride (PVC) plastic degrading fungal isolates. J Basic Microbiol 2014; 54(1): 18-27.
[http://dx.doi.org/10.1002/jobm.201200496] [PMID: 23686796]

[112] Anwar MS, Kapri A, Chaudhry V, *et al.* Response of indigenously developed bacterial consortia in progressive degradation of polyvinyl chloride. Protoplasma 2016; 253(4): 1023-32.
[http://dx.doi.org/10.1007/s00709-015-0855-9] [PMID: 26231814]

[113] Giacomucci L, Raddadi N, Soccio M, Lotti N, Fava F. Polyvinyl chloride biodegradation by *Pseudomonas citronellolis* and *Bacillus flexus*. N Biotechnol 2019; 52: 35-41.
[http://dx.doi.org/10.1016/j.nbt.2019.04.005] [PMID: 31026607]

[114] Eisaku O, Linn K, Takeshi E, Taneaki O, Yoshinobu I. Isolation and characterization of polystyrene degrading microorganisms for zero emission treatment of expanded polystyrene. Environ Eng Res 2003; 40: 373-9.

[115] Mor R, Sivan A. Biofilm formation and partial biodegradation of polystyrene by the actinomycete *Rhodococcus ruber*. Biodegradation 2008; 19(6): 851-8.
[http://dx.doi.org/10.1007/s10532-008-9188-0] [PMID: 18401686]

[116] Yang Y, Yang J, Wu WM, *et al.* Biodegradation and mineralization of polystyrene by plastic-eating mealworms: Part 2. Role of gut microorganisms. Environ Sci Technol 2015; 49(20): 12087-93.
[http://dx.doi.org/10.1021/acs.est.5b02663] [PMID: 26390390]

[117] Liu B, He L, Wang L, *et al.* Protein crystallography and site direct mutagenesis analysis of the poly (ethylene terephthalate) hydrolase PETase from *Ideonella sakaiensis*. ChemBioChem 2018; 19(14): 1471-5.
[http://dx.doi.org/10.1002/cbic.201800097] [PMID: 29603535]

[118] Chen CC, Han X, Ko TP, Liu W, Guo RT. Structural studies reveal the molecular mechanism of PET ase. FEBS J 2018; 285(20): 3717-23.
[http://dx.doi.org/10.1111/febs.14612] [PMID: 30048043]

[119] Peng YH, Shih Y, Lai YC, Liu YZ, Liu YT, Lin NC. Degradation of polyurethane by bacterium isolated from soil and assessment of polyurethanolytic activity of a *Pseudomonas putida* strain. Environ Sci Pollut Res Int 2014; 21(16): 9529-37.
[http://dx.doi.org/10.1007/s11356-014-2647-8] [PMID: 24633845]

[120] Muhamad WNAW, Othman R, Shaharuddin RI, Irani MS. Microorganism as plastic biodegradation agent towards sustainable environment. Adv Environ Biol 2015; 9: 8-14.

[121] Kasirajan S, Ngouajio M. Polyethylene and biodegradable mulches for agricultural applications: a review. Agron Sustain Dev 2012; 32(2): 501-29.
[http://dx.doi.org/10.1007/s13593-011-0068-3]

[122] Shimao M. Biodegradation of plastics. Curr Opin Biotechnol 2001; 12(3): 242-7.
[http://dx.doi.org/10.1016/S0958-1669(00)00206-8] [PMID: 11404101]

[123] Mahdi MS, Ameen RS, Ibrahim HK. Study on Degradation of Nylon 6 by thermophilic bacteria *Anoxybacillus rupiensis* Ir3 (JQ912241). Int J Adv Res Biol Sci 2016; 3(9): 200-9.
[http://dx.doi.org/10.22192/ijarbs.2016.03.09.027]

CHAPTER 14

Characteristic Features of Plastic Microbial Degradation

Soumyaranjan Senapati[1], **Sreelipta Das**[1] and **Alok Kumar Panda**[1,*]

[1] *School of Applied Sciences, Kalinga Institute of Industrial Technology, Deemed to be University, Bhubaneswar-751024, India*

Abstract: The increase in the amount of plastic waste, especially microplastics and the environmental pollution caused by it has diverted the research focus of the world into plastic recycling and degradation. Hence in the last decade, different strategies have been adopted to combat this problem. Albeit many physiochemical technologies are there for the degradation of plastics, they give rise to harmful chemicals as by-products. This has shifted the priority of our research to the biodegradation of plastics by microbes. In fact, in the last decade, many microorganisms have been discovered with the ability to degrade many conventional plastics with moderate efficiency but longer duration. The initial part of this chapter discusses the various kinds of plastics present and the methods adopted for the degradation of plastics, with special emphasis on the factors affecting plastic degradation. In the subsequent section, the microbial degradation of different plastics by bacteria and fungi, along with a mechanism, has been outlined. Furthermore, this chapter also briefly discusses the role of enzymes in the degradation of different plastics by microbes and the future of plastic biodegradation.

Keywords: Plastics, Plastics biodegradation, Microbes, Bacteria, Fungi, Microbial consortia, Microbial biofilm, Biodegradation mechanism, Enzymes, PETase.

INTRODUCTION

Polymers have been associated with human society since 1600 BC, when natural rubber was processed to form different items [1]. The first synthetic plastic was revealed by Alexander Parkes in the year 1862 and Leo Hendrik Baekland in the year 1907, who accidentally discovered a new synthetic polymer that came to be known as Bakelite. Two years later, Baekland coined the name "plastics," which represented a new material [2]. Plastic, derived from the Greek word "plastikos,"

---

* **Corresponding author Alok Kumar Panda:** School of Applied Sciences, Kalinga Institute of Industrial Technology, Deemed to be University, Bhubaneswar-751024, India; E-mail: alok.pandafch@kiit.ac.in means things that can be molded into different shapes. Polyvinyl chloride or PVC was the first plastic to be patented and registered in 1914. However, the boom in plastics started just after World War II, when polyurethane, silicones, polypropylene, polyester joined the race with polyvinyl chloride and polystyrene. It is estimated that there is an increase in plastic production of about 30-40% worldwide due to the application of plastic for different packing purposes, as a result of which the increase rate in the garbage production would rise by around twelve percent per annum [3]. In 1950, the production of plastic per year was 2 million tonnes worldwide. Since then, the annual production of plastic has increased nearly 200-fold and reached 381 million tonnes in 2015 [4]. Due to this large increase in the production of plastic, there is a persistent accumulation of plastics in the environment, which causes severe water and land pollution. A large amount of plastics is found littering the agricultural lands than in the ocean waters, which causes land pollution and hampers agricultural productivity. The plastics released from the household activities end up in the waste treatment plant, which is then accumulated in the agricultural soils, which leads to the transportation of the invasive species [5]. Apart from this, the swallowing of the disposed of plastic materials by both terrestrial and marine animals leads to entanglement which may result in the death of these species [6]. While landfilling is a short-term solution for managing plastic waste, the long-term solution for the disposal of plastic waste needs a lot of work and still has a long way to go. In addition to this, many chemical and physical degradation processes such as hydrolysis, ammonolysis, oxidants, physical and UV stress have been employed to degrade plastics, but these processes lead to the production of toxic substances [7]. Conventional plastics like polyethylene terephthalate (PET) and polyethylene (PE), are very resistant to degradation as they are made of a highly stable carbon backbone and hence are durable and consumer-friendly plastics. The plastic waste generation per person per day in India in 2010 was about 0.01 kilograms, which is less than 10 times as compared to the United States, Kuwait, Germany, the Netherlands, *etc.* In recent times the average per capita plastic usage in India is eleven kg which is less than half of the global consumption which stands at 28 kg [8]. Albeit the consumption of plastics in India is less as compared with that of the world, but India faces a lot of problems in recycling or degrading these plastics in an eco-friendly manner. Therefore, biodegradation is one of the major solutions which can be employed to deal with the plastics wastes from both the world and Indian perspectives.

As conventional plastics are difficult to degrade and generate a lot of pollution, biodegradable plastics are being used in many fields [9 - 13]. But due to the high cost and less durability of biodegradable plastics, their commercial application is limited [14]. Hence, in recent times a lot of efforts and research has been directed to the biodegradation of conventional plastics which is eco-friendly and negates the release and accumulation of toxic and harmful by-products [15]. The biodegradation of plastics is mainly carried out by various micro-organisms, but the degree of biodegradability by different microbes depends on various properties of the plastic especially the chemical and the physical properties [16]. The polymer bonds in the plastic are mainly degraded by the microbes with the aid of enzymatic hydrolysis and the biofilms generation on the plastic surface [17]. In this chapter, the various kind of plastics used and their present method of disposal and degradation has been outlined. Thereafter, the various methods adopted for the biodegradation of selected conventional plastics and the factors affecting their biodegradation and the extensive list of the microbes utilized for the degradation of various plastics have been discussed in detail. Finally, the biodegrading enzymes, their molecular aspects, and the future scope of biodegradation of plastics have been summarized.

CLASSIFICATION AND CATEGORIES OF PLASTICS

Polymers are synthesized by the polymerization of the monomers which may be derived from fossil fuels or nature. In general, there are around twenty different types of polymers and within each group, there are numerous grades. Plastics are divided into thermosetting, elastomers, and thermoplastics and the classification is based on their physical characteristics. But the majority of the plastic that we use is thermoplastic *i.e* it can reform repeatedly. In context to the biodegradability of the plastics, it can be classified as follows:

1. Natural Plastics
2. Biodegradable Synthetic Plastics
3. Non-biodegradable Synthetic Plastics

Natural Plastics

Natural plastics are easily available in nature in the form of biopolymers and dry materials and are named natural bioplastics. These types of plastics are mainly synthesized by plants, fungi, crustaceans, insects or algae, *etc* [18, 19]. The cell wall of different plants have a variable composition and usually, lignin, cellulose and hemicellulose are the main constituents of dry material that endows rigidity to the plant cell wall [18]. So, these lignin, cellulose, and hemicellulose, are the building block of natural polymers [20]. Chitin is another most important and abundant natural polymer [19]. It is the structural element in many organisms, like fungi, insects and algae, *etc* [21].

Biodegradable Synthetic Plastics

Biodegradable synthetic polymers are those which can be degraded by using certain microbes through a different pathway. The pathways through which the degradation proceeds may be enzymatic and non-enzymatic hydrolysis. Also, various other processes are available for the degradation of synthetic plastics. Different factors also affect the degradation of plastic-like surface morphology of plastics, the nature of bonds present in the plastics, *etc.* There are mainly three categories of biodegradable synthetic plastics and they are

Bio-based

Fossil based

Biodegradable Polymer Blends

Bio-Based Biodegradable Plastics

PHA (Polyhydroxyalkanoate) and PLA (Polylactic acid) are bio-based biodegradable plastics that are commonly used.

PHA is polyhydroxyalkanoate which is produced naturally by bacterial fermentation of carbohydrates and lipids [22]. PHA polymer has wide use in different fields of the pharmaceutical industry and medical [23]. Also, it is used in the packaging of different substances like fast food materials, packaging goods, medical tools, *etc* [24]. In the pharmaceutical industry, PHA is extensively used since it shows a good degree of bio-acceptance in patients. Microorganisms can degrade PHA under the condition of limited energy and utilize carbon sources to generate CO_2 and H_2O [25]. Some bacterial genera like *Bacillus, Burkholderia, Nocardiopsis, and Cupriavidus* are known to assimilate PHA by both aerobic and anaerobic mechanisms as well as some fungal genera [26].

PLA stands for polylactic acid and the source of this polymer is corn starch, sugarcane, and tapioca roots. The monomeric unit of PLA is lactic acid or hydroxycarboxylic acid. Due to PLA's availability, biodegradability, and good mechanical attribute, it is used extensively in the field of medicine [27, 28]. Many microbes lead to the easy hydrolytic degradation of PLA [29]. *B. licheniformis* and *Amycolatopsis sp.*, are example of microbes isolated from the soil which can degrade PLA [29, 30].

Fossil Based Biodegradable Plastics

Fossil based biodegradable plastics are used in different sector such as the packing industry but the majority of these types of plastics are not biodegradable and cause concern in the management of this waste product [31, 32]. The degradation process for this type of plastic is very slow, and different enzymes, as well as microorganisms, are involved in the degradation process [25, 33 - 35].

Currently, the research in the biodegradation of plastic focuses on the microorganisms that can degrade fossil-based plastics. In addition to this, the focus is also on the development of the enzymes which can lead to the degradation of plastics and then encoding these enzymatic genes inside a bacteria to synthesize these enzymes in larger quantities [36]. The fossil-based biodegradable polymers are polyethylene succinate (PES) and polycaprolactone (PCL)

PES stands for polyethylene succinate which is a thermoplastic polyester. PES is synthesized from either ethylene glycol and succinic acid or succinic anhydride and ethylene oxide with the aid of co-polymerization [37]. PES is utilized in the field of agriculture, carrying bags, paper coating agents, *etc* [38]. *Pseudomonas sp. AKS2* a mesophilic bacterium is capable of degrading PES polymer [39]. As compared to PCL, there are less number of microbes that can degrade PES [38]. There is another thermophilic bacterial strain isolated from the soil for the degradation of PES namely *Bacillus sp. TT96* [40].

PCL stands for polycaprolactone which is a fossil-based plastic and it is seen that both aerobic and anaerobic microorganisms have the ability to degrade this polymer. This plastic finds extensive application in biomedical materials such as blood bags, catheters, *etc* [41]. It is highly flexible and biodegradable [41]. The enzymes that are involved in the biodegradation of PCL are microbial lipases and esterases [42]. *Aspergillus sp. ST-01* can degrade PCL and generate a wide range of products such as succinic, caproic, butyric, and valeric acids [43].

Biodegradable Polymer Blends

Polymer blends are the new class of materials formed by the blending of two different polymers. It is one of the cheap methods to synthesize biodegradable polymers with desirable properties [40]. Synthesis of polymer blends is cost-effective and easier to prepare than co-polymerization [40]. In between polymer blend and fossil-based polymer, the polymer blend finds more commercial applications [44, 45]. Each component of this type of plastic is completely degraded therefore it is considered a completely biodegradable polymer. Some of the polymer blends such as starch/polyester blends and starch/PVA blends are biodegradable. By increasing the starch concentration, the rate of degradation of blended PCL increases [46]. There is an increase in the characteristics of degradation of starch and polyester blend in the presence of anhydride polyesters [47]. There are many Rhizopus microbial species that aid in the hydrolysis of many polyester blends with the help of a lipase enzyme [48]. Starch/PVA blends can also be biodegraded easily. PVA is known as polyvinyl alcohol and is water-soluble. Several microbes have been isolated which can hydrolytically biodegrade starch/PVA with the aid of various enzymes. *Alcaligenes faecalis T1* efficiently degrades this starch/PVA blend [40, 49]

Non-biodegradable Synthetic Plastics

Non-biodegradable plastics generally have a high molecular weight and cannot be degraded easily and they are mainly derived from fossil fuels. They have usually a long chain of monomeric repeating units and are highly stable to harsh environmental factors. These plastics are first broken down into microplastics which are then accessible to the microbes for biodegradation to some extent. Some of the non-biodegradable synthetic plastics are as follows:

1. Polyethylene Terephthalate (PET)

2. Polyurethane (PUR)

3. Polyethylene (PE)

4. Polyamide (PA)

5. Polystyrene (PS)

6. Polyvinyl Chloride (PVC)

7. Polypropylene (PP)

8. Polyvinyl Alcohol (PVA)

9. Polyethylene Glycol (PEG)

PET stands for polyethylene terephthalate polymer which is stable (both thermally and chemically) with semi-crystalline properties, having a molecular weight range of 30000- 80000 gram per mole [50]. PET is a linear and polar polymer made up of ethylene glycol and aromatic terephthalic acid [51]. PET is an example of a thermoplastic polymer and shows partial crystallization. Esterase is responsible for PET degradation [52]. Researchers have reported some bacteria and fungi that are responsible for the partial degradation of PET polymer to monomers and oligomers [53].

The monomer of the PET polymer is bis (2-hydroxyethyl) terephthalate [51]. Serine hydrolases are the enzymes assumed to degrade this polymer to some extent [54]. The presence of the disulfide bonds due to cysteine residues promotes the thermal stability of the enzyme and also leads to the specific binding of PET to the enzyme [54]. In addition to this, the bacterial hydrolases, and fungal cutinases also exhibit the ability to degrade PET [54]. PETases represent the best explored and studied class of enzymes to degrade this polymer effectively by hydrolysis [54].

PUR stands for polyurethane. Polyether or polyester polyols can be used to synthesize PUR [54]. This is a thermosetting polymer and is widely used for foam generation, textile coating, and paints [54]. Some bacteria and fungi are reported to degrade this polymer [54]. PUR stands for polyurethane. Polyether or polyester polyols can be used to synthesize PUR [54]. This polymer is an example of thermosetting polymers *i.e.* they cannot melt and are irreversible. It is widely used for the production of foam, textile coating, and paint to prevent corrosion [54]. Some bacteria and fungi were reported to degrade this polymer. Moreover, a large number of fungal species are found to degrade this polymer with some PUR-active bacteria [54]. Fungi such as *Aureobasidium pullulans* and *Fusarium solani* can be degraded PUR [54].

PE stands for polyethylene and is synthesized by polymerization of ethene. These polymers are hydrophobic and are used extensively for packaging. PE has a very long chain of ethylene monomers and has a high molecular weight. Mostly PE is not biodegradable, but there are a few microbes that can degrade PE [55]. PE is mainly of two types; HDPE (High-density) and LDPE (Low-density) polymers. These polymers are non-degradable by microorganisms. But some microorganisms can degrade PE. For example, polyethylene of low molecular weight can be degraded by *Actinobacter sp* [55]. *Penicilillium simplicissimum YK* is another fungus strain that can degrade PE [56]. It is mainly of two types of PE; HDPE (High-density) and LDPE (Low-density). The biotic and abiotic degradation process plays an important role to degrade this HDPE and LDPE polyethylene. Surprisingly, a large number of bacterial and some fungal species are reported to degrade PE.

PA stands for polyamides and it has many different repeating units such as aromatic, aliphatic, and semi-aromatic molecules which can be connected through amide bonds [54]. Hence, these polymers are mainly used in textile, carpets, sportswear, and automotive applications [54]. Various microorganisms are isolated to degrade nylon. *Vibrio furnisii, Bacillus cereus, Brevundimonas vesicularis,* and *Bacillus sphaericus* are microorganisms that can degrade nylon [54].

PS stands for polystyrene which consists of styrene monomers. Polystyrenes are high molecular weight and hydrophobic polymers. PS polymer is used in packing industries and in daily use materials such as petri dishes, CD cases, plastic cutlery *etc* [54]. The gut bacteria of mealworms *i.e. Exiguobacterium sp. Strain YT2* can degrade polystyrene [57].

PVC stands for polyvinyl chloride and its repeating unit constitutes of vinyl chloride monomer. This polymer is the third most-produced and used [58]. Microorganisms like *Micrococcus species* bacteria are found to play a major role in the degradation of PVC [59]. Also, white-rot fungi are responsible for PVC degradation [60].

PP stands for polypropylene and is built from the monomeric unit propylene. It is produced by chain-growth polymerization. It is the second most used and produced after PE. Bacteria like *Vibrio and Pseudomonas* and fungi such as *Aspergillus niger* have the capacity to biodegrade polypropylene [61].

PVA stands for polyvinyl alcohol and is built from the monomeric unit of vinyl alcohol. Since PVA is biocompatible it is used in an array of medical applications. This polymer can be degraded by using the consortia of microbes which has been discovered from sludge samples. PVA can be degraded by some extracellular and intracellular enzymes. PVA polymers can be biodegraded by different types of microorganisms [62]. *Alcaligenes faecalis T1* which secretes the enzyme depolymerase can efficiently degrade starch/PVA blend [40, 49].

PEG stands for polyethylene glycol and the monomer unit of this polymer is ethylene glycol. It has many applications in biotechnological, biomedical, and industries. A bacterial strain namely *Pelobacterpropionicus*. Strain KoB58 can degrade PEG [63].

PLASTIC DEGRADATION

Degradation of plastic or polymers means the complete depolymerization and mineralization of the plastic materials. The polymer degradation can take place majorly in three ways *i.e.* chemical degradation, physiochemical degradation, and biodegradation (Fig. **1**). Chemical and physiochemical degradation are conventional degradation methods and most of these processes release toxic gases that lead to environmental pollution. Biodegradation methods are eco-friendly processes but have a long way to go to be exploited on a commercial scale.

Conventional Degradation Methods

The method of chemical degradation of plastics involves oxidation and hydrolysis of the polymer chains while the physiochemical degradation of plastics mainly includes a thermal, photo, and mechanical degradation of the polymers as presented in Fig. (**1**). Furthermore, based on the agents causing polymer degradation, the degradation process can be categorized as thermal, photo-oxidative, ozone-induced, catalytic, and mechanochemical. Photodegradation and thermal degradation are the major processes that are used to degrade conventional plastics. Thermal degradation involves high temperature and is carried out in an aerobic atmosphere which leads to the polymeric bond rupture resulting in the creation of reactive and radical sites in the polymer and a change in its molecular weight. Photodegradation as the name suggests is initiated by photons absorbance by the polymer which results in the excision of the long polymer chains. This leads to the conversion of the larger polymer chain into smaller portions [2]. But the process of

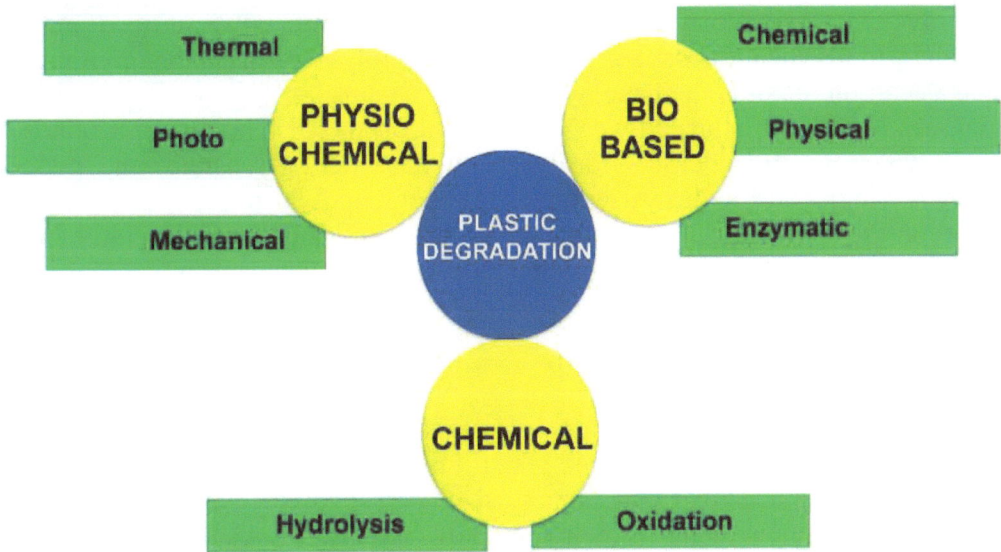

Fig. (1). Polymer degradation under various environmental conditions. The above figure depicts the various ways utilized to degrade polymers including biodegradation.

photodegradation of the plastics requires the side chain, functional groups of the polymer, to be photoresponsive [2]. But many synthetic plastics are resistant to both physical and chemical degradation, which leaves biodegradation as an alternative option to degrade conventional plastic.

Biodegradation Methods

The process of biodegradation involves the breaking down of organic and inorganic materials by living organisms. In the case of plastic, biodegradation is mainly carried out by the microorganisms under the influence of various chemical, physical, and enzymatic actions (Fig. **1**). Plastic degradation, in general, is a slow process, where at first several environmental factors such as temperature, pH and ultraviolet rays contribute towards breaking the long polymer chains into smaller fragments, and to do this the microorganisms in the process of degradation have developed different strategies. The major microbes which play a vital role in the degradation of plastics are bacteria and fungi and each of their biodegradation mechanism is different from each other. The biodegradation of plastic may be enzyme-catalyzed or non-enzymatic reaction [64]. The knowledge aggregated to date on the degradation of plastics, reveals that microorganisms secrete exoenzymes which leads to the disintegration of the complex polymeric chain into smaller oligomers, dimers, and monomers. These smaller polymer particles then seep through the microbial membranes and are utilized by the microbial enzymes [65]. The biodegradation of the polymers may take different pathways that are majorly dependent on the environmental conditions and the complete mineralization or degradation of the polymers generates compounds such as carbon dioxide, methane, water, organic acids, *etc* [2]. Biodegradation of microbes is majorly induced by enzymatic hydrolysis [2]. The different bonds in the polymer which undergo hydrolysis are the ester, glycosidic, and peptide linkages. Enzymatic hydrolysis of the polymer bonds also depends to a great extent on the structure of the polymer. The details of polymer biodegradation by the microbes and enzymes as well as the most probable mechanism is discussed in the subsequent sections.

Microbial Biodegradation of Plastics

Among the biodegradation methods, microbial degradation of plastics mainly by bacteria and fungi is vastly explored but still many questions remain unanswered. Microorganisms due to their small size can evolve rapidly and utilize environmental contaminants as their source of food. This gives rise to the concept of microbial bioremediation of many environmental pollutants. Microbes can use the organic pollutants for their propagation and expansion because these organic pollutants act as a major source of carbon which is exploited by the microorganisms as their building block to build their cells and they also provide energy to the cells [2]. In this section, the degradation of plastic especially non-biodegradable plastics will be covered comprehensively.

FACTORS INFLUENCING THE MICROBIAL BIODEGRADATION OF PLASTIC

Exposure Conditions (Abiotic and Biotic)

The degradation of plastic by microbes is judged by several factors which are shown in Fig. (**2**). The biodegradation of plastics/ micro-nano plastics can be affected by various factors and these factors usually affects the growth of microorganisms. The characteristics of plastics include many of their physical and chemical properties such as polymer density, polymer's molecular weight, crystallinity contained inside the plastic, type of functional group, type of substituents present in its structure, *etc* [33]. In addition to the preceding characteristics of the plastics, the environmental factors like ultraviolet radiation, high temperature, relative humidity, presence of chemicals in the plastic dumping site, *etc* affect the biodegradation of the plastics. The plasticizer is an entity that is used as additive to enhance the plastic properties and it significantly affects the biodegradability of plastics by lowering the glass transition temperature of the plastic [66]. Other processes and chemical reactions that affect the process of plastic biodegradation are methanolysis, ammonolysis, and hydrolysis [67, 68]. The properties of the plastic, the environmental factors, and the chemical reactions affect the extent of the biodegradable capability of the microorganisms as well as the biodegradability of the plastics. The resistance of synthetic plastics towards biodegradation is majorly due to their properties like crystallinity, density, and molecular weight, functional groups, *etc*. Plastics with highly ordered and more densely packed crystalline structure with high molecular weight are difficult to degrade and has more resistance power towards biodegradation. Different type of bonding and linkages present in the polymer affect its degradation property. The presence of hydrolysable bonds in the polymer backbone eases the biodegradation process of PET and PUR as compared to PE, PP, and PS.

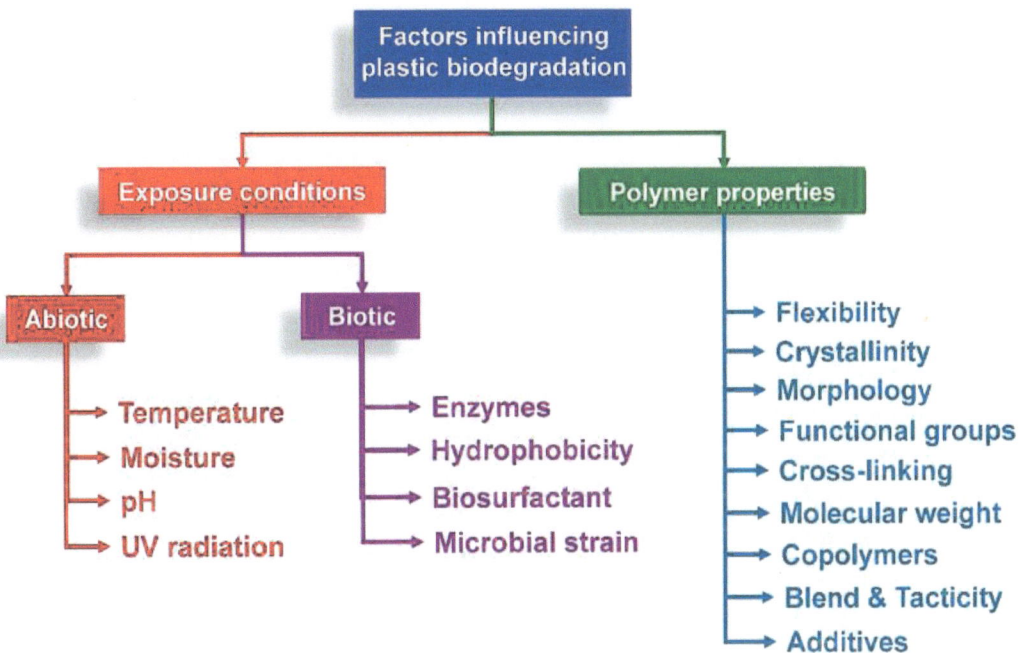

Fig. (2). Factors involved in biodegradation of plastics. The above figure shows the various environmental exposure conditions and the properties of the plastics that influence the degree of biodegradability of the plastics.

The external environment (abiotic and biotic) can affect two aspects of microbial biodegradation. The abiotic and the biotic components of the environment can affect the growth and metabolic behavior of the microbes and it can also lead to the aging and destruction of the polymeric structure by reactive oxidative processes. Plastic degradation under the influence of biotic and abiotic factors is a long-term process. Ultraviolet radiation, an increase in temperature, mechanical stress, and other external environmental effects can affect the plastic degradation process and lead to the conversion of the larger plastic particles into brittle and disintegrated micro and nano-size particles [69]. These micro and nano-size particles lead to the surface area increment and improve the bioavailability of the polymer particles to the plastic degrading microorganisms. Such type of degrading process is especially relevant to marine and beach environments but does not hold much relevance to the soil. Photooxidation is one of the important environmental effects which leads to the rise in the hydrophilicity of the polymers and also surface oxidation that leads to the fragmentation of the polymer backbone [70 - 72]. In line with photo-oxidation, high temperature also induces similar reactions in the degradation of the plastics. Oxidation plays a vital role in plastic degradation. Oxo-biodegradable plastics have been developed containing additives that are pro-oxidant in nature. These additives are majorly metals like cobalt, manganese, iron, *etc* and they generate free radicals on exposure to UV radiations or elevated temperatures. As a result, there is a scission of longer polymeric chains into smaller oligomers which then can be utilized by microbes for degradation [69]. Plastics with chromophore groups lead to enhanced photo-oxidation, which in turn increases microbial degradation [70, 73, 74]. Apart from this, the presence of photoresponsive groups in the polymer also has a similar effect [2]. Furthermore, oxidation decreases the surface hydrophobicity of the polymer and favors biological degradation [70]. Thermal degradation in the presence of oxygen is known as thermo-oxidative degradation which involves rupture of the polymeric bonds resulting in radical sites [2]. Plastic recycling is mainly done by mechanical and chemical methods. Mechanical recycling involves collecting, sorting, washing, grinding, and pelletizing. It is carried out in two modes *i.e.* primary and secondary [75]. Primary mode helps in reprocessing the plastics into products [75]. Chemical recycling includes gasification, pyrolysis, fluid-catalyzed cracking, and so on [76]. Chemical recycling leads to the disintegration of the polymeric chains into their chemical constituents which are then put back into products. Chemical recycling requires a large amount of energy and hence is not practiced [75]. The rate of biotransformation of plastics is very low due to their stable carbon-based backbone and the presence of cross-linkages [8, 69]. This makes conventional plastic hard to bio-degrade [33, 40, 69]. The polymers with a hydrolyzable backbone have a high degree of biodegradability due to which all the natural polymers are biodegradable [69].

Polymer Properties

The macromolecular structure of plastics is another obstacle to the microbial degradation of plastics. Microbes secrete extracellular enzymes that initiate the breaking of the plastics into smaller fragments which are consequently taken up by the microbes for metabolization. Hydrolyzable bonds in polymers are easier to attack by various extracellular hydrolases but polymers having non-hydrolyzable bonds such as PE, PP, PS, and PVC are difficult to break by the microbes without the help of external factors [69, 77, 78]. Polymers made up of only carbon-carbon backbone are first fragmented by a redox reaction and consequently mineralized by microorganisms [8]. Smooth surface and hydrophobicity restrict the biodegradation process of the polymer by the microorganisms [8]. When microbes attack and colonize the surface of the plastic, they hydrolyze the ester and other similar types of bonds like acetals and amide by enzyme action [68, 75, 79, 80]. So, the order in which different bonds are hydrolyzed by microbial enzymes is, ester > ether > amide > urethane [81]. Plastics harbor functional groups which lead to an increase in the hydrophilic nature of the plastic, enhancing the biodegradability of the polymer. On the other hand, hydrophobicity hinders the attachment of the microorganism to the plastic surface which decreases plastic biodegradation. This problem of hydrophobicity can be overcome by the formation of microbial biofilm where the bacteria attach themselves to the non-polar surface by fimbrial appendages [80]. Plastics with high melting temperatures are hard to biodegrade because depolymerization only happens above the ceiling temperature (T_c) [75, 81]. Usually, polymer chains having a lack of segmental mobility result in high thermal stability [70]. Aromatic polymers tend to resist degradation while the presence of carboxyl leads to a decrease in their thermo-oxidative stability [70].

Crystallinity

It is one of the most important aspects of polymer biodegradation, as highly crystalline polymers resist hydrolysis while amorphous polymers facilitate hydrolysis and enzymatic action. This limitation of crystallinity can be overcome by many biological processes [69]. Due to the difference in the crystallinity, colonization of bacteria takes place faster on the low-density polyethylene (LDPE) than on high-density polyethylene (HDPE). LDPE structure is softer, flexible, and not perfectly crystalline as HDPE [82]. Usually, microbes attack the amorphous region first as compared to other parts of the plastic [8]. Thus, amorphous polymers are more prone to enzymatic action and faster biodegradation than crystalline polymers [76]. Usually, the degradation of crystalline plastic occurs by increasing the temperature and exposing them to water [76]. In addition to this, the longer side chain of the plastic makes it more amorphous [75]. The glass transition temperature (T_g) is also an important parameter for amorphous plastic materials which determines their biodegradability. Usually, commercially biodegradable polymers have low T_g. The glass transition temperature (T_g) is mainly influenced by the structure of the polymers and it is seen that plastics having larger aromatic structures have a higher T_g value [81].

Plasticizer

The plasticizer is an additive that makes the material softer and flexible and increases its plasticity, thereby decreasing its viscosity and friction. Plasticizers increase the free volume and lower the glass transition temperature. It increases the strength of the plastic which makes it less biodegradable. The presence of other additives such as anti-oxidants, flame retardants, and light stabilizers can also affect the biodegradation of the polymers [83].

Biosurfactants

Surfactants are surface-active compounds that lower the interfacial tension between liquid-liquid, liquid-gas, and solid-gas due to the presence of amphiphilic compounds. Similar to surfactant, biosurfactants are surface-active biomolecules that are amphiphilic. The biosurfactants have several benefits as they can biodegrade the nonbiodegradable polymers like fossil-based polymers and non-biodegradable synthetic polymers and also, they can lower the toxicity as well as impart better environmental compatibility [84]. The reason for biosurfactant to facilitate the process of biodegradation is because of the presence of the specific functional groups which allow the biosurfactant to be active under high temperatures and in different pH as well as salinity conditions [85, 86].

Moisture

It can influence the biodegradation of plastic as microorganism growth and multiplication requires water and help in breaking the polymer bonds by the process of hydrolysis. So, the greater the amount of moisture present, the greater the chances of polymer biodegradation [38]. In moisture-rich conditions, the degree of polymer hydrolysis increases which leads to a greater number of chain scission. This makes the sites in the polymer chain more accessible to microbes to attack for faster biodegradation [2].

Temperature and pH

The pH value can also affect polymer biodegradation. At pH 5, the conditions are optimal for the hydrolysis of PLA capsules [38]. Due to the degradation of the polymer, several products come out which results in a change of pH value, affectingthe degradation process and microbial growth. Also, temperature affects the biodegradation of polymers, such as polymers with a high melting point have less affinity towards microbial biodegradation which may be attributed to the diminished enzymatic activity at elevated temperatures. But polymers having low melting points are easily biodegraded by the microbes. For example, a purified lipase from *R. delemar* hydrolyzed polyester (PCL) at a low melting temperature.

Enzymes

Enzymes have different active sites that aid in the degradation of different types of plastics. For example, diacid monomers generating straight-chain polyesters having a carbon chain length of six to twelve are degraded easily with the aid of the enzymes secreted by several fungal species [38].

Molecular Weight

The biodegradability is inversely proportional to the polymer's molecular mass. For instance, PCL with a molecular weight of greater than 4000, is easily degraded by the enzyme lipase isolated from the strain *Rhizopus delemar* [40]. Hence, microbial enzymes are more efficient in degrading lower molecular weight polymers [38]. Also, molecular composition (blend) leads to an increase in the biodegradation of plastics [87].

Shape and Size

Plastics with a larger surface area tend to biodegrade more quickly than plastics with a smaller surface area as the microbes have a larger area to adhere themselves to the plastics and degrade them. Polymers possessing side chains are more difficult to biodegrade than polymers without side chains [88].

Tacticity and Flexibility

Tacticity is the relative stereochemistry of the adjacent chiral centre within a macromolecule. This tacticity can affect polymer properties. The regularity of the polymeric structure influences the degree to which it is rigid, crystalline, or amorphous, which in turn becomes the deciding factor for the biodegradability of the polymer. Conformation flexibility is another factor that plays a role in the degradability of the polymer. Flexible polymers are easily accessible to the microbes and their enzymes, which enhances microbial biodegradation [2].

Blend

Blending is the mixing of one or more substances and as a result, the resultant substance differs in its properties from their individual ones. So, in a polymer blend two or more polymer mix with each other and generates a polymeric system with interesting properties [89]. It is a less cost-effective method to produce biodegradable polymers with desirable properties [40]. It is also a faster and easier method as compared to copolymerization.

ROLE OF DIFFERENT MICROBES IN PLASTIC DEGRADATION

Different types of microorganisms and groups of microorganisms are involved in the degradation of plastic as well as microplastic like bacteria, fungi, bacterial consortia, and biofilms. Although plastics are less susceptible to microbial attacks than other environmental pollutants, nevertheless, fragmented plastics provide an environment for microbial growth and serve as a carbon source for the microorganisms. In this section, the different type of microbes or consortia of microbes used in the degradation of plastic is discussed in detail.

Bacteria

Bacteria is one of the major categories of microbes that is used in environmental bioremediation and also in the biodegradation of plastics. Till date, a vast number of investigations have been carried out to study the role of bacteria in the degradation of plastics. These studies are majorly focussed on using pure bacterial cultures in the laboratory which are generally isolated from environmental dumping sites such as landfills, wastewater, sludge, *etc.* A large variety of bacterial strains aiding the degradation of various kinds of plastics has been isolated from these dumping sites which are enlisted in Table **1**. Auta *et al.* have shown that bacterial culture of *Rhodococcus sp.* and *Bacillus sp.* strain 36 and 27, respectively, degraded the plastic polypropylene by an amount of 6.4 and 4.0 percent, respectively [90]. In the same study, they found that pure bacterial cultures adhere to the plastic efficiently which leads to the degradation of the plastics. In another study, Auta and his co-workers found that *Bacillus cereus* led to the degradation of the plastics such as PS, PET and PE by 7.4, 6.6 and 1.6, respectively while the bacteria *Bacillus gottheilii* degraded the plastics, PS, PP, PET, and PE by 5.8, 3.6, 3.0 and 6.2, respectively [91]. In both the studies conducted by Auta and his co-workers, the surface of the plastic developed many irregularities, pores, and grooves. It has been found in many studies that along with the surface modifications, many bacteria species can modify or change the functional group on the polymer chain. The analysis of PE biodegraded by *Bacillus cereus* by FTIR spectroscopy revealed the disappearance of the carbonyl bands and the appearance of N-H and O-H bands at 3738 and 3419 cm^{-1}, respectively. Jeon *et al.* showed that bacteria *S. maltophilia* LB decreased the tensile strength and the molecular weight of PLA [92]. Uscateugui *et al.* carried out the degradation of polyurethanes.

Table 1. Bacterial strains involved in the polymer degradation. The table enlists the various bacteria extracted from different environmental sites and involved in the degradation of the different plastics.

S. No.	Name of the Bacteria	Types of Plastic Degrade	References
1	*Comamonas acidovorans* TB-35	Polyester polyurethane (PUR)	[93 - 95]
2	*Schlegelella thermodepolymerans* and *Pseudomonas indica* K2	Poly(3-hydroxybutyrate-co-3-mercaptopropionate), Poly(3-hydroxyalkanoate) (PHA	[96]
3	*Brevibacillus borstelensis*	Polyethylene	[97]
4	*Rhodococcus ruber* C208	Polyethylene	[98]
5	*Streptomyces sp.* SNG9	Poly(3-hydroxybutyrate) (PHB), Poly(3-hydroxybutyrate-co-3-hydroxyvalerate)	[99]
6	*Acidovorax sp.* DP5	Polyhydroxyalkanoates (PHA)	[100]
7	*Alcaligenes faecalis*	Poly(ε-caprolactone) (PCL) Film,	[101, 102]
8	*Pseudomonas stutzeri*	Polyethylene glycols (PEGs)	[103]
9	*Amycolatopsis sp.*	Polylactide (PLA), PE	[104]
10	*Schlegelella sp.* KB1a	Poly(3-hydroxybutyrate)	[105]
11	*Streptomyces sp.*	Polyethylene	[106]
12	*Corynebacterium sp.*	Polyester-polyurethane	[107]
13	*Enterobacter agglomerans*	Polyester-polyurethane	[107]
14	*Serratia rubidaea*	Polyester-polyurethane	[107]
15	*Pseudomonas aeruginosa*	Polyester-polyurethane	[107]
16	*Staphylococcus epidermidis* KH 11	Polyurethanes	[108]
17	*Acinetobacter johnsonii*	Poly-3-hydroxybutyrate	[109]
18	*Burkholderia cepacia*	Poly-3-hydroxybutyrate	[109]
19	*Comamonas testosteroni*	Poly-3-hydroxybutyrate	[109]
20	*Comamonas sp.*	Poly-3-hydroxybutyrate	[109, 110]
21	*Flavobacterium johnsoniae*	Poly-3-hydroxybutyrate	[109]
22	*Ochrobactrum anthropi*	Poly-3-hydroxybutyrate	[109]
23	*Pseudoalteromonas haloplanktis*	Poly-3-hydroxybutyrate	[109]
24	*Pseudomonas alcaligenes*	Poly-3-hydroxybutyrate	[109]

(Table 1) cont.....

S. No.	Name of the Bacteria	Types of Plastic Degrade	References
25	*Pseudomonas mendocina*	Poly-3-hydroxybutyrate	[109]
26	*Pseudomonas pseudoalcaligenes*	Poly-3-hydroxybutyrate	[109]
27	*Pseudomonas pseudomallei*	Poly-3-hydroxybutyrate	[109]
28	*Pseudomonas vesicularis*	Poly-3-hydroxybutyrate	[109]
29	*Ralstonia pickettii*	Poly-3-hydroxybutyrate	[109, 111]
30	*Stenotrophomonas maltophilia*	Poly-3-hydroxybutyrate	[109, 112]
31	*Variovorax paradoxus*	Poly-3-hydroxybutyrate	[109, 113]
32	*Vibrio ordalii*	Poly-3-hydroxybutyrate	[109]
33	*Zoogloea ramigera*	Poly-3-hydroxybutyrate	[109, 110]
34	*Arthrobacter ilicis*	Poly-3-hydroxybutyrate	[109]
35	*Bacillus circulans*	Poly-3-hydroxybutyrate	[109]
36	*Bacillus laterosporus*	Poly-3-hydroxybutyrate	[109]
37	*Bacillus megaterium*	Poly-3-hydroxybutyrate	[109]
38	*Bacillus sp.*	Poly-3-hydroxybutyrate	[109, 110]
39	*Clavibacter michiganensis* subsp.	Poly-3-hydroxybutyrate	[109]
40	*Staphylococcus aureus*	Poly-3-hydroxybutyrate	[109]
41	*Clostridium sp.*	Poly-3-hydroxybutyrate	[109]
42	*Paenibacillus polymyxa*	Poly-3-hydroxybutyrate	[109]
43	*Staphylococcus epidermidis*	Poly-3-hydroxybutyrate	[109]
44	*Bacillus sphericus*	LDPE	[114]
45	*Arthrobacter sp.* GMB5	HDPE	[115]
46	*Pseudomonas sp.* GMB7	HDPE	[115]
47	*Pseudomonas sp.* AKS2	LDPE	[116]
48	*Bacillus subtilis* H1584	LDPE	[117]
49	*Enterobacter asburiae* YT1 and *Bacillus* sp. YP1	PE	[118]
50	*Serratia marcescens* subsp. marcescens	LDPE	[119]
51	*Achromobacter xylosoxidans* PE-1	HDPE	[120]
52	*Alcanivorax borkumensis*	LDPE	[121]
53	*Xanthomonas sp.* and *Sphingobacterium sp.*	PS (Polystyrene)	[122]

(Table 1) cont.....

S. No.	Name of the Bacteria	Types of Plastic Degrade	References
54	*Exiguobacterium sp.* YT2	PS (Polystyrene)	[57]
55	*Brevibacillus sps.* and *Aneurinibacillus sp.*	PP, LDPE, HDPE	[123]
56	*Pseudomonas citronellolis* and *Bacillus flexus*	PVC	[124]
57	*Pseudomonas fluorescens* and *Bacillus subtilis*	Poly(ether urethanes)	[125]
58	*Bacillus sp.* strain 27 and *Rhodococcus sp.* strain 36	Polypropylene	[90]
59	*Chelatococcus sp.* E1	Low molecular-weight polyethylene (LMWPE)	[126]
60	*Pseudomonas sp.* MYK1 and *Bacillus sp.* MYK2	Polylactic acid (PLA)	[127]
61	*Ideonella sakaiensis* 201-F6	PET	[128]
62	*Stenotrophomonas maltophilia*	PLA	[92]
63	*Bacillus cereus*	PET, PE, PS	[91]
64	*Bacillus pumilus* VRKPC1	HDPE	[129]
65	*Thermobifida fusca* Cutinase	PET	[130]
66	*Pseudomonas aeruginosa* E7	PE	[131]
67	*Rhodococcus ruber* C208,	PE	[132]
68	*Alcaligenes faecalis* T1	Poly-3-hydroxybutyrate	[111]
69	*Pseudomonas lemoignei* ATCC17989	Poly-3-hydroxybutyrate	[111]
70	*Comamonas testosteroni* YM1004	Poly-3-hydroxybutyrate	[111]
71	*Pseudomonas stutzeri* YM1006	Poly-3-hydroxybutyrate	[111]
72	*Variovorax paradoxus* ATCC17718	Poly-3-hydroxybutyrate	[111]
73	*Acidovorax delafieldii* ATCC17505	Poly-3-hydroxybutyrate	[111]
74	*Bacillus megaterium* JFRL92	Poly-3-hydroxybutyrate	[111]
75	*Pseudomonas alcaligenes* YM1409	Poly-3-hydroxybutyrate	[111]
76	*Pseudomonas stutzeri* YMI414	Poly-3-hydroxybutyrate	[111]
77	*Pseudomonas vesicular* YM1423	Poly-3-hydroxybutyrate	[111]

(Table 1) cont.....

S. No.	Name of the Bacteria	Types of Plastic Degrade	References
78	*Comamonas acidovorans* YM1609	Poly-3-hydroxybutyrate	[111]
79	*Pseudomonas cepacia* YM1614	Poly-3-hydroxybutyrate	[111]
80	*Alcaligenes sp.* YM1623	Poly-3-hydroxybutyrate	[111]
81	*Alcaligenes paradoxus* DSM 66	PHB	[110]
82	*Bacillus megaterium* DSM 32	PHB	[110]
83	*Bacillus sp.* A6	PHB	[110]
84	*Bacillus sp.* K5	PHB	[110]
85	*Comamonas testosteroni* LMG 1393	PHB	[110]
86	*Comamonas testosteroni* LMG 1394	PHB	[110]
87	*Pseudomonas facilis* DSM 50181	PHB	[110]
88	*Pseudomonas facilis* DSM 649	PHB	[110]
89	*Pseudomonas delafieldii* DSM 50403	PHB	[110]
90	*Pseudomonas lemoignei* LMG 2207	PHB	[110]
91	*Pseudomonas sp.* DSM 1455	PHB	[110]
92	*Zoogloea ramigera* strain ATCC25935	PHB	[110]
93	*Kocuria palustris* M16	LDPE	[117]
94	*Bacillus pumilus* M27	LDPE	[117]
95	*Bacillus gottheilii*	PE, PET, PP, PS	[91]
96	*Pseudomonas aeruginosa* VRKPC5	HDPE	[129]
97	*Leucobacter sp.* VRKPC22	HDPE	[129]
98	*Pseudomonas aeruginosa* VRKPCH4	HDPE	[129]
99	*Bacillus licheniformis* VRKPCH23	HDPE	[129]
100	*Bacillus aryabhattai* VRKPV15	HDPE	[129]
101	*Bacillus cereus* VRKPK25	HDPE	[129]

(Table 1) cont.....

S. No.	Name of the Bacteria	Types of Plastic Degrade	References
102	*Bacillus subtilis* VRKPP1	HDPE	[129]
103	*Bacillus sp.* VRKPP17	HDPE	[129]
104	*Bacillus amyloliquefaciens* VRKPR13	HDPE	[129]
105	*Corynebacterium* and *Pseudomonas sp.*	Polyester Polyurethane	[107]

in the presence of *E. coli* and found that the degradation rate after three days is around 2 percent [133]. In a study done by Shimpi *et al.*, the biodegradability of PS-PLA by *P. aeruginosa* is around 10 percent [134]. The *Bacillus* strain isolated by Mohan *et al.* degraded the brominated high impact polystyrene by around 23 percent [135]. Apart from the environmental samples, bacterial strains are isolated by Yang *et al. i.e. Bacillus sp. YP1* and *Enterobacter asburiae YT1*, from waxworm, help in the degradation of PE [118]. Oda's group from Kyoto, Japan discovered a bacteria *Ideonella sakainesis* which leads to PET degradation and utilizes the polymer as its carbon source [128].

A single bacterium usually leads to the generation of toxins as end products which may inhibit the microbial growth and prevent the biodegradation of plastics [136]. Hence, using a bacterial consortium stabilizes the microbial community where the toxic metabolites produced by one microorganism are used by another for its growth and survival. Therefore, bacterial consortia act as a sustainable source for the biodegradation of plastic. But if the bacteria present in the consortia are symbiotic and have synergistic interaction with each other then the bacteria consortia will lead to tolerance against the process of bioremediation [137]. So, nowadays many researchers have focused their work on utilizing bacterial consortia for plastic biodegradation. Park et.al have shown that a mesophilic bacterial consortium of two bacteria species *Bacillus sp.* and *Paenibacillus sp.* reduces both the weight as well as the size of the PE [138]. The advent of new techniques and development in the field of molecular biology, *in vitro* transcription, *etc.*, has led to the opportunity to study different bacterial consortia for the bioremediation of plastics [139]. However, plastic degradation by the bacterial cluster is often a complicated process due to the multi-network reactions between the microbes and their enzymes.

In addition to the bacterial consortia, biofilms created by multiple bacteria on the surface of the plastic also contribute to the biodegradation of the plastics. Biofilms are a typical conglomeration of various types of microorganisms on the surface of the materials. Almost all types of microorganisms can form biofilms on various surfaces. Usually in a water environment, plastic and microplastics come in contact with different types of microorganisms like algae, protists, fungi, bacteria, and even viruses that adhere themselves to the plastic surface leading to the creation of the biofilm [140]. Biofilm is often a complex ecosystem that is primarily composed of microorganisms and their metabolites [141]. After the microorganisms attach themselves to the plastic surface and form the biofilm, it changes the structure of the polymer by modulating its surface properties, degrading the additives added to the polymer, and fragmenting the plastics into smaller polymeric chains [142]. The type of biofilm formed on the surface of the plastics often influences the fate of the microplastics formed by changing their physical property and enhancing their chances of degradation [143]. Lobelle *et al.* have revealed that early matured biofilm formation takes place on the surface of PE after three weeks. The formation of the initial biofilm took place in the first week which kept on increasing till the third week. The analysis of the biofilm revealed that the heterotrophic bacterial population on the plastic surface is around 4000 cells cm^{-2} in the first week which increased to 120000 cells cm^{-2} by the end of the third week. In the same study, it is also seen that with an increase in the bacteria population, the hydrophilicity of the plastic also increased significantly [144]. There are several studies in the literature that pertains to the mechanism involved in the biofilm-mediated degradation of plastic and microplastics. Generally, it is categorized into four phases. The first phase involves the attachment of the micro-organism onto the surface of the plastic and altering its surface properties such as hydrophilicity, adhesion *etc.* In the second phase the microorganism leaches the additives in the plastics and fragments them into smaller pieces with the aid of the enzymes. The third phase involves the attack on these fragmented plastics by microbial enzymes and radicals which leads to the loss in the mechanical stability of the plastic fragments. In the final phase, microbial filaments and water molecules penetrate the fragmented plastic, leading to the degradation and utilization of the fragmented plastic as a nutritional source by the microorganism [141]. But, the biofilm-mediated plastic degradation is a budding field and still many of its complex processes are still in dark and need to be systematically studied.

Although a large number of investigations have been done to explore the role of bacteria in the microbial degradation of plastic, still there is a vast array of questions that still need to be answered. The bacterial degradation of plastic is relatively long and hence optimizing and improving the bacterial strains would decrease the duration of the biodegradation process and can make it commercially viable.

Fungi

Fungi also have the ability to degrade plastics and microplastics [145]. A list of fungi utilized to degrade different kinds of plastics is shown in Table **2**. Bacteria are the pioneer invaders and play important role in the biofouling formation and they entrap the fungi onto their surface. The presence of fungi in the biofilm accelerates the biodegradation process of the plastic [84]. Fungi enhance the biodegradation process by decreasing the hydrophobicity and increasing the hydrophilicity on the surface of the plastic. Also, both bacteria and fungi in the biofilm share each other metabolites and accelerate their degradation process [84]. Fungi have the potential for the formation of various chemical bonds in the polymer-like carboxyl, carbonyl, and ester functional groups, and these groups reduce the hydrophobicity of plastics and increase the nature of hydrophilicity of the surface of the plastic. Fungi also for its growth uses plastic and microplastic as source of carbon. Yamada-Onodera and co-workers showed that *Penicillium simplicissimum* YK isolated for the biodegradation of polyethylene alters the morphology and properties of PE [56]. In another study, it is found that *Fusarium* and similar fungal species make the surface of the plastic more hydrophilic in nature by decreasing the hydrophobicity nature into hydrophilic nature after their attachment [33]. A single bacterium usually generates toxic metabolites which inhibit the microbial growth due to which the biodegradation of plastic is restricted by a single microorganism [136]. So, the use of a cluster of bacteria generally known as the bacterial consortium is used to produce a sustainable microbial community in which toxic metabolites produced by one microorganism can be used by another microorganism for their growth and causes no harm to the microbial growth. So, bacterial consortia positively contribute to the degradation of plastic. Another important point is that the symbiotic and synergistic interaction between the bacteria in the consortia increases the activity and tolerance of the microbial community towards the pollutants [137]. So, nowadays researchers focus on the bacterial consortia for the plastic as well as microplastic degradation. Several bacterial consortia studies carried out by different researchers have shown good results in the biodegradation of the plastic. For example, a study by Park *et al.*, in 2019 showed that, the degradation of polyethylene by a bacterial consortium of two bacteria, reduces the weight as well as the diameter of PE [138]. With the advent of modern molecular techniques in the field of bioscience, several bacterial consortia aiding the degradation of plastics have been investigated in various environmental conditions [139].

Table 2. Fungal strains involved in polymer degradation. The table enlists the various fungi extracted from different environmental sites and involved in the degradation of the different plastics.

S. No.	Name of Microorganism (Fungi/Fungal Strain)	Type of Plastics Degrade	References
1	*Penicillium simplicissimum* YK	Polyethylene	[56]
2	*Curvularia senegalensis*	Ester based Polyurethane	[146]
3	*Fusarium solani*	Poly(butylene succinate)	[147]
4	*Aspergillus niger*	Poly(vinyl alcohol) (PVA)	[148]
5	*Aspergillussp.* Strain S45	Polyester polyurethane	[149]
6	*Aspergillus niger* van Tieghem F-1119	Poly vinyl chloride	[150]
7	*Phanerochaete chrysosporium*	Polyethylene	[106]
8	*Fusarium*	PCL	[151]
9	*Aspergillus japonicus*	Polythene	[152]
10	*Aspergillus niger*	Polythene	[152]
11	*Phanerochaete chrysosporium* PV1	PVC	[153]
12	*Lentinus tigrinus* PV2	PVC	[153]
13	*Aspergillus niger* PV3	PVC	[153]
14	*Aspergillus sydowii* PV4	PVC	[153]
15	*Acremonium sp.*	Poly-3-hydroxybutyrate	[109]
16	*Alternaria sp.* PURDK2	Ether PUR (Ether type polyurethane)	[154]
17	*Cladosporium pseudocladosporioides* strain T1.PL.1	Polyurethane	[155]
18	*Xepiculopsis graminea*	Polyurethane	[156]
19	*Penicillium griseofulvum*	Polyurethane	[156]
20	*Leptosphaeria sp.*	Polyurethane	[156]
21	*Agaricus bisporus*	Polyurethane	[156]
22	*Marasmius oreades*	Polyurethane	[156]
23	*Bjerkandera adusta* MZKI G-84	Nylon-6	[157]
24	*Ulocladium sp.* MZKI B-1085	Nylon-6	[157]
25	*Aspergillus puniceus* MZKI A-518	Nylon-6	[157]
26	*Penicillium sp.* MZKI P-261	Nylon-6	[157]
27	*Mucor hiemalis* MZKI B-1078	Nylon-6	[157]
28	*Aureobasidium pullulans* MZKI B-802	Nylon-6	[157]
29	*Fusarium sp.* MZKI B-1083	Nylon-6	[157]
30	*Trichotecium roseum* MZKI B-1080	Nylon-6	[157]
31	*Cephalosporium sp*	poly-3-hydroxybutyrate	[109]

(Table 2) cont.....

S. No.	Name of Microorganism (Fungi/Fungal Strain)	Type of Plastics Degrade	References
32	*Cladosporium sp*	poly-3-hydroxybutyrate	[109]
33	*Eupenicillium sp.*	poly-3-hydroxybutyrate	[109]
34	*Gerronema postii*	poly-3-hydroxybutyrate	[109]
35	*Mucor sp.*	poly-3-hydroxybutyrate	[109]
36	*Penicillium adametzii*	poly-3-hydroxybutyrate	[109]
37	*Penicillium chermisinum*	poly-3-hydroxybutyrate	[109]
38	*Penicillium daleae*	poly-3-hydroxybutyrate	[109]
39	*Penicillium funiculosum*	poly-3-hydroxybutyrate	[109]
40	*Penicillium ochrochloron*	poly-3-hydroxybutyrate	[109]
41	*Penicillium restrictum*	poly-3-hydroxybutyrate	[109]
42	*Penicillium simplicissimum*	poly-3-hydroxybutyrate	[109]
43	*Polyposur circinatus*	poly-3-hydroxybutyrate	[109]
44	*Verticillium sp*	poly-3-hydroxybutyrate	[109]
45	*Verticillium leptobactrum*	poly-3-hydroxybutyrate	[109]
46	*Pestalotiopsis microspora*	polyester polyurethane	[158]
47	*Aspergillus tubingensis* VRKPT1	HDPE	[159]
48	*Aspergillus niger* and *Penicillium pinophilum*	Thermo-oxidized-LDPE	[160]
49	*Zalerion maritimum*	PE (Polyethylene)	[161]
50	*Candida rugosa*	Polyester PUR	[158]
51	*Engyodontium album* MTP091 (F2)	PP (Polypropylene	[77]
52	*Cladosporium cladosporioides*	polyurethane	[156]
53	*Pestalotiopsis microspore*	polyurethane	[156]
54	*Ulocladium sp.* MZKI B-1082	Nylon-6	[157]
55	*Phanerochaete chrysosporium* MZKI B-223	Nylon-6	[157]
56	*Phanerochaete chrysosporium* MZKI B-186	Nylon-6	[157]
57	*Aureobasidium pullulans* MZKI B-241	Nylon-6	[157]
58	*Aspergillus flavus* VRKPT2	HDPE	[159]

Biofilms are a collection of multiple types of microorganisms that can symbiotically grow on several surfaces. Many microorganisms such as fungi, bacteria, algae, *etc* form both homogenous or heterogeneous biofilm. Usually, in an aqueous environment, plastic and microplastics come in contact with different types of microorganisms that can attach to the surface of the plastic and microplastic [140]. So, this attachment produces a colony which is called biofilm and this biofilm has a complex ecosystem that is primarily composed of microorganisms, their secretions, primary and secondary metabolites, *etc* [141]. The surface of the plastic and microplastic serve as a substrate for the formation of biofilm. So, after the microbes attach themselves to the surface of the plastic, they change the structure and surface properties of plastics in various ways like masking, degrading plastic into microplastic and degrading the microplastics with enzymes, and releasing metabolites [142]. Biofilms influence the fate of MPs by modifying their physical properties and degrading them [143].

It is shown that *Aspergillus niger* and *Penicillium pinophilum* decreased the mass of thermo-oxidized LDPE by around 0.57 and 0.37 percent, respectively after a period of 31 months [160]. In another study, Russell *et al.* showed that *Pestaltiopsis microspore* degraded polyurethane with the aid of the enzyme serine hydrolase [158]. Devi *et al.* showed that *Aspergillus tubingensis VRKPT1 and Aspergillus flavus VRKPT2* decreased the weight of HDPE by 6.02 and 8.51 percent, respectively [159]. In another study, it is shown that *Trametes versicolor* and *Phanerochaete chrysosporium* degrades nylon-6,6 by the oxidative attack. Raaman *et al.* have shown that *A. niger and A. japonicus* degraded LDPE by 5.8 and 11.11 percent, respectively in one month [152]. A particular strain of *A. niger* (ITCC 6052) has been seen to decrease the weight of PE by 3.44 percent and reduce its tensile strength by 61 percent after one month [162]. *P. chyrosogenum and P. oxalicum* degraded both HDPE and LDPE by an amount of 55.60 and 34.35 percent after an incubation period of 90 days [163]. In the same study, they observed that the degradation of plastic was followed by a decrease in pH which may be due to the oxidation of the carbon double bonds to carboxylic acids by the fungal enzymes. *Penicillium sp.* in consortia with other microbes increases the degradation rate of the PE by decreasing the incubation period by half [164]. These studies show that fungal strains can be exploited to efficiently degrade many types of plastics but still a lot of work needs to be done to make the degradation process by fungi economically viable.

Mechanism of Degradation of Plastic and Microplastic by Microorganisms

Microorganisms such as fungi, bacteria, *etc.* are the heterotrophs, and the main motivation for them to degrade plastic is to get nutrients (energy and carbon) from these polymers [165]. Microorganisms or microbes can attack the polymer surface either directly or indirectly to assimilate the nutrients from the plastic [166]. Bacteria and fungi exist in a synergistic relationship on the polymer's surface which ultimately degrades the plastic. In the microbial degradation of plastic, TCA cycle in the organism plays a vital role in the energy generation from plastic [167, 168]. Microbial degradation of microplastic and plastic goes through multiple chemical reactions but there are a few differences in the degradation of different plastics. The major chemical reaction involved in breaking the polymer backbone is the oxidation-reduction reaction which involves electron transfer [169]. Many microorganisms can degrade the chemical bonds in the polymer through anaerobic respiration *i.e.* in absence of oxygen. So, in anaerobic respiration, sulphate, nitrate, iron, and manganese, and even CO_2 can accept the electrons from the biodegradable contaminants [169]. The by-products of anaerobic respiration typically depend on the electron acceptor and are mainly hydrogen sulfide, nitrogen, reduced metals, and methane. But oxidative degradation is the main mechanism for the degradation of plastics [169]. As discussed before, the biodegradation of plastics depends on the exposure conditions as well as the polymer properties (Fig. **2**). Usually, it is seen that microbial degradation increases in the presence of abiotic factors like UV radiations which leads to the photooxidation of the polymer backbone [97, 170]. Biotic factors often play an important role in the absence of the abiotic factor and vice versa. The first step in the biodegradation of the plastic is the fragmentation of the larger polymeric backbone. Abiotic factor plays an important role in degrading the longer polymeric chain into short polymers like oligomer, dimers, and monomers. Thereafter, these fragmented polymers are attacked by microbes. Hence, abiotic degradation is preceded by biotic degradation [67]. Then the biotic degradation changes the physicochemical properties of the plastic which leads to the complete depolymerization and mineralization of the plastics. The addition of limiting nutrients to the microbial culture is known as bio-stimulation and the addition of live microbes capable of degradation is known as bio-augmentation and is used to enhance the biodegradation process of the polymers [171]. Therefore, the whole biodegradation process involves four consecutive steps namely biodeterioration, bio-fragmentation, assimilation, and mineralization [172]. Biodeterioration involves a change in the physical, chemical, and mechanical properties of the polymer by biological agents and it depends on both the exposure conditions and polymer properties as discussed before [173]. Bio-fragmentation mainly entails the enzymatic or non-enzymatic breaking of polymer into simpler oligomers. The third step is assimilation which is the incorporation of the molecules by microorganisms and the last step is mineralization which can proceed through either aerobic and anaerobic conditions to produce different end products like CO_2, H_2O or CO_2, H_2O, CH_4, H_2S, respectively. These biomineralized products can be utilized by the microbes as their carbon source [65, 174]. A schematic representation of the microbial biodegradation is shown in Fig. (**3**).

During the biodegradation of both plastic and microplastic, the polymer is converted to monomer, after which these monomers are mineralized. Usually, microplastics and plastics are too large as compared to the pore size of the cellular membrane of the microorganisms. Hence, to cross through the cellular membrane, they are first converted into monomers, dimers and oligomers. Thereafter, they are absorbed and biodegraded within the cells and that used by microbes as carbon and energy sources [175]. Plastics are usually depolymerized into monomers by microbial enzymes as well as microbes themselves [83]. During the depolymerization of plastic, enzymatic hydrolysis plays an important role in which the enzyme binds to the plastic and causes hydrolytic cleavage. Also, oxidative degradation plays another important role in the degradation of plastic and it can degrade both hydrolyzable and nonhydrolyzable plastics/polymers [83]. In the degradation of plastic two significant mechanisms play an important role called intracellular and extracellular degradation [83, 176]. Extracellular degradation involves the secretion of extracellular enzymes or exoenzyme such as hydrolases which degrade complex polymers into simpler molecules and this mainly happens in the periplasmic space or the cell membrane [33, 83, 177]. Consequently these short sized oligomers- monomer and dimers are transported to the cytoplasmic membrane [33]. Then intracellular degradation caused by the use of intracellular enzyme is exploited to generate carbon as an energy source [177]. Biosurfactants are also helpful in directly internalizing the oligomers and after internalizing the different processes like beta-oxidation and TCA cycle help in mineralizing these smaller oligomers. TCA cycle plays an important role in central carbon metabolism. The biodegradation of styrene generates molecules like acetaldehyde, pyruvate, 2-vinylmuconate, and 2-phenyl ethanol [178]. These molecules are further metabolized to phenyl-acetyl-CoA which enters the central carbon metabolism or tricarboxylic acid (TCA) cycle [83]. So, both intracellular and extracellular depolymerases are required in microorganisms [83]. Intracellular degradation mainly utilizes the endoenzyme to hydrolyze the endogenous carbon whereas extracellular degradation utilizes extracellular enzymes to degrade the exogenous carbon source since extracellular enzymes of microorganisms break down plastic into short chains of oligomers, dimers, and monomers [83]. Another important thing is that when a polymer is made up of two or more monomers then the degradation becomes more difficult. So, in this case a new engineered pathway is developed by the microbe to degrade co-polymers. It has been reported that during PET degradation, *E. coli* BL21 synthesizes the enzyme LC-Cutinase which generates terephthalate and ethylene glycol as products of hydrolysis reaction [179]. Thereafter, ethylene glycol is mineralized to carbon dioxide and water by *E. coli* BL21. In some instances, the genetically engineered strain of *E. coli i.e.* BL21 can synthesize LC-Cutinase which can hydrolyze PET to terephthalate and ethylene glycol as two principal monomers [179]. Then, the ethylene glycol is mineralized by *E. coli* BL21 to CO_2 and H_2O. However, to enhance the process of biodegradation, bacterial strains can be genetically modified and inserted with other genes to carry out plastic degradation in a different pathway [179]. These genetically engineered bacterial strains harbours all the necessary enzymes required for the biodegradation of the plastic [180].

Also, there is possibility that the polymer can disintegrate into monomers outside the microbial cell, or the engineered microbial strains possessing transporters can transport the polymer inside the cell and degrade [179]. Hence, the plastic degrading microbial species often harbours a complete set of enzymes required for biodegradation of platsics [180]. Also, now-a-days with the aid of genetic engineering and systems biology, new strains of microbes with better degradability of the plastics have been developed which can overcome the limitations associated with the conventional methods of plastic disposal [181].

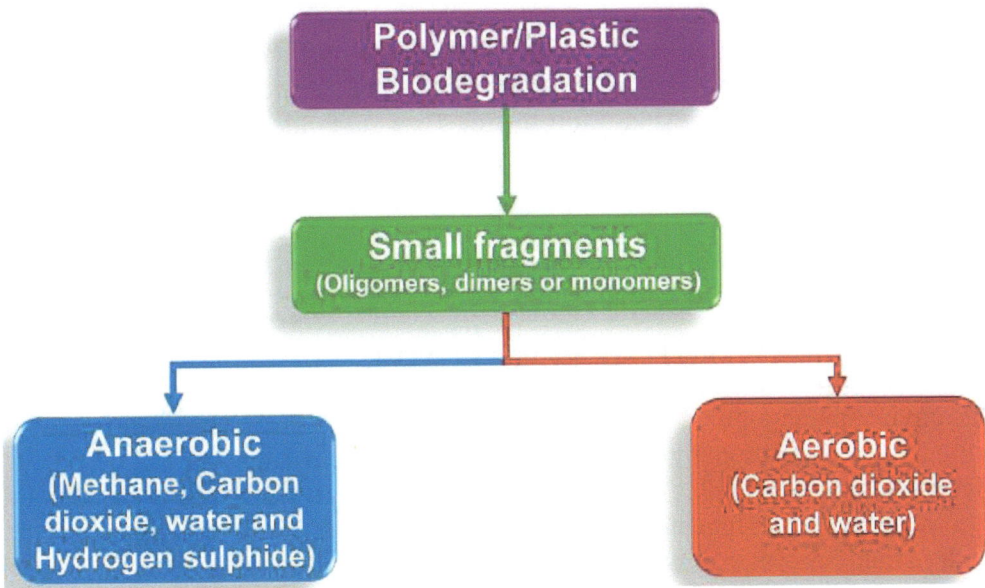

Fig. (3). General mechanism of biodegradation of plastics. The above figure shows a very simple and general mechanism adopted by the microbes to degrade plastics.

These short-chain monomer molecules are then mineralized into an end product like CO_2, H_2O, CH_4 *etc*, and this process is called mineralization. Another important point about the microorganism is that to survive in a harsh or extreme environment, microbes form a protective layer to protect themselves from toxic substances and gather nutrients by producing extracellular polymers [143, 182].

Different Enzymes in Microbial Degradation of Plastic

Enzymes are present in every living cell and play a vital role in the survival of the microbes. Enzymes' action is very specific toward their substrates and the relative amount of enzymes increases or decreases in accordance with the need of the microbes. It is found that laccase helps in the oxidation of PE and reduces the molecular weight of PE by around 20 percent [132].This enzyme is mostly present in the fungi which are used to degrade lignin. There are many other enzymes such as cutinases, lipase, and hydrolase from many different microorganisms which play a very important role in the degradation of plastics. The microbe *I. sakaiensis* in the process of PET degradation generates two enzymes PETase and MHETase which helps in the degradation of the PET. PETase hydrolyses the polymer into MHET [mono (2-hydroxyethyl) terephthalic acid] and BHET [(2-hydroxyethyl) terephthalic acid]. Further, MHETase hydrolyses MHET and BHET into TPA [Terepthalic acid] and EG (Ethylene glycol) [128]. TPA after undergoing a series of the enzymatic reaction is converted into pyruvate and oxaloacetate and EG is assimilated through glyoxylate. Similarly, *Thermobifida fusca* also helps in degrading PET and contains a two enzyme system *i.e.* polyester hydrolase and carboxyl esterase [130]. In addition to this *Pseudomonas spp.* can also degrade LDPE by utilizing a different set of enzymes [53]. Although a few enzymes have been characterized by the microbes still a lot of work needs to be done to understand the structure and function of these plastic degrading enzymes which will allow us to employ protein engineering to these enzymes and make them more efficient in degrading plastics.

CONCLUSION AND FUTURE PERSPECTIVE

The environmental pollution caused due to plastic is alarming and it needs immediate intervention by quality research. The main cause of concern is the microplastics which are lesser than 5mm in size and can easily find their way into the food chain when ingested by small marine or terrestrial organisms. This may result in many health issues. Therefore, the detection of microplastics and filtering them out of the environment should be one of the major areas which can be answered with the aid of biotechnological intervention. Thereafter, in the arena of plastic degradation by microbes, genetically engineered microbes, and engineered enzymes can be developed to increase the efficiency and lower the duration of microbial degradation of plastics. Thirdly, more emphasis should be given to developing commercially viable biodegradable polymers and every country should opt for a circular polymer economy so that plastic waste can be reduced drastically, and recycling can be promoted. Although, to date, a lot of work has been done in the characterisation of plastic degrading microbes from various places, still a deeper understanding of the microbes and their metabolic cycles and enzymes used for plastic degradation will open new avenues and strategies for faster and efficient degradation of the plastics.

CONSENT FOR PUBLICATION

Not applicable.

CONFLICT OF INTEREST

The authors declare no conflict of interest, financial or otherwise.

ACKNOWLEDGEMENT

Declared none.

REFERENCES

[1]   Hosler D, Burkett SL, Tarkanian MJ. Prehistoric polymers: rubber processing in ancient mesoamerica. Science 1999; 284(5422): 1988-91.
        [http://dx.doi.org/10.1126/science.284.5422.1988] [PMID: 10373117]

[2]   Devi RS, Kannan VR, Natarajan K, *et al.* The role of microbes in plastic degradation. In: Chandra R (Ed.) Environmental Waste Management. CRC Press: Boca Raton 2015; pp. 341-70.

[3]   Clukey KE, Lepczyk CA, Balazs GH, *et al.* Persistent organic pollutants in fat of three species of Pacific pelagic longline caught sea turtles: Accumulation in relation to ingested plastic marine debris. Sci Total Environ 2018; 610-611: 402-11.
        [http://dx.doi.org/10.1016/j.scitotenv.2017.07.242] [PMID: 28806556]

[4]   Ritchie H, Roser M. Plastic Pollution 2018. Published online at OurWorldInData.org.

[5]   Nizzetto L, Futter M, Langaas S. Are agricultural soils dumps for microplastics of urban origin? Environ Sci Technol 2016; 50(20): 10777-9.
        [http://dx.doi.org/10.1021/acs.est.6b04140]

[6]   Rios LM, Moore C, Jones PR. Persistent organic pollutants carried by synthetic polymers in the ocean environment. Mar Pollut Bull 2007; 54(8): 1230-7.
        [http://dx.doi.org/10.1016/j.marpolbul.2007.03.022] [PMID: 17532349]

[7]   Hauenstein O, Agarwal S, Greiner A. Bio-based polycarbonate as synthetic toolbox. Nat Commun 2016; 7(1): 11862.
        [http://dx.doi.org/10.1038/ncomms11862] [PMID: 27302694]

[8]   Kalita NK, Kalamdhad A, Katiyar V. Recent Trends and Advances in the Biodegradation of Conventional Plastics.Advances in Sustainable Polymers. Springer 2020; pp. 389-404.
        [http://dx.doi.org/10.1007/978-981-15-1251-3_17]

[9]   Dhar P, Bhasney SM, Kumar A, Katiyar V. Acid functionalized cellulose nanocrystals and its effect on mechanical, thermal, crystallization and surfaces properties of poly (lactic acid) bionanocomposites films: A comprehensive study. Polymer (Guildf) 2016; 101: 75-92.
        [http://dx.doi.org/10.1016/j.polymer.2016.08.028]

[10]   Valapa R, G P, Katiyar V. Hydrolytic degradation behaviour of sucrose palmitate reinforced poly(lactic acid) nanocomposites. Int J Biol Macromol 2016; 89: 70-80.
        [http://dx.doi.org/10.1016/j.ijbiomac.2016.04.040] [PMID: 27095433]

[11]   Gupta A, Pal AK, Woo EM, Katiyar V. Effects of Amphiphilic Chitosan on Stereocomplexation and Properties of Poly(lactic acid) Nano-biocomposite. Sci Rep 2018; 8(1): 4351.
        [http://dx.doi.org/10.1038/s41598-018-22281-1] [PMID: 29531341]

[12]   Pal AK, Katiyar V. Nanoamphiphilic Chitosan Dispersed Poly(lactic acid) Bionanocomposite Films with Improved Thermal, Mechanical, and Gas Barrier Properties. Biomacromolecules 2016; 17(8): 2603-18.
        [http://dx.doi.org/10.1021/acs.biomac.6b00619] [PMID: 27332934]

[13]   Pradhan R, Misra M, Erickson L, Mohanty A. Compostability and biodegradation study of PLA–wheat straw and PLA–soy straw based green composites in simulated composting bioreactor. Bioresour Technol 2010; 101(21): 8489-91.

[http://dx.doi.org/10.1016/j.biortech.2010.06.053] [PMID: 20594827]

[14]   Wei R, Zimmermann W. Microbial enzymes for the recycling of recalcitrant petroleum☐based plastics: how far are we? Microb Biotechnol 2017; 10(6): 1308-22.
[http://dx.doi.org/10.1111/1751-7915.12710] [PMID: 28371373]

[15]   Restrepo-Flórez JM, Wood JA, Rehmann L, Thompson M, Bassi A. Effect of biodiesel on biofilm biodeterioration of linear low density polyethylene in a simulated fuel storage tank. J Energy Resour Technol 2015; 137(3): 032211.
[http://dx.doi.org/10.1115/1.4030107]

[16]   Influence of cell surface hydrophobicity in colonization and biofilm formation on LDPE biodegradation. Int J Pharm Pharm Sci 2013; 5(4): 690-4.

[17]   Uchida H, Nakajima-Kambe T, Shigeno-Akutsu Y, Nomura N, Tokiwa Y, Nakahara T. Properties of a bacterium which degrades solid poly(tetramethylene succinate)-co-adipate, a biodegradable plastic. FEMS Microbiol Lett 2000; 189(1): 25-9.
[http://dx.doi.org/10.1111/j.1574-6968.2000.tb09201.x] [PMID: 10913861]

[18]   Pathak VM, Navneet . Review on the current status of polymer degradation: a microbial approach. Bioresour Bioprocess 2017; 4(1): 15.
[http://dx.doi.org/10.1186/s40643-017-0145-9]

[19]   Beier S, Bertilsson S. Bacterial chitin degradation—mechanisms and ecophysiological strategies. Front Microbiol 2013; 4: 149.
[http://dx.doi.org/10.3389/fmicb.2013.00149] [PMID: 23785358]

[20]   Pérez J, Muñoz-Dorado J, de la Rubia T, Martínez J. Biodegradation and biological treatments of cellulose, hemicellulose and lignin: an overview. Int Microbiol 2002; 5(2): 53-63.
[http://dx.doi.org/10.1007/s10123-002-0062-3] [PMID: 12180781]

[21]   Gooday GW. The ecology of chitin degradation.Advances in microbial ecology. Springer 1990; pp. 387-430.
[http://dx.doi.org/10.1007/978-1-4684-7612-5_10]

[22]   Shimao M. Biodegradation of plastics. Curr Opin Biotechnol 2001; 12(3): 242-7.
[http://dx.doi.org/10.1016/S0958-1669(00)00206-8] [PMID: 11404101]

[23]   Philip S, Keshavarz T, Roy I. Polyhydroxyalkanoates: biodegradable polymers with a range of applications. J Chem Tech Biotech 2007; 82(3): 233-47.

[24]   Flieger M, Kantorová M, Prell A, Řezanka T, Votruba J. Biodegradable plastics from renewable sources. Folia Microbiol (Praha) 2003; 48(1): 27-44.
[http://dx.doi.org/10.1007/BF02931273] [PMID: 12744074]

[25]   Chen GQ, Patel MK. Plastics derived from biological sources: present and future: a technical and environmental review. Chem Rev 2012; 112(4): 2082-99.
[http://dx.doi.org/10.1021/cr200162d] [PMID: 22188473]

[26]   Boyandin AN, Prudnikova SV, Karpov VA, *et al.* Microbial degradation of polyhydroxyalkanoates in tropical soils. Int Biodeterior Biodegradation 2013; 83: 77-84.
[http://dx.doi.org/10.1016/j.ibiod.2013.04.014]

[27]   Liu L, Li S, Garreau H, Vert M. Selective enzymatic degradations of poly(L-lactide) and poly(ε-caprolactone) blend films. Biomacromolecules 2000; 1(3): 350-9.
[http://dx.doi.org/10.1021/bm000046k] [PMID: 11710123]

[28]   Ikada Y, Tsuji H. Biodegradable polyesters for medical and ecological applications. Macromol Rapid Commun 2000; 21(3): 117-32.
[http://dx.doi.org/10.1002/(SICI)1521-3927(20000201)21:3<117::AID-MARC117>3.0.CO;2-X]

[29]   Fukushima K, Abbate C, Tabuani D, Gennari M, Camino G. Biodegradation of poly(lactic acid) and its nanocomposites. Polym Degrad Stabil 2009; 94(10): 1646-55.

[http://dx.doi.org/10.1016/j.polymdegradstab.2009.07.001]

[30] Anderson JM, Shive MS. Biodegradation and biocompatibility of PLA and PLGA microspheres. Adv Drug Deliv Rev 2012; 64: 72-82.
[http://dx.doi.org/10.1016/j.addr.2012.09.004]

[31] Vert M, Santos ID, Ponsart S, *et al.* Degradable polymers in a living environment: where do you end up? Polym Int 2002; 51(10): 840-4.
[http://dx.doi.org/10.1002/pi.903]

[32] Hoshino A, Tsuji M, Ito M, *et al.* Study of the aerobic biodegradability of plastic materials under controlled compost.Biodegradable polymers and plastics. Springer 2003; pp. 47-54.
[http://dx.doi.org/10.1007/978-1-4419-9240-6_3]

[33] Shah AA, Hasan F, Hameed A, Ahmed S. Biological degradation of plastics: A comprehensive review. Biotechnol Adv 2008; 26(3): 246-65.
[http://dx.doi.org/10.1016/j.biotechadv.2007.12.005] [PMID: 18337047]

[34] Mir S, Asghar B, Khan AK, *et al.* The effects of nanoclay on thermal, mechanical and rheological properties of LLDPE/chitosan blend. Journal of Polymer Engineering 2017; 37(2): 143-9.
[http://dx.doi.org/10.1515/polyeng-2015-0350]

[35] Chen G-Q. Introduction of Bacterial Plastics PHA, PLA, PBS, PE, PTT, and PPP. In: Chen, GQ (Ed.) Plastics from Bacteria. Microbiology Monographs, vol 14. Springer, Berlin, Heidelberg 2010; 1-16.
[http://dx.doi.org/10.1007/978-3-642-03287-5_1]

[36] Vijaya C, Reddy RM. Impact of soil composting using municipal solid waste on biodegradation of plastics. Ind J Biotech 2008; 7: 235-9.

[37] Hoang KC, Tseng M, Shu WJ. Degradation of polyethylene succinate (PES) by a new thermophilic Microbispora strain. Biodegradation 2007; 18(3): 333-42.
[http://dx.doi.org/10.1007/s10532-006-9067-5] [PMID: 17109189]

[38] Ahmed T, Shahid M, Azeem F, *et al.* Biodegradation of plastics: current scenario and future prospects for environmental safety. Environ Sci Pollut Res Int 2018; 25(8): 7287-98.
[http://dx.doi.org/10.1007/s11356-018-1234-9] [PMID: 29332271]

[39] Tribedi P, Sil AK. Cell surface hydrophobicity: a key component in the degradation of polyethylene succinate by *Pseudomonas* sp. AKS2. J Appl Microbiol 2014; 116(2): 295-303.
[http://dx.doi.org/10.1111/jam.12375] [PMID: 24165295]

[40] Tokiwa Y, Calabia B, Ugwu C, Aiba S. Biodegradability of Plastics. Int J Mol Sci 2009; 10(9): 3722-42.
[http://dx.doi.org/10.3390/ijms10093722] [PMID: 19865515]

[41] Wu CS. A comparison of the structure, thermal properties, and biodegradability of polycaprolactone/chitosan and acrylic acid grafted polycaprolactone/chitosan. Polymer (Guildf) 2005; 46(1): 147-55.
[http://dx.doi.org/10.1016/j.polymer.2004.11.013]

[42] Karakus K, Mengeloglu F. Polycaprolactone (PCL) based polymer composites filled wheat straw flour. Kastamonu Univ J Forest Facul 2016; 16(1): 264-8.
[http://dx.doi.org/10.17475/kujff.03251]

[43] Sanchez JG, Tsuchii A, Tokiwa Y. Degradation of polycaprolactone at 50° C by a thermotolerant *Aspergillus* sp. Biotechnol Lett 2000; 22(10): 849-53.
[http://dx.doi.org/10.1023/A:1005603112688]

[44] Jayasekara R, Harding I, Bowater I, Lonergan G. Biodegradability of a selected range of polymers and polymer blends and standard methods for assessment of biodegradation. J Polym Environ 2005; 13(3): 231-51.
[http://dx.doi.org/10.1007/s10924-005-4758-2]

[45]   Garg S, Jana AK. Studies on the properties and characteristics of starch–LDPE blend films using cross-linked, glycerol modified, cross-linked and glycerol modified starch. Eur Polym J 2007; 43(9): 3976-87.
[http://dx.doi.org/10.1016/j.eurpolymj.2007.06.030]

[46]   Ratto JA, Stenhouse PJ, Auerbach M, Mitchell J, Farrell R. Processing, performance and biodegradability of a thermoplastic aliphatic polyester/starch system. Polymer (Guildf) 1999; 40(24): 6777-88.
[http://dx.doi.org/10.1016/S0032-3861(99)00014-2]

[47]   Mani R, Bhattacharya M. Properties of injection moulded blends of starch and modified biodegradable polyesters. Eur Polym J 2001; 37(3): 515-26.
[http://dx.doi.org/10.1016/S0014-3057(00)00155-5]

[48]   Tokiwa Y, Calabia BP. Biodegradability and biodegradation of polyesters. J Polym Environ 2007; 15(4): 259-67.
[http://dx.doi.org/10.1007/s10924-007-0066-3]

[49]   Jayasekara R, Harding I, Bowater I, Christie GBY, Lonergan GT. Biodegradation by composting of surface modified starch and PVA blended films. J Polym Environ 2003; 11(2): 49-56.
[http://dx.doi.org/10.1023/A:1024219821633]

[50]   Webb H, Arnott J, Crawford R, Ivanova E. Plastic degradation and its environmental implications with special reference to poly (ethylene terephthalate). Polymers (Basel) 2012; 5(1): 1-18.
[http://dx.doi.org/10.3390/polym5010001]

[51]   Gubbels E, Heitz T, Yamamoto M, *et al.* Polyesters. In: Ullmann F (Ed.) Ullmann's Encyclopedia of Industrial Chemistry. Wiley: New York 6th Edition 2000; pp. 1-30.

[52]   Sharon C, Sharon M. Studies on biodegradation of polyethylene terephthalate: A synthetic polymer. J Microbiol Biotechnol Res 2012; 2(2): 248-57.

[53]   Wei R, Zimmermann W. Microbial enzymes for the recycling of recalcitrant petroleum□based plastics: how far are we? Microb Biotechnol 2017; 10(6): 1308-22.
[http://dx.doi.org/10.1111/1751-7915.12710] [PMID: 28371373]

[54]   Danso D, Chow J, Streit WR. Plastics: microbial degradation, environmental and biotechnological perspectives. Appl Environ Microbiol 2019; 85(19).

[55]   Ghosh SK, Pal S, Ray S. Study of microbes having potentiality for biodegradation of plastics. Environ Sci Pollut Res Int 2013; 20(7): 4339-55.
[http://dx.doi.org/10.1007/s11356-013-1706-x] [PMID: 23613206]

[56]   Yamada-Onodera K, Mukumoto H, Katsuyaya Y, Saiganji A, Tani Y. Degradation of polyethylene by a fungus, *Penicillium simplicissimum* YK. Polym Degrad Stabil 2001; 72(2): 323-7.
[http://dx.doi.org/10.1016/S0141-3910(01)00027-1]

[57]   Yang Y, Yang J, Wu WM, *et al.* Biodegradation and mineralization of polystyrene by plastic-eating mealworms: part 2. Role of gut microorganisms. Environ Sci Technol 2015; 49(20): 12087-93.
[http://dx.doi.org/10.1021/acs.est.5b02663] [PMID: 26390390]

[58]   Allsopp MW, Vianello G. Poly(vinyl chloride). Ullmann's Encyclopedia of Industrial Chemistry, 2000.
[http://dx.doi.org/10.1002/14356007.a21_717]

[59]   Patil R, Bagde U. Isolation of polyvinyl chloride degrading bacterial strains from environmental samples using enrichment culture technique. Afr J Biotechnol 2012; 11(31): 7947-56.

[60]   y ZKL, Keskin N, Güner A. Biodegradation of polyvinylchloride (PVC) by white rot fungi. Bull Environ Contam Toxicol 1999; 63(3): 335-42.
[http://dx.doi.org/10.1007/s001289900985] [PMID: 10475911]

[61]   Arutchelvi J, Sudhakar M, Arkatkar A, Doble M, Bhaduri S, Uppara P V. Biodegradation of polyethylene and polypropylene. Ind J Biotech 2008; 7(1): 9-22.

[62]   Park E, George E, Muldoon M, Flammino M. Thermoplastic Starch Blends With Poly (Vinyl Alcohol): Processability, Physical Properties And Biodegradability. Polymer News 1994; 19(8): 230-8.

[63]   Wagener S, Schink B. Fermentative degradation of nonionic surfactants and polyethylene glycol by enrichment cultures and by pure cultures of homoacetogenic and propionate-forming bacteria. Appl Environ Microbiol 1988; 54(2): 561-5.
[http://dx.doi.org/10.1128/aem.54.2.561-565.1988] [PMID: 3355141]

[64]   Wackett LP, Hershberger CD. Biocatalysis and biodegradation: microbial transformation of organic compounds. ASM Press, Washington, D.C., 2001.

[65]   Gu JD. Microbiological deterioration and degradation of synthetic polymeric materials: recent research advances. Int Biodeterior Biodegradation 2003; 52(2): 69-91.
[http://dx.doi.org/10.1016/S0964-8305(02)00177-4]

[66]   Gradin P, Howgate PG, Seldén R, Brown R. "Dynamic--Mechanical Properties," *Pergamon Press plc, Comprehensive Polymer Science: the Synthesis, Characterization*. Reactions & Applications of Polymers 1989; 2: 533-69.

[67]   Gewert B, Plassmann MM, MacLeod M. Pathways for degradation of plastic polymers floating in the marine environment. Environ Sci Process Impacts 2015; 17(9): 1513-21.
[http://dx.doi.org/10.1039/C5EM00207A] [PMID: 26216708]

[68]   Kamini NR, Iefuji H. Lipase catalyzed methanolysis of vegetable oils in aqueous medium by Cryptococcus spp. S-2. Process Biochem 2001; 37(4): 405-10.
[http://dx.doi.org/10.1016/S0032-9592(01)00220-5]

[69]   Krueger MC, Harms H, Schlosser D. Prospects for microbiological solutions to environmental pollution with plastics. Appl Microbiol Biotechnol 2015; 99(21): 8857-74.
[http://dx.doi.org/10.1007/s00253-015-6879-4] [PMID: 26318446]

[70]   Fotopoulou KN, Karapanagioti HK. Degradation of various plastics in the environment.Hazardous Chemicals Associated with Plastics in the Marine Environment. Springer 2017; pp. 71-92.
[http://dx.doi.org/10.1007/698_2017_11]

[71]   Arkatkar A, Juwarkar AA, Bhaduri S, Uppara PV, Doble M. Growth of *Pseudomonas* and *Bacillus* biofilms on pretreated polypropylene surface. Int Biodeterior Biodegradation 2010; 64(6): 530-6.
[http://dx.doi.org/10.1016/j.ibiod.2010.06.002]

[72]   Cooper DA, Corcoran PL. Effects of mechanical and chemical processes on the degradation of plastic beach debris on the island of Kauai, Hawaii. Mar Pollut Bull 2010; 60(5): 650-4.
[http://dx.doi.org/10.1016/j.marpolbul.2009.12.026] [PMID: 20106491]

[73]   Albertsson AC, Andersson SO, Karlsson S. The mechanism of biodegradation of polyethylene. Polym Degrad Stabil 1987; 18(1): 73-87.
[http://dx.doi.org/10.1016/0141-3910(87)90084-X]

[74]   David C, Trojan M, Daro A, Demarteau W. Photodegradation of polyethylene: comparison of various photoinitiators in natural weathering conditions. Polym Degrad Stabil 1992; 37(3): 233-45.
[http://dx.doi.org/10.1016/0141-3910(92)90165-2]

[75]   Hatti-Kaul R, Nilsson LJ, Zhang B, Rehnberg N, Lundmark S. Designing biobased recyclable polymers for plastics. Trends Biotechnol 2020; 38(1): 50-67.
[http://dx.doi.org/10.1016/j.tibtech.2019.04.011] [PMID: 31151764]

[76]   Jenkins S, Quer AMi, Fonseca C, Varrone C. Microbial Degradation of Plastics: New Plastic Degraders, Mixed Cultures and Engineering Strategies. Soil Microenvironment for Bioremediation and Polymer Production 2019; pp. 213-38.

[77]   Jeyakumar D, Chirsteen J, Doble M. Synergistic effects of pretreatment and blending on fungi mediated biodegradation of polypropylenes. Bioresour Technol 2013; 148: 78-85.

[http://dx.doi.org/10.1016/j.biortech.2013.08.074] [PMID: 24045194]

[78]   Restrepo-Flórez JM, Bassi A, Thompson MR. Microbial degradation and deterioration of polyethylene – A review. Int Biodeterior Biodegradation 2014; 88: 83-90.
[http://dx.doi.org/10.1016/j.ibiod.2013.12.014]

[79]   Ghosh S, Qureshi A, Purohit HJ. Microbial degradation of plastics: Biofilms and degradation pathways. Contaminants in Agriculture and Environment: Health Risks and Remediation 2019; 1: 184-99.
[http://dx.doi.org/10.26832/AESA-2019-CAE-0153-014]

[80]   Jaiswal S, Sharma B, Shukla P. Integrated approaches in microbial degradation of plastics. Environmental Technology & Innovation 2020; 17: 100567.
[http://dx.doi.org/10.1016/j.eti.2019.100567]

[81]   Haben Fesseha M. Degradation of Plastic Materials Using Microorganisms: A review. Public Health Open J 2019; 4(2): 57-63.
[http://dx.doi.org/10.17140/PHOJ-4-136]

[82]   Tiwari N, Santhiya D, Sharma JG. Microbial remediation of micro-nano plastics: Current knowledge and future trends. Environ Pollut 2020; 265(Pt A)115044
[http://dx.doi.org/10.1016/j.envpol.2020.115044] [PMID: 32806397]

[83]   Yuan J, Ma J, Sun Y, Zhou T, Zhao Y, Yu F. Microbial degradation and other environmental aspects of microplastics/plastics. Sci Total Environ 2020; 715: 136968.
[http://dx.doi.org/10.1016/j.scitotenv.2020.136968]

[84]   Orr IG, Hadar Y, Sivan A. Colonization, biofilm formation and biodegradation of polyethylene by a strain of *Rhodococcus ruber*. Appl Microbiol Biotechnol 2004; 65(1): 97-104.
[PMID: 15221232]

[85]   Kawai F, Watanabe M, Shibata M, Yokoyama S, Sudate Y. Experimental analysis and numerical simulation for biodegradability of polyethylene. Polym Degrad Stabil 2002; 76(1): 129-35.
[http://dx.doi.org/10.1016/S0141-3910(02)00006-X]

[86]   Kawai F, Watanabe M, Shibata M, Yokoyama S, Sudate Y, Hayashi S. Comparative study on biodegradability of polyethylene wax by bacteria and fungi. Polym Degrad Stabil 2004; 86(1): 105-14.
[http://dx.doi.org/10.1016/j.polymdegradstab.2004.03.015]

[87]   Alshehrei F. Biodegradation of synthetic and natural plastic by microorganisms. J Appl Environ Microbiol 2017; 5(1): 8-19.

[88]   Trivedi P, Hasan A, Akhtar S, Siddiqui MH, Sayeed U, Khan MKA. Role of microbes in degradation of synthetic plastics and manufacture of bioplastics. J Chem Pharm Res 2016; 8(3): 211-6.

[89]   La Mantia FP, Morreale M, Botta L, Mistretta MC, Ceraulo M, Scaffaro R. Degradation of polymer blends: A brief review. Polym Degrad Stabil 2017; 145: 79-92.
[http://dx.doi.org/10.1016/j.polymdegradstab.2017.07.011]

[90]   Auta HS, Emenike CU, Jayanthi B, Fauziah SH. Growth kinetics and biodeterioration of polypropylene microplastics by *Bacillus* sp. and *Rhodococcus* sp. isolated from mangrove sediment. Mar Pollut Bull 2018; 127: 15-21.
[http://dx.doi.org/10.1016/j.marpolbul.2017.11.036] [PMID: 29475646]

[91]   Auta HS, Emenike CU, Fauziah SH. Screening of *Bacillus* strains isolated from mangrove ecosystems in Peninsular Malaysia for microplastic degradation. Environ Pollut 2017; 231(Pt 2): 1552-9.
[http://dx.doi.org/10.1016/j.envpol.2017.09.043] [PMID: 28964604]

[92]   Jeon HJ, Kim MN. Biodegradation of poly(l-lactide) (PLA) exposed to UV irradiation by a mesophilic bacterium. Int Biodeterior Biodegradation 2013; 85: 289-93.
[http://dx.doi.org/10.1016/j.ibiod.2013.08.013]

[93]   Nomura N, Shigeno-Akutsu Y, Nakajima-Kambe T, Nakahara T. Cloning and sequence analysis of a

polyurethane esterase of Comamonas acidovorans TB-35. J Ferment Bioeng 1998; 86(4): 339-45.
[http://dx.doi.org/10.1016/S0922-338X(99)89001-1]

[94]   Akutsu Y, Nakajima-Kambe T, Nomura N, Nakahara T. Purification and properties of a polyester polyurethane-degrading enzyme from Comamonas acidovorans TB-35. Appl Environ Microbiol 1998; 64(1): 62-7.
[http://dx.doi.org/10.1128/AEM.64.1.62-67.1998] [PMID: 16349494]

[95]   Nakajima-Kambe T, Onuma F, Akutsu Y, Nakahara T. Determination of the polyester polyurethane breakdown products and distribution of the polyurethane degrading enzyme of Comamonas acidovorans strain TB-35. J Ferment Bioeng 1997; 83(5): 456-60.
[http://dx.doi.org/10.1016/S0922-338X(97)83000-0]

[96]   Elbanna K, Lütke-Eversloh T, Jendrossek D, Luftmann H, Steinbüchel A. Studies on the biodegradability of polythioester copolymers and homopolymers by polyhydroxyalkanoate (PHA)-degrading bacteria and PHA depolymerases. Arch Microbiol 2004; 182(2-3): 212-25.
[http://dx.doi.org/10.1007/s00203-004-0715-z] [PMID: 15340783]

[97]   Hadad D, Geresh S, Sivan A. Biodegradation of polyethylene by the thermophilic bacterium *Brevibacillus borstelensis*. J Appl Microbiol 2005; 98(5): 1093-100.
[http://dx.doi.org/10.1111/j.1365-2672.2005.02553.x] [PMID: 15836478]

[98]   Sivan A, Szanto M, Pavlov V. Biofilm development of the polyethylene-degrading bacterium *Rhodococcus ruber*. Appl Microbiol Biotechnol 2006; 72(2): 346-52.
[http://dx.doi.org/10.1007/s00253-005-0259-4] [PMID: 16534612]

[99]   Mabrouk MM, Sabry SA. Degradation of poly (3-hydroxybutyrate) and its copolymer poly (3-hydroxybutyrate-co-3-hydroxyvalerate) by a marine *Streptomyces* sp. SNG9. Microbiol Res 2001; 156(4): 323-35.
[http://dx.doi.org/10.1078/0944-5013-00115] [PMID: 11770850]

[100] Vigneswari S, Lee T, Bhubalan K, Amirul A. Extracellular polyhydroxyalkanoate depolymerase by *Acidovorax* sp. DP5 Enzyme Res 2015; 2015: 212159.
[http://dx.doi.org/10.1155/2015/212159]

[101] Khatiwala VK, Shekhar N, Aggarwal S, Mandal UK. Biodegradation of poly (ε-caprolactone)(PCL) film by Alcaligenes faecalis. J Polym Environ 2008; 16(1): 61-7.
[http://dx.doi.org/10.1007/s10924-008-0104-9]

[102] Oda Y, Oida N, Urakami T, Tonomura K. Polycaprolactone depolymerase produced by the bacterium Alcaligenes faecalis. FEMS Microbiol Lett 1997; 152(2): 339-43.
[http://dx.doi.org/10.1111/j.1574-6968.1997.tb10449.x] [PMID: 9273313]

[103] Obradors N, Aguilar J. Efficient biodegradation of high-molecular-weight polyethylene glycols by pure cultures of *Pseudomonas stutzeri*. Appl Environ Microbiol 1991; 57(8): 2383-8.
[http://dx.doi.org/10.1128/aem.57.8.2383-2388.1991] [PMID: 1768106]

[104] Pranamuda H, Tokiwa Y, Tanaka H. Polylactide degradation by an Amycolatopsis sp. Appl Environ Microbiol 1997; 63(4): 1637-40.
[http://dx.doi.org/10.1128/aem.63.4.1637-1640.1997] [PMID: 16535586]

[105] Romen F, Reinhardt S, Jendrossek D. Thermotolerant poly(3-hydroxybutyrate)-degrading bacteria from hot compost and characterization of the PHB depolymerase of Schlegelella sp. KB1a. Arch Microbiol 2004; 182(2-3): 157-64.
[http://dx.doi.org/10.1007/s00203-004-0684-2] [PMID: 15340791]

[106] Lee B, Pometto AL III, Fratzke A, Bailey TB Jr. Biodegradation of degradable plastic polyethylene by *phanerochaete* and *streptomyces* species. Appl Environ Microbiol 1991; 57(3): 678-85.
[http://dx.doi.org/10.1128/aem.57.3.678-685.1991] [PMID: 16348434]

[107] Kay M, Morton L, Prince E. Bacterial degradation of polyester polyurethane. Int Biodeteriorat 1991; 27: 205-22.

[http://dx.doi.org/10.1016/0265-3036(91)90012-G]

[108] Jansen B, Schumacher-Perdreau F, Peters G, Pulverer G. Evidence for degradation of synthetic polyurethanes by *Staphylococcus epidermidis*. Zentralbl Bakteriol 1991; 276(1): 36-45.
[http://dx.doi.org/10.1016/S0934-8840(11)80216-1] [PMID: 1789899]

[109] Mergaert J, Swings J. Biodiversity of microorganisms that degrade bacterial and synthetic polyesters. J Ind Microbiol 1996; 17(5-6): 463-9.

[110] Jendrossek D, Knoke I, Habibian RB, Steinbüchel A, Schlegel HG. Degradation of poly (3-hydroxybutyrate), PHB, by bacteria and purification of a novel PHB depolymerase fromComamonas sp. J Environ Polym Degrad 1993; 1(1): 53-63.
[http://dx.doi.org/10.1007/BF01457653]

[111] Mukai K, Yamada K, Doi Y. Efficient hydrolysis of polyhydroxyalkanoates by *Pseudomonas stutzeri* YM1414 isolated from lake water. Polym Degrad Stabil 1994; 43(3): 319-27.
[http://dx.doi.org/10.1016/0141-3910(94)90002-7]

[112] Ramsay BA, Saracovan I, Ramsay JA, Marchessault RH. A method for the isolation of microorganisms producing extracellular long-side-chain poly (β-hydroxyalkanoate) depolymerase. J Environ Polym Degrad 1994; 2(1): 1-7.
[http://dx.doi.org/10.1007/BF02073481]

[113] Delafield FP, Doudoroff M, Palleroni NJ, Lusty CJ, Contopoulos R. Decomposition of poly--hydroxybutyrate by pseudomonads. J Bacteriol 1965; 90(5): 1455-66.
[http://dx.doi.org/10.1128/jb.90.5.1455-1466.1965] [PMID: 5848334]

[114] Sudhakar M, Doble M, Murthy PS, Venkatesan R. Marine microbe-mediated biodegradation of low- and high-density polyethylenes. Int Biodeterior Biodegradation 2008; 61(3): 203-13.
[http://dx.doi.org/10.1016/j.ibiod.2007.07.011]

[115] Balasubramanian V, Natarajan K, Hemambika B, *et al.* High-density polyethylene (HDPE)-degrading potential bacteria from marine ecosystem of Gulf of Mannar, India. Lett Appl Microbiol 2010; 51(2): no.
[http://dx.doi.org/10.1111/j.1472-765X.2010.02883.x] [PMID: 20586938]

[116] Tribedi P, Sil AK. Low-density polyethylene degradation by *Pseudomonas* sp. AKS2 biofilm. Environ Sci Pollut Res Int 2013; 20(6): 4146-53.
[http://dx.doi.org/10.1007/s11356-012-1378-y] [PMID: 23242625]

[117] Harshvardhan K, Jha B. Biodegradation of low-density polyethylene by marine bacteria from pelagic waters, Arabian Sea, India. Mar Pollut Bull 2013; 77(1-2): 100-6.
[http://dx.doi.org/10.1016/j.marpolbul.2013.10.025] [PMID: 24210946]

[118] Yang J, Yang Y, Wu WM, Zhao J, Jiang L. Evidence of polyethylene biodegradation by bacterial strains from the guts of plastic-eating waxworms. Environ Sci Technol 2014; 48(23): 13776-84.
[http://dx.doi.org/10.1021/es504038a] [PMID: 25384056]

[119] Azeko ST, Etuk-Udo GA, Odusanya OS, Malatesta K, Anuku N, Soboyejo WO. Biodegradation of linear low density polyethylene by *Serratia marcescens* subsp. marcescens and its cell free extracts. Waste Biomass Valoriz 2015; 6(6): 1047-57.
[http://dx.doi.org/10.1007/s12649-015-9421-0]

[120] Kowalczyk A, Chyc M, Ryszka P, Latowski D. Achromobacter xylosoxidans as a new microorganism strain colonizing high-density polyethylene as a key step to its biodegradation. Environ Sci Pollut Res Int 2016; 23(11): 11349-56.
[http://dx.doi.org/10.1007/s11356-016-6563-y] [PMID: 27072033]

[121] Delacuvellerie A, Cyriaque V, Gobert S, Benali S, Wattiez R. The plastisphere in marine ecosystem hosts potential specific microbial degraders including Alcanivorax borkumensis as a key player for the low-density polyethylene degradation. J Hazard Mater 2019; 380 120899.
[http://dx.doi.org/10.1016/j.jhazmat.2019.120899] [PMID: 31326835]

[122] K. T. Linn, T. ENDO, T. OIKAWA, and Y. ISHIBASHI, "Isolation and characterization of polystyrene degrading microorganisms for zero emission treatment of expanded polystyrene,". Environ Eng Res 2003; 40: 373-9.

[123] Skariyachan S, Patil AA, Shankar A, Manjunath M, Bachappanavar N, Kiran S. Enhanced polymer degradation of polyethylene and polypropylene by novel thermophilic consortia of *Brevibacillus* sps. and *Aneurinibacillus* sp. screened from waste management landfills and sewage treatment plants. Polym Degrad Stabil 2018; 149: 52-68.
[http://dx.doi.org/10.1016/j.polymdegradstab.2018.01.018]

[124] Giacomucci L, Raddadi N, Soccio M, Lotti N, Fava F. Polyvinyl chloride biodegradation by *Pseudomonas citronellolis* and *Bacillus flexus*. N Biotechnol 2019; 52: 35-41.
[http://dx.doi.org/10.1016/j.nbt.2019.04.005] [PMID: 31026607]

[125] Stepien AE, Zebrowski J, Piszczyk Ł, *et al*. Assessment of the impact of bacteria *Pseudomonas denitrificans*, *Pseudomonas fluorescens*, *Bacillus subtilis* and yeast *Yarrowia lipolytica* on commercial poly(ether urethanes). Polym Test 2017; 63: 484-93.
[http://dx.doi.org/10.1016/j.polymertesting.2017.08.038]

[126] Jeon HJ, Kim MN. Isolation of a thermophilic bacterium capable of low-molecular-weight polyethylene degradation. Biodegradation 2013; 24(1): 89-98.
[http://dx.doi.org/10.1007/s10532-012-9560-y] [PMID: 22661062]

[127] Kim MY, Kim C, Moon J, Heo J, Jung SP, Kim JR. Polymer film-based screening and isolation of polylactic acid (PLA)-degrading microorganisms. J Microbiol Biotechnol 2017; 27(2): 342-9.
[http://dx.doi.org/10.4014/jmb.1610.10015] [PMID: 27840398]

[128] Yoshida S, Hiraga K, Takehana T, *et al*. A bacterium that degrades and assimilates poly(ethylene terephthalate). Science 2016; 351(6278): 1196-9.
[http://dx.doi.org/10.1126/science.aad6359] [PMID: 26965627]

[129] Sangeetha Devi R, Ramya R, Kannan K, Robert Antony A, Rajesh Kannan V. Investigation of biodegradation potentials of high density polyethylene degrading marine bacteria isolated from the coastal regions of Tamil Nadu, India. Mar Pollut Bull 2019; 138: 549-60.
[http://dx.doi.org/10.1016/j.marpolbul.2018.12.001] [PMID: 30660306]

[130] Barth M, Oeser T, Wei R, Then J, Schmidt J, Zimmermann W. Effect of hydrolysis products on the enzymatic degradation of polyethylene terephthalate nanoparticles by a polyester hydrolase from Thermobifida fusca. Biochem Eng J 2015; 93: 222-8.
[http://dx.doi.org/10.1016/j.bej.2014.10.012]

[131] Jeon HJ, Kim MN. Functional analysis of alkane hydroxylase system derived from *Pseudomonas aeruginosa* E7 for low molecular weight polyethylene biodegradation. Int Biodeterior Biodegradation 2015; 103: 141-6.
[http://dx.doi.org/10.1016/j.ibiod.2015.04.024]

[132] Santo M, Weitsman R, Sivan A. The role of the copper-binding enzyme – laccase – in the biodegradation of polyethylene by the actinomycete *Rhodococcus ruber*. Int Biodeterior Biodegradation 2013; 84: 204-10.
[http://dx.doi.org/10.1016/j.ibiod.2012.03.001]

[133] Uscátegui YL, Arévalo FR, Díaz LE, Cobo MI, Valero MF. Microbial degradation, cytotoxicity and antibacterial activity of polyurethanes based on modified castor oil and polycaprolactone. J Biomater Sci Polym Ed 2016; 27(18): 1860-79.
[http://dx.doi.org/10.1080/09205063.2016.1239948] [PMID: 27654066]

[134] Shimpi N, Borane M, Mishra S, Kadam M. Biodegradation of polystyrene (PS)-poly(lactic acid) (PLA) nanocomposites using *Pseudomonas aeruginosa*. Macromol Res 2012; 20: 181-7.
[http://dx.doi.org/10.1007/s13233-012-0026-1]

[135] Mohan AJ, Sekhar VC, Bhaskar T, Nampoothiri KM. Microbial assisted high impact polystyrene

(HIPS) degradation. Bioresour Technol 2016; 213: 204-7.
[http://dx.doi.org/10.1016/j.biortech.2016.03.021] [PMID: 26993201]

[136] Dobretsov S, Abed RMM, Teplitski M. Mini-review: Inhibition of biofouling by marine microorganisms. Biofouling 2013; 29(4): 423-41.
[http://dx.doi.org/10.1080/08927014.2013.776042] [PMID: 23574279]

[137] Singh L, Wahid ZA. Methods for enhancing bio-hydrogen production from biological process: A review. J Ind Eng Chem 2015; 21: 70-80.
[http://dx.doi.org/10.1016/j.jiec.2014.05.035]

[138] Park SY, Kim CG. Biodegradation of micro-polyethylene particles by bacterial colonization of a mixed microbial consortium isolated from a landfill site. Chemosphere 2019; 222: 527-33.
[http://dx.doi.org/10.1016/j.chemosphere.2019.01.159] [PMID: 30721811]

[139] Sangwan P, Wu DY. New insights into polylactide biodegradation from molecular ecological techniques. Macromol Biosci 2008; 8(4): 304-15.
[http://dx.doi.org/10.1002/mabi.200700317] [PMID: 18383571]

[140] Oberbeckmann S, Löder MGJ, Labrenz M. Marine microplastic-associated biofilms – a review. Environ Chem 2015; 12(5): 551-62.
[http://dx.doi.org/10.1071/EN15069]

[141] Flemming HC. Relevance of biofilms for the biodeterioration of surfaces of polymeric materials. Polym Degrad Stabil 1998; 59(1-3): 309-15.
[http://dx.doi.org/10.1016/S0141-3910(97)00189-4]

[142] Miao L, Wang P, Hou J, *et al.* Distinct community structure and microbial functions of biofilms colonizing microplastics. Sci Total Environ 2019; 650(Pt 2): 2395-402.
[http://dx.doi.org/10.1016/j.scitotenv.2018.09.378] [PMID: 30292995]

[143] Rummel CD, Jahnke A, Gorokhova E, Kühnel D, Schmitt-Jansen M. Impacts of biofilm formation on the fate and potential effects of microplastic in the aquatic environment. Environ Sci Technol Lett 2017; 4(7): 258-67.
[http://dx.doi.org/10.1021/acs.estlett.7b00164]

[144] Lobelle D, Cunliffe M. Early microbial biofilm formation on marine plastic debris. Mar Pollut Bull 2011; 62(1): 197-200.
[http://dx.doi.org/10.1016/j.marpolbul.2010.10.013] [PMID: 21093883]

[145] Mitik-Dineva N, Wang J, Truong VK, *et al. Escherichia coli, Pseudomonas aeruginosa,* and *Staphylococcus aureus* attachment patterns on glass surfaces with nanoscale roughness. Curr Microbiol 2009; 58(3): 268-73.
[http://dx.doi.org/10.1007/s00284-008-9320-8] [PMID: 19020934]

[146] Crabbe JR, Campbell JR, Thompson L, Walz SL, Schultz WW. Biodegradation of a colloidal ester-based polyurethane by soil fungi. Int Biodeterior Biodegradation 1994; 33(2): 103-13.
[http://dx.doi.org/10.1016/0964-8305(94)90030-2]

[147] Abe M, Kobayashi K, Honma N, Nakasaki K. Microbial degradation of poly(butylene succinate) by *Fusarium solani* in soil environments. Polym Degrad Stabil 2010; 95(2): 138-43.
[http://dx.doi.org/10.1016/j.polymdegradstab.2009.11.042]

[148] Stoica-Guzun A, Jecu L, Gheorghe A, *et al.* Biodegradation of poly (vinyl alcohol) and bacterial cellulose composites by *Aspergillus niger.* J Polym Environ 2011; 19(1): 69-79.
[http://dx.doi.org/10.1007/s10924-010-0257-1]

[149] Osman M, Satti SM, Luqman A, Hasan F, Shah Z, Shah AA. Degradation of polyester polyurethane by *Aspergillus* sp. strain S45 isolated from soil. J Polym Environ 2018; 26(1): 301-10.
[http://dx.doi.org/10.1007/s10924-017-0954-0]

[150] Mogil'nitskii G, Sagatelyan R, Kutishcheva T, Zhukova S, Kerimov S, Parfenova T. Disruption of the protective properties of the polyvinyl chloride coating under the effect of microorganisms. 1987.

[151] Muhamad W, Othman R, Shaharuddin RI, Irani MS. Microorganism as plastic biodegradation agent towards sustainable environment. Adv Environ Biol 2015; 9: 8-14.

[152] Raaman N, Rajitha N, Jayshree A, Jegadeesh R. Biodegradation of plastic by *Aspergillus* spp. isolated from polythene polluted sites around Chennai. J Acad Ind Res 2012; 1(6): 313-6.

[153] Ali MI, Ahmed S, Robson G, *et al.* Isolation and molecular characterization of polyvinyl chloride (PVC) plastic degrading fungal isolates. J Basic Microbiol 2014; 54(1): 18-27.
[http://dx.doi.org/10.1002/jobm.201200496] [PMID: 23686796]

[154] Matsumiya Y, Murata N, Tanabe E, Kubota K, Kubo M. Isolation and characterization of an ether-type polyurethane-degrading micro-organism and analysis of degradation mechanism by Alternaria sp. J Appl Microbiol 2010; 108(6): 1946-53.
[PMID: 19912428]

[155] Álvarez-Barragán J, Domínguez-Malfavón L, Vargas-Suárez M, González-Hernández R, Aguilar-Osorio G, Loza-Tavera H. Biodegradative activities of selected environmental fungi on a polyester polyurethane varnish and polyether polyurethane foams. Appl Environ Microbiol 2016; 82(17): 5225-35.
[http://dx.doi.org/10.1128/AEM.01344-16] [PMID: 27316963]

[156] Brunner I, Fischer M, Rüthi J, Stierli B, Frey B. Ability of fungi isolated from plastic debris floating in the shoreline of a lake to degrade plastics. PLoS One 2018; 13(8) e0202047.
[http://dx.doi.org/10.1371/journal.pone.0202047] [PMID: 30133489]

[157] Friedrich J, Zalar P, Mohorčič M, Klun U, Kržan A. Ability of fungi to degrade synthetic polymer nylon-6. Chemosphere 2007; 67(10): 2089-95.
[http://dx.doi.org/10.1016/j.chemosphere.2006.09.038] [PMID: 17257652]

[158] Russell JR, Huang J, Anand P, *et al.* Biodegradation of polyester polyurethane by endophytic fungi. Appl Environ Microbiol 2011; 77(17): 6076-84.
[http://dx.doi.org/10.1128/AEM.00521-11] [PMID: 21764951]

[159] Sangeetha Devi R, Rajesh Kannan V, Nivas D, Kannan K, Chandru S, Robert Antony A. Biodegradation of HDPE by *Aspergillus* spp. from marine ecosystem of Gulf of Mannar, India. Mar Pollut Bull 2015; 96(1-2): 32-40.
[http://dx.doi.org/10.1016/j.marpolbul.2015.05.050] [PMID: 26006776]

[160] Volke-Sepúlveda T, Saucedo-Castañeda G, Gutiérrez-Rojas M, Manzur A, Favela-Torres E. Thermally treated low density polyethylene biodegradation by *Penicillium pinophilum* and *Aspergillus niger*. J Appl Polym Sci 2002; 83(2): 305-14.
[http://dx.doi.org/10.1002/app.2245]

[161] Paço A, Duarte K, da Costa JP, *et al.* Biodegradation of polyethylene microplastics by the marine fungus Zalerion maritimum. Sci Total Environ 2017; 586: 10-5.
[http://dx.doi.org/10.1016/j.scitotenv.2017.02.017] [PMID: 28199874]

[162] Mathur G, Mathur A, Prasad R. Colonization and Degradation of Thermally Oxidized High-Density Polyethylene by *Aspergillus niger* (ITCC No. 6052) Isolated from Plastic Waste Dumpsite. Bioremed J 2011; 15(2): 69-76.
[http://dx.doi.org/10.1080/10889868.2011.570281]

[163] Ojha N, Pradhan N, Singh S, *et al.* Evaluation of HDPE and LDPE degradation by fungus, implemented by statistical optimization. Sci Rep 2017; 7(1): 39515.
[http://dx.doi.org/10.1038/srep39515] [PMID: 28051105]

[164] Mahalakshmi V, Siddiq SA. Enhanced biodegradation of polyethylene by development of a consortium. Pelagia Research Library 2015; 6: 183-9.

[165] Uttiya Dey UD, Mondal NK, Das K, Dutta S. An approach to polymer degradation through microbes. IOSR J Pharm 2012; 2(3): 385-8.
[http://dx.doi.org/10.9790/3013-0230385388]

[166] Shalini R, Sasikumar C. Biodegradation of Low Density Polythene Materials Using Microbial Consortium–An Overview. International Journal of Pharmaceutical and Chemical Sciences 2015; 4(4): 507-14.

[167] Ghosh S, Qureshi A, Purohit HJ. Enhanced expression of catechol 1,2 dioxygenase gene in biofilm forming Pseudomonas mendocina EGD-AQ5 under increasing benzoate stress. Int Biodeterior Biodegradation 2017; 118: 57-65.
[http://dx.doi.org/10.1016/j.ibiod.2017.01.019]

[168] Upreti M, Srivastava R. A potential *Aspergillus* species for biodegradation of polymeric materials. Curr Sci 2003; 84(11): 1399-402.

[169] Council NR. *In situ* bioremediation: When does it work? National Academies Press 1993.

[170] Sharma M, Sharma P, Sharma A, Chandra S. Microbial Degradation of Plastic—A Brief Review. CIBTech Journal of Microbiology 2015; 4(1): 85-9.

[171] Yu KSH, Wong AHY, Yau KWY, Wong YS, Tam NFY. Natural attenuation, biostimulation and bioaugmentation on biodegradation of polycyclic aromatic hydrocarbons (PAHs) in mangrove sediments. Mar Pollut Bull 2005; 51(8-12): 1071-7.
[http://dx.doi.org/10.1016/j.marpolbul.2005.06.006] [PMID: 16023146]

[172] Dussud C, Ghiglione J-F. Bacterial degradation of synthetic plastics CIESM Workshop Monogr. vol. 46: 49-54.

[173] Okada M. Chemical syntheses of biodegradable polymers. Prog Polym Sci 2002; 27(1): 87-133.
[http://dx.doi.org/10.1016/S0079-6700(01)00039-9]

[174] Yoon JS, Chin IJ, Kim MN, Kim C. Degradation of microbial polyesters: a theoretical prediction of molecular weight and polydispersity. Macromolecules 1996; 29(9): 3303-7.
[http://dx.doi.org/10.1021/ma950314k]

[175] Raziyafathima M, Praseetha P, Rimal I. Microbial degradation of plastic waste: a review. Chemical and Biological Sciences 2016; 4: 231-42.

[176] Wilkes RA, Aristilde L. Degradation and metabolism of synthetic plastics and associated products by *Pseudomonas* sp.: capabilities and challenges. J Appl Microbiol 2017; 123(3): 582-93.
[http://dx.doi.org/10.1111/jam.13472] [PMID: 28419654]

[177] Koutny M, Lemaire J, Delort AM. Biodegradation of polyethylene films with prooxidant additives. Chemosphere 2006; 64(8): 1243-52.
[http://dx.doi.org/10.1016/j.chemosphere.2005.12.060] [PMID: 16487569]

[178] Mooney A, Ward PG, O'Connor KE. Microbial degradation of styrene: biochemistry, molecular genetics, and perspectives for biotechnological applications. Appl Microbiol Biotechnol 2006; 72(1): 1-10.
[http://dx.doi.org/10.1007/s00253-006-0443-1] [PMID: 16823552]

[179] http://2016.igem.org/Team:BGU_ISRAEL/Description

[180] Boronat A, Caballero E, Aguilar J. Experimental evolution of a metabolic pathway for ethylene glycol utilization by *Escherichia coli*. J Bacteriol 1983; 153(1): 134-9.
[http://dx.doi.org/10.1128/jb.153.1.134-139.1983] [PMID: 6336729]

[181] Satlewal A, Soni R, Zaidi M, Shouche Y, Goel R. Comparative biodegradation of HDPE and LDPE using an indigenously developed microbial consortium. J Microbiol Biotechnol 2008; 18(3): 477-82.
[PMID: 18388465]

[182] Junge K, Eicken H, Deming JW. Bacterial Activity at -2 to -20 degrees C in Arctic wintertime sea ice. Appl Environ Microbiol 2004; 70(1): 550-7.
[http://dx.doi.org/10.1128/AEM.70.1.550-557.2004] [PMID: 14711687]

SUBJECT INDEX

A

Acid 16, 20, 28, 31, 53, 122, 123, 139, 146, 148, 161, 162, 163, 164, 169, 170, 182, 184, 189, 203, 208, 216, 218, 241, 257, 302, 353, 354, 438, 454, 476
   carboxylic 184, 302, 353, 354, 476
   chlorendic 182
   chlorobenzoic 203
   citric 20, 241
   clofibric 216
   galactosamine uronic 123
   galacturonic 123
   glutamic 53
   humic 257
   hyaluronic 28
   lactic 454
   muconic 208
   mycolic 122
   nitric 438
   oxalic 218
   sulfanilic 163, 164
Acinetobacter 23, 304
   baumannii 304
   junii 23
Actinomycetes 107, 109
Aeromonas veronii 415
Agents, lubricating 119
Alcanivorax borkumensis 359, 439, 468
Algae-mediated degradation 413
Ammonolysis 434, 452, 461
Anaemia 3, 75
Artificial 4, 110
   release of heavy metals 4
   restriction enzymes 110
Aspergillus 15, 16, 17, 28, 29, 54, 55, 85, 122, 170, 307, 308, 346, 412, 439, 440, 441, 474

flavus 17, 122, 170, 307, 346, 412, 439, 441
fumigatus 17, 441

niger 15, 16, 28, 29, 54, 55, 85, 307, 308, 440, 474
terreus 17, 307
Asthma 195, 409
   occupational 409
Atherosclerosis 195
Atomic force microscopy (AFM) 253, 258, 259, 260, 261, 417
ATP phosphorylation process 379

B

Bacillus 14, 15, 122, 163, 279, 304, 305, 410, 440, 457, 466, 469
   brevis 410, 440
   cereus 14, 15, 163, 279, 304, 305, 457, 466, 469
   subtillis 122
Bacteria 13, 15, 53, 97, 105, 109, 111, 220, 286, 302, 303, 305, 306, 307, 337, 356, 357, 359, 365, 368, 441, 466, 467, 473
   ammonia-oxidizing 337
   biofilm dwelling 359
   endophytic 105
   heterotrophic 286
   nitrogen cycling 337
   oil-degrading 368
   rhizospheric 53, 105
Biodegradation 161, 351, 415
   dye 161
   microplastic 415
   oxidative 351
Biological oxygen demand (BOD) 22, 66, 67, 138, 193, 380
Biomass production 24, 277, 341, 343, 379
   reduced microbial 24
Bioremediation 5, 6, 7, 10, 18, 22, 23, 25, 50, 51, 52, 54, 58, 65, 78, 87, 89, 97, 102, 103, 104, 109, 110, 111, 181, 196, 197, 200, 201, 202, 273, 274, 275, 284, 300, 301, 311, 312, 319, 366, 372, 376, 378, 381

leaching 86
metabolic 211
nitrification 337
purification 256
Production 19, 71, 75, 124, 141, 183, 185,
      187, 189, 200, 205, 317, 319, 353, 355,
      433
   biofuel 19
   cosmetics 124
   industrial fabric 141
Proteins 3, 8, 13, 20, 22, 28, 44, 50, 120, 123,
      218, 219, 362, 377, 440
   bacterial cell wall 13
   glycosylated 362
   organic 28
   release hydrophobic 440

Q

Quantum environmental technologies 311
Quorum sensing (QS) 360

R

Reactions 14, 16, 24, 44, 47, 48, 74, 84, 212,
      214, 343, 344, 345, 352, 355, 356, 366,
      375, 477
   biodegradative 375
   microbial-catalysed 344
   microbial-mediated 345
   oxidation-reduction 74, 366, 477
Reactive oxygen species (ROS) 207, 208, 215
Redox reactions, enzymatic activities 32
Reduction 5, 8, 20, 29, 82, 84, 104, 214, 239,
      241, 313, 333, 334, 335, 336, 337, 363
   anaerobic 84
   cometabolic 363
   enzymatic 29
   manganese 313
Reductive 208, 341, 344
   biotransformation 341, 344
   dehalogenases 208
Reverse osmosis (RO) 47, 77, 78, 239, 249,
      250, 251

RNA template 111
Roxithromycin 190

S

Saccharomyces cerevisiae 17, 22, 51, 55, 86,
      161
Sewage waste 192
Skin 3, 45, 74, 182, 186, 188
   cancer 186
   inflammation 182
   irritation 188
Sleeplessness 75
Sludge 65, 160, 185, 198, 317, 318, 355, 408
   bioremediation 317
   digesters 355
   oily 317, 318
   sewage 160, 185, 198, 408
   toxic 65
Soil 7, 452
   agricultural 452
   harsh 7
Stress, metallic 8
Sugar(s) 45, 79, 254, 304, 363
   abnormal blood 45
   industry 254
   rhamnose 304
Synthetic 127, 139, 141, 165, 166, 167, 187,
      189, 192, 204, 207, 211
   dyes 127, 139, 141, 187, 189, 192, 204,
      207, 211
   wastewater 165, 166, 167
Synthetic plastics 184, 453, 454, 456, 459,
      461
   biodegradable 453, 454
   non-biodegradable 453, 456
System 4, 75, 84, 104, 120, 166, 168, 186,
      255, 278, 282, 308, 311, 409
   aerobic bacterial 278
   aerobic treatment 166
   digestive 75, 409
   endocrine 4
   heavy metal transportation 84

* 9 7 8 9 8 1 5 1 2 3 5 1 7 *